Operations Research Proceedi

GOR (Gesellschaft für Operations Research e.V.)

More information about this series at http://www.springer.com/series/722

Marco Lübbecke · Arie Koster
Peter Letmathe · Reinhard Madlener
Britta Peis · Grit Walther
Editors

Operations Research Proceedings 2014

Selected Papers of the Annual International
Conference of the German Operations
Research Society (GOR), RWTH Aachen
University, Germany, September 2–5, 2014

 Springer

Editors
Marco Lübbecke
Operations Research
RWTH Aachen University
Aachen, Nordrhein-Westfalen
Germany

Arie Koster
Lehrstuhl II für Mathematik
RWTH Aachen University
Aachen, Nordrhein-Westfalen
Germany

Peter Letmathe
Management Accounting
RWTH Aachen University
Aachen, Nordrhein-Westfalen
Germany

Reinhard Madlener
E.ON Energy Research Center
RWTH Aachen University
Aachen, Nordrhein-Westfalen
Germany

Britta Peis
Management Science
RWTH Aachen University
Aachen, Nordrhein-Westfalen
Germany

Grit Walther
Operations Management
RWTH Aachen University
Aachen, Nordrhein-Westfalen
Germany

ISSN 0721-5924   ISSN 2197-9294   (electronic)
Operations Research Proceedings
ISBN 978-3-319-28695-2   ISBN 978-3-319-28697-6   (eBook)
DOI 10.1007/978-3-319-28697-6

Library of Congress Control Number: 2015960228

Printed on acid-free paper

This Springer imprint is published by SpringerNature
The registered company is Springer International Publishing AG Switzerland

Preface

OR2014, the German Operations Research Society's (GOR) annual international scientific conference was held at RWTH Aachen University on September 2–5, 2014. The conference's general theme was "business analytics and optimization" that reflected the growing importance of data-driven applications and the underlying decision support by operations research (OR) models and methods. This volume contains a selection of extended abstracts of papers presented at OR2014. Each of the 124 submissions was reviewed by two or three editors and we decided to accept 90 of them, solely on the basis of scientific merit. In particular, the GOR Master's thesis and dissertation price winners summarize their work in this volume. The convenient EasyChair system was used for submission and paper handling.

OR2014 was a truly interdisciplinary conference, as is OR itself, and it reflected also the great interest from the natural sciences. Researchers and practitioners from mathematics, computer science, business and economics, and engineering (in decreasing order of their share) attended. The conference Web site www.or2014.de contains additional materials such as slides and videos of several of the presentations, in particular the talks given by practitioners at the business day.

We would like to thank the many people who made the conference a tremendous success, in particular the program committee, the 35 stream chairs, our 12 invited plenary and semi-plenary speakers, our exhibitors and sponsors, the 60 persons organizing behind the scenes, and, last but not least, the 881 participants from 47 countries. We hope that you enjoyed the conference as much as we did.

Aachen
November 2015

Marco Lübbecke
Arie Koster
Peter Letmathe
Reinhard Madlener
Britta Peis
Grit Walther

Contents

Experimental Validation of an Enhanced System Synthesis Approach

Lena C. Altherr, Thorsten Ederer, Ulf Lorenz, Peter F. Pelz and Philipp Pöttgen

Abstract Planning the layout and operation of a technical system is a common task for an engineer. Typically, the workflow is divided into consecutive stages: First, the engineer designs the layout of the system, with the help of his experience or of heuristic methods. Secondly, he finds a control strategy which is often optimized by simulation. This usually results in a good operating of an unquestioned system topology. In contrast, we apply Operations Research (OR) methods to find a cost-optimal solution for both stages simultaneously via mixed integer programming (MILP). Technical Operations Research (TOR) allows one to find a provable global optimal solution within the model formulation. However, the modeling error due to the abstraction of physical reality remains unknown. We address this ubiquitous problem of OR methods by comparing our computational results with measurements in a test rig. For a practical test case we compute a topology and control strategy via MILP and verify that the objectives are met up to a deviation of 8.7 %.

1 Introduction

Mixed-integer linear programming (MILP) [4] is the outstanding modeling technique for computer-aided optimization of real-world problems, e.g. logistics, flight or production planning. Regarding the successful application in other fields, it is desirable

L.C. Altherr (✉) · U. Lorenz (✉) · P.F. Pelz (✉) · P. Pöttgen (✉)
Chair of Fluid Systems, TU Darmstadt, Darmstadt, Germany
e-mail: lena.altherr@fst.tu-darmstadt.de

U. Lorenz
e-mail: ulf.lorenz@fst.tu-darmstadt.de

P.F. Pelz
e-mail: peter.pelz@fst.tu-darmstadt.de

P. Pöttgen
e-mail: philipp.poettgen@fst.tu-darmstadt.de

T. Ederer (✉)
Discrete Optimization, TU Darmstadt, Darmstadt, Germany
e-mail: ederer@mathematik.tu-darmstadt.de

© Springer International Publishing Switzerland 2016
M. Lübbecke et al. (eds.), *Operations Research Proceedings 2014*,
Operations Research Proceedings, DOI 10.1007/978-3-319-28697-6_1

to transfer Operations Research (OR) methods to the optimization of technical systems.

The design process of a technical system is typically divided into two consecutive stages: First, the engineer designs the layout of the system, with the help of his experience or of heuristic methods. Secondly, he finds a control strategy which is often optimized by simulation. This usually results in a good operating of an unquestioned system topology.

In order to provide engineers with a methodical procedure for the design of new technical systems, we strive to establish Technical Operations Research (TOR) in engineering sciences. The TOR approach allows one to find an optimal solution for both the topology decision and the usage strategy simultaneously via MILP [3]. While this formulation enables us to prove global optimality and to assess feasible solutions using the global optimality gap, the modeling error often cannot be quantified.

Our aim in this paper is to quantify the modelling error for a MILP of a booster station with accumulators based on [1, 2]. We examine a practical test case and compare the computed results with measurements in a test rig.

2   Problem Description

We replicate MILP predictions for the topology and operating of a technical system in a test rig and compare the computed optimal solution to experimental results. A manageable test case is a water-conveying system, in which a certain amount of water per time has to be pumped from the source to the sink. Such a time-dependent volume flow demand can for example be observed when people shower in a multistory building. To fulfill this time-varying load, a system designer may choose one single speed-controlled pump dimensioned to meet the peak demand.

Another option is a booster station. It consists of an optional accumulator and a set of pumps which are able to satisfy the peak load in combined operation. Compared to the single pump, this set-up allows for a more flexible operating that may lead to lower energy consumption. The speed of each active pump can be adjusted according to the demand, so that they may operate near their optimal working point and thus with higher efficiency. The designer's challenging task is to trade off investment costs and energy efficiency while considering all possible topology and operating options.

3   Mixed Integer Linear Program

Our model consists of two stages: First, find a low-priced investment decision in an adequate set of pumps, pipes, accumulators and valves. Secondly, find energy-optimal operating settings for the selected components. The goal is to compare all possible systems that fulfill the load and to minimize the sum of investment and energy costs over a given depreciation period.

All possible systems can be modelled by a graph $G = (V, E)$ with edges E corresponding to possible components, and vertices V representing connection points between these components. A binary variable $p_{i,j}$ for each optional component $(i, j) \in V$ indicates the purchase decision. Since accumulators can store volume, we generate a time-expansion $\mathcal{G} = (\mathcal{V}, \mathcal{E})$ of the system graph G by copying it once for every time step [1]. Each edge $(i, t_i, j, t_j) \in \mathcal{E}$ connects vertices $(i, t_i) \in \mathcal{V}$ at time t_i and $(j, t_j) \in \mathcal{V}$ at time t_j. An accumulator is represented by edges in time, connecting one point in time with the next, while the other components are edges in space, representing quasi-static behavior. Binary variables a_{i, t_i, j, t_j} for each edge of the expanded graph allow to deactivate purchased components during operation. The conservation of the volume flow Q_{i, t_i, v, t_v} in space and time is given by

$$\forall v \in \mathcal{V}: \qquad \sum_{(i, t_i, v, t_v) \in \mathcal{E}} Q_{i, t_i, v, t_v} \cdot \Delta t = \sum_{(v, t_v, j, t_j) \in \mathcal{E}} Q_{v, t_v, j, t_j} \cdot \Delta t \qquad (1)$$

with time step Δt. An additional condition with an adequate upper limit $Q_{\max}$ makes sure that only active components contribute to the volume flow conservation:

$$\forall e \in \mathcal{E}: \qquad\qquad Q_{i, t_i, j, t_j} \leq Q_{\max} \cdot a_{i, t_i, j, t_j} \qquad (2)$$

Another physical constraint is the pressure propagation

$$\forall (i, t_i, j, t_i) \in \mathcal{E}: \qquad p_{j, t_i} \leq p_{i, t_i} + \Delta p + M \cdot a_{i, t_i, j, t_i} \qquad (3)$$

$$p_{j, t_i} \geq p_{i, t_i} + \Delta p - M \cdot a_{i, t_i, j, t_i} \qquad (4)$$

which has to be fulfilled along each edge in each time step, if the component is active. Regarding pumps, the resulting increase of pressure depends on the rotational speed of the pump and on the volume flow that is conveyed, cf. Fig. 1b. For pipes and valves, pressure loss increasing with the volume flow is observed, cf. Fig. 1a, c and d. All of the measured characteristic curves were linearly approximated and included in the model by a convex combination formulation [5].

4 Experimental Validation

To validate our mathematical model, we consider three test cases with different time-dependent demand profiles. To assess the modeling error, the computed optimal combination of the available components is replicated in an experimental setup, and the settings of the system (e.g. the speed of the used pumps or the valve lift) are adjusted according to the computed optimal variable assignment. Subsequently, we verify if the demand profiles are met in each time step. Moreover, the energy consumption of the setup is measured and the resulting energy costs are calculated and compared to the objective value of the mathematical model.

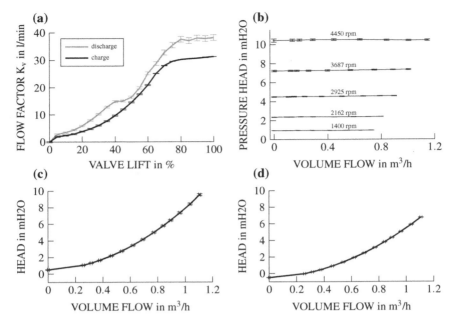

Fig. 1 Input data for the model are the measured characteristic curves of the components of the fluid system. Each data point is the mean value of 10,000 samples. The error bars depict the corresponding standard deviation. **a** Characteristic curves of the valve. Discharging the accumulator causes more pressure loss than charging it. **b** Characteristic curve of the most powerful of the available three pumps with a speed range of 1400–4500 rpm. **c** Characteristic curve of the system for the section from the source to the junction, cf. Fig. 2. A geodetic offset of around 0.5 m has to be overcome from the source to the junction. **d** Characteristic curve from junction to sink. Since the geodetic height of the sink is around 0.5 m lower than that of the junction, this curve starts with negative pressure values

4.1 The Test Rig

Figure 2 shows the modular test rig used for validation measurements. It consists of a combination of up to three speed-controlled centrifugal pumps in a row and an optional acrylic barrel which serves as volume and pressure accumulator. The three pumps differ in their maximum rotating speed (S: 2800 rpm, M: 3400 rpm, L: 4450 rpm) and power consumption. Figure 1b depicts the characteristic curves of pump L. The accumulator has a maximum volume of 50 l and a maximum storable pressure of ≈ 0.2 bar. The barrel can be charged and discharged via a controllable valve, cf. Fig. 1a. Closing the ball valve allows to charge the accumulator without conveying water to the sink. The volume flow is measured by a magnetic flow meter with a tolerance of ± 0.1 l/min $= \pm 0.006$ m^3/h. Pressure measurements are performed by manometers with a tolerance range of ± 0.01 bar $\approx \pm 0.1$ mH2O. All data points represent the mean value of 10,000 samples, collected within 10 s.

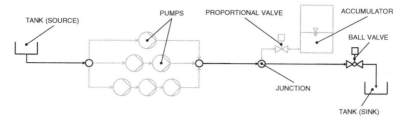

Fig. 2 The test rig consists of a combination of up to three out of three different speed-controlled centrifugal pumps. An optional accumulator can be used to fulfill the demand at the sink. It can be charged and discharged via a controllable valve

Fig. 3 Test case 1. One pump fulfills the load

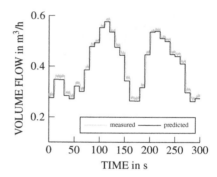

4.2 Comparison of Optimization Results and Measurements

Three different load profiles are given as an input to the optimization program. We built every calculated first-stage solution on our test rig, set up the control strategy and measured the volume flows at the sink and the power consumption of the pumps. The measurement results are given in Figs. 3, 4 and 5.

The time-varying flow demand of the first test case is between $0.25\,\mathrm{m^3/h}$ and $0.6\,\mathrm{m^3/h}$. It can be fulfilled by pump M and pump L, but not by pump S. As pump M is at a lower price than pump L, the optimal result via MILP is to buy pump M. The measured flow is in good agreement with the demand profile, cf. Fig. 3, if the pump is driven with the predicted control settings. The computed total energy consumption for a recovery period of 10 years is $1.9126 \times 10^3\,\mathrm{kWh}$, corresponding to energy costs of €478.14, and total costs of €923.14. The measured energy consumption for one repetition of the load cycle is $(663 \pm 11.2) \times 10^3\,\mathrm{kWh}$, which sums up to $(1.9369 \pm 0.1117) \times 10^3\,\mathrm{kWh}$ and €(484.23 ± 9.57) within 10 years.

The second test case contains higher flow demands than the first one: $0.4\,\mathrm{m^3/h}$ to $0.9\,\mathrm{m^3/h}$. The optimization result is to use pumps S and M to cover the load. The demanded and measured volume flow rates match, cf. Fig. 4. During a recovery period

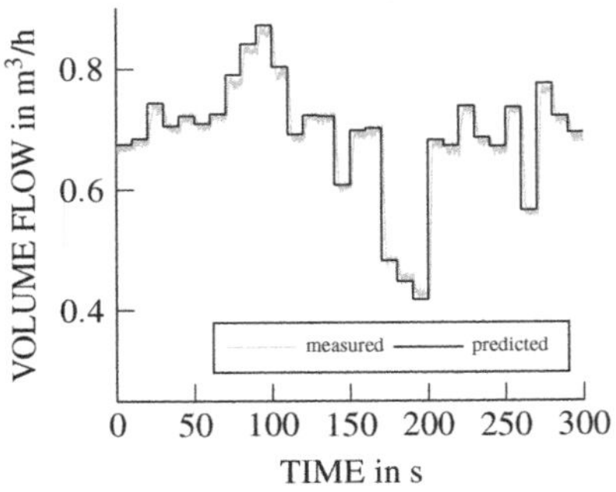

Fig. 4 Test case 2. Two
pumps fulfill the load

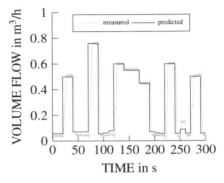

Fig. 5 Test case 3. One
pump in combination with an
accumulator fulfills the load

of 20 years the pumps consume 1.0565×10^4 kWh according to the optimization
result, compared to $(1.0765 \pm 0.0239) \times 10^4$ kWh derived from the measurements.
This leads to total optimal costs of €3436.27, compared to €(3486.32 ± 59.85).
Pump L could have also been used, but its energy consumption is higher for flow
demands around $0.7\,\mathrm{m}^3/\mathrm{h}$.

The flow demands in the third test case range from $0.1\,\mathrm{m}^3/\mathrm{h}$ to $0.8\,\mathrm{m}^3/\mathrm{h}$. The
optimal topology consists of pump L, the accumulator and the valve. Pump L cannot
convey volume flows as low as $0.05\,\mathrm{m}^3/\mathrm{h}$ in the test rig configuration. The optimization model correctly predicts the usage of the accumulator during time steps with
these small demands. The accumulator starts with a positive water level. To satisfy
the conservation of energy, the water level at the last time step has to be equal to this
starting value.

In Fig. 5 the measured data is in satisfying agreement with the time-varying
demand. The optimal energy costs are €537.57. Compared to €(584.27 ± 27.91)
derived from the measurements this corresponds to the highest observed deviation
of 8.7 %. For all test cases a delayed step response of around 5–10 s to the changed
rotational speed settings is observed.

5 Conclusion

In this paper, we presented a MILP model for a system synthesis problem. We are able to find the best combination out of a set of pumps, valves and accumulators to satisfy a given time-dependent flowrate demand with minimal weighted purchase and energy costs. The predicted topology and operating decisions were validated in an experimental setup for three different load demands. The measured volume flows and the power consumption of the pumps match the predicted values with satisfying accuracy. The observed deviations could be caused by the delayed response of the pumps when changing their speed settings. We plan to investigate the influence of the time step size on the modeling error in a future research project. This will allow us to determine to which degree the components' start-up characteristics and deferred adaptation should be included into our model formulation.

Acknowledgments This work is partially supported by the German Federal Ministry for Economic Affairs and Energy funded project "Entwicklung hocheffizienter Pumpensysteme" and by the German Research Foundation (DFG) funded SFB 805.

References

1. Dörig, B., Ederer, T., Hedrich, P., Lorenz, U., Pelz, P. F., Pöttgen, P.: Technical operations research (TOR) exemplified by a hydrostatic power transmission system. In: 9. IFK-Proceedings, vol. 1 (2014)
2. Pelz, P. F., Lorenz, U., Ederer, T., Lang, S., Ludwig, G.: Designing pump systems by discrete mathematical topology optimization: the artificial fluid systems designer (AFSD). In: IREC, Düsseldorf, Germany (2012)
3. Pelz, P. F., Lorenz, U., Ederer, T., Metzler, M., Pöttgen, P.: Global system optimization and scaling for turbo systems and machines. In: ISROMAC-15, Honolulu, USA (2014)
4. Schrijver, A.: Theory of Linear and Integer Programming. Wiley (1998)
5. Vielma, J. P., Ahmed, S., Nemhauser, G.: Mixed-integer models for nonseparable piecewise-linear optimization: unifying framework and extensions. Oper. Res. **58**(2), 303–315 (2010)

Stochastic Dynamic Programming Solution of a Risk-Adjusted Disaster Preparedness and Relief Distribution Problem

Ebru Angün

Abstract This chapter proposes a multistage stochastic optimization framework that dynamically updates the purchasing and distribution decisions of emergency commodities in the aftermath of an earthquake. Furthermore, the models consider the risk of exceeding the budget levels at any stage through chance constraints, which are then converted to Conditional Value-at-Risk constraints. Compared to the previous papers, our framework provides the flexibility of adjusting the level of conservativeness to the users by changing risk related parameters. Under some conditions, the resulting linear programming problems are solved through the Stochastic Dual Dynamic Programming algorithm. The preliminary numerical results are encouraging.

1 Introduction

This chapter proposes a dynamic and stochastic methodology to generate a risk-averse disaster preparedness and logistics plan that can mitigate demand and road capacity uncertainties. More specifically, we apply multistage stochastic optimization for dynamically purchasing and distributing emergency commodities with time dependent demands and road capacities. Several authors have dealt with problems similar to ours, but [2, 4] are the most related papers. In many cases, our approach can give less conservative solutions than [2], which considers a robust dynamic optimization framework. Furthermore, our approach gives more conservative solutions than [4], which considers a risk-neutral dynamic stochastic optimization framework with a finite number of scenarios.

This research with the project number 13.402.005 has been financially supported by Galatasaray University Research Fund.

E. Angün (✉)
Department of Industrial Engineering, Galatasaray University,
Ciragan Cad. Ortaköy, 34349 Istanbul, Turkey
e-mail: eangun@gsu.edu.tr; ebru.angun@gmail.com

© Springer International Publishing Switzerland 2016   9
M. Lübbecke et al. (eds.), *Operations Research Proceedings 2014*,
Operations Research Proceedings, DOI 10.1007/978-3-319-28697-6_2

The structure of the chapter is as follows. In Sect. 2, we introduce multistage stochastic programming models that take risk into account. Section 3 presents the novelty in our application of the risk-averse Stochastic Dual Dynamic Programming (SDDP) algorithm, and Sect. 3.1 presents some preliminary numerical results. Finally, Sect. 4 summarizes the chapter and presents a few future research directions.

2 Risk-Adjusted Multistage Stochastic Programming Model

We formulate the problem through a risk-adjusted, T-stage stochastic programming model, where the decisions at the first-stage belong to the preparedness phase, and the decisions at later stages belong to the response phase of a disaster. The risk adjustments are achieved by adding probabilistic constraints to the risk-neutral formulation at stages $t = 1, \ldots, T - 1$. A risk-neutral formulation and solution of this problem is given in [1].

We make the following two assumptions for the random vector $\boldsymbol{\xi}_t$ whose components are the demands and the road capacities: i—The distribution P_t of $\boldsymbol{\xi}_t$ is known, and this P_t is supported on a set $\Xi_t \subset \mathbb{R}^{d_t}$; ii—The random process $\{\boldsymbol{\xi}_t\}_{t=2}^{T}$ is stage-wise independent.

We formulate the T-stage problem through the following dynamic programming equations. At stage $t = 1$, the problem is

$$
\begin{aligned}
\text{Min} \quad & \sum_{i \in I} \left[\sum_{l \in L} f_{il} y_{il} + \sum_{k \in K} q_1^k r_{1i}^k \right] + \mathbb{E}\left[Q_2\left(\mathbf{x}_1, \boldsymbol{\xi}_2\right) \right] \\
\text{s.t} \quad & \sum_{k \in K} b^k r_{1i}^k \leq \sum_{l \in L} M_l y_{il} \;\; \forall i \in I \\
& \sum_{l \in L} y_{il} \leq 1 \qquad \forall i \in I \\
& \text{Prob}\left\{ Q_2\left(\mathbf{x}_1, \boldsymbol{\xi}_2\right) \leq \eta_2 \right\} \geq 1 - \alpha_2 \\
& y_{il} \in \{0, 1\}, r_{1i}^k \geq 0, \forall i \in I, l \in L, k \in K
\end{aligned}
\tag{1}
$$

where I, L, and K are the set of potential nodes to open storage facilities, the set of size categories of the facilities, and the set of commodity types, respectively, f_{il} is the fixed cost of opening a facility of size l in location i, q_t^k is the unit acquisition cost of commodity k at stage t, b^k is the unit space requirement for commodity k, M_l is the overall capacity of a facility of size l, r_{ti}^k is the amount of commodity k purchased at stage t in location i, y_{il} is the location i and the size l of a facility, η_t and α_t are the known budget limit and the significance level at stage t, respectively, and $\mathbf{x}_1$ is the vector with the components y_{il}'s and r_{1i}^k's. Furthermore, in (1), the first set of constraints limits the capacity of a facility, the second set of constraints restricts the number of facilities per node, and the chance constraint ensures that the second-stage cost-to-go function $Q_2\left(\mathbf{x}_1, \boldsymbol{\xi}_2\right)$ does not exceed the budget limit η_2 with high probability.

For later stages $t = 2, \ldots, T - 1$ and for a realization $\boldsymbol{\xi}_t^s$ of $\boldsymbol{\xi}_t$, the cost-to-go functions $Q_t\left(\mathbf{x}_{t-1}, \boldsymbol{\xi}_t^s\right)$ are given by

$$
\begin{aligned}
\text{Min} \sum_{k \in K} &\left[\sum_{i \in I} q_t^k r_{ti}^k + \sum_{(i',j') \in A} c_{ti'j'}^k m_{ti'j'}^k + \sum_{j \in J} p_t^k w_{tj}^k \right] + \mathbb{E}\left[Q_{t+1}\left(\mathbf{x}_t, \boldsymbol{\xi}_{t+1}\right) \right] \\
\text{s.t} \quad z_{ti}^k + &\sum_{(i,j') \in A} m_{tij'}^k - \sum_{(j',i) \in A} m_{tj'i}^k = r_{t-1,i}^k + z_{t-1,i}^k \quad \forall i \in I, k \in K \\
&\sum_{(i',j) \in A} m_{ti'j}^k - \sum_{(j,i') \in A} m_{tji'}^k + w_{tj}^k = v_{tj}^{ks} \qquad\qquad \forall j \in J, k \in K \\
&\sum_{k \in K} b^k \left(m_{ti'j'}^k + m_{tj'i'}^k \right) \qquad\quad \leq \kappa_{ti'j'}^s \qquad \forall \left(i', j'\right) \in A \\
&\sum_{k \in K} b^k \left(z_{ti}^k + r_{ti}^k \right) \qquad\qquad\quad \leq \sum_{l \in L} M_l y_{il} \qquad \forall i \in I \\
&\text{Prob}\left\{ Q_{t+1}\left(\mathbf{x}_t, \boldsymbol{\xi}_{t+1}\right) \leq \eta_{t+1} \right\} \geq 1 - \alpha_{t+1} \\
&r_{ti}^k, m_{ti'j'}^k, w_{tj}^k, z_{ti}^k \geq 0 \forall i \in I, j \in J, k \in K, \left(i', j'\right) \in A
\end{aligned}
\tag{2}
$$

where J and A are the set of nodes that represent shelters and the set of arcs that represent roads in the network, respectively, $c_{ti'j'}^k$ is the unit transportation cost of commodity k through arc $\left(i', j'\right)$, p_t^k is the unit shortage cost of commodity k, $m_{ti'j'}^k$ is the amount of commodity k transported through arc $\left(i', j'\right)$, w_{tj}^k and z_{ti}^k are the shortage amount of commodity k in shelter j and the amount of commodity k stored in location i, respectively, v_{tj}^{ks} and $\kappa_{ti'j'}^s$ are the demand for the commodity k in shelter j and the road capacity of arc $\left(i', j'\right)$ for a realization s, respectively, and $\mathbf{x}_t$ is the vector with components r_{ti}^k's and z_{ti}^k's; all values depend on stage t. Moreover, in (2), the first set of constraints represents the flow conservation with $z_{1,i}^k = 0 \forall i \in I, k \in K$, the second set of constraints is for the demand satisfaction, and the third set of constraints is for the road capacity. The stage T problem has the same three sets of constraints as in (2), but there are no more acquisition decisions and the remaining inventories are penalized through a unit holding cost h_T^k. Hence, the objective function at $t = T$ becomes

$$
\text{Min} \sum_{k \in K} \left[\sum_{i \in I} h_T^k z_{Ti}^k + \sum_{(i',j') \in A} c_{Ti'j'}^k m_{Ti'j'}^k + \sum_{j \in J} p_T^k w_{Tj}^k \right].
$$

It was suggested in [5] to replace the chance constraint by the $\mathsf{CV@R}_\alpha$-type constraint, where $\mathsf{CV@R}_\alpha$ is given by

$$
\mathsf{V@R}_\alpha \left[Q_t\left(\mathbf{x}_{t-1}, \boldsymbol{\xi}_t\right) \right] + \alpha^{-1} \mathbb{E}\left[Q_t\left(\mathbf{x}_{t-1}, \boldsymbol{\xi}_t\right) - \mathsf{V@R}_\alpha \left[Q_t\left(\mathbf{x}_{t-1}, \boldsymbol{\xi}_t\right) \right] \right]_+ \tag{3}
$$

where the Value-at-Risk $\left(\mathsf{V@R}_\alpha\right)$ in (3) is, by definition, the left-side $(1 - \alpha)$-quantile of the distribution of $Q_t\left(\mathbf{x}_{t-1}, \boldsymbol{\xi}_t\right)$, and

$$
\left[Q_t - \mathsf{V@R}_\alpha \left(Q_t\right) \right]_+ = \max \left\{ Q_t - \mathsf{V@R}_\alpha \left(Q_t\right), 0 \right\}.
$$

A problem with a CV@R-type constraint is that it can make the problem infeasible. Consequently, it could be convenient to move the CV@R-type constraint into the objective function; that is, we redefine the cost-to-go function as

$$
V_{\lambda_t}\left[Q_t\left(\mathbf{x}_{t-1},\boldsymbol{\xi}_t\right)\right] := (1-\lambda_t)\,\mathbb{E}\left[Q_t\left(\mathbf{x}_{t-1},\boldsymbol{\xi}_t\right)\right] + \lambda_t \mathsf{CV@R}_{\alpha_t}\left[Q_t\left(\mathbf{x}_{t-1},\boldsymbol{\xi}_t\right)\right]
\tag{4}
$$

where $\lambda_t \in [0, 1]$ is a parameter that can be tuned for a tradeoff between minimizing on average and risk control.

The expectation and the CV@R in (4) usually make the problem analytically untractable. A possible way to deal with this problem is to use Sample Average Approximation (SAA). That is, sample $\boldsymbol{\xi}_t$ from its distribution P_t to obtain $\mathscr{S}_t := \left\{\boldsymbol{\xi}_t^1, \ldots, \boldsymbol{\xi}_t^{N_t}\right\}$, where N_t is the sample size at stage t. Then, setting $\lambda_t = 0$ in (4) and for a fixed $\overline{\mathbf{x}}_{t-1}$, solve the stage t problem to obtain the N_t optimal values $Q_t\left(\overline{\mathbf{x}}_{t-1},\boldsymbol{\xi}_t^1\right), \ldots, Q_t\left(\overline{\mathbf{x}}_{t-1},\boldsymbol{\xi}_t^{N_t}\right)$. Let $Q_{t,(1)} < Q_{t,(2)} < \cdots < Q_{t,(\iota)} < \cdots < Q_{t,(N_t)}$ be the order statistics obtained from these optimal values, and ι be the smallest integer that satisfies $\iota \geq N_t(1-\alpha_t)$. This $Q_{t,(\iota)}$ is an estimate of $\mathsf{V@R}\left[Q_t\left(\overline{\mathbf{x}}_{t-1},\boldsymbol{\xi}_t\right)\right]$ so that (4) is estimated through

$$
\frac{(1-\lambda_t)}{N_t}\sum_{s=1}^{N_t} Q_t\left(\overline{\mathbf{x}}_{t-1},\boldsymbol{\xi}_t^s\right) + \lambda_t Q_{t,(\iota)} + \frac{\lambda_t}{N_t\alpha_t}\sum_{s=1}^{N_t}\left[Q_t\left(\overline{\mathbf{x}}_{t-1},\boldsymbol{\xi}_t^s\right) - Q_{t,(\iota)}\right]_+ .
$$

3 Stochastic Dual Dynamic Programming Applications

The Stochastic Dual Dynamic Programming (SDDP) algorithm was introduced in [3], and the risk-averse SDDP algorithm was applied to an SAA problem in [6]. Furthermore, a detailed description of the risk-neutral SDDP algorithm applied to an SAA problem was given in [1]. We do not give further detail on the SDDP algorithm, but refer to the papers above.

The novelty in our application of the risk-averse SDDP follows from the following proposition.

Proposition 1 *For a realization $\boldsymbol{\xi}_t^s$ of $\boldsymbol{\xi}_t$ and at a given $\overline{\mathbf{x}}_{t-1}$, a subgradient $\mathbf{g}_t^s$ of $V_{\lambda_t}\left[Q_t\left(\mathbf{x}_{t-1},\boldsymbol{\xi}_t\right)\right]$ is computed through*

$$
\mathbf{g}_t^s = \begin{cases} -\left(1-\lambda_t+\lambda_t\alpha_t^{-1}\right)\mathbf{B}_t^{sT}\boldsymbol{\pi}_t^s - \left(\lambda_t-\lambda_t\alpha_t^{-1}\right)\mathbf{B}_t^{(\iota)T}\boldsymbol{\pi}_t^{(\iota)} & \text{if } Q_t\left(\overline{\mathbf{x}}_{t-1},\boldsymbol{\xi}_t^s\right) > Q_{t,(\iota)} \\ -(1-\lambda_t)\mathbf{B}_t^{sT}\boldsymbol{\pi}_t^s - \lambda_t\mathbf{B}_t^{(\iota)T}\boldsymbol{\pi}_t^{(\iota)} & \text{if } Q_t\left(\overline{\mathbf{x}}_{t-1},\boldsymbol{\xi}_t^s\right) \leq Q_{t,(\iota)} \end{cases}
$$

where $\boldsymbol{\pi}_t^s$ is the vector of dual variables corresponding to the first set of constraints for $t = 3, \ldots, T$, and to the first and the second set of constraints for $t = 2$, $\mathbf{B}_t^s$ is the *matrix whose entries are given by the coefficients of $r_{t-1,i}^k$ and $z_{t-1,i}^k$ for $t = 3, \ldots, T$, and by the coefficients of $r_{t-1,i}^k$ and y_{il} for $t = 2$, and $\mathbf{B}_t^{(i)}$ and $\boldsymbol{\pi}_t^{(i)}$ correspond to $Q_{t,(i)}$.*

Then, a subgradient $\hat{\mathbf{g}}_t$ of (4) is estimated through $\hat{\mathbf{g}}_t = \frac{1}{N_t} \sum_{s-1}^{N_t} \mathbf{g}_t^s$.

3.1 Numerical Results

We consider three consumable emergency commodity types, 10 potential locations for facilities, and 30 shelters in the two boroughs of Istanbul. The data for costs and volumes of commodities, the data for costs and capacities of facilities, and the population data are the same as in [1]. Furthermore, [7] estimated the total numbers of buildings that are prone to be damaged at various levels for an earthquake of magnitude 7.3 on the Richter scale; these data are also summarized in [1].

We model the random demand v_{tj}^k for commodity k at shelter j and random capacity $\kappa_{ti'j'}$ for any arc (i', j') at stage t $(t = 2, \ldots, T)$ as follows:

$$v_{tj}^k = \delta_t^k \left(\varsigma_{t-1,j} + \varsigma_{t,j} \right) \forall j \in J \text{ and } \kappa_{ti'j'} = \eta * \frac{\tau(t)}{\omega(i',j')/\gamma_{ti'j'}} \forall (i', j') \in A$$

where δ_t^k is the amount of commodity k needed by a single individual during stage t, $\varsigma_{t-1,j}$ is the number of evacuees who were expected to arrive at shelter j by the end of stage $(t-1)$, and $\varsigma_{t,j}$ is the random additional number of evacuees who arrive at shelter j at stage t. Moreover, η is the capacity of a single vehicle, $\tau(t)$ is the length of stage t, $\omega(i', j')$ is the actual distance between nodes i' and j', and $\gamma_{ti'j'}$ is the random speed of the vehicle. Both $\varsigma_{t,j}$ and $\gamma_{ti'j'}$ are assumed to be normal; see [1].

We consider $T = 6$ stages, and concentrate on the first 72 h in the aftermath of an earthquake. The stopping criterion of the SDDP algorithm is the maximum number of iterations, which is 100. All computational experiments are conducted on a workstation with Windows 2008 Server, three Intel(R) Xeon(R) CPU E5-2670 CPUs of 2.60 GHz, and 4 GB RAM. The linear programming problems are solved by ILOG CPLEX Callable Library 12.2.

So far we have only experimented with risk-related parameters, namely λ and α. Values of λ closer to 1 and values of α closer to 0 make the 6-stage problems more risk-averse. In Fig. 1, for $\alpha = 1\%$ (on the left) the lower bounds on the 6-stage costs for $\lambda = 0.4$ and $\lambda = 0.5$ stabilize at almost the same value. For $\alpha = 5\%$ (on the right), however, the lower bound for the more risk-averse case ($\lambda = 0.5$) stabilizes at a value which is much lower than the lower bound of the less risk-averse case ($\lambda = 0.4$); this is due to the fact that for the $\lambda = 0.5$ case, facilities store more emergency commodities, and hence the shortage amounts and the penalty costs are much lower compared to the $\lambda = 0.4$ case.

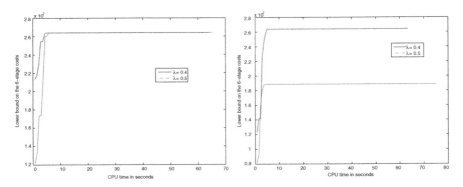

Fig. 1 Changes in the lower bound on the 6-stage costs for $\alpha = 1\%$ on the *left* and $\alpha = 5\%$ on the *right*

4 Conclusions

In this chapter, we formulate a short-term disaster management problem through a multistage stochastic programming model. The model takes the risk of exceeding the budget level at that stage into account through a chance constraint, which is then converted into a CV@R-type constraint. Because the CV@R-type constraint can make the problem infeasible, that constraint is further added to the objective function. Under some assumptions, the resulting problem is solved through the Stochastic Dual Dynamic Programming (SDDP) algorithm. The numerical results are very preliminary, but nevertheless encouraging; the model responds to the risk factors, namely λ and α, as it should. Furthermore, the solution time of a risk-adjusted problem is not worse than a risk-neutral one.

Future research should include the derivation of a stopping rule for the risk-adjusted SDDP. Moreover, more numerical experiments should be done concerning the risk factors and testing the model sensitivities to various cost parameters.

References

1. Angun, E.: Stochastic dual dynamic programming solution of a short-term disaster management problem. In: Dellino, G., Meloni, C. (eds.) Uncertainty Management in Simulation-Optimization of Complex Systems: Algorithms and Applications. Operations Research/Computer Science Series. Springer, New York, 2015, pp. 225–250
2. Ben-Tal, A., Chung, B.D., Mandala, S.R., Yao, T.: Robust optimization for emergency logistics planning: risk mitigation in humanitarian relief supply chains. Transp. Res. Part B: Methodol. **45**, 1177–1189 (2011)
3. Pereira, M.V.F., Pinto, L.M.V.G.: Multi-stage stochastic optimization applied to energy planning. Math. Program. **52**, 359–375 (1991)
4. Rawls, C.G., Turnquist, M.A.: Pre-positioning and dynamic delivery planning for short-term response following a natural disaster. Socio. Econ. Plan. Sci. **46**, 46–54 (2012)

5. Rockafellar, R.T., Uryasev, S.P.: Optimization of conditional value-at-risk. J. Risk **2**, 21–41 (2000)
6. Shapiro, A., Tekaya, W., da Costa, J.P., Soares, M.P.: Risk neutral and risk averse stochastic dual dynamic programming method. Eur. J. Oper. Res. **224**, 375–391 (2013)
7. Unpublished data obtained through private communication from Ansal, A., Boğaziçi University Kandilli Observatory and Earthquake Research Center, Istanbul, Turkey

Solution Approaches for the Double-Row Equidistant Facility Layout Problem

Miguel F. Anjos, Anja Fischer and Philipp Hungerländer

Abstract We consider the Double-Row Equidistant Facility Layout Problem and show that the number of spaces needed to preserve at least one optimal solution is much smaller compared to the general double-row layout problem. We exploit this fact to tailor exact integer linear programming (ILP) and semidefinite programming (SDP) approaches that outperform other recent methods for this problem. We report computational results on a variety of benchmark instances showing that the ILP is preferable for small and medium instances whereas the SDP yields better results on large instances with up to 60 departments.

1 Introduction

An instance of the Double-Row Equidistant Facility Layout Problem (DREFLP) consists of d one-dimensional departments with equal lengths and pairwise non-negative weights w_{ij}. The objective is to find an assignment $r : [d] \to \{1, 2\}$ of the departments to the rows and feasible horizontal positions p for the centers of the departments minimizing the total weighted sum of the center-to-center distances between all pairs of departments:

M.F. Anjos
Canada Research Chair in Discrete Nonlinear Optimization in Engineering,
GERAD & École Polytechnique de Montréal, Montreal, QC H3C 3A7, Canada
e-mail: anjos@stanfordalumni.org

A. Fischer (✉)
Department of Mathematics, TU Dortmund, Dortmund, Germany
e-mail: anja.fischer@mathematik.tu-dortmund.de

P. Hungerländer
Department of Mathematics, Alpen-Adria Universität Klagenfurt,
Klagenfurt am Wörthersee, Austria
e-mail: philipp.hungerlaender@uni-klu.ac.at

© Springer International Publishing Switzerland 2016
M. Lübbecke et al. (eds.), *Operations Research Proceedings 2014*,
Operations Research Proceedings, DOI 10.1007/978-3-319-28697-6_3

$$\min_{r,p} \sum_{\substack{i,j\in[d]\\i<j}} w_{ij}|p(i) - p(j)|, \quad \text{s.t.} \quad |p(i) - p(j)| \geq 1 \text{ if } i \neq j \text{ and } r(i) = r(j).$$

In this short paper we summarize our new results on the structure of the optimal layouts, present the ILP and SDP models that follow from those results and provide representative computational results. Formal proofs and detailed computational results are omitted due to space limitations and are provided in the full paper [3].

2 The Structure of Optimal Layouts

The definition of the (DREFLP) suggests that the spaces between the departments can be of arbitrary lengths in a feasible layout. Hence it is intuitive to model the problem using continuous variables. But in fact the (DREFLP) has some hidden underlying combinatorial structure that makes it possible to model it using only binary variables. The following theorem is a special case of Theorem 2 from [8]:

Theorem 1 *There is always an optimal solution to the (DREFLP) on the grid.*

Note that for such layouts all departments and spaces have equal size. An illustration of this result is provided in Fig. 1:

Note that the grid property is implicitly fulfilled for layouts corresponding to the graph version of the (DREFLP), i.e. the extension of the linear arrangement problem where two or more nodes can be assigned to the same position. Hence in particular the Minimum Duplex Arrangement Problem considered by Amaral [2] is a special case of the (DREFLP) by Theorem 1.

For layouts fulfilling the grid property we say that department i lies in column j if the center of department i is located at the jth grid point. For example, department 5 lies in column 4 in Fig. 1.

In the next theorem we make three assumptions.

Assumption 1:   Columns that contain only spaces can be deleted.

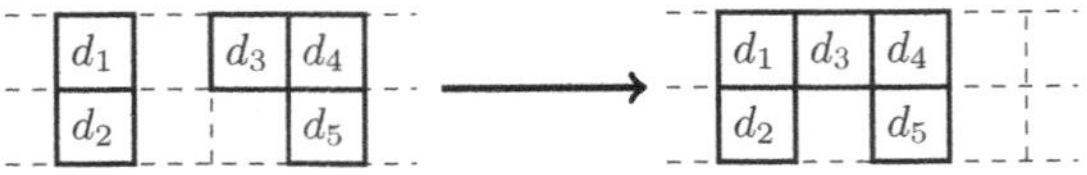

Assumption 2:   If two non-empty adjacent columns both contain only one department, then the two departments can be assigned to the left column and the right column can be deleted.

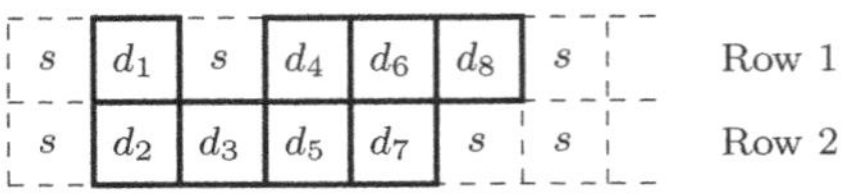

Fig. 1 Illustration of the grid property of the (DREFLP)

Table 1 Minimum number of columns needed for small values of d

d	1	2	3	4	5	6	7	8	9	10	11	12	13	14	15	16
# columns	1	1	2	2	3	4	4	5	5	6	7	7	8	9	9	10

Assumption 3:   If $d > 4$ and the first column and the third column contain in total at most 2 departments, then both departments can be assigned to the third column and the first column can be deleted. The same holds for the last and third-to-last columns.

Note that the local changes described in the three assumptions cannot worsen the objective value of the layout. The next theorem shows that under these assumptions, it is possible to determine in advance the minimum number of columns needed, and hence the minimum number of spaces needed, to compute an optimal solution for the (DREFLP).

Theorem 2 *If Assumptions 1, 2, and 3 hold, then the number of columns needed to preserve at least one optimal layout for an instance of the (DREFLP) with d departments is* $\lceil \frac{2d}{3} \rceil - 1$ *for* $d \geq 9$.

For small values of d, Table 1 gives the minimum number of columns needed so that at least one optimal solution for the (DREFLP) is preserved.

Theorem 2 allows us to reduce the number of spaces needed to formulate the (DREFLP). It also helps to eliminate some of the symmetries in the problem (e.g. the position of columns containing no department) and hence to obtain stronger global bounds independently of the choice of the relaxation.

3 Formulations

To simplify notation we let n denote the total number of departments (original ones plus spaces) such that $n = 2c$ for some c and the number of spaces is $s = n - d$. After the insertion of a suitable number of spaces we have a space-free optimization problem as introduced by Hungerländer and Anjos [7], and Theorems 1 and 2 ensure that the optimal solution of the space-free problem solves the corresponding (DREFLP).

The MILP Formulation of Amaral. To the best of our knowledge the paper by Amaral [2] contains the only approach tailored specifically to the (DREFLP).

We briefly recall his MILP formulation. Let us introduce the binary variables $y_{ip} \in \{0, 1\}$, $i \in [d]$, $p \in [c]$, with the interpretation

$$y_{ip} = \begin{cases} 1, & \text{department } i \text{ is assigned to column } p, \\ 0, & \text{otherwise.} \end{cases}$$

Using these variables we can write the objective function of the (DREFLP) as:

$$\sum_{\substack{i,j \in [d] \\ i < j}} \sum_{\substack{p,q \in [c] \\ p < q}} w_{ij}(q - p)y_{ip}y_{jq}.$$

This quadratic objective function can now be linearized in a standard way, and adding appropriate constraints yields an MILP formulation of the (DREFLP).

The ILP Formulation. Next we outline our ILP formulation for the (DREFLP). This is an extension of the model of Amaral [1] for the Single-Row Facility Layout Problem. Our model uses the binary betweenness variables $b_{ijk} = b_{kji} \in \{0, 1\}$, $i, j, k \in [n]$, $i < k$, $i \neq j \neq k$, with the interpretation

$$b_{ijk} = \begin{cases} 1, & \text{if department } j \text{ lies between departments } i \text{ and } k, \\ 0, & \text{otherwise,} \end{cases}$$

as well as the binary overlap variables $a_{ij} = a_{ji} \in \{0, 1\}$, $i, j \in [n]$, $i < j$, with the interpretation

$$a_{ij} = \begin{cases} 1, & \text{if departments } i \text{ and } j \text{ are assigned to the same column,} \\ 0, & \text{otherwise.} \end{cases}$$

Using these variables the objective function of (DREFLP) can be expressed as:

$$\min \sum_{i,j \in N, i < j} \frac{w_{ij}}{2} \left(\sum_{k \in N \setminus \{i,j\}} b_{ikj} + 2(1 - a_{ij}) \right)$$

We count all departments that lie between the departments i and j and because of the double-row structure the corresponding sum is divided by two. Furthermore we add w_{ij} if departments i and j do not lie in the same column. Using standard constraints and also some additional symmetry-breaking conditions, we obtain an ILP formulation for the (DREFLP).

The SDP Formulation. The ILP formulation can be reformulated as a quadratic program by expressing the variables b_{ijk} and a_{ij} in terms of the following bivalent ordering variables x_{ij}, $i, j \in [n]$, $i \neq j$:

$$x_{ij} = \begin{cases} 1, & \text{if department } i \text{ lies left of department } j, \\ -1, & \text{otherwise,} \end{cases}$$

$$b_{ikj} = \tfrac{1}{4}(x_{ik,kj} + x_{jk,ki} + x_{ik} + x_{kj} + x_{jk} + x_{ki}) + \tfrac{1}{2}, \quad i, j, k \in [n], \ i < j,$$

$$a_{ij} = -\tfrac{1}{2}(x_{ij} + x_{ji}), \quad i, j \in [n], \ i < j.$$

We can further rewrite the resulting quadratic program as an SDP problem with a rank-one condition and use standard techniques for relaxing it. Hence we obtain a basic semidefinite relaxation for the (DREFLP) that can be tightened by adding several classes of valid constraints. Our final SDP relaxation can be interpreted as a generalization of the SDP relaxations for the (SRFLP) [9].

4 Computational Experiments

We implemented a simple as well as an SDP-based construction heuristic and a 3-OPT improvement heuristic in order to obtain good upper bounds on the optimal solution, i.e., integer solutions that describe feasible layouts of the departments. The MILP and ILP formulations were solved using Gurobi 5.6.3 [4], partially with our own separators. The SDP approach used a spectral bundle method [5, 6] in conjunction with primal cutting plane generation. Hence, we always obtain both a feasible solution and an optimality gap showing how far this solution can be from the true optimum. Exploiting these lower and upper bounds allows us to solve (DREFLP) instances of reasonable size to optimality or near-optimality.

All experiments were conducted on an Intel Core i7 CPU 920 with 2.67 GHz and 12 GB RAM in single processor mode using openSUSE Linux 12.2.

We denote by MILP I the original approach proposed by Amaral [2] using d columns, and by MILP II the approach to solve the same model using only the smallest number of columns according to Theorem 2. Table 2 shows that reducing the number of columns (and hence the number of variables) in the MILP model reduces the running times considerably. Nonetheless this improvement is not enough to create a competitive solution approach. Although the instances are small ($d = 11, 12$), MILP II cannot solve these instances within 1 hour, and the optimality gaps are significant. On the other hand, the ILP and SDP approaches solved all instances to optimality. For this reason we do not present results for MILP I and II in the next tables.

Table 2 CPU times (sec, min:sec, or h:min:sec) and gaps for small instances

	Optimal	Gap (%)		Time			
	Solution	MILP I	MILP II	MILP I	MILP II	ILP	SDP
Y-11_t	2008	41.2	3.0	1:00:00	1:00:00	0.63	3:59
Y-12_t	2342	91.3	16.1	1:00:00	1:00:00	1.99	1:33
S-12_t	2167	84.4	20.7	1:00:00	1:00:00	9.85	1:40

Table 3 CPU times (sec, min:sec, or h:min:sec) and gaps for medium-sized instances

Inst.	Best	Gap (%)		Time	
	UB	ILP	SDP	ILP	SDP
S-13_t	2940	0.0	0.0	24.59	18:15
S-14_t	3608	0.0	0.0	3:03	57:00
S-15_t	4466	0.0	0.3	2:51	1:00:00
S-16_t	5446	0.0	0.0	27:29	12:50
S-17_t	6577	0.0	0.3	44:48	1:00:00
S-18_t	7788	0.2	0.1	1:00:00	1:00:00
S-19_t	9343	0.5	0.5	1:00:00	1:00:00
S-20_t	10841	8.3	0.1	1:00:00	1:00:00
Y-13_t	2730	0.0	0.0	12.63	25:41
Y-14_t	3164	0.0	0.0	1:52	14:44
Y-15_t	3676	0.0	0.3	2:26	1:00:00
Y-20_t	6046	9.5	0.0	1:00:00	1:00:00

Table 4 SDP gaps for large instances

Inst.	Best	Gap (%)	Gap (%)
	UB	(1h)	(5h)
S-21_t	12431	0.6	0.2
S-22_t	14208	0.1	0.0
S-23_t	16521	0.8	0.4
S-24_t	18658	0.3	0.1
S-25_t	21172	0.8	0.4
Y-25_t	10170	0.8	0.3
Y-30_t	13790	0.8	0.1
Y-35_t	19087	1.3	0.3
Y-40_t	23747	1.6	0.4
Y-45_t	31442	1.9	0.6
Y-50_t	41517	5.3	0.9
Y-60_t	55996	9.8	1.9

The results in Table 3 show that for instances with up to 17 departments the ILP model is faster than the SDP and all instances can be solved to optimality within 1 hour. But for large instances, particularly for $d \geq 20$, the SDP approach provides much better lower bounds, although we are only considering the SDP root node relaxation. Accordingly Table 4 gives the SDP results for large instances with more than 20 departments after 1 hour and 5 hours of computation. This shows that SDP is well-suited for providing high-quality lower bounds for large-scale instances in adequate running times.

One direction for future work is to use these bounds within a branch-and-bound scheme. Furthermore, it remains an open question to what extent the results in this paper can be generalized to the non-equidistant case.

References

1. Amaral, A.R.S.: A new lower bound for the single row facility layout problem. Discret. Appl. Math. **157**(1), 183–190 (2009)
2. Amaral, A.R.S.: On duplex arrangement of vertices. Technical report, Dep. de Informtica, Univ. Federal do Esprito Santo, Brazil (2011)
3. Anjos, M.F., Fischer, A., Hungerländer, P.: Solution approaches for equidistant double- and multi-row facility layout problems. Technical report, submitted (2015)
4. Gurobi Optimization Inc: Gurobi optimizer reference manual (2014). http://www.gurobi.com
5. Helmberg, C.: ConicBundle 0.3.11. TU Chemnitz, Germany (2012). http://www.tu-chemnitz.de/~helmberg/ConicBundle
6. Helmberg, C., Rendl, F.: A spectral bundle method for semidefinite programming. SIAM J. Optim. **10**(3), 673–696 (1999)
7. Hungerländer, P., Anjos, M.F.: A semidefinite optimization approach to space-free multi-row facility layout. Cahier du GERAD G-2012-03, GERAD, Montreal, QC, Canada (2012)
8. Hungerländer, P., Anjos, M.F.: Semidefinite optimization approaches to multi-row facility layout. Technical report, submitted (2014)
9. Hungerländer, P., Rendl, F.: A computational study and survey of methods for the single-row facility layout problem. Comput. Optim. Appl. **55**(1), 1–20 (2013)

Simulation of the System-Wide Impact of Power-to-Gas Energy Storages by Multi-stage Optimization

Christoph Baumann, Julia Schleibach and Albert Moser

Abstract In this paper a simulation method for the European power and natural gas system based on an optimization is introduced. The mathematical formulation of the problem represents a minimization of total costs subject to the coverage of demand and reserve requirements as well as the adherence of technical constraints. Due to the complexity of the problem, especially due to binary decisions and non-linearities resulting from restrictions of thermal power as well as Power-to-Gas plants, a closed-loop formulation is not practicable. Thus, this paper presents a simulation method consisting of a multi-stage optimization with the use of Lagrangian Relaxation and decomposition techniques.

1 Introduction

The transformation of the European energy system with a strong expansion of Renewable Energy Sources (RES) leads to an increasing coupling of the systems of natural gas and power. On the one hand, flexible thermal power plants are needed to provide backup for the intermittent feed-in of RES, on the other hand energy surpluses in situations with high RES feed-in and low demand will trigger the need for long-term energy storages. For the necessary thermal capacities, especially gas-fired power plants are suitable due to their high flexibility and efficiency as well as low greenhouse gas emissions. One of the most promising options for long-term storage is the Power-to-Gas technology (PtG), which transforms electrical energy into methane or hydrogen. The produced gas can be stored in the natural gas infrastructure afterwards. In order to simulate the impact of the increased coupling in general and PtG storages in particular, a combined simulation of the European energy system regarding the elements presented in Fig. 1 is necessary. Assuming an efficient energy market the simulation can be formulated as a cost minimization problem. This problem and its solution are described in the following.

C. Baumann (✉) · J. Schleibach · A. Moser
Institute of Power Systems and Power Economics (IAEW), RWTH Aachen University,
Schinkelstrasse 6, 52062 Aachen, Germany
e-mail: bm@iaew.rwth-aachen.de

© Springer International Publishing Switzerland 2016
M. Lübbecke et al. (eds.), *Operations Research Proceedings 2014*,
Operations Research Proceedings, DOI 10.1007/978-3-319-28697-6_4

25

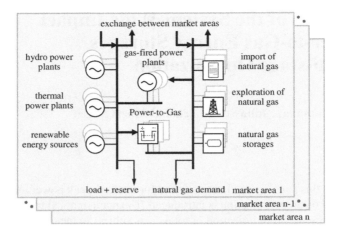

Fig. 1 Scope of model

2 Mathematical Model

The European power and natural gas system is modeled as several single gas and power market areas. Each power market area contains one or more gas market areas. Between the market areas cross-border exchange of gas and electrical energy, which is limited by maximum transfer capacities, is possible. Each power market area includes an hourly power demand and feed-in of renewable energy. The residual load, which is defined as the difference between power demand and non-dispatchable generation (especially RES), has to be covered by the hydraulic and thermal generation system.

In analogy to the power system, each gas market area has a daily demand for natural gas, which can either be covered by indigenous production, import from other European countries or import from countries outside of Europe. These different sources are defined by their flexibility of supply and production costs. Beside this, natural gas storages can adjust short-term and seasonal fluctuations in demand.

PtG and gas-fired power plants are coupling elements between the power and the natural gas system. PtG plants produce gas from electrical energy whereas in gas-fired power plants gas is used for electricity generation.

2.1 Objective Function

The problem of simulating the energy system for natural gas and power is formulated as minimization of the total costs C_{tot} for natural gas and power supply in N market areas and for T time intervals under consideration of load coverage and technical constraints. The usual simulation period is one year in hourly time steps for the power system and daily time steps for the natural gas system.

$$min \; C_{tot} = \sum_{n=1}^{N} \sum_{t=1}^{T} \left(C_{gas,n,t} + C_{power,n,t} \right) \tag{1}$$

The costs $C_{gas,n,t}$ include costs for gas transport and supply whereas the costs $C_{power,n,t}$ compose costs for power generation $\dot{C}_{i,stat}$ and transport. Thereby, the costs for the power output $p_{t,i}$ of a power plant i with the primary energy carrier source's costs $c_{i,prim}$, the heat consumption $\dot{Q}(p_{t,i})$ and the additional costs $c_{i,add}$ yield to:

$$\dot{C}_{i,stat} = c_{i,prim} \cdot \dot{Q}(p_{t,i}) + c_{i,add} \cdot p_{t,i} \tag{2}$$

The heat consumption is either modeled by a linearization in pieces or by a second-degree polynomial. Thereby, the convex heat consumption $\dot{Q}(p_{t,i})$ subject to the output power and the heat consumption coefficients hcc_0, hcc_1 and hcc_2 has the form:

$$\dot{Q}(p_{t,i}) = hcc_0 + hcc_1 \cdot p_{t,i} + hcc_2 \cdot p_{t,i}^2 \tag{3}$$

2.2 Constraints

In order to ensure safe operation of the grid and to guarantee security of supply, the electrical load has to be covered and reserve power has to be provided for each power market area. These two requirements are added to the minimization problem as linear constraints. The load is covered, if the feed-in of hydraulic and thermal power plants minus the consumption of electrical energy by PtG plants plus import balance of all exchange capacities (NTC) equals the residual load $d_{t,power}$:

$$d_{t,power} = \sum_{i=1}^{I} p_{t,i} + \sum_{j=1}^{J} p_{t,j} - \sum_{k=1}^{K} p_{t,k} + \sum_{m=1}^{M} \left(s_{t,m} - w_{t,m} \right) + \sum_{n=1}^{NTC} \left(im_{t,n} - ex_{t,n} \right) \tag{4}$$

In the stated formula I indicates the number of non-gas-fired thermal power plants, J the number of gas-fired power plants, K the number of PtG plants and M the number of hydro power plants. p_t represents the current power output or consumption, $s_{t,j}$ the feed-in and $w_{t,j}$ the power consumption of hydro power plants. The power exchange is represented by the imports im and exports ex.

The natural gas demand, which has to be covered in the natural gas system, consists of a fixed and a variable part. The gas demand of households, businesses and industry is predetermined in the model as a function of temperature and forms the fixed part $d_{t,gas,fix}$. The variable part arises from the demand or feed-in x_t of gas-fired power plants and PtG plants and results endogenously from the simulation. The gas load $d_{t,gas}$ is covered when fixed and variable demand equals the sum of domestic gas production $x_{t,prod}$, balance of exchanges of all market area interconnection points ($MAIP$) and gas storages ($STOR$) injection in or withdrawal wi.

$$d_{t,gas} = \sum_{j=1}^{J} x_{t,j} - \sum_{k=1}^{K} x_{t,k} + d_{t,gas,fix}$$

$$= x_{t,prod} + \sum_{o=1}^{MAIP} \left(im_{t,o} - ex_{t,o}\right) + \sum_{p=1}^{STOR} \left(wi_{t,p} - in_{t,p}\right) \qquad (5)$$

Besides load coverage, especially technical constraints have to be regarded in the model. PtG and power plants have dispatch limitations such as start-up procedures, minimum up- and down-times and minimum and maximum output. These limitations generate binary variables in the optimization problem.

For hydro power and gas storages, technical restrictions and time couplings need to be considered. The supply of natural gas splits into import and domestic production. Both sources underlie quantitative restrictions, which can be yearly or short-term maximum and minimum amounts of gas. Also, for the import of liquefied natural gas (LNG) the maximum capacity of LNG terminals has to be observed.

3   Developed Method

The mathematical optimization task forms a highly complex mixed integer quadratic programming problem. For this reason, a closed-loop solution by using commercially available solver software is not possible. Therefore, a method based on a decomposition approach is developed to solve the problem. The procedure, which is divided into three stages, is shown in Fig. 2. In the first stage, the power and natural gas system is optimized by Linear Programming (LP), while integer decisions and non-linearities are neglected. The resulting power exchanges and price indicators for natural gas, which are derived from the dual variables of the gas load coverage constraints ("shadow prices"), are passed on to the second stage. In the second stage, the PtG and power plant dispatch is optimized under consideration of all constraints for the power system described in Sect. 2.2 using Lagrangian Relaxation (LG). The Lagrangian equation $\mathscr{L}$ includes coordinators λ for load balance (LB) and μ for reserve balance (RB) [1]:

$$min \; \mathscr{L} = \max_{\lambda,\mu}\{ \min_{p,s,w,x} \left(C_{power} + \lambda(LB) + \mu(RB)\right)\} \qquad (6)$$

The dual problem of $\mathscr{L}$ can be decomposed into subpoblems for thermal and hydraulic units which can be solved by Dynamic Programming and Network Flow Optimization, respectively. The aim of maximizing the dual problem is achieved iteratively by adjusting λ and μ using gradient descent method. Finally, in the third stage the power and natural gas system is optimized by LP again by adopting the dispatch decisions found in the second stage and with linearized heat consumption curves.

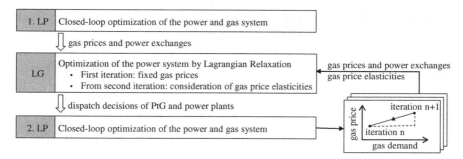

Fig. 2 Three-stage optimization

In the second stage of the optimization, gas prices are fixed. Since certain constraints are neglected in the first stage of the optimization, gas demand and thus gas prices might be different after the second LP. Therefore, new gas prices and power exchanges are derived from the last stage and passed on again to the second stage in an iterative process. In further developments of the method, gas price elasticities for specific market areas and time intervals shall be determined and considered in the LG stage if gas prices differ between iterations. Since the multi-stage approach does not guarantee finding the optimal solution inherently, close attention has to be paid to the parameterization of the LG stage and the development of the objective value between iterations.

4 Exemplary Results

In this section, first exemplary results computed with the developed method for the power and natural gas system simulation are presented. The underlying scenario represents an RES dominated scenario of the European energy system for the year 2050. The power system scenario is based on the German grid development plan [2] and the natural gas system is designed according to the outlook in ENTSO-G TYNDP [3]. In order to make use of power surpluses, PtG plants with a total capacity of 20 GW are considered in Germany. Overall, the scenario contains 992 thermal and 424 hydraulic units in the power system as well as 146 storages and 30 LNG terminals in the natural gas system.

The computer used for the calculations has 16 Intel Xeon processor cores and a main memory of 256 GB. All linear problems are solved with IBM ILOG Optimization Studio using barrier algorithm. The resulting computation time for an annual simulation and the development of total costs between iterations for a monthly simulation is shown in Fig. 3.

It can be seen that the computing time for the LG relaxation is about twice as long as the time for each LP. With each iteration of the LG and second LP stage the computational time increases correspondingly. The total costs after the first iteration

Process	Duration
Import	0.7 h
First LP	5.9 h
LG Relaxation	9.6 h
Second LP	5.3 h
Output	1.1 h

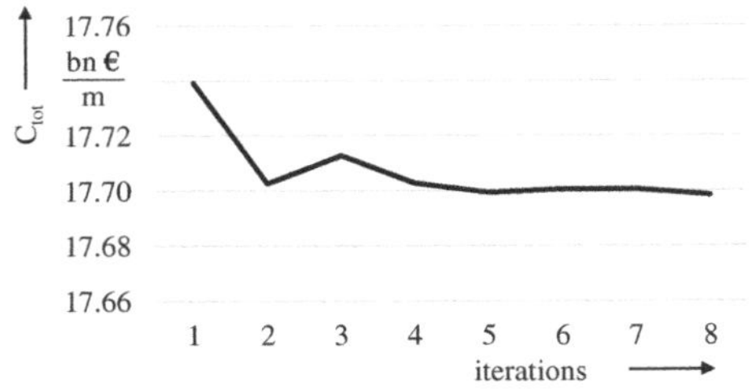

Fig. 3 Computation time and development of total costs

are the highest, because the determination of gas prices in this iteration does not include e.g. starting processes of power plant due to their negligence in the first LP. In the following iterations the total costs converge and only vary less than 0.06 %. An implementation of the consideration of gas price elasticities in the second optimization stage should further improve convergence.

5 Conclusions

The introduced method allows to simulate the European natural gas and power system under consideration of relevant constraints. In order to keep the complex problem solvable and the computational time manageable, a multi stage optimization is necessary. After implementing final improvements the method will be applied to assess the impact of PtG on the energy system by simulating different scenarios for the future.

References

1. Drees, T., Schuster, R., Moser, A.: Proximal bundle methods in unit commitment optimization. In: Operations Research Proceedings 2012, pp. 469–474. Springer International Publishing, Heidelberg (2013)
2. German TSO: Netzentwicklungsplan Strom 2013. Berlin (2013)
3. ENTSO-G: Ten-Year Network Development Plan 2013–2022. Brussels (2013)

An Approximation Result for Matchings in Partitioned Hypergraphs

Isabel Beckenbach and Ralf Borndörfer

Abstract We investigate the matching and perfect matching polytopes of hypergraphs having a special structure, which we call partitioned hypergraphs. We show that the integrality gap of the standard LP-relaxation is at most $2\sqrt{d}$ for partitioned hypergraphs with parts of size $\leq d$. Furthermore, we show that this bound cannot be improved to $\mathcal{O}(d^{0.5-\varepsilon})$.

1 Introduction

It is well known that the set packing problem is $\mathcal{NP}$-hard and that it admits no constant factor approximation algorithm unless $\mathcal{P} = \mathcal{NP}$. Here, we look at the equivalent problem of finding a maximum weight matching in a hypergraph. There exists a lot of work characterizing classes of hypergraphs for which the matching problem can be solved in polynomial time. For example, this is the case for balanced hypergraphs introduced by Berge in [2]. As for bipartite graphs the standard LP-Relaxation of the matching polytope of a balanced hypergraph is integral. So the polynomial time solvability of the matching problem in balanced hypergraphs follows from the fact that one can optimize over the LP-Relaxation in polynomial time. In general, the integrality gap of the standard LP-Relaxation can be arbitrarily high. In the following we look at hypergraphs in which the hyperedges have a special structure. We call this class partitioned hypergraphs as in [3]. We investigate the matching and perfect matching polytope of partitioned hypergraphs and their LP-Relaxations. Furthermore, we show that the integrality gap of the standard LP-relaxation of a partitioned hypergraph is bounded by its part size.

I. Beckenbach (✉) · R. Borndörfer (✉)
Zuse Institut Berlin, Takustraße 7, 14195 Berlin, Germany
e-mail: beckenbach@zib.de

R. Borndörfer
e-mail: borndoerfer@zib.de

© Springer International Publishing Switzerland 2016
M. Lübbecke et al. (eds.), *Operations Research Proceedings 2014*,
Operations Research Proceedings, DOI 10.1007/978-3-319-28697-6_5

2 Definitions

In this section we introduce some basic definitions and notations that we use in the remainder. First, we give two definitions that show the close connection between matchings in hypergraphs and the set packing problem and perfect matching and the set partitioning problem, respectively.

Every hypergraph can be represented by a 0, 1 matrix in the following way:

Definition 1 Let $H = (V, E)$ be a hypergraph, The *incidence matrix* of H is the matrix $A = (a_{v,e})_{v \in V, e \in E} \in \{0, 1\}^{V \times E}$ defined by

$$a_{v,e} = \begin{cases} 1, & \text{if } v \in e \\ 0, & \text{else} \end{cases} \tag{1}$$

Now, we define the four polytopes that we investigate in the next section.

Definition 2 Let $H = (V, E)$ be a hypergraph the *matching polytope*, the *fractional matching polytope*, the *perfect matching polytope*, and the *fractional perfect matching polytope* are defined by:

$$\mathrm{IP}_M(H) = \mathrm{conv}\left(\left\{x \in \{0, 1\}^E \,|\, Ax \leq 1\right\}\right), \tag{2}$$

$$\mathrm{LP}_M(H) = \mathrm{conv}\left(\left\{x \in \mathbb{R}^E \,|\, Ax \leq 1, \ x \geq 0\right\}\right), \tag{3}$$

$$\mathrm{IP}_{PM}(H) = \mathrm{conv}\left(\left\{x \in \{0, 1\}^E \,|\, Ax = 1\right\}\right), \tag{4}$$

$$\mathrm{LP}_{PM}(H) = \mathrm{conv}\left(\left\{x \in \mathbb{R}^E \,|\, Ax = 1, \ x \geq 0\right\}\right). \tag{5}$$

The extreme points of $\mathrm{IP}_M(H)$ are exactly the incidence vectors of matchings in H. So, finding a maximum weight matching is equivalent to optimizing over $\mathrm{IP}_M(H)$ which is hard. However, we can optimize over $\mathrm{LP}_M(H)$ to obtain an upper bound. Therefore, if we can bound the integrality gap of $\mathrm{LP}_M(H)$ we obtain an approximation result for the maximum weight of a matching in H.

Borndörfer and Heismann introduced in [3] the hypergraph assignment problem which is a generalization of the assignment problem to hypergraphs. The hypergraph assignment problem can also be seen as a perfect matching problem in a hypergraph having the following special structure:

Definition 3 Let $H = (V \cup W, E)$ be a hypergraph with $|V| = |W|$, $V \cap W = \emptyset$, and $|e \cap V| = |e \cap W|$ for all $e \in E$. A nonempty set $P \subseteq V$ or $P \subseteq W$ is called a *part* of H if for all $e \in E$ either $e \cap V \subseteq P$ or $(e \cap V) \cap P = \emptyset$ holds or in the case $P \subseteq W$ either $e \cap W \subseteq P$ or $(e \cap W) \cap P = \emptyset$ holds. H is a *partitioned hypergraph* with *maximum part size* d if there are disjoint parts $P_1, \ldots, P_r \subseteq V$ and $Q_1, \ldots, Q_s \subseteq W$ that form a partition of V and W, respectively, and $|P_i| \leq d, |Q_j| \leq d$ for $1 \leq i \leq r, 1 \leq j \leq s$.

It is easy to see that the intersection of two parts is empty or again a part. So, there exists a unique finest partition $P_1, \ldots, P_r$ of V and a unique finest partition

$Q_1, \ldots, Q_s$ of W into parts. We always assume that we have a partitioned hypergraph with its finest partition into parts. Under this assumption the part size of a partitioned hypergraphs is the maximum size of one of its parts.

In the next section we deal with the following "complete" partitioned hypergraph with parts of size two:

Definition 4 Let $n \in \mathbb{N}$ be an even number. The partitioned hypergraph D_n consists of two disjoint vertex sets $V_n = \{v_1, \ldots, v_n\}$ and $W_n = \{w_1, \ldots, w_n\}$. Each of the two vertex sets is partitioned into $\frac{n}{2}$ parts of size two, say $V_n^i = \{v_{2i-1}, v_{2i}\}$, $W_n^i = \{w_{2i-1}, w_{2i}\}$ for all $1 \leq i \leq \frac{n}{2}$. The set of hyperedges E_n of D_n consists of n^2 edges $\{v_i, w_j\}$ for all $1 \leq i, j \leq n$ and $\frac{n^2}{4}$ hyperedges of the form $V_n^i \cup W_n^j$ for all $1 \leq i, j \leq \frac{n}{2}$.

3 Polyhedral Investigations

In this section we give some results on the matching polytope, the perfect matching polytope, and their fractional variants. We begin with the dimension of these polytopes.

Theorem 1 $\mathrm{IP}_M(H)$ and $\mathrm{LP}_M(H)$ have full dimension.

Proof $\{\chi_\emptyset\} \cup \{\chi_{\{e\}} | e \in E(H)\}$ is a set of $|E| + 1$ affinely independent vectors in $\mathrm{IP}_M(H)$ and $\mathrm{LP}_M(H)$, so $\mathrm{IP}_M(H), \mathrm{LP}_M(H) \subseteq \mathbb{R}^E$ have full dimension. $\square$

The dimension of the perfect matching polytope is more difficult to calculate, as it is $\mathcal{NP}$-hard to decide whether a hypergraph has a perfect matching (i.e. $\mathrm{IP}_{PM}(H)$ is non-empty). However, for D_n it is possible to calculate the dimension of the perfect matching polytope and the fractional perfect matching polytope.

Theorem 2 *The dimension of* $\mathrm{IP}_{PM}(D_n)$ *and* $\mathrm{LP}_{PM}(D_n)$ *is* $\frac{5}{4}n^2 - 2n + 1$.

Proof As every valid equation for $\mathrm{LP}_{PM}(D_n)$ is a linear combination of the rows of $Ax = 1$, the dimension of $\mathrm{LP}_{PM}(D_n)$ is $|E_n| - \mathrm{rank}(A)$. Let a_e be a column of A corresponding to a hyperedge of the form $V_n^i \cup W_n^j$. Then e is the disjoint union of the two edges $e_1 = \{v_{2i-1}, w_{2i-1}\}$ and $e_2 = \{v_{2i}, w_{2i}\}$ and a_e is the sum of the two column vectors corresponding to e_1 and e_2. So we can delete column a_e from A without changing the rank of A. Doing this for all columns corresponding to hyperedges of size four, shows that the rank of A is the same as the rank of the incidence matrix of $K_{n,n}$ which is $2n - 1$. It follows that $\dim(\mathrm{LP}_{PM(D_n)}) = |E_n| - 2n + 1 = \frac{5}{4}n^2 - 2n + 1$.

To see that $\dim(\mathrm{IP}_{PM}(D_n)) = \dim(\mathrm{LP}_{PM}(D_n))$, we construct $\frac{5}{4}n^2 - 2n + 2$ affinely independent vectors in $\mathrm{IP}_{PM}(D_n)$. First, observe that every fixed hyperedge of size four can be completed to a perfect matching of D_n by adding edges. Clearly, the incidence vectors of these $\frac{n^2}{4}$ perfect matchings are affinely independent. The matching polytope of $K_{n,n}$ has dimension $n^2 - 2n + 1$. Thus, there are $n^2 - 2n + 2$ perfect matchings in $K_{n,n}$ such that their incidence vectors are affinely independent. These vectors can be lifted to vectors in $\mathrm{IP}_{PM}(D_n)$ by setting all entries corresponding to hyperedges of size four to 0. The $\frac{n^2}{4}$ first vectors and these $n^2 - 2n + 1$ new vectors are affinely independent.   □

Now, we state some results on valid inequalities and facets of the matching polytope and the perfect matching polytope (see [1] for proofs).

Theorem 3 *Every trivial inequality $x_e \geq 0$ defines a facet of* $\mathrm{IP}_M(H)$.

In the case of the perfect matching polytope it is even difficult to decide when a trivial inequality is facet defining. So we restrict ourselves to the hypergraphs D_n.

Theorem 4 *The trivial inequality $x_e \geq 0$ defines a facet of* $\mathrm{IP}_{PM}(D_n)$

A clique in a hypergraph is a set $Q \subseteq E$ of hyperedges such that every two elements of Q intersect. Clearly, every matching contains at most one edge from a clique. So $x(Q) \leq 1$ is a valid inequality for $\mathrm{IP}_M(H)$.

Theorem 5 *A clique inequality $x(Q) \leq 1$ defines a facet of* $\mathrm{IP}_M(H)$ *if and only if Q is a maximal clique.*

Heismann also generalized the odd set inequalities that are valid for the matching polytope of a graph to valid inequalities for the (perfect) matching polytope of a hypergraph. See references [1, 9] for more details.

4   Integrality Gap

In this section we prove our main result concerning the multiplicative integrality gap of maximizing over the matching polytope of a partitioned hypergraph. This gap can be bounded in contrast to the unbounded integrality gap of the perfect matching polytope (see [1]).

Füredi, Kahn and Seymour show in [5] that the integrality gap of

$$\max \, w^t x \tag{6}$$
$$\text{s.t.} \ \ x \in \mathrm{LP}_M(H)$$

is at most $k - 1 + \frac{1}{k}$ for k-uniform hypergraphs. For k-partite hypergraphs the result can be strengthen to $k - 1$. The proofs of [5] are non-algorithmic, however, in [4] an iterative rounding algorithm with approximation factor $k - 1$ is given for the maximum weight matching problem in k-partite hypergraphs. For the analysis of their algorithm Chan and Lau consider the following linear program for fixed degree bounds $0 \leq B_v \leq 1$:

$$\max \ \ w^t x \tag{7}$$
$$\text{s.t.} \ \ x(\delta(v)) \leq B_v \ \forall v \in V(H)$$
$$x_e \ \ \geq 0 \ \forall e \in E(H)$$

Let $N[e] := \{e' : e \cap e' \neq \emptyset\}$ be the set of all hyperedges intersecting e. The crucial point of their proof for an integrality gap of $k - 1$ for (6) is that for every extreme point x of (7) with $x > 0$ there exists a hyperedge $e \in E(H)$ with $x(N[e]) \leq k - 1$. The further analysis of the algorithm in [4] does not use the k-partiteness of the hypergraph. If we can show that for every extreme point x with $x > 0$ there exists a hyperedge $e \in E(H)$ with $x(N[e]) \leq \alpha$ for H in some class $\mathcal{C}$ of hypergraphs, then the result of [4] directly gives an α-approximation algorithm for the weighted matching problem in $\mathcal{C}$. For partitioned hypergraphs we can proof the following bound:

Lemma 1 *Let H be a partitioned hypergraph with maximum part size d and x be an extreme point of (7) with $x_e > 0$ for all $e \in E(H)$. There exists a hyperedge $e^* \in E(H)$ with $x(N[e^*]) \leq 2\sqrt{d}$.*

Proof 1. Case: There exists a hyperedge e^* of size less than $2\sqrt{d}$. Then

$$\sum_{e \in N[e^*]} x(e) \leq \sum_{v \in e^*} \sum_{e:v \in e} x(e) \leq |e^*| < 2\sqrt{d}. \tag{8}$$

2. Case: $|e| \geq 2\sqrt{d}$ for all $e \in E(H)$. We choose $e^* \in E$ arbitrarily. Let P and P' be the two parts of H such that $e^* \subseteq P \cup P'$. Summing over all inequalities $x(\delta(v)) \leq 1$ for $v \in P$ gives

$$\sum_{e \in \delta(P)} \sqrt{d} x(e) \leq \sum_{e \in \delta(P)} \frac{|e|}{2} x(e) \leq d, \tag{9}$$

and the same inequality holds for $e \in \delta(P')$. Thus we get

$$\sum_{e \in N[e^*]} x(e) \leq \sum_{e \in \delta(P)} x(e) + \sum_{e \in \delta(P')} x(e) \leq 2\sqrt{d}. \tag{10}$$

$\square$

Now, we can proof that (6) has an integrality gap $\leq 2\sqrt{d}$ for partitioned hypergraphs with maximum part size d. The proof is based on the ideas used in [4] for the analysis of the k-dimensional matching algorithm.

Theorem 6 *The multiplicative integrality gap of (6) is at most $2\sqrt{d}$ for a partitioned hypergraph H with maximum part size d.*

Proof Let x be an extreme point of (6). We have to show that there exists a matching M of H such that $w^t x \leq 2\sqrt{d} \times w(M)$.

We use induction on the number of hyperedges $e \in E(H)$ with positive weight. If $w(e) = 0$ for all hyperedges $e \in \mathcal{E}$ the claim trivially holds. Otherwise, there exists a hyperedge e^* of positive weight with $x(N[e^*]) \leq 2\sqrt{d}$.

Define a weight function w^1 by $w^1(e) := w(e^*)$ for all $e \in N[e^*]$ and $w^1(e) := 0$ for all other $e \in E(H)$. Furthermore, set $w^2(e) := w(e) - w^1(e)$ for all $e \in E(H)$.

The weight function w^2 has fewer hyperedges with positive weight then w. By induction there exists a matching M' of H with $(w^2)^t x \leq 2\sqrt{d} \times w^2(M')$. If $M' \cup \{e^*\}$ is a matching we set $M := M' \cup \{e^*\}$, otherwise we set $M := M'$. In both cases, we have $w^2(M) = w^2(M')$ and $w^1(M) = w(e^*)$, because $w^2(e^*) = 0$ and $N[e^*] \cap M \neq \emptyset$. It follows that:

$$2\sqrt{d}w(M) = 2\sqrt{d}w^2(M) + 2\sqrt{d}w^1(M) = 2\sqrt{d}w^2(M') + 2\sqrt{d}w(e^*) \quad (11)$$
$$\geq (w^2)^t x + w(e^*)x(N[e^*]) = (w^2)^t x + (w^1)^t x = w^t x. \quad (12)$$

$\square$

For general hypergraphs with hyperedges of size k Hazan, Safra and Schwartz proved in [8] that there is no $\mathcal{O}(\frac{k}{\ln k})$ approximation algorithm for the maximum matching problem unless $\mathcal{P} = \mathcal{N}\mathcal{P}$. So, if the maximum part size of a partitioned hypergraph is ck for some constant $c \in \mathbb{Q}_+$ we get a $\mathcal{O}(\sqrt{k})$-approximation algorithm which is better than $\mathcal{O}(\frac{k}{\ln k})$.

Furthermore, there exists a $2|V(H)|^{0.5}$ approximation algorithm for the maximum weight matching problem in hypergraphs with hyperedges of unbounded size (see [6]). This approximation factor cannot be improved to $\mathcal{O}(|V(H)|^{0.5-\varepsilon})$ in the unweighted case (see [7]). Every hypergraph H can be transformed into a partitioned hypergraph H_P with maximum part size $\leq |V(H)|$ by setting $V(H_P) := V(H) \times \{0, 1\}$ and $E(H_P) := \{\{(v, 0), (v, 1) : v \in e\} : e \in E(H)\}$. This shows that (6) cannot have an integrality gap of $\mathcal{O}(d^{\frac{1}{2}-\varepsilon})$, so Theorem 6 is nearly tight.

Acknowledgments The first author was supported by BMBF Research Campus MODAL.

References

1. Beckenbach, I.: Special cases of the hypergraph assignment problem. Master Thesis, TU Berlin (2013)
2. Berge, C.: Sur certains hypergraphes généralisant les graphes bipartites. In: Combinatorial Theory and its Applications, I (Proc. Colloq., Balatonfred, 1969), pp. 119–133. North-Holland, Amsterdam (1970)
3. Borndörfer, R., Heismann, O.: The Hypergraph Assignment Problem. Technical Report, Zuse Institut, Berlin (2012)
4. Chan, Y., Lau, L.: On linear and semidefinite programming relaxations for hypergraph matching. Math. Prog. **135**(1–2), 123–148 (2012)
5. Füredi, Z., Kahn, J., Seymour, P.D.: On the fractional matching polytope of a hypergraph. Combinatorica **13**(2), 167–180 (1993)
6. Halldórsson, M.M.: Approximations of weighted independent set and hereditary subset problems. Lect. Notes Comput. Sci. **1627**, 261–270 (1999)
7. Halldórsson, M.M., Kratochvíl, J., Telle, J.A.: Independent sets with domination constraints. Discret. Appl. Math. **99**, 39–54 (2000)
8. Hazan, E., Safra, M., Schwartz, O.: On the complexity of approximating k-set packing. Comput. Complex. **15**(1), 20–29 (2006)
9. Heismann, O.: The Hypergraph Assignment Problem. Ph.D. Thesis, TU Berlin (2014)

On Class Imbalance Correction for Classification Algorithms in Credit Scoring

Bernd Bischl, Tobias Kühn and Gero Szepannek

Abstract Credit scoring is often modeled as a binary classification task where defaults rarely occur and the classes generally are highly unbalanced. Although many new algorithms have been proposed in the recent past to mitigate this specific problem, the aspect of class imbalance is still underrepresented in research despite its great relevance for many business applications. Within the "Machine Learning in R" (mlr) framework methods for imbalance correction are readily available and can be integrated into a systematic classifier optimization process. Different strategies are discussed, extended and compared.

1 Introduction

Credit scoring denotes the assignment of ordered values (scores) to individuals that are supposed to be decreasing with risk. Here, risk is interpreted as the probability of a lender to default in the future. Business application scoring models are a major element in credit decisions and the IRBA Basel capital framework [1, 19].

In order to model credit risk, typically a binary random variable with two outcomes (default and non-default) and classification algorithms are used. The most widely-used technique is logistic regression, but in the recent past several new models have been proposed and studies have compared performances in different data situations [17] as well as on real world credit scoring data [2, 13].

The opinions expressed in this paper are those of the authors and do not reflect views of any organization or employer.

B. Bischl (✉) · T. Kühn (✉) · G. Szepannek (✉)
LMU München, Munich, Germany
e-mail: bernd_bischl@gmx.net; bischl@statistik.uni-dortmund.de

T. Kühn
e-mail: tobi.kuehn@gmx.de

G. Szepannek
Stralsund University of Applied Sciences, Stralsund, Germany
e-mail: gero.szepannek@web.de

© Springer International Publishing Switzerland 2016
M. Lübbecke et al. (eds.), *Operations Research Proceedings 2014*,
Operations Research Proceedings, DOI 10.1007/978-3-319-28697-6_6

One complicating factor in credit scoring is that classes typically follow a highly unbalanced distribution, i.e., the default class is much smaller. The effect of this on the performance of different classification algorithms has been investigated by Brown and Mues [7]. Vincotti and Hand [20] discuss both the introduction of misclassification costs at either scorecard construction or classification stage and preprocessing of the training sample by over- or undersampling of the classes. The effect of over- and undersampling in relation to effective class sizes has been extensively investigated by Crone and Finlay [9].

In this article, we study different strategies for imbalance correction together with different classifiers in a comprehensive setting, introducing several new aspects:

- Joint framework for tuning of classifiers and imbalance correction.
- Newer techniques like SMOTE [8] and *overbagging* are investigated.
- Extension of SMOTE through the Gower distance for categorical data.
- Iterated F-racing instead of grid search [13] for tuning within mlr [5, 16].
- Large data base of real world data sets and validation on credit scoring data.
- Realistic evaluation of logistic regression using coarse classed data.

2 Methodology

Imbalance Correction: A standard approach for class imbalance correction consists in sampling [9]: In *undersampling* a random subset of the majority class is used for training, whereas *oversampling* randomly duplicates instances of the minority class. Some classifiers allow *weighting* of observations during training, which is a straight-forward, alternative *intrinsic imbalance correction* to sampling, if one downweights majority and upweights minority class observations. Oversampling can be extended to *overbagging*, where the oversampling of the minority class is repeated several times. Majority class instances are bootstrapped in each iteration and for the new training sets we fit a bagging predictor in order to reduce prediction variance. The popular *synthetic minority over-sampling (SMOTE)* [8] generates new observations of the minority class as random convex combinations of neighboring observations. As categorical features occur in many real-world problems, we use the Gower distance in this mixed space to identify neighbors and sample a new category for each categorical feature from the respective two entries of the neighbors during the convex combination step.

The **mlr R package** [6] offers an interface to more than 50 classification, regression and survival analysis models, and most standard resampling and evaluation procedures. Models can be chained and extended with, e.g., preprocessing operations and jointly optimized. The package allows for different optimization/configuration techniques, from simple random search, to iterated F-racing and sequential model based optimization. The latter two are arguably among the most popular and successful approaches for algorithm configuration nowadays.

Iterated F-racing [12, 14] builds upon the simpler racing technique, where algorithm candidate configurations are sequentially evaluated on a stream of instances. After each iteration, a statistical test is performed—usually the non-parametric Friedman test—to identify outperformed candidates, which are eliminated from the candidate set. In our case, candidates are joint hyperparameter settings for classifiers and imbalance correction and instances are subsampled versions of the training data set. Iterated F-racing samples one set of candidate configurations from a joint distribution over the parameter space, performs a usual F-race to reduce the current candidates to number of elite configurations and adapts the distribution by centering it around the elites as well as reducing its spread. The latter results in exploration in the beginning and exploitation in the later stages of the optimization.

3   Experiments

A typical problem in credit scoring research is the availability of data so that most studies are based on only a few data sets [2]. In order to obtain general results we follow a two-fold approach: First, all methods are evaluated on a large set of public unbalanced data sets,[1] among them the popular German credit data, which is not very representative due to the low degree of imbalance (30 %) and the low number of observations. In a second step, we validate the results on two more realistic real world credit scoring problems: gmsc[2] and glc [15].

Industrial standard includes preliminary coarse classing of the data. We address this by generating additional binned data sets (suffix "nom") using decision trees with varying complexity parameters [18] and subsequent manual investigation of bins concerning numbers of defaults and default rates using binomial tests. The manual step implies a loss in scientific rigor, but allows to assess the results with respect to industrial practice.

As a general preprocessing step, constant features are removed from the data sets. Afterwards, five classification techniques, *logistic regression (logreg), rpart decision tree (rpart), random forest (RF), gradient boosting (gbm)* and *support vector machines (ksvm)* with a Gaussian kernel, are applied to each data set. We combine all classifiers with all mentioned imbalance correction techniques. In this context, we study the following variants: classifiers without tuning or imbalance correction (bl = baseline), normal tuning of hyperparameters (tune), and joint tuning of the classifier and an imbalance correction method like class weighting (cw), undersampling (us), oversampling (os), SMOTE (sm) and 10 iterations of overbagging (ob) (Tables 1, 2, 3 and 4).

---

[1] http://www.cs.gsu.edu/~zding/research/benchmark-data.php.
[2] http://www.kaggle.com/c/GiveMeSomeCredit.

Table 1 Overview of tuning parameters (arguments of corresponding R functions) for each learner

Learner	Tuning parameters with range (lower, upper)
gbm	n.trees (100, 5000)/interaction.depth (1, 5)/shrinkage (1e-05, 0.1)/bag.fraction (0.7, 1)
ksvm	C $(2^{-12}, 2^{12})$ / sigma $(2^{-12}, 2^{12})$
logreg	–
RF	ntree (10, 500)/mtry (1, 10)
rpart	cp (0.0001, 0.1)/minsplit (1, 50)

Table 2 Effect of tuning versus imbalance correction: top ten AUC improvements of the baseline by tuning (left, columns 1 to 4) as well as improvements of tuning by additional imbalance correction together with the best strategy (right, columns 5 to 8)

Data	Learner	Base	Tuning	Data	Learner	Tuning	Imbal	Method
balance	gbm	0.29	0.89	poker	rpart	0.47	0.76	sm
poker	gbm	0.53	1.00	abalone19	rpart	0.56	0.81	ob
balance	ksvm	0.68	0.92	balance	rpart	0.50	0.73	ob
mammography	gbm	0.71	0.94	balance	logreg	0.29	0.50	us
gmsc	gbm	0.66	0.87	solar flare m0	ksvm	0.62	0.82	sm
satellite image	gbm	0.78	0.97	ozone level	rpart	0.67	0.84	ob
abalone7	rpart	0.50	0.67	poker	logreg	0.34	0.52	us
vehicle	gbm	0.69	0.86	abalone7	rpart	0.67	0.83	os
coil2000	rpart	0.50	0.66	oil spill	rpart	0.70	0.85	ob
glc	rpart	0.70	0.85	balance	RF	0.36	0.50	ob

Table 3 Effect of tuning versus imbalance correction: Mean improvements per learner by tuning across data sets (left) and further improvements by imbalance correction across data sets, averaged over all sampling methods and best sampling method on average (right)

Learner	Tuning mean	Imbal mean	Imbal max	Method
gbm	0.14	0.02	0.04	ob
ksvm	0.05	0.04	0.06	us
logreg	0.00	0.05	0.13	us
RF	0.00	0.04	0.14	ob
rpart	0.04	0.13	0.29	sm

The parameter controlling the upsampling ratio/minority class upweighting is tuned in the range of 1 and $1.5 \times IR$, where IR is the class imbalance ratio. For undersampling we use a range of $0.67 \times IR^{-1}$ and 1.

For all set-ups (except the baseline) tuning is performed via iterated F-racing and a budget of 400 evaluations. During the inner resampling for tuning (in each racing step) we use 80 % of the observations for training and 20 % for testing. The whole tuning/model selection process is embedded into an outer loop of stratified 5-fold

Table 4 AUC of the best algorithm (bold) and learner per dataset

Data	IR	N	Feat	Learner	Base	Tuning	Weights	Sampling	Method	Rate
vehicle	2.90	846	18	ksvm	0.868	0.926	0.489	**0.935**	os	2.90
satellite image	9.28	6435	36	ksvm	0.935	0.967	0.965	**0.968**	sm	9.29
abalone7	9.68	4177	10	ksvm	0.774	0.85	0.865	**0.87**	os	8.72
balance	11.76	625	20	ksvm	0.68	**0.917**	0.857	0.765	os	8.13
us crime	12.29	1994	100	ksvm	0.87	0.927	0.925	**0.928**	sm	4.17
yeast ml8	12.58	2417	103	ksvm	0.592	0.605	**0.619**	0.605	sm	9.80
scene	12.60	2407	294	RF	0.763	0.783	0.804	**0.815**	os	9.64
coil2000	15.76	9822	85	gbm	0.685	0.758	0.762	**0.762**	os	15.64
solar flare m0	19.43	1389	32	ksvm	0.628	0.619	0.814	**0.818**	sm	18.00
oil spill	21.85	937	48	RF	0.93	0.923	0.907	**0.948**	sm	10.59
yeast2vs8	23.10	482	8	RF	0.927	0.847	0.906	**0.929**	os	14.54
wine quality4	25.77	4898	11	RF	0.898	0.898	0.874	**0.907**	sm	4.15
yeast uci me2	28.10	1484	8	RF	0.93	0.923	0.915	**0.934**	os	16.44
ozone level	33.74	2536	72	ksvm	0.845	0.886	0.915	**0.916**	os	17.07
yeast6	41.40	1484	8	gbm	0.903	0.945	0.944	**0.954**	sm	33.08
mammography	42.01	11183	6	gbm	0.708	0.943	**0.953**	0.949	os	22.18
poker	58.40	1485	10	gbm	0.525	0.998	**1**	1	os	26.38
abalone19	129.53	4177	10	logreg	0.816	0.816	0.816	**0.842**	os	153.70
gcd	2.33	1000	19	RF	**0.798**	0.792	0.781	0.787	os	2.98
gcd nom	2.33	1000	20	logreg	**0.787**	0.787	0.787	0.784	os	2.78
glc	11.91	28882	26	gbm	0.788	0.922	**0.924**	0.922	os	9.49
glc nom	11.91	28882	19	logreg	**0.909**	0.909	0.909	0.909	os	8.94
gmsc	13.96	150000	10	gbm	0.656	0.865	**0.866**	0.864	os	8.94
gmsc nom	13.96	150000	10	logreg	0.86	0.86	0.86	**0.861**	os	13.72

In comparison AUC of the same learner without tuning or imbalance correction (Base), with tuning of hyperparameters only (Tuning), —and class weights (Weights)—and the best sampling method (Sampling). Method gives the name of the sampling strategy, the best found sampling rate/class upweighting rate/class upweighting parameter for this method is shown in Rate

cross-validation to ensure unbiased performance estimation. As it represents a standard for credit scoring applications the area under the ROC curve (AUC) is used both as a measure for tuning and performance evaluation [4, 11]. We parallelize our experiments via the BatchJobs and BatchExperiments R packages [3].

4   Results and Summary

The tables show the results of the conducted experiments. Often strong improvements are achieved, mostly using upsampling strategies—which are unfortunately the computationally most expensive ones. Also, these improvements are only observed in combination with proper hyperparameter tuning, especially for SVMs and boosting which reflects their strong dependence on parameterization. Decision trees are most strongly affected by imbalance correction, followed by random forests and logistic regression. Note that in some rare cases the results after imbalance correction worsen, which might be due to an overfitting on the validation sets. The results do not uniquely favor a single combination of methods and the picture is much less clear than in [10], where only decision trees and no tuning was considered. Nevertheless, boosting and upsampling (sm or os) seem to be a good choice in many cases.

For credit scoring, the established pre-binned logistic regression shows good results, but improvements by the proposed integrated tuning and imbalance correction framework are visible. This is especially noteworthy, as the pre-binning comes with a substantial time investment for the human expert, while the automated one does not.

References

1. Baesens, B., van Gestel, T.: Credit Risk Management—Basic Concepts. Oxford University Press, Oxford (2009)
2. Baesens, B., Van Gestel, T., Viaene, S., Stepanova, M., Suykens, J., Vanthienen, J.: Benchmarking state of the art classification algorithms for credit scoring. J. Oper. Res. Soc. **54**(6), 627–635 (2003)
3. Bischl, B., Lang, M., Mersmann, O., Rahnenführer, J., Weihs, C.: BatchJobs and BatchExperiments: abstraction mechanisms for using R in batch environments (ACCEPTED). J. Stat. Soft. (2015)
4. Bischl, B., Schiffner, J., Weihs, C.: Benchmarking local classification methods. Comput. Stat. **28**(6), 2599–2619 (2013)
5. Bischl, B., Schiffner, J., Weihs, C.: Benchmarking classification algorithms on high-performance computing clusters. In: Spiliopoulou, M., Schmidt-Thieme, L., Janning, R. (eds.) Data Analysis, Machine Learning and Knowledge Discovery, Studies in Classification, Data Analysis, and Knowledge Organization, pp. 23–31. Springer, Heidelberg (2014)
6. Bischl, B., Lang, M., Richter, J., Judt, L.: mlr: Machine Learning in R. R package version 2.0. http://CRAN.R-project.org/package=mlr (2014)
7. Brown, I., Mues, C.: An experimental comparison of classification algorithms for imbalanced credit scoring data sets. Expert Syst. Appl. **39**(3), 3446–3453 (2012)

8. Chawla, N.V., Bowyer, K.W., Hall, L.O., Kegelmeyer, W.P.: SMOTE: synthetic minority over-sampling technique. J. Artif. Intell. Res. **16**, 321–357 (2002)
9. Crone, S., Finlay, S.: Instance sampling in credit scoring: an empirical study of sample size and balancing. Int. J. Forecast. **28**(1), 224–238 (2012)
10. Galar, M., Fernandez, A., Barrenechea Tartas, E., Bustince Sola, H., Herrera, F.: A review on ensembles for the class imbalance problem: Bagging-, Boosting-, and Hybrid-Based Approaches. IEEE Trans. Syst. Man Cybern. Part C **42**(4), 463–484 (2012)
11. Koch, P., Bischl, B., Flasch, O., Bartz-Beielstein, T., Weihs, C., Konen, W.: Tuning and evolution of support vector kernels. Evol. Intell. **5**(3), 153–170 (2012)
12. Lang, M., Kotthaus, H., Marwedel, P., Weihs, C. Rahnenführer, J., Bischl, B.: Automatic model selection for high-dimensional survival analysis. J. Stat. Comput. Simul. (2014)
13. Lessmann S., Seow H.-V., Baesens, B., Thomas, L.C.: Benchmarking state-of-the-art classification algorithms for credit scoring: A ten-year update. http://www.business-school.ed.ac.uk/waf/crc_archive/2013/42.pdf (2013)
14. Lopez-Ibanez, M., Dubois-Lacoste, J., Stützle, T., Birattari, M.: The irace Package: iterated racing for automatic algorithm configuration, Technical report TR/IRIDIA/2011-004. IRIDIA, Bruxelles (2011)
15. Strackeljahn, J., Jonscher, R., Prieur, S., Vogel, D., Deslaers, T., Keysers, D., Mauser, A., Bezrukov, I., Hegerath, A.: GfKl Data mining competition 2005—predicting liquidity crisis of companies. In: Spiliopoulou, M., Kruse, R., Borgelt, C., Nürnberger, A., Gaul, W. (eds.) From Data and Information Analysis to Knowledge Engineering, pp. 748–758. Springer (2005)
16. Szepannek, G., Gruhne, M., Bischl, B., Krey, S., Harczos, T., Klefenz, F., Dittmar, C., Weihs, C.: Perceptually based phoneme recognition in popular music. In: Locarek-Junge, H., Weihs, C. (eds.) Classification as a Tool for Research, pp. 751–758. Springer, Heidelberg (2010)
17. Szepannek, G., Schiffner, J., Wilson, J.C., Weihs, C.: Local modelling in classification. In: Perner, P. (ed.) Advances in Data Mining: Medical Applications, E-Commerce, Marketing, and Theoretical Aspects, pp. 153–164. Springer LNAI 5077, Berlin (2008)
18. Therneau, T., Atkinson, E.: In introduction to recursive partitioning using RPART routines, TR 61, Mayo Foundation. http://www.mayo.edu/hsr/techrpt/61.pdf (1997)
19. Thomas, L.C., Edelman, D.B., Crook, J.N.: Credit scoring and its applications. SIAM (2002)
20. Vincotti, T., Hand, D.: Scorecard construction with unbalanced class sizes. J. Iran. Stat. Soc. **2**, 189–205 (2002)

The Exact Solution of Multi-period Portfolio Choice Problem with Exponential Utility

Taras Bodnar, Nestor Parolya and Wolfgang Schmid

Abstract In the current paper we derive the exact analytical solution of the multi-period portfolio choice problem for an exponential utility function. It is assumed that the asset returns depend on predictable variables and that the joint random process of the asset returns follows a vector autoregression. We prove that the optimal portfolio weights depend on the covariance matrices of the next two periods and the conditional mean vector of the next period. The case without predictable variables and the case of independent asset returns are partial cases of our solution.

1 Introduction

Nowadays, the investment analysis and portfolio choice theory are very important and challenging topics in finance, economics and management. Since Markowitz [7] presented his mean-variance paradigm modern portfolio theory has become a fundamental tool for understanding the interactions of systematic risk and reward.

It is well known that the mean-variance optimization problem of Markowitz [7] is equivalent to the expected exponential utility optimization under the assumption of normal distribution (see Merton [8]). Unfortunately, his approach only gives an answer to the single-period portfolio choice problem in discrete time but it says nothing about the multi-period setting. Therefore, it is important to investigate the multi-period (dynamic) portfolio optimization problem which is of great relevance for an investor as well.

T. Bodnar
University of Stockholm, Stockholm, Sweden
e-mail: taras.bodnar@math.su.se

N. Parolya (✉)
Leibniz University Hannover, Hannover, Germany
e-mail: nestor.parolya@ewifo.uni-hannover.de

W. Schmid
European University Viadrina, Frankfurt, Germany
e-mail: schmid@europa-uni.de

© Springer International Publishing Switzerland 2016
M. Lübbecke et al. (eds.), *Operations Research Proceedings 2014*,
Operations Research Proceedings, DOI 10.1007/978-3-319-28697-6_7

45

The continuous case has already been solved for many types of utility functions in the single- and multi-period case by Merton [8]. On the other hand, the discrete time setting is more useful and seems to be more realistic for the investors because they observe the underlying data-generating process discretely and make their decisions only at discrete points of time. In other words, they would prefer to deal with the time series models rather than the stochastic differential equations for modelling the asset returns. Moreover, the continuous time portfolio strategies are often inadmissible in discrete time because they may cause a negative wealth (see, Brandt [4]). In general, there are not many closed-form solutions provided in the literature to the discrete-time multi-period case. Moreover, the most of them contain the assumption of independence of the asset returns in time. This assumption, however, is unfortunately not fulfilled in many practical situations. As a result, analytical solutions of the discrete-time multi-period optimal portfolio choice problems are not easy to obtain.

In the present paper we consider an investor who invests into k risky assets and one riskless asset with an investment strategy based on the exponential utility function

$$U(W_t) = -e^{-\alpha W_t} . \tag{1}$$

Here W_t denotes the investor's wealth at period t and $\alpha > 0$ stands for the coefficient of absolute risk aversion (ARA), which is a constant over time for the exponential utility (CARA utility). The application of the exponential utility function is more plausible than the use of the quadratic utility since the last one possesses an increasing risk aversion coefficient. That is why the exponential utility function is commonly used in portfolio selection theory. Moreover, the optimization of the expected exponential utility function leads to the well known mean-variance utility maximization problem and consequently its solution lays on the mean-variance efficient frontier. This is also true when the uncertainty of the asset returns is taken into account (see, Bodnar et al. [3]).

We extend the previous findings on the exponential utility function in three ways: (i) we do not assume the independence of the asset returns in time; (ii) taking into account the untradable predictable variables, we work in the presence of the incomplete market; and (iii) we derive a closed-form solution of the discrete-time multi-period portfolio choice problem with the exponential utility function (1) under the assumption that the joint process consists of the asset returns and the predictable variables and it is assumed to follow a vector autoregressive (VAR) process.

2 Multi-period Portfolio Problem for an Exponential Utility

Let $\mathbf{X}_t = \left(X_{t,1}, X_{t,2}, \ldots, X_{t,k}\right)'$ denote the vector of the returns (e.g., log-returns) of k risky assets and let $r_{f,t}$ be the return of the riskless asset at time t. Let $\mathbf{z}_t$ be a p-dimensional vector of predictable variables. We assume that $\mathbf{Y}_t = (\mathbf{X}_t', \mathbf{z}_t')'$ follows a VAR(1) process given by

$$\mathbf{Y}_t = \tilde{\mathbf{v}} + \tilde{\mathbf{\Phi}}\mathbf{Y}_{t-1} + \tilde{\boldsymbol{\varepsilon}}_t \tag{2}$$

where the $k + p$-dimensional vector $\tilde{\mathbf{v}}$ and the $(k + p) \times (k + p)$-dimensional matrix $\tilde{\mathbf{\Phi}}$ contain the model parameters and $\tilde{\boldsymbol{\varepsilon}}_t \sim \mathcal{N}(\mathbf{0}, \tilde{\mathbf{\Sigma}}(t))$, where $\tilde{\mathbf{\Sigma}}(t)$ is a $(k + p) \times (k + p)$-dimensional positive definite deterministic matrix function. Let $\mathscr{F}_t$ denote the information set available at time t. Then $\mathbf{Y}_t|\mathscr{F}_{t-1} \sim \mathcal{N}_{k+p}(\tilde{\boldsymbol{\mu}}_t, \tilde{\mathbf{\Sigma}}(t))$, i.e., the conditional distribution of $\mathbf{Y}_t$ given $\mathscr{F}_{t-1}$ is a $k + p$ dimensional normal distribution with mean vector $\tilde{\boldsymbol{\mu}}_t = E(\mathbf{Y}_t|\mathscr{F}_{t-1}) = E_{t-1}(\mathbf{Y}_t)$ and covariance matrix $\mathrm{Var}(\mathbf{Y}_t|\mathscr{F}_{t-1}) = \tilde{\mathbf{\Sigma}}(t)$.

The stochastic model (2) is described in detail by Campbell et al. [5] who argued that the application of VAR(1) is not a restrictive assumption because every vector autoregression can be presented as a VAR(1) process through an expansion of the vector of state (predictable) variables. Let $\mathbf{v} = \mathbf{L}\tilde{\mathbf{v}}$ and $\mathbf{\Phi} = \mathbf{L}\tilde{\mathbf{\Phi}}$ with $\mathbf{L} = [\mathbf{I}_k \ \mathbf{O}_{k,p}]$ where $\mathbf{I}_k$ is a $k \times k$ identity matrix and $\mathbf{O}_{k,p}$ is a $k \times p$ matrix of zeros. From (2) we obtain the following model for $\mathbf{X}_t$ expressed as

$$\mathbf{X}_t = \mathbf{L}\mathbf{Y}_t = \mathbf{L}\tilde{\mathbf{v}} + \mathbf{L}\tilde{\mathbf{\Phi}}\mathbf{Y}_{t-1} + \mathbf{L}\tilde{\boldsymbol{\varepsilon}}_t = \mathbf{v} + \mathbf{\Phi}\mathbf{Y}_{t-1} + \boldsymbol{\varepsilon}_t \,. \tag{3}$$

Consequently, the conditional distribution of $\mathbf{X}_t$ given $\mathscr{F}_{t-1}$ is a multivariate normal distribution with conditional mean vector $\boldsymbol{\mu}_t = E(\mathbf{X}_t|\mathscr{F}_{t-1}) = \mathbf{v} + \mathbf{\Phi}\mathbf{Y}_{t-1}$ and conditional covariance matrix $\mathbf{\Sigma}(t) = \mathrm{Var}(\mathbf{X}_t|\mathscr{F}_{t-1}) = \mathbf{L}\tilde{\mathbf{\Sigma}}(t)\mathbf{L}'$, that is $\mathbf{X}_t|\mathscr{F}_{t-1} \sim \mathcal{N}_k(\boldsymbol{\mu}_t, \mathbf{\Sigma}(t))$.

Let $\mathbf{w}_t = (w_{t,1}, w_{t,2}, \ldots, w_{t,k})'$ denote the vector of the portfolio weights of the k risky assets at period t. Then the evolution of the investor's wealth is expressed as

$$W_t = W_{t-1}\left(1 + r_{f,t} + \mathbf{w}_{t-1}'(\mathbf{X}_t - r_{f,t}\mathbf{1})\right) = W_{t-1}\left(R_{f,t} + \mathbf{w}_{t-1}'\breve{\mathbf{X}}_t\right), \tag{4}$$

where $R_{f,t} = 1 + r_{f,t}$ and $\breve{\mathbf{X}}_t = \mathbf{X}_t - r_{f,t}\mathbf{1}$ with $\breve{\boldsymbol{\mu}}_t = E_{t-1}(\breve{\mathbf{X}}_t) = \mathbf{v} + \mathbf{\Phi}\mathbf{Y}_{t-1} - r_{f,t}\mathbf{1}$. The aim of the investor is to maximize the expected utility of the final wealth. The optimization problem is given by

$$V(0, W_0, \mathscr{F}_0) = \max_{\{\mathbf{w}_s\}_{s=0}^{T-1}} E_t[U(W_T)] \tag{5}$$

with the terminal condition

$$U(W_T) = -\exp(-\alpha W_T) \quad \text{for } \alpha > 0. \tag{6}$$

Following Pennacchi [9] the optimization problem (5) can be solved by applying the following Bellman equation at time point $T - t$

$$\begin{aligned} V(T - t, W_{T-t}, \mathscr{F}_{T-t}) &= \max_{\mathbf{w}_{T-t}} E_{T-t}\left[\max_{\{\mathbf{w}_s\}_{s=T-t+1}^{T-1}} E_{T-t+1}[U(W_T)]\right] \\ &= \max_{\mathbf{w}_{T-t}} E_{T-t}\left[V\left(T - t + 1, W_{T-t}\left(r_{f,T-t} + \mathbf{w}_{T-t+1}^{*'}\breve{\mathbf{X}}_{T-t+1}\right), \mathscr{F}_{T-t+1}\right)\right] \end{aligned} \tag{7}$$

subject to (6), where $\mathbf{w}_{T-t+1}^*$ are the optimal portfolio weights at period $T - t + 1$. Note that in contrast to the static case now the vector of optimal portfolio weights $\mathbf{w}_{T-t}$ is a function of the weights of the next periods, i.e., of $\mathbf{w}_{T-t+1}, \mathbf{w}_{T-t+2}, \ldots, \mathbf{w}_{T-1}$, what is the consequence of the backward recursion method (see, e.g. Pennacchi [9]).

For the period $T - 1$ we get

$$
\begin{aligned}
&V(T-1, W_{T-1}, \mathscr{F}_{T-1}) \\
&= E_{T-1}\left[-\exp(-\alpha W_{T-1}(R_{f,T} + \mathbf{w}_{T-1}'\check{\mathbf{X}}_T))\right] \\
&= -\exp(-\alpha W_{T-1}R_{f,T})E_{T-1}[\exp(-\alpha W_{T-1}\mathbf{w}_{T-1}'\check{\mathbf{X}}_T)] \\
&= \exp(-\alpha W_{T-1}R_{f,T})\left(-\exp\left[-\alpha(W_{T-1}\mathbf{w}_{T-1}'\check{\mathbf{\mu}}_T - \frac{\alpha}{2}\mathbf{w}_{T-1}'\mathbf{\Sigma}(T)\mathbf{w}_{T-1}W_{T-1}^2)\right]\right) \rightarrow \max.
\end{aligned}
\tag{8}
$$

The last optimization problem is equivalent to

$$
W_{T-1}\mathbf{w}_{T-1}'\check{\mathbf{\mu}}_T - \frac{\alpha}{2}\mathbf{w}_{T-1}'\mathbf{\Sigma}(T)\mathbf{w}_{T-1}W_{T-1}^2 \rightarrow \max \quad \text{over } \mathbf{w}_{T-1}.
\tag{9}
$$

Taking the derivative and solving (9) with respect to $\mathbf{w}_{T-1}$ we get the classical solution for the period $T - 1$

$$
\mathbf{w}_{T-1}^* = \frac{1}{\alpha W_{T-1}}\mathbf{\Sigma}^{-1}(T)\check{\mathbf{\mu}}_T = \frac{1}{\alpha W_{T-1}}(\mathbf{L}\tilde{\mathbf{\Sigma}}(T)\mathbf{L}')^{-1}(\check{\mathbf{v}}_T + \mathbf{\Phi}\mathbf{Y}_{T-1}) \text{ with } \check{\mathbf{v}}_T = \mathbf{v} - r_{f,T}\mathbf{1}.
\tag{10}
$$

2.1 Multi-period Portfolio Weights

In Theorem 1 the multi-period portfolio weights for all periods from 0 to $T - 1$ are given (cf. Bodnar et al. [2]).

Theorem 1 Let $\mathbf{X}_\tau = (X_{\tau,1}, X_{\tau,2}, \ldots, X_{\tau,k})'$ be a random return vector of k risky assets. Suppose that $\mathbf{X}_\tau$ and the vector of p predictable variables $\mathbf{z}_\tau$ jointly follow a VAR(1) process as defined in (2). Let $r_{f,\tau}$ be the return of the riskless asset. Then the optimal multi-period portfolio weights are given by (10) for period $T - 1$,

$$
\mathbf{w}_{T-2}^* = \frac{1}{\alpha W_{T-2}R_{f,T}}\left(\mathbf{L}\tilde{\mathbf{\Sigma}}^{-1}(T-1)\tilde{\mathbf{\mu}}_{T-1}^* - \mathbf{L}\mathbf{\Phi}'\mathbf{\Sigma}^{-1}(T)(\check{\mathbf{v}}_T + r_{f,T}\mathbf{\Phi}\mathbf{L}'\mathbf{1})\right),
\tag{11}
$$

and

$$
\mathbf{w}_{T-t}^* = C_{T-t}\left(\mathbf{L}\tilde{\mathbf{\Sigma}}^{-1}(T-t+1)\tilde{\mathbf{\mu}}_{T-t+1}^* - \mathbf{L}\tilde{\mathbf{\Phi}}'\tilde{\mathbf{\Sigma}}^{-1}(T-t+2)(\check{\mathbf{v}}_{T-t+3}^* + r_{f,T-t+2}\tilde{\mathbf{\Phi}}\mathbf{L}'\mathbf{1})\right)
\tag{12}
$$

for $t = 3, \ldots, T$ *with*

$$C_{T-t} = \frac{1}{\alpha W_{T-t}} \left(\prod_{i=T-t+2}^{T} R_{f,i} \right)^{-1}, \quad \tilde{\mu}_{T-t+1}^{*} = \tilde{\mu}_{T-t+1} - r_{f,T-t+2} \mathbf{L}' \mathbf{1} \quad and \quad (13)$$

$$\check{\mathbf{v}}_{T-t+3}^{*} = \check{\mathbf{v}} - r_{f,T-t+3} \mathbf{L}' \mathbf{1}. \tag{14}$$

The results of Theorem 1 show us that the optimal portfolio weights at every period of time except the last one depend on the covariance matrices of the next two periods and the conditional mean vector of the next period. This property turns out to be very useful if we want to calculate the optimal portfolio weights for a real data set. Moreover, Theorem 1 presents a solution to the discrete-time multi-period optimal portfolio choice problem with the CARA utility in the presence of an incomplete market.

Note that the case without predictable variables (the case with a complete market) is a special case of Theorem 1. In this case the following expressions are obtained.

Corollary 1 *Let* $\mathbf{X}_{\tau} = \left(X_{\tau,1}, X_{\tau,2}, \ldots, X_{\tau,k} \right)'$ *be a random return vector of k risky assets which follows a VAR(1) process as defined in (2) but without a vector of predictable variables* z_{τ}. *Let* $r_{f,\tau}$ *be the return of the riskless asset. Then the optimal multi-period portfolio weights for period* $T - 1$ *are given by*

$$\mathbf{w}_{T-1}^{*} = \frac{1}{\alpha W_{T-1}} \boldsymbol{\Sigma}^{-1}(T) \check{\mu}_{T} = \frac{1}{\alpha W_{T-1}} \boldsymbol{\Sigma}^{-1}(T)(\check{\mathbf{v}}_{T} + \boldsymbol{\Phi} \mathbf{Y}_{T-1}) \text{ with } \check{\mathbf{v}}_{T} = \mathbf{v} - r_{f,T} \mathbf{1}$$
$$\tag{15}$$

and for $t = 2, \ldots, T$ *by*

$$\mathbf{w}_{T-t}^{*} = C_{T-t} \left(\boldsymbol{\Sigma}^{-1}(T - t + 1) \check{\mu}_{T-t+1} - \boldsymbol{\Phi}' \boldsymbol{\Sigma}^{-1}(T - t + 2)(\check{\mathbf{v}}_{T-t+2} + r_{f,T-t+2} \boldsymbol{\Phi} \mathbf{1}) \right), \tag{16}$$

with $C_{T-t} = \dfrac{1}{\alpha W_{T-t} \displaystyle\prod_{i=T-t+2}^{T} R_{f,i}}.$

In Corollary 2 the return vectors are assumed to be independent.

Corollary 2 *Let* $\mathbf{X}_{\tau} = \left(X_{\tau,1}, X_{\tau,2}, \ldots, X_{\tau,k} \right)'$ *be a sequence of the independently and identically normally distributed vectors of k risky assets, i.e.,* $\mathbf{X}_{\tau} \sim \mathcal{N}(\boldsymbol{\mu}, \boldsymbol{\Sigma})$. *Let* $r_{f,\tau}$ *be the return of the riskless asset. We assume that* $\boldsymbol{\Sigma}$ *is positive definite. Then for all* $t = 1, \ldots, T$ *the optimal multi-period portfolio weights for period* $T - t$ *are given by*

$$\mathbf{w}_{T-t}^{*} = \frac{1}{\alpha W_{T-t} \displaystyle\prod_{i=T-t+2}^{T} R_{f,i}} \boldsymbol{\Sigma}^{-1} \check{\mu} \text{ with } \check{\mu} = \boldsymbol{\mu} - r_{f,T-t+2} \mathbf{1}. \tag{17}$$

The results of Corollary 2 can be obtained as a partial case of Çanakoğlu and Özekici [6], where the stochastic market was presented by a discrete time Markov chain. In that case the asset returns depend on the present state of the market and not on the previous ones which implies the independence of the asset return over time.

It is noted that the dynamics of the optimal portfolio weights in Corollary 2 is hidden in the coefficient of the absolute risk aversion α which is given by $\alpha_\tau =$

$$\left(\alpha W_{T-\tau} \prod_{i=T-\tau+2}^{T} R_{f,i} \right)^{-1}.$$

3 Summary

There are only a few results on closed-form solutions of the multi-period portfolio choice problem in the discrete time available in literature (see, e.g., Bodnar et al. [1]). Unfortunately, most of them are derived under the assumption that the asset returns are independently distributed in time.

In the present paper we derive an exact solution of the multi-period portfolio selection problem for an exponential utility function which is obtained under the assumption that the asset returns and the vector of predictable variables follow a vector autoregressive process of order 1. Under the assumption of independence the obtained expressions of the weights are proportional to the weights of the tangency portfolio obtained as a solution in the case of a single-period optimization problem. We show that only the coefficient of absolute risk aversion depends on the dynamics of the asset returns in this case. The weights of the optimal portfolio derived without a vector of predictable variables are obtained as a partial case of the suggested general solution.

The obtained results can be further extended by taking into account the uncertainties about the parameters of the data generating process. The analytical expressions of the weights can be used to derive the expected mean vector and the covariance matrix of the estimated weights which provide us the starting point for the detailed analysis of their distributional properties.

References

1. Bodnar, T., Parolya, N., Schmid, W.: A closed-form solution of the multi-period portfolio choice problem for a quadratic utility function. To appear in Ann. Oper. Res. (2015)
2. Bodnar, T., Parolya, N., Schmid, W.: On the exact solution of the multi-period portfolio choice problem for an exponential utility under return predictability. Accept/minor revision in Eur. J. Oper. Res. (2014). arXiv:1207.1037
3. Bodnar, T., Parolya, N., Schmid, W.: On the equivalence of quadratic optimization problems commonly used in portfolio theory. Eur. J. Oper. Res. **229**, 637–644 (2013)
4. Brandt, M.: Portfolio choice problems. In: Aït-Sahalia, Y., Hansen, L.P. (eds.) Handbook of Financial Econometrics: Tools and Techniques, vol. 1, pp. 269–336. North Holland (2010)

5. Campbell, J.Y., Chan, Y.L., Viceira, L.M.: A multivariate model of strategic asset allocation. J. Financ. Econ. **67**, 41–80 (2003)
6. Çanakoğlu, E., Özekici, S.: Portfolio selection in stochastic markets with exponential utility functions. Ann. Oper. Res. **166**, 281–297 (2009)
7. Markowitz, H.: Portfolio selection. J. Finance **7**, 77–91 (1952)
8. Merton, R.C.: Lifetime portfolio selection under uncertainty: the continuous time case. Rev. Econ. Stat. **50**, 247–257 (1969)
9. Pennacchi, G.: Theory of Asset Pricing. Pearson/Addison-Wesley, Boston (2008)

On the Discriminative Power of Tournament Solutions

Felix Brandt and Hans Georg Seedig

Abstract Tournament solutions constitute an important class of social choice functions that only depend on the pairwise majority comparisons between alternatives. Recent analytical results have shown that several concepts with appealing axiomatic properties tend to not discriminate at all when the tournaments are chosen from the uniform distribution. This is in sharp contrast to empirical studies which have found that real-world preference profiles often exhibit Condorcet winners, i.e., alternatives that all tournament solutions select as the unique winner. In this work, we aim to fill the gap between these extremes by examining the distribution of the number of alternatives returned by common tournament solutions for empirical data as well as data generated according to stochastic preference models.

1 Introduction

A key problem in social choice theory is to identify functions that map the preference relations of multiple agents over some abstract set of alternatives to a socially acceptable alternative. Whenever the social choice function is required to be impartial towards alternatives and voters, it may be possible that several alternatives qualify equally well to be chosen. It is typically understood that such ties will eventually be broken by some procedure that is independent of the agents' preferences. In general, it seems desirable to narrow down the choice as much as possible based on the preferences of the voters alone. The goal of this paper is to study the discriminative power of various social choice functions—i.e., how many tied alternatives are returned—when preferences are drawn from common distributions that have been proposed in the literature.

An important class of social choice functions only depends on the pairwise majority relation between alternatives. When the pairwise majority relation is asymmetric,

F. Brandt (✉) · H.G. Seedig (✉)
Institut für Informatik, Technische Universität München, München, Germany
e-mail: brandtf@in.tum.de

H.G. Seedig
e-mail: seedigh@in.tum.de

© Springer International Publishing Switzerland 2016
M. Lübbecke et al. (eds.), *Operations Research Proceedings 2014*,
Operations Research Proceedings, DOI 10.1007/978-3-319-28697-6_8

as is the case when there is an odd number of agents with linear preferences, these functions are known as *tournament solutions*. The tradeoff between discriminative power and axiomatic foundations is especially evident for tournament solutions as many of them can be axiomatically characterized as the *most discriminating* functions that satisfy certain desirable properties.[1]

Analytical results about the discriminative power of tournament solutions for realistic distributions of preferences are very difficult to obtain. To the best of our knowledge, all existing papers explicitly or implicitly consider a uniform distribution over all tournaments of a fixed size. Under this assumption, it was shown that the Banks set and the minimal covering set almost always selects all alternatives as the number of alternatives goes to infinity [5, 14]. For the bipartisan set, a more precise result by Fisher and Reeves [6] implies that on average, it returns half of the alternatives for odd $|T|$.

These analytical results stand in sharp contrast to empirical observations that Condorcet winners exist in many real-world situations, implying that tournament solutions very frequently return singletons.

Simulations with stochastic preference models have been used for the analysis of several problems in (computational) social choice. See Laslier [8] and McCabe-Dansted and Slinko [11] for examples. In comparison, we consider tournaments of larger sizes because several tournament solutions are known to always coincide when there are only few alternatives [3].

2 Methodology

2.1 Preference Profiles and Tournament Solutions

Let A be a set of alternatives and $N = \{1, \ldots, n\}$ a set of voters. The preferences of voter $i \in N$ are represented by a complete and antisymmetric *preference relation* $R_i \subseteq A \times A$. The interpretation of $(a, b) \in R_i$, usually denoted by $a \, R_i \, b$, is that voter i values alternative a at least as much as alternative b. When individual preferences are transitive, we also speak of *rankings*. A *preference profile* $R = (R_1, \ldots, R_n)$ is an n-tuple containing a preference relation R_i for each agent $i \in N$. The majority relation $\succ_R$ of a given preference profile is defined as $a \succ_R b \Leftrightarrow |\{i \mid a \, R_i \, b\}| > |\{i \mid b \, R_i \, a\}|$.

A *tournament* T is a pair $(A, \succ)$, where $\succ$ is an asymmetric and connex binary relation on A. Whenever the number of voters n is odd, $(A, \succ_R)$ constitutes a tournament. If there is an alternative a such that $a \succ_R b$ for all $b \in A \setminus \{b\}$, a is a *Condorcet winner* according to R. A *tournament solution* is a function that maps a tournament to a nonempty subset of its alternatives, the *choice set*, and uniquely chooses a Condorcet winner whenever one exists. The simplest tournament solution is *COND* which

---

[1]See, e.g., [2], Chap. 6, Sect. 2.2.2.

Fig. 1 Set-theoretic relationships between tournament solutions. If the ellipses of two tournament solutions S and S' intersect, then $S(T) \cap S'(T) \neq \emptyset$ for all tournaments T. If the ellipses for S an S' are disjoint, however, this signifies that $S(T) \cap S'(T) = \emptyset$ for some tournament T. The exact locations of *BP* and *TEQ* in this diagram are unknown

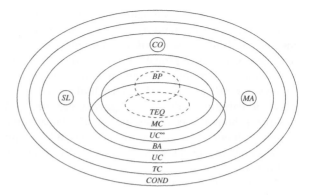

chooses the set of all alternatives whenever there is no Condorcet winner. The other tournament solutions considered in this paper are the *top cycle* (*TC*), the *uncovered set* (*UC*), the *iterated uncovered sets* (*UC$^\infty$*), the *Copeland set* (*CO*), the *bipartisan set* (*BP*), the *Markov set* (*MA*), the *Banks set* (*BA*), the *Slater set* (*SL*), the *minimal covering set* (*MC*), and the *tournament equilibrium set* (*TEQ*).

For definitions and discussions on most of these concept, we refer to the excellent overview by Laslier [7]. The set-theoretic relationships of these concepts are depicted in Fig. 1.

2.2 Empirical Data

In the preference library PREFLIB [10], scholars have contributed data sets from a variety of real world scenarios. At the time of writing, PREFLIB contained 354 tournaments induced from pairwise majority comparisons. Out of these, all except 9 exhibit a Condorcet winner. The remaining tournaments are still very structured as the uncovered set never contains more than 4 alternatives. This is in line with earlier observations that real-world majority relations tend to be close to linear orders and often have Condorcet winners [13].

2.3 Stochastic Models

As the available empirical data does not allow to draw conclusions about the differences in discriminative power of tournament solutions, we now consider stochastic models to generate tournaments.

The *uniform random tournament* model was used in the previous analysis of the discriminative power of tournament solutions (e.g., [5, 6]). It assigns the same probability to each *labeled* tournament of a fixed size.

In the next models, we sample preference profiles and work with the tournament induced by the majority relation. One of the most widely-studied models is the *impartial culture model* (IC), where for each voter, every possible ranking of the alternatives has equal probability. If we add anonymity by having indistinguishable voters, the set of profiles is partitioned into equivalence classes. Under the *impartial anonymous culture model* (IAC), each of these equivalence classes is chosen with equal probability. In the Pólya-Eggenberger *urn model*, each possible preference ranking is thought to be represented by a ball in an urn from which individual preferences are drawn. After each draw, the chosen ball is put back and $\alpha \in \mathbb{N}_0$ new balls of the same kind are added to the urn [1]. The urn model subsumes both IC ($\alpha = 0$) and IAC ($\alpha = 1$).

A very different kind of model is the *spatial model* used frequently in political and social choice (see, e.g., [12]). Here, alternatives and voters are uniformly at random placed in a multi-dimensional space and the voters' preferences are determined by the (Euclidian) distanced to the alternatives.

In *distance-based* models, it is assumed that agents report noisy estimates of a pre-existing truth as their preferences and such models are usually parameterized by a homogeneity parameter. In its arguably simplest form, every agent provides (possibly intransitive) preferences R where each pairwise preference $a\ R\ b$ is 'correct', with a probability p. We will call this the *Condorcet noise model*. In *Mallows-ϕ m*odel [9], the probability of a ranking is determined by its Kendall-tau distance to a reference ranking, i.e., the number of pairwise disagreements. Obviously, one can define a number of such distance-based models. See [4] for a discussion. To overcome the bias of distance-based models to transitive majority relations, *mixtures* of models have been considered. We consider uniform mixtures over k Mallows-ϕ models with a shared parameter ϕ and refer to this as *Mallows k-mixtures*.

For a more detailed exposition of these models and additional results, we refer to Seedig [15].

3 Experimental Results and Discussion

In our experimental setup, we generated tournament instances according to the aforementioned models and computed the different choice sets for them. For the sampling step, we built on the implementations from Mattei and Walsh [10] to generate preference profiles of which we considered the majority relation. The computation of the various tournament solutions was done by our own implementations.

We examined the ability of the various solutions to rule out alternatives. Our informal measure for discriminative power of a tournament solution on a specific model is the distance of its average choice set size to the average size of *COND* which, by definition, is the least discriminative tournament solution. In our comparisons, we provide *COND* not only as a baseline but also as an indicator for the frequency of tournaments with a Condorcet winner. We examined the average choice set sizes of the aforementioned tournament solutions for a fixed number of voters $n = 51$. The

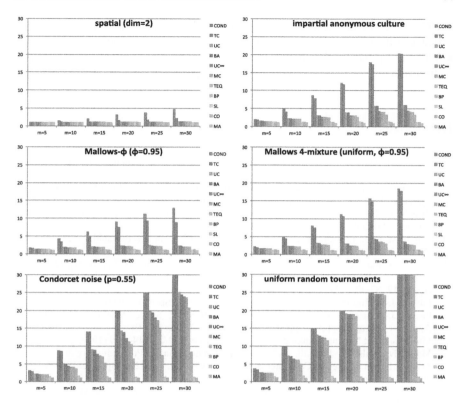

Fig. 2 Comparison of average absolute choice set sizes for various stochastic preference models. The number of alternatives is on the horizontal axis, the number of voters is $n = 51$. Averages are taken over 100 samples. The Slater set (*SL*) is omitted whenever its computation was infeasible

results are shown in Fig. 2 where the graphics for IC and the urn model ($\alpha = 10$) have been omitted due to space constraints. They are very similar to the graphic for IAC.

The following conclusions can be drawn from our results.

- *TC* is almost as undiscriminating as *COND*.
- All other tournament solutions are much more discriminating than the analytical results for uniform random tournaments suggest. In fact, for all reasonable parameterizations of the considered models with transitive individual preferences and at least 10 alternatives (including impartial culture) all tournament solutions except *TC* discarded at least 75 % of the alternatives on average.
- All tournament solutions except *TC* behave similarly in terms of discriminative power. One may conclude that the decision which one to use in practice should not be based on discriminative power, but rather on axiomatic properties.
- Tournament solutions based on scoring (*SL*, *CO*, and *MA*) are more discriminating than all other tournament solutions.

- UC^∞ (and thereby also MC) discriminates more than BA. This could not be deduced from the set-theoretic relationships between tournament solutions.
- Within the group of tournament solutions with appealing axiomatic properties, BP discriminates the most (and is efficiently computable).

References

1. Berg, S.: Paradox of voting under an urn model: the effect of homogeneity. Public Choice **47**, 377–387 (1985)
2. Brandt, F., Conitzer, V., Endriss, U.: Computational social choice. In: Weiß, G. (ed.) Multiagent Systems, chapter 6, 2nd edn., pp. 213–283. MIT Press (2013)
3. Brandt, F., Dau, A., Seedig, H.G.: Bounds on the disparity and separation of tournament solutions. Discrete Appl. Math. **187**, 41–49 (2015)
4. Critchlow, D.E., Fligner, M.A., Verducci, J.S.: Probability models on rankings. J. Math. Psychol. **35**, 294–318 (1991)
5. Fey, M.: Choosing from a large tournament. Soc. Choice Welfare **31**(2), 301–309 (2008)
6. Fisher, D.C., Reeves, R.B.: Optimal strategies for random tournament games. Linear Algebra Appl. **217**, 83–85 (1995)
7. Laslier, J.-F.: Tournament Solutions and Majority Voting. Springer (1997)
8. Laslier, J.-F.: In silico voting experiments. In: Laslier, J.-F., Sanver, M.R. (eds.) Handbook on Approval Voting, chapter 13, pp. 311–335. Springer-Verlag (2010)
9. Mallows, C.L.: Non-null ranking models. Biometrika **44**(1/2), 114–130 (1957)
10. Mattei, N., Walsh, T.: PrefLib: A library for preference data. In: Proceedings of 3rd ADT, vol. 8176 of Lecture Notes in Computer Science (LNCS), pp. 259–270. Springer (2013). http://www.preflib.org
11. McCabe-Dansted, J.C., Slinko, A.: Exploratory analysis of similarities between social choice rules. Group Decis. Negot. **15**(1), 77–107 (2006)
12. Ordeshook, P. C.: The spatial analysis of elections and committees: four decades of research. Technical report, California Institute of Technology. Mimeo (1993)
13. Regenwetter, M., Grofman, B., Marley, A.A.J., Tsetlin, I.M.: Behavioral Social Choice: Probabilistic Models, Statistical Inference, and Applications. Cambridge University Press (2006)
14. Scott, A., Fey, M.: The minimal covering set in large tournaments. Soc. Choice Welfare **38**(1), 1–9 (2012)
15. Seedig, H.G.: Majority Relations and Tournament Solutions: A Computational Study. Ph.D. thesis, Technische Universität München (2015)

An Actor-Oriented Approach to Evaluate Climate Policies with Regard to Resource Intensive Industries

Patrick Breun, Magnus Fröhling and Frank Schultmann

Abstract Metal production is responsible for a large part of greenhouse gas (GHG) emissions in Germany. While some political stakeholders call for a more restrictive climate policy to force further reductions of GHG emissions, the exceptions made for the metal industry increased so far to guarantee its global competitiveness. The question rises how a more restrictive climate policy would affect industrial GHG emissions and the profitability. To estimate the impact of political instruments the actor-oriented approach presented focuses on the simulation of plant-specific investment decisions. First, a detailed database of the internal material and energy flows of all relevant iron, steel and aluminium producing plants together with the best available techniques (BAT) for GHG emission reduction is developed. In the subsequent simulation, the plants, modelled as actors, decide on the implementation of these techniques dependent on the political conditions which are varied in scenarios. The results show, that there are only minor GHG emission reduction potentials due to already implemented high efficiency standards. Nevertheless, more restrictive climate policies can lead to significant cost increases influencing global competitiveness.

1 Introduction

The GHG emissions of the German metal production constitute 27 % of the manufacturing industrie's GHG emissions in 2011. A large part of this amount can be allocated to a comparably small number of plants. Hence, there is a special interest of political stakeholders to focus on the decarbonization of those industrial activities. At

P. Breun (✉) · M. Fröhling (✉) · F. Schultmann (✉)
French-German Institute for Environmental Research (DFIU), Karlsruhe Institute
of Technology (KIT), Hertzstraße 16, Bldg. 06.33, Karlsruhe, Germany
e-mail: patrick.breun@kit.edu

M. Fröhling
e-mail: magnus.froehling@kit.edu

F. Schultmann
e-mail: frank.schultmann@kit.edu

© Springer International Publishing Switzerland 2016
M. Lübbecke et al. (eds.), *Operations Research Proceedings 2014*,
Operations Research Proceedings, DOI 10.1007/978-3-319-28697-6_9

the same time, about 740,000 employees work in the metal producing and processing sectors which consequently form an important economic factor. Furthermore, numerous industries downstream the value chain, as e.ġ. the automotive industry, depend on the metal supply. Taking into account these two perspectives, national climate policies need to be formulated to incentivize the utilization of available energy efficient techniques in order to reach environmental long-term targets without harming the global competitiveness due to increasing production costs.

As various possible ecological and economic impacts of different climate policies together with the given industry-specific technical restrictions have to be investigated, decision support for political stakeholders is required. So we developed the national integrated assessment model DECARBONISE, funded by the Federal Ministry of Education and Research (BMBF), which focuses on the decarbonization of the German metal industry taking into account plant-specific investment decisions as well as macroeconomic developments. The actor-oriented simulation model is part of this project and helps to gain insight into the economic and technical reduction potentials which are currently available and how these can be made use of by the German metal industry.

Before the approach is sketched in Sects. 3 and 2 is concerned with a short classification of the approach and the used data which is required to determine the status quo of the metal industry. Section 4 shows exemplary model results whereas Sect. 5 gives a short conclusion.

2 Previous Works and Estimation of Plant Configurations

Ilsen [4] developed an actor-oriented approach for environmental policy evaluation regarding emission relevant industrial processes carried out in Germany. Therein, every plant is modelled as an actor and reacts on imposed political instruments by deciding on the implementation of emission reduction measures based on the net present value (NPV). As this approach is capable to capture the affects of climate policies on different industries, it has been used and modified in our described works. Modifications are necessary to calculate plant- and facility-specific efficiencies more precisely in order to estimate carbonaceous inputs, GHG emissions and reduction potentials. Especially gaseous by-products of facilities, whose energy content can be used at other stages of the process, as well as recovered waste heat determine the overall efficiency of a plant (cf. [2]).

Thus, the initial internal material and energy flows of all simulated plants have to be estimated for the starting year of the simulation. This is done by a nonlinear programming model (NLP), which uses the reported CO_2 emissions of each facility [3] as well as plant-specific production volumes [6], capacities and technical restrictions [2] that are combined via a carbon balance (The detailed NLP can be found in [1]). The calculations are carried out for eight (partly) integrated steelworks, 20 electric arc furnaces, four aluminium smelters and 14 aluminium refiners and remelters. Figure 1 shows exemplary results for one integrated steelworks.

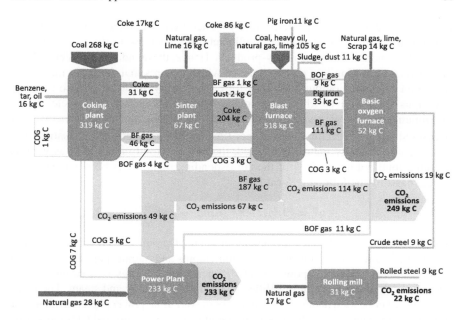

Fig. 1 Sankey diagram of carbon flows for an exemplary integrated steelworks (COG: coke oven gas, BF gas: blast furnace gas, BOF gas: basic oxygen furnace gas)

3 Actor-Oriented Simulation Model

The modelling of the actors, i.e. the simulated plants, follows an input-output approach. Initially, the input and output coefficients of the simulated facilities are obtained from the results of the NLP and translated into a technology matrix T^p, where t_{ij}^p represents the input (<0) or output (>0) of material i to produce the intermediate product j of the corresponding facility. This matrix, which contains 52 materials (columns), reflects the utilized technology for each facility (rows) of a plant p and is the basis for the calculation of plant-specific emissions and production costs.

In the simulation, plants decide on the implementation of in total 22 efficiency increasing techniques for the iron and steel industry and seven for the aluminium industry, as these techniques have been identified by screening the related latest BAT documents [2]. Every modelled efficiency increasing technique *tech* influences at least one entry t_{ij}^p in the technology matrix, which in case of an implementation has to be adapted according to the input coefficients of this technique Δt_{ij}^{tech}. A distinction is made between onetime and continuous measures to incorporate also techniques which can be extended continuously (i.e. over several years) as e.g. the direct injection of reducing agents into the blast furnace.

Before the plant-specific investment decisions are annually taken, all possible alternatives are implemented seperately for a test entity of the regarded plant to compare possible GHG emission reductions and cost changes induced by the respective techniques and to calculate the related NPVs. Thereby, an implementation sometimes not only affects the aforementioned coefficients Δt_{ij}^{tech} but also further material flows as e.g. the generation of gaseous by-products. Hence, these indirectly affected flows are adapted according to a carbon balance which ensures that the input mass flow of carbon equals the output mass flow for every facility. In this way, also the CO_2 emissions can be derived while other non-carbonaceous GHGs as PFC—relevant for aluminium smelters—are adapted directly. For every adaptation of coefficients t_{ij}^p, it is ensured that given technological upper t_{ij}^{max} and lower bounds t_{ij}^{min} are satisfied.

Once a technology matrix is adapted, the required total inputs (>0) as well as accruing total outputs (<0) g of the corresponding plant p can be obtained on an annual basis by the following focal Eq. 1:

$$g^p = (T^p)^{-1} \cdot \pi^p + in^{PP} - out^{PP} \,, \tag{1}$$

where the vector π^p denotes the annual amount of products to be sold while the vectors in^{PP} and out^{PP} indicate the absolute inputs and outputs per year (each >0) of a potentially connected power plant PP incinerating energy-rich surplus gases, which are not utilized in other facilities, to produce electricity and steam. The latter ones can either be used at other process stages or sold. If no power plant is available at the production site, the surplus gases are flared leading to additional CO_2 emissions without energy recovery and in^{PP} and out^{PP} are set to zero.

To calculate the NPVs of all efficiency increasing techniques *tech*, the investments as well as the costs and revenues have to be computed to obtain the corresponding technique- and plant-specific cash flows $CF^{p,tech}$. While the costs for raw materials and supplies can directly be calculated using g^p, the costs for depreciation, labor, maintenance, administration, etc. are derived from Peters et al. [5] and depend on e.g. overall investments or capacities. The costs induced by political instruments are emphasized here and computed dependent on the configuration of ecotaxes, the trade with GHG certificates and the EEG reallocation charge, which all are varied in scenarios. With this information the cash flows CF can be obtained. Finally, the calculation of the NPVs is carried out using Eq. 2.

$$NPV^{p,tech} = -Inv^{tech} + \sum_{t=1}^{LS} \frac{CF_t^{p,tech} - CF_t^p}{(1+e)^t} \,, \tag{2}$$

where Inv^{tech} denotes the investments required for technique *tech*, LS the life span of this technique and e the assumed interest on capital. For the investments, mainly data of the BAT documents [2] are used while economies of scale are incorporated. In short, Eq. 2 calculates the net effect of an investment in efficiency increasing techniques compared to no additional investments.

Now, every plant chooses the highest positive $NPV^{p,tech}$ and the corresponding technique is implemented or, if there are no positive NPVs, no investments are carried out. Subsequently, the technology matrices T^p as well as the carbon balance are updated and the simulation run is repeated for the next year.

With this approach, the ecological and economic impacts of individual political instruments—or climate policies containing a set of different instruments—on the regarded industries can be quantified. The model is implemented in MATLAB and connected to an input-output model of the German economy, which allows for an estimation of the resulting macroeconomic impacts and the calculation of GHG emissions accruing in industry sectors upstream the value chain.

4 Results of an Example Application

Simulation runs are carried out up to the year 2030. The scenarios contain probable raw material price changes, especially for fossil fuels, and various configurations of economic political instruments. The latter ones are modelled in detail to analyze different free allocation rules for and different prices of GHG certificates, different ecotaxes for used energy carriers or different EEG reallocation charges, which depend on the respective electricity consumption.

First of all, the plant-specific GHG emission reduction potentials are estimated. For this purpose, the standard decision routine of the simulation using NPVs is replaced by another objective function to *maximize GHG savings*. The results are shown in Fig. 2. These results are compared to the standard scenario where the plants act economically and a reduction potential of in total 2.2 m. t CO_2-eq. is obtained, which represents only about 4 % of the GHG emissions of all simulated plants.

As it becomes obvious from Fig. 2, these further GHG mitigations are possible at the cost of reduced (but still positive) profits. This leads to the question, how the calculated remaining reduction potential can be achieved when plants decide according to the NPV by changing political instruments. With high prices for CO_2 certificates, a large part of the reduction potential is used. But this leads to comparably high windfall profits for some plants if the current free allocation rules are maintained.

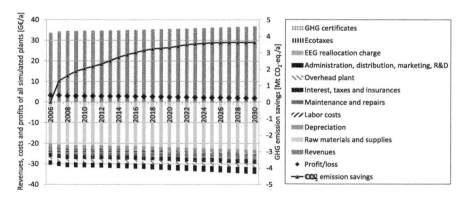

Fig. 2 Exemplary results for the scenario *maximize GHG savings*

If the actual financial reliefs of the energy intensive industries are repealed by eliminating the free allocation of certificates as well as by increasing the ecotaxes and EEG reallocation charges up to the level of the manufacturing industry, the GHG emission reductions are comparably low while notable financial losses are recorded.

5 Conclusions

To estimate the affects of climate policies, the reactions of the affected actors (here: the metal industry) have to be incorporated. These reactions depend on given technical restrictions and financial conditions. Thus, the presented approach focuses on a detailed modelling of possible efficiency gains on facility and plant level as well as on a detailed calculation of costs induced by imposing political instruments. This level of detail goes beyond most approaches in the field of policy evaluation and allows for a realistic estimation of the future developments in the regarded sectors. The results show that the possible contribution of the metal industry to achieve ambitious national long-term GHG reduction targets is comparably low and that more restrictive climate policies rather lead to unprofitable production than to additional GHG emission savings.

References

1. Breun, P., Fröhling, M., Schultmann, F.: Ein nichtlineares Optimierungsmodell zur Bestimmung der Stoffflüsse in der deutschen Eisen- und Stahlindustrie. In: Kunze, R., Fichtner, W. (ed.) Einsatz von OR-Verfahren zur Analyse von Fragestellungen im Umweltbereich (2013)
2. European Commission: Best Available Techniques (BAT) Reference Document for Iron and Steel Production, Industrial Emissions Directive 2010/75/EU (Integrated Pollution Prevention and Control) (2012)
3. European Environment Agency: The European Pollutant Release and Transfer Register (E-PRTR), Member States reporting under Article 7 of Regulation (EC) No 166/2006 (2012)
4. Ilsen, R.: Ein Beitrag zur modellgestützten Analyse der Auswirkungen umweltpolitischer Instrumente in den Bereichen Luftreinhaltung und Klimawandel. Dissertation, Karlsruhe (2011)
5. Peters, M.S., Timmerhaus, K.D., West, R.E.: Plant Design and Economics for Chemical Engineers. McGraw-Hill, New York (2004)
6. Wirtschaftsvereinigung Stahl, Stahlinstitut VDEh: Statistisches Jahrbuch der Stahlindustrie 2009/2010. Düsseldorf (2009)

Upper Bounds for Heuristic Approaches to the Strip Packing Problem

Torsten Buchwald and Guntram Scheithauer

Abstract We present an algorithm for the two-dimensional strip packing problem (SPP) that improves the packing of the FFDH heuristic, and we state theoretical results of this algorithm. We also present an implementation of the FFDH heuristic for the three-dimensional case which is used to construct the COMB-3D heuristic with absolute worst-case performance ratio of 5. We also show, that this heuristic has absolute worst-case performance ratio of at most 4.25 for the z-oriented three-dimensional SPP. Based on this heuristic, we derive a general upper bound for the optimal height which depends on the continuous and the maximum height lower bound. We prove that the combination of both lower bounds also has an absolute worst-case performance ratio of at most 5 for the standard three-dimensional SPP. We also show that the layer-relaxation has a worst-case performance ratio of at most 4.25 for the z-oriented three-dimensional SPP.

1 Introduction

In this paper, we consider the two- and three-dimensional *strip packing problem* (SPP) with rectangular items. In the two-dimensional case (SPP-2), a list $\mathscr{L} := (R_1, \ldots, R_n)$ of small rectangles R_i (items) of width $w_i \leq W$ and height $h_i \leq H$, $i \in K := \{1, \ldots, n\}$ is given. The items have to be packed into a strip of given width W and minimal height OPT such that the items do not overlap each other. In the three-dimensional case (SPP-3), the items (boxes) R_i additionally have length $l_i \leq L, i \in K$, and they have to be packed into a container of given length L and width W such that minimal height is used. For standard SPP-3, rotation of items is not permitted, but we also consider a version of SPP-3 where the length and width of items can be

T. Buchwald (✉) · G. Scheithauer
Institute of Numerical Mathematics, Dresden University of Technology,
Dresden, Germany
e-mail: torsten.buchwald@tu-dresden.de

G. Scheithauer
e-mail: guntram.scheithauer@tu-dresden.de

© Springer International Publishing Switzerland 2016   65
M. Lübbecke et al. (eds.), *Operations Research Proceedings 2014*,
Operations Research Proceedings, DOI 10.1007/978-3-319-28697-6_10

interchanged. Without loss of generality, we assume that all input-data are positive integers.

In this paper, we present new results based on those given in [2]. There, a new algorithm for SSP-2 is introduced which improves the well-known FFDH heuristic (cf. [3]). Theoretical results are also given. Furthermore, in [2] an implementation of the FFDH heuristic for SPP-3 is proposed. Based on this heuristic, in this paper a new upper performance bound for some special case is proposed. Using this heuristic, in [2] a new heuristic for SPP-3 is created which enables us to prove a new general upper bound for the optimal value of this packing problem.

2   Two-Dimensional Approaches

A simple heuristic for the two-dimensional SPP is the FFDH heuristic. In [3] it is proved that this heuristic has an asymptotic worst-case performance ratio of 1.7 and an absolute worst-case performance bound of 2.7 . One disadvantage of this heuristic is the fact that all items of every single strip has decreasing heights. It might be more useful to have strips with decreasing heights and with increasing heights. Based on this idea, Schiermeyer provided an algorithm, called Reverse Fit [6], with an absolute worst-case performance bound of 2. The idea that we analyzed is a post-reduction procedure FFDH* (cf. [2]) for a packing created by the FFDH heuristic.

Algorithm FFDH*

1. Pack all items of K using the FFDH heuristic.
2. Choose two strips $t, u \in \{1, \ldots, s\}, t < u$.
3. Rearrange the order of the strips, such that t is the second-highest strip and u is the highest one.
4. Rotate the whole strip u by 180° (cf. Fig. 1).
5. Drop down strip u as much as possible.
6. If possible, repeat this method with two of the not already considered strips.

For this procedure we proved the following statement (cf. [2]):

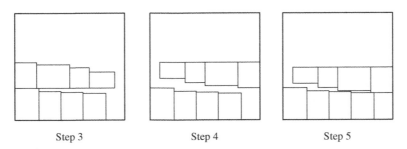

Step 3   Step 4   Step 5

Fig. 1 Algorithm FFDH*

Theorem 1 *Let $p \geq 2$ be a natural number, $w_i \leq W/p$ for all items of K and let $A(K)$ denote the area of all items. Furthermore, let $\widetilde{h}_j$ denote the height of strip j created by the FFDH heuristic. In the case that Algorithm FFDH* cannot achieve any improvement, we have*

$$FFDH^*(K) < \frac{p+1}{pW}A(K) + \widetilde{h}_1$$

$$- \left(\frac{1}{2} + \frac{p-2}{2p^2}\right)\left(\widetilde{h}_1 - \max\{0, 2 \max_{j \in \{1,\dots,s\}}\{\widetilde{h}_j - \widetilde{h}_{j+1}\} - \widetilde{h}_1\}\right).$$

This result also provides that the FFDH* procedure improves the height of the packing created by the FFDH heuristic, if this height is greater than the bound stated in Theorem 1. Another algorithm with absolute worst-case performance bound of 2 was provided by Steinberg in [7]. This algorithm is based on a reduction procedure and packs all items in polynomial time.

3 Three-Dimensional Approaches

In this section, we consider an implementation of the FFDH heuristic for the three-dimensional case, called *FFDH-3D heuristic*. The FFDH-3D heuristic is then combined with the algorithm of Steinberg [7] to formulate a new three-dimensional packing heuristic, called *COMB-3D heuristic* [2].

Using the layer-based FFDH-3D heuristic, the items are sorted according to non-increasing height and each item is placed in the lowest layer, where the item can be packed. A set of items can be packed in one layer, if all items of this set can be placed next to each other or a packing can be guaranteed by the algorithm of Steinberg [7]. The general absolute worst-case performance ratio of the FFDH-3D heuristic is unlimited, but we get a finite absolute worst-case performance ratio in some special cases:

Theorem 2 *Let a container with base area $L \times W$ and a natural number $p \in [2, W]$ be given. Furthermore, let K be a set of items of size $l_i \times w_i \times h_i$ with $w_i \leq W/p$ for all $i \in K$. Then we have for the absolute worst-case performance ratio of the FFDH-3D heuristic:*

$$\sup_E \frac{FFDH\text{-}3D(E)}{OPT(E)} = \frac{2(p+1)}{p} + 1.$$

We use the fact that an instance of the three-dimensional SPP with $w_i > W/2$ for all items can be reduced to an instance of the two-dimensional SPP, since it is not possible to pack two items next to each other in the width-direction. For the COMB-3D heuristic we partition the set of items into the set of items with $w_i \leq W/2$ and the set of items with $w_i > W/2$ and pack both sets separately. We pack all items with $w_i \leq W/2$ using the FFDH-3D heuristic. For the items with $w_i > W/2$ we reduce the instance of SPP-3 to an instance of the two-dimensional SPP and use the algorithm of Steinberg [7] to pack these items. We also analyzed the worst-case performance ratio of the COMB-3D heuristic.

Theorem 3 *Let a container with base area $L \times W$ be given. Furthermore, let K be a set of items of size $l_i \times w_i \times h_i$. Then we have for the absolute worst-case performance ratio of the COMB-3D heuristic:*

$$\sup_E \frac{COMB\text{-}3D(E)}{OPT(E)} = 5.$$

More detailed examinations show that this result can be further improved.

Theorem 4 *Let a container with base area $L \times W$ be given. Furthermore, let K be a set of items of size $l_i \times w_i \times h_i$ and let $K_2 := \{i \in K : w_i > \frac{1}{2}W, l_i \leq \frac{2}{3}L\} \neq \emptyset$. Let the constant C be defined by $\sum_{j \in K_2} l_j h_j = CL \max\{h_i : i \in K_2\}$. Then we have for the absolute worst-case performance ratio of the COMB-3D heuristic:*

$$\sup_E \frac{COMB\text{-}3D(E)}{OPT(E)} = \max\{5 - \frac{3}{2}C, \frac{9}{2}\}.$$

Theorem 5 *Let a container with base area $L \times W$ be given. Furthermore, let K be a set of items of size $l_i \times w_i \times h_i$ and let $K_2 := \{i \in K : w_i > \frac{1}{2}W, l_i \leq \frac{2}{3}L\} = \emptyset$. Then we have for the absolute worst-case performance ratio of the COMB-3D heuristic:*

$$\sup_E \frac{COMB\text{-}3D(E)}{OPT(E)} = 4.$$

Moreover, the idea of the COMB-3D heuristic is not only applicable for the three-dimensional SPP without rotation. We also consider the case that items can be rotated so that l_i and w_i can be interchanged but h_i is fixed. For this problem we assume $\max\{l_i, w_i\} \leq \min\{L, W\}$ for all items $i \in K$, so that each item can be packed in both orientations. This kind of three-dimensional SPP, called z-oriented three-dimensional SPP, was analyzed in [5] and [4]. The authors provided algorithms which are focused on the asymptotic worst-case performance bounds. We can show that the COMB-3D heuristic has an absolute worst-case performance bound for the z-oriented three-dimensional strip packing problem, that is even better than the absolute worst-case performance bound for the three-dimensional SPP without rotation.

Theorem 6 *Let a container with base area $L \times W$ with $L \leq W$ be given. Furthermore, let K be a set of items of size $l_i \times w_i \times h_i$ with $l_i \geq w_i$. For the COMB-3D heuristic, we get an absolute worst-case performance bound of*

$$\sup_E \frac{COMB\text{-}3D(E)}{OPT(E)} \leq 17/4 = 4.25.$$

Since the condition $l_i \geq w_i$ can be fulfilled by rotation, this absolute worst-case performance bound holds also for the considered SPP with rotation. For the special case $L = W$, we get an even better result.

Theorem 7 *Let a container with square base area $L \times W$ with $L = W$ be given. Furthermore, let K be a set of items of size $l_i \times w_i \times h_i$ with $l_i \geq w_i$. For the COMB-3D heuristic adapted to the z-oriented SPP-3, we get an absolute worst-case performance bound of*

$$\sup_{E} \frac{COMB\text{-}3D(E)}{OPT(E)} \leq 4.$$

The COMB-3D heuristic is also useful to analyze the absolute worst-case performance ratios of lower bounds of the three-dimensional SPP:

Theorem 8 *Let $E = (L, W, K)$ be any instance of the three-dimensional strip packing problem. Let $V(K)$ denote the volume of all items in K. For the absolute worst-case performance ratio of the combined lower bound $b_0 := \max\{\lceil \frac{1}{LW} V(K) \rceil,$ $\max_{j \in K} h_j\}$ holds*

$$\sup_{E} \frac{OPT(E)}{b_0(E)} \leq \sup_{E} \frac{COMB\text{-}3D(E)}{b_0(E)} \leq 5.$$

Considering the z-oriented three-dimensional SPP we can show a similar result for a class of lower bounds.

Theorem 9 *Let $E = (L, W, K)$ be any instance of the z-oriented three-dimensional strip packing problem with $L \leq W$ and $l_i \geq w_i$ for all $i \in K$. Furthermore, let $\overline{K}_1 := \{i \in K : w_i > W/2\}$. Let b be any lower bound for the z-oriented three-dimensional strip packing problem with*

$$b \geq \max \left\{ \left\lceil \frac{1}{LW} V(K) \right\rceil, \max_{j \in K} h_j, \sum_{j \in \overline{K}_1} h_j \right\}.$$

For the absolute worst-case performance ratio of the lower bound b holds

$$\sup_{E} \frac{OPT(E)}{b(E)} \leq \sup_{E} \frac{COMB\text{-}3D(E)}{b(E)} \leq 17/4 = 4.25.$$

An example for a lower bound of this class is the layer-relaxation. This relaxation replaces each item i with h_i items with same length and width but height 1. Two items which are created by the same original item are not allowed to be placed at the same layer.

The proofs of the Theorems of this section can be found in [1].

4   Conclusions and Outlook

In this paper, we presented a new algorithm for the two-dimensional strip packing problem which improves packings of the FFDH heuristic and we stated an improved upper bound of this algorithm for an important special case. Furthermore, we showed new upper bounds for two special cases of the three-dimensional strip packing problem. By using these special cases we succeeded to create a new general upper bound for the optimal value of this problem. We used this upper bound to improve the upper bound of the absolute worst-case performance ratio of a well known natural lower bound of the three-dimensional strip packing problem. Furthermore, we adapted the new heuristic to the z-oriented three-dimensional SPP and proved absolute worst-case performance bounds for the adapted heuristic. Moreover, we used this heuristic to improve the absolute worst-case performance bound of a class of lower bounds of the z-oriented SPP-3.

It will be part of our future research to prove a stronger general bound for the new algorithm for the two-dimensional strip packing problem. Furthermore, we will try to improve the bounds of the worst-case performance ratio mentioned above.

References

1. Buchwald, T., Scheithauer, G.: Improved Performance Bounds of the COMB-3D Heuristic for the three-dimensional Strip Packing Problem. Preprint MATH-NM-01-2015, Technische Universität Dresden
2. Buchwald, T., Scheithauer, G.: Upper bounds for heuristic approaches to the strip packing problem. Int. Trans. Oper. Res. **23/1-2**, 93–119 (2016)
3. Coffman Jr., E.G., et al.: Performance bounds for level-oriented two-dimensional packing algorithms. SIAM J. Comput. **9**, 808–826 (1980)
4. Epstein, L., Van Stee, R.: This side up!. Approximation and Online Algorithms, pp. 48–60. Springer, Berlin Heidelberg (2005)
5. Miyazawa, F.K., Wakabayashi, Y.: Approximation algorithms for the orthogonal z-oriented three-dimensional packing problem. SIAM J. Comput. **29**, 1008–1029 (2000)
6. Schiermeyer, I.: Reverse-fit: A 2-optimal algorithm for packing rectangles. In: Algorithms-ESA'94, pp. 290–299 (1994)
7. Steinberg, A.: A strip-packing algorithm with absolute performance bound 2. SIAM J. Comput. **26**, 401–409 (1997)

An Approximative Lexicographic Min-Max Approach to the Discrete Facility Location Problem

Ľuboš Buzna, Michal Koháni and Jaroslav Janáček

Abstract We propose a new approximative approach to the discrete facility location problem that provides solutions close to the lexicographic minimax optimum. The lexicographic minimax optimum is concept that allows to find equitable location of facilities. Our main contribution is the approximation approach, which is based on the rules allowing: (i) to take into account the multiplicities assigned to different customers; (ii) to detect whether for a given distance active customers can reach higher, equal or smaller distance to the closest located facility; and (iii) to use methods customized for solving the p-median problem. Customized methods can handle larger problems than state-of-the-art general purpose integer programming solvers. We use the resulting algorithm to perform extensive study using the well-known benchmarks and benchmarks derived from the real-world road network data. We demonstrate that our algorithm allows to solve larger problems than existing algorithms and provides high-quality solutions. The algorithm found the optimal solution for all tested benchmarks, where we could compare the results with the exact algorithm.

1 Introduction

Our study is motivated by problems faced by public authorities when locating facilities, such as schools, branch offices and ambulance, police or fire stations. These systems are typically paid from public money and they should account for equitable access to services. Previous approaches to the equitable location of facilities [1, 2], result in a specific form of the mathematical model that is supposed to be solved by a general purpose solver. Our initial experience with the algorithm [1], implemented on the state of the art solver XPRESS, indicated that we are able to solve problems up to 900 customers and 900 candidate facility locations, while restricting the distances to integers and measuring them in kilometres. This limitation might be too tight for some real-world applications and therefore it is of interest to elaborate

Ľ. Buzna (✉) · M. Koháni · J. Janáček
Department of Mathematical Methods and Operations Research, University of Zilina,
Univerzitna 8215/5, 01026 Zilina, Slovakia
e-mail: Lubos.Buzna@fri.uniza.sk

© Springer International Publishing Switzerland 2016
M. Lübbecke et al. (eds.), *Operations Research Proceedings 2014*,
Operations Research Proceedings, DOI 10.1007/978-3-319-28697-6_11

algorithms, which can provide high-quality solutions to larger problems. Building on the concept of unique classes of distances [1], we propose approximation algorithm providing high quality solutions for large instances of solved problems. We use the resulting algorithm to perform extensive study using the well-known benchmarks and two new large benchmarks derived from the real-world data.

2 Problem Formulation

We consider a set of potential locations of facilities I and a set of aggregate customers J. Each aggregate customer $j \in J$ is characterized by a unique geographical position and by an integer weight b_j. The weight b_j represents the number of individual customers situated in the location j. The decisions to be made can be represented by a set of binary variables. The variable y_i equals to 1 if the location $i \in I$ is used as a facility location and 0 otherwise. Allocation decisions are modelled by variables x_{ij} for $i \in I$ and $j \in J$, whereas $x_{ij} = 1$ if location i is serving the customer j and $x_{ij} = 0$ otherwise. In order to obtain a feasible solution, the decision variables have to satisfy the following set of constraints:

$$\sum_{i \in I} y_i = p, \tag{1}$$

$$\sum_{i \in I} x_{ij} = 1 \qquad \text{for all } j \in J, \tag{2}$$

$$x_{ij} \le y_i \qquad \text{for all } i \in I, j \in J, \tag{3}$$

$$x_{ij} \in \{0, 1\} \qquad \text{for all } i \in I, j \in J, \tag{4}$$

$$y_i \in \{0, 1\} \qquad \text{for all } i \in I, \tag{5}$$

where the Eq. (1) specifies that the number of located facilities equals to p. The constraints (2) make sure that each customer is assigned to exactly one facility, and the constraints (3) allow to assign a customer only to the located facilities. Following Ref. [1], we denote the set of all feasible location patterns, which satisfy the constraints (1)–(5), by the symbol Q.

We order the set of all feasible distance values d_{ij} into the descending sequence of unique distance values D_k, for $k = 1, \ldots, k_{max}$. Each feasible solution in the set Q can be associated with a sequence of subsets $[J_1, J_2, \ldots, J_{k_{max}}]$ and with a vector $[B_1, B_2, \ldots, B_{k_{max}}]$. The distance between customers in the set J_k and the assigned facility is exactly D_k. The component B_k is a number defined as $B_k = \sum_{j \in J_k} b_j$. If the set J_k is empty, then the associated value B_k is zero. The lexicographically minimal solution in the set Q is a solution that corresponds to the lexicographically minimal vector $[B_1, B_2, \ldots, B_{k_{max}}]$ [3].

3 The Algorithm A-LEX

Similarly to the algorithm [1], our algorithm solves optimization problems in stages corresponding to the distance values. For each $k = 2, \ldots, k_{max}$, we consider a partitioning of the set J into the system of subsets $\{J_1, \ldots, J_{k-1}, C_k\}$, where C_k is a set of active customers. The subset $J_k \subseteq C_k$ is determined as the minimal subset of customers (i.e. the set, where the sum of multiplicities $\sum_{j \in J_k} b_j$ is the smallest), whose distance from the closest facility location equals to the value D_k. For a given value of D_k, we find the minimal set J_k by solving the problem P_k:

$$\text{Minimize} \qquad g^k(\mathbf{x}) = \sum_{i \in I} \sum_{j \in J} r_{ij}^k x_{ij} \qquad (6)$$

$$\text{Subject to} \qquad (\mathbf{x}, \mathbf{y}) \in Q, \qquad (7)$$

where r_{ij}^k are the costs defined for $j \in C_k$ and $i \in I$ in the following way:

$$r_{ij}^k = \begin{cases} 0, & \text{if } d_{ij} < D_k, \\ b_j, & \text{if } d_{ij} = D_k, \\ (1 + \sum_{u \in C_k} b_u), & \text{if } d_{ij} > D_k, \end{cases} \qquad (8)$$

and for $j \in J_l$ where $l = 1, \ldots, k - 1$ and $i \in I$ according to the following prescription:

$$r_{ij}^k = \begin{cases} 0, & \text{if } d_{ij} \leq D_l, \\ (1 + \sum_{u \in C_k} b_u), & \text{otherwise.} \end{cases} \qquad (9)$$

Knowing the optimal solution $(\mathbf{x}^k, \mathbf{y}^k)$ of the problem P_k, the following implications can be derived:

1. If $g^k(\mathbf{x}^k) = 0$, then each customer $j \in C_k$ can be assigned to a facility whose distance from j is less than D_k.
2. If $0 < g^k(\mathbf{x}^k) < 1 + \sum_{u \in C_k} b_u$, then each customer $j \in C_k$ can be assigned to a facility, whose distance from j is less or equal to D_k. The minimal subset of customers $J_k \subseteq C_k$, whose distance from the closest facility locations equals to the value D_k can be defined as $\{j \in C_k | \sum_{i \in I} r_{ij}^k x_{ij}^k = b_j\}$.
3. If $g^k(\mathbf{x}^k) > \sum_{u \in C_k} b_u$, then this case indicates non-existence of a solution $(\mathbf{x}, \mathbf{y})$ to the problem P_k, for which $\sum_{i \in I} d_{ij} x_{ij} \leq D_l$ for $j \in J_l$, where $l = 1, \ldots, k$.

We formulate the algorithm A-LEX, where we identify the customers whose distance from the closest facility location cannot be shorter than D_k, by embedding the problem P_k:

Step 0: Set $k = 1$ and $C_1 = J$.
Step 1: Solve the problem P_k and denote the solution by $(\mathbf{x}^k, \mathbf{y}^k)$.
Step 2: If $g^k(\mathbf{x}^k) = 0$, set $C_{k+1} = C_k$ and go to Step 4, otherwise go to Step 3.

Step 3: Set $J_k = \{ j \in C_k | \sum_{i \in I} r^k_{ij} x^k_{ij} = b_j \}$; $C_{k+1} = C_k - J_k$.

Step 4: If $C_{k+1} = \emptyset$, then terminate and return $(\mathbf{x}^k, \mathbf{y}^k)$ as the solution, otherwise set $k = k + 1$ and continue with the Step 1.

Correctness and finiteness of the algorithm A-LEX, including the optimality conditions, are in more details analysed in Ref. [4].

4 Numerical Experiments

Our two main goals are to evaluate the quality of solutions provided by the algorithm A-LEX by comparing them to the exact algorithm (algorithm O-LEX hereafter) [1] and to test the limits of the algorithm regarding the size of solvable problems. To be able to compare our results with the exact algorithm O-LEX, we implemented algorithm A-LEX (algorithm A-LEX^X hereafter) in the XPRESS-Mosel language (version 3.4.0) and we ran it using the XPRESS-Optimizer (version 23.01.05). To explore the properties of the algorithm A-LEX beyond the limits of the general purpose integer solvers, we implemented the algorithm A-LEX in the Microsoft Visual C++ 2010 using the algorithm ZEBRA, the state-of-the-art solver for the p-median problem [5] (algorithm A-LEX^Z hereafter). Two sets of testing problems organized by the size were used to perform the computational study. In all cases, customers' sites are considered to be also possible facility locations, i.e. sets I and J are identical. As there are no standard test problems for the facility location problem with the lexicographic minimax objective, we used the problems originally proposed for the capacitated p-median problem while interpreting the demands as b_j values (multiplicities of customers). Three problems $SJC2$ ($|I| = 200$), $SJC3$ ($|I| = 300$) and $SJC4$ ($|I| = 402$) are taken from the Ref. [6] and together with two instances derived from the network of $|I| = 737$ Spanish cities [7] constitute the medium sized instances. Large test problems include the problem $p3038$ ($|I| = 3038$) originally proposed for the TSP [8] and later adjusted to the capacitated p-median problem [6]. Furthermore, considering the population data as b_j values, we created large-sized benchmarks from the interurban road network of the Slovak Republic ($|I| = 2928$) [9] and the interurban road network of six south-eastern U.S. states: Tennessee, North Carolina, South Carolina, Georgia, Alabama and Mississippi ($|I| = 2398$) [10].

We summarized the computational results in Table 1. Due to problems with the computer memory XPRESS solver was not able to solve large instances successfully. Therefore, for large instances we show in Table 1 the results obtained by the solver ZEBRA only. Comparison of results reveals that the algorithm A-LEX^X outperforms the algorithm O-LEX in terms of the computational time. A-LEX^X computed all medium instances in 47.6 % of the time needed by the algorithm O-LEX. In order to compare the quality of the solution, we evaluated the Manhattan distance between the location vectors $\mathbf{y}$. Occasionally, we found small differences in the location of facilities. However, we checked, and we found that customers have the same distance

Table 1 Computational results for the algorithm A-LEX

Medium instances				Large instances		
Instance	p	O-LEX Time [s]	A-LEX^X Time [s]	Instance	p	A-LEX^Z Time [s]
$SJC2$	10	131,4	50,9	$p3038$	2000	4204,3
$SJC2$	20	64,4	37,4	$p3038$	1500	8915,9
$SJC2$	30	32,2	17,4	$p3038$	900	190092,9
$SJC2$	40	20,3	9,7	$p3038$	700	17902,9
$SJC3$	15	461,6	357,7	$p3038$	100	*
$SJC3$	30	145,1	68,8	$p3038$	50	*
$SJC3$	45	71,1	37,9	$p3038$	10	201165,3
$SJC3$	60	53,3	29,8	SR	2000	1021,6
$SJC4$	20	1371,2	1205,8	SR	1500	1083,4
$SJC4$	40	1207,5	1052,5	SR	900	1988,5
$SJC4$	60	158,7	87,2	SR	700	2954,2
$SJC4$	80	144,9	56,2	SR	100	9624,7
$Spain\_737\_1$	37	116838	81185,1	SR	50	10509,8
$Spain\_737\_1$	50	196000	27296,2	SR	10	11888,4
$Spain\_737\_1$	185	12367,4	279,5	US	2000	1006,6
$Spain\_737\_1$	259	430,4	32,2	US	1500	1203,7
$Spain\_737\_2$	37	35590,7	29185,6	US	900	1694,7
$Spain\_737\_2$	50	64005,7	27806,1	US	700	11022,5
$Spain\_737\_2$	185	3182,3	232,4	US	100	*
$Spain\_737\_2$	259	72,5	43,2	US	50	*
				US	10	9038,2

The symbol "*" denotes the cases when the algorithm did not terminate within 3 days

to the closest located facility in all tested instances. Thus, the algorithm A-LEX found an optimal solution in all cases where we were able to compare it with the algorithm O-LEX.

5 Conclusions

The proposed algorithm A-LEX is competitive with the state-of-the-art algorithm O-LEX:

- it allows to solve large instances of the problem,
- the algorithm found the optimal solution for all instances, where we could compare the results with the exact algorithm,
- the algorithm A-LEX computed all medium instances in 47.6 % of the time needed by the algorithm O-LEX.

The proposed approximation approach is also applicable to other types of similar combinatorial optimization problems with lexicographic minimax objective. Example of a problem, where this approach could be used, is the maximum generalized assignment problem.

Acknowledgments This work was supported by the research grants VEGA 1/0339/13 and VEGA 1/0463/16.

References

1. Ogryczak, W.: On the lexicographic minmax approach to location problems. Eur. J. Oper. Res. **100**, 566–585 (1997)
2. Ogryczak, W., Śliwiński, T.: On direct methods for lexicographic min-max optimization. In: Cigarillo, M., Gervasi, O., Kumar, V., Tan, C., Taniar, D., Laganá, A., Mun, Y., Choo, H. (eds.) Computational Science and Its Applications—ICCSA 2006, Lecture Notes in Computer Science, pp. 802–811. Springer, Berlin (2006)
3. Luss, H.: Equitable Resources Allocation: Models, Algorithms and Applications, Wiley (2012). ISBN: 978-1-118-05468-0
4. Buzna, Ľ., Koháni, M., Janáček, J.: An approximation algorithm for the facility location problem with lexicographic minimax objective. J. Appl. Math. **2014** (562373) (2014)
5. García, S., Labbé, M., Marín, A.: Solving large p-median problems with a radius formulation. INFORMS J. Comput. **23**(4), 546–556 (2011)
6. Lorena, L.A., Senne, E.L.: A column generation approach to capacitated p-median problems. Comput. Oper. Res. **31**(6), 863–876 (2004)
7. Díaz, J.A., Fernández, E.: Hybrid scatter search and path relinking for the capacitated p-median problem. Eur. J. Oper. Res. **169**(2), 570–585 (2006)
8. Reinelt, G.: The traveling salesman: Computational Solutions for TSP Applications. Springer, Berlin, Heidelberg (1994). ISBN: 3-540-58334-3
9. Data describing the road network of the Slovak Republic were purchased from the publisher MAPA Slovakia Plus s.r.o www.mapaslovakia.sk. Accessed 22 June 2014
10. United States Census Database: http://www.nationalatlas.gov/mld/ce2000t.html (2000) Accessed 22 June 2014

Effects of Profit Taxes in Matrix Games

Marlis Bärthel

Abstract It is consensus in different fields of practical relevance that the introduction of taxes will—in one way or the other—affect the playing behavior of actors. However, it is not clear what the effects might really look like. Here, a game theoretic model is considered that concentrates on effects of relative or constant taxes on transferred monetary volume. For matrix games it is asked: How do taxes change the behavior of players and the expected transacted volume? Analyzing this basic research model clearly shows: One has to be careful in considering taxes as a powerful instrument to confine aggressive playing behavior. Taxes might encourage increased expected transfers.

1 Introduction

In some economic applications not only the real payoffs of the actors, but the transferred amount of money plays an important role. When for instance a financial transaction tax is introduced, the traded financial volume would be crucial to determine resulting tax revenues. Offerers of bets or gambling (e.g. online portals of sports betting, poker platforms or casinos) are faced with a similar situation. It is not important for the providers who of the participating actors has which payoff; instead the monetary transfer is interesting.

In this context the concept of (bi) matrix games offers an interesting modelling framework. Commonly, for a 2-person-zerosum matrix game Γ with payoff matrix A for player 1 (and $-A$ for player 2) and mixed strategies p and q of the players, the expected payoff for player 1 is defined by $v_1(A; p, q) = \sum \sum p_i q_j a_{ij}$. We define the **expected transfer** (cf. [1])

$$ET(A; p, q) := \sum \sum p_i q_j |a_{ij}|.$$

M. Bärthel (✉)
Department of Mathematics, Friedrich-Schiller-University Jena,
Ernst-Abbe-Platz 2, 07743 Jena, Germany
e-mail: marlis.baerthel@uni-jena.de

© Springer International Publishing Switzerland 2016
M. Lübbecke et al. (eds.), *Operations Research Proceedings 2014*,
Operations Research Proceedings, DOI 10.1007/978-3-319-28697-6_12

The only difference between expected payoff and expected transfer is, that the absolute values of the entries a_{ij} instead of the real, signed payoffs are considered in our setting. If the matrix game has a unique Nash-equilibrium $(\bar{p}, \bar{q})$, we call $ET(\Gamma) := ET(A; \bar{p}, \bar{q})$ the expected transfer of the game. We investigate how the expected transfer changes, when taxes have to be paid to a third, uninvolved party. We basically consider four different tax scenarios, depending on **who** (winner or loser) is taxed and **how** (relative or constant tax) the amount of the taxes is determined.

In Sect. 2 we explain the model. An example and some general results are presented in Sect. 3. We conclude by summarizing the main facts and listing some open questions in Sect. 4.

2 The Model

Our basic model can be described in six steps. We always assume fully informed, risk neutral and perfectly rational players.

(I) Consider a 2-person-zerosum matrix game Γ with payoff matrix A for player 1 and $-A$ for player 2.

(II) Determine the optimal strategies, i.e. the Nash-equilibrium $(\bar{p}, \bar{q})$ (cf. [3]). To ensure that the game has a unique Nash-equilibrium, consider only non-degenerate games in step (I) (cf. [2] for definition and properties of non-degenerate matrix games).

(III) The game theoretic value of the game (cf. [4]) is given by the expected payoff of player 1: $v(\Gamma) = \sum \sum \bar{p}_i \bar{q}_j a_{ij}$. Determine the expected transfer of the game (cf. [1])

$$ET(\Gamma) := \sum \sum \bar{p}_i \bar{q}_j |a_{ij}|.$$

(IV) Change the payoffs of the players by charging a tax with tax-parameter x and receive a bimatrix game $\Gamma_{\text{tax}}(x)$. This step involves the essential idea of modelling a tax and will be explained in more detail below. Different scenarios are considered.

(V) Determine the optimal strategies, i.e. the Nash-equilibrium $(\bar{p}(x), \bar{q}(x))$ of the taxed game. The Nash-equilibrium need not be unique. In this case, consider all possible Nash-equilibria.

(VI) The expected transfer of the taxed game with respect to the equilibrium $(\bar{p}(x), \bar{q}(x))$ is

$$ET(\Gamma_{\text{tax}}(x); \bar{p}(x), \bar{q}(x)) := \sum \sum \bar{p}_i(x) \bar{q}_j(x) |a_{ij}|.$$

In case there is a unique equilibrium in step (V), we shortly write

$$ET(\Gamma_{\text{tax}}(x)) := ET(\Gamma_{\text{tax}}(x); \bar{p}(x), \bar{q}(x)).$$

In step (IV) the taxation of the matrix game $\Gamma = (A, -A)$ is modelled. Basically we define four different tax scenarios. The scenarios arise by combining two options of *who* is taxed (*winner* or *loser*) and two options of *how* is taxed (*relative* or *constant* *tax*).

Assume for instance a *winner tax* with a *relative tax* rate of 10%, i.e. $x = 0.1$ (**or** a *constant tax* of $y = 0.1$). When row i and column j are played, then $|a_{ij}|$ is transferred. For $a_{ij} > 0$, *winner tax* means that player 1 receives only $0.90 \cdot a_{ij}$ in the *relative* case (**or** $a_{ij} - 0.1$ in the *constant* case), whereas player 2 has to pay the full a_{ij}. For $a_{ij} < 0$, player 1 has to pay the full $|a_{ij}|$, but player 2 only gets $0.90 \cdot |a_{ij}|$ (**or** $|a_{ij}| - 0.1$). For $a_{ij} = 0$, nothing at all is transferred and no taxes (**or** $\frac{1}{2}y = 0.05$ in the *constant* case) are taken from both players. We denote the new game by $\Gamma_{\text{WiT}}(x)$, where the index "WiT" is short for *"Winner Tax"*.

Introducing such a tax changes the character of the game. It is no longer a zerosum, but now a bimatrix game. The resulting bimatrix game may have several equilibria, with different expected payoffs and transfers. However, at least in the case where also the taxed game has a unique equilibrium, one may compare the expected transfers of $\Gamma_{\text{WiT}}(x)$ and Γ. We do this comparison with respect to the original $|a_{ij}|$. So, when for instance in the taxed game player 1 gets $(1 - x) \cdot a$ (**or** $a - y$) and player 2 has to pay $1 \cdot a$, then we count this as transfer of size $1 \cdot a$. I.e. our transfer means "transfer before tax".

The payoff changes and comparisons of the expected transfers in the other tax scenarios are realized analogously. We investigate the scenarios

- *relative winner tax*: the winner only gets $(1 - x) \cdot |a_{ij}|$ instead of $|a_{ij}|$, the loser has to pay $|a_{ij}|$ (where $x \in [0, 1)$).
- *relative loser tax*: the loser has to pay $(1 + x) \cdot |a_{ij}|$ instead of simply $|a_{ij}|$, the winner gets $|a_{ij}|$ (where $x \in [0, 1)$).
- *constant winner tax*: the winner only gets $|a_{ij}| - y$ instead of $|a_{ij}|$, the loser has to pay $|a_{ij}|$ (where $y \geq 0$).
- *constant loser tax*: the loser has to pay $|a_{ij}| + y$ instead of simply $|a_{ij}|$, the winner gets $|a_{ij}|$ (where $y \geq 0$).

Naive common sense might suggest that such a tax should discourage high transfers, resulting in a smaller expected transfer. But analysis tells a different story, at least for small, fair matrix games. Rather often, a tax increases the expected transfer. There are interesting differences between relative and constant taxes.

3 Results for Relative and Constant Profit Taxes

In this article, we only consider non-degenerate, *fair* games, i.e. games with game theoretic value $v(\Gamma) = 0$ according to step (III) in Sect. 2. In the initial situation the equilibrium payoffs are 0 for both players.

Table 1 The four scenarios for the fair 2×2 example game with fixed relative tax rate $x = 0.1$ or constant tax $y = 0.1$

Scenario	A	B	$\bar{p}$	$\bar{q}$	$\bar{v}_1$	$\bar{v}_2$	ET
Matrix game: Γ	$\begin{pmatrix} 1 & -2 \\ -3 & 6 \end{pmatrix}$	$\begin{pmatrix} -1 & 2 \\ 3 & -6 \end{pmatrix}$	$\begin{pmatrix} \frac{3}{4} \\ \frac{1}{4} \end{pmatrix}$	$\begin{pmatrix} \frac{2}{3} \\ \frac{1}{3} \end{pmatrix}$	0	0	2
Relative winner tax: $\Gamma_{\text{WiT}}(0.1)$	$\begin{pmatrix} 0.9 & -2 \\ -3 & 5.4 \end{pmatrix}$	$\begin{pmatrix} -1 & 1.8 \\ 2.7 & -6 \end{pmatrix}$	$\begin{pmatrix} \frac{87}{115} \\ \frac{28}{115} \end{pmatrix}$	$\begin{pmatrix} \frac{74}{113} \\ \frac{39}{113} \end{pmatrix}$	$-\frac{57}{565}$ ≈ -0.1009	$-\frac{57}{575}$ ≈ -0.0991	$2 + \frac{2}{12995}$ ≈ 2.0002
Relative loser tax: $\Gamma_{\text{LoT}}(0.1)$	$\begin{pmatrix} 1 & -2.2 \\ -3.3 & 6 \end{pmatrix}$	$\begin{pmatrix} -1.1 & 2 \\ 3 & -6.6 \end{pmatrix}$	$\begin{pmatrix} \frac{96}{127} \\ \frac{31}{127} \end{pmatrix}$	$\begin{pmatrix} \frac{82}{125} \\ \frac{43}{125} \end{pmatrix}$	$-\frac{63}{625}$ ≈ -0.1008	$-\frac{63}{635}$ ≈ -0.0992	$2 + \frac{2}{15875}$ ≈ 2.0001
Constant winner tax: $\Gamma_{\text{WiT}}(0.1)$	$\begin{pmatrix} 0.9 & -2 \\ -3 & 5.9 \end{pmatrix}$	$\begin{pmatrix} -1 & 1.9 \\ 2.9 & -6 \end{pmatrix}$	$\begin{pmatrix} \frac{89}{118} \\ \frac{29}{118} \end{pmatrix}$	$\begin{pmatrix} \frac{79}{118} \\ \frac{39}{118} \end{pmatrix}$	$-\frac{69}{1180}$ ≈ -0.0585	$-\frac{49}{1180}$ ≈ -0.0415	$2 - \frac{54}{3481}$ $\approx 1,9845$
Constant loser tax: $\Gamma_{\text{LoT}}(0.1)$	$\begin{pmatrix} 1 & -2.1 \\ -3.1 & 6 \end{pmatrix}$	$\begin{pmatrix} -1.1 & 2 \\ 3 & -6.1 \end{pmatrix}$	$\begin{pmatrix} \frac{91}{122} \\ \frac{31}{122} \end{pmatrix}$	$\begin{pmatrix} \frac{81}{122} \\ \frac{41}{122} \end{pmatrix}$	$-\frac{51}{1220}$ ≈ -0.0418	$-\frac{71}{1220}$ ≈ -0.0582	$2 + \frac{56}{3721}$ ≈ 2.0150

Example 1 As an example, we look at the non-degenerate 2×2 matrix game $\Gamma = (A, -A)$ with

$$A = \begin{pmatrix} 1 & -2 \\ -3 & 6 \end{pmatrix}.$$

This game has the unique equilibrium $(\bar{p}, \bar{q}) = \left(\left(\frac{3}{4}, \frac{1}{4}\right)^\top, \left(\frac{2}{3}, \frac{1}{3}\right)^\top \right)$. The value of the game is $v(\Gamma) = \bar{p}^\top A \bar{q} = 0$, so it is a fair game. The expected transfer of Γ is

$$ET(\Gamma) = \begin{pmatrix} \frac{3}{4} & \frac{1}{4} \end{pmatrix} \begin{pmatrix} 1 & 2 \\ 3 & 6 \end{pmatrix} \begin{pmatrix} \frac{2}{3} \\ \frac{1}{3} \end{pmatrix} = \frac{3 \cdot 2 \cdot 1 + 3 \cdot 1 \cdot 2 + 1 \cdot 2 \cdot 3 + 1 \cdot 1 \cdot 6}{12} = 2.$$

Now we assume that a third party (for instance the state, a platform provider etc.) enters the scene and collects a profit tax. Table 1 gives an overview of all four scenarios if the relative tax rate is $x = 0.1$ or the constant tax is $y = 0.1$, respectively. It contains the payoff matrices (A, B), the equilibrium strategies $(\bar{p}, \bar{q})$, and the expected payoffs (v_1, v_2) of the two players, as well as the resulting expected transfer (ET).

An interesting observation is: The expected transfer in the scenarios *relative winner tax*, *relative loser tax* and *constant loser tax* has increased, while it has decreased in the scenario *constant winner tax*. Figure 1 shows the effects for all relative tax rates $x \in [0, 1)$ and all constant taxes $y \in [0, 3)$.

This phenomenon of monotonically increasing or decreasing expected transfers is not an artifact of the special 2×2 game with matrix A from above, as shown by Theorem 1 (cf. [1]) and Theorem 2: Resulting effects, represented by Fig. 1, occur for all non-degenerate, fair 2×2 matrix games.

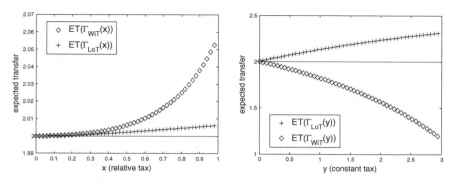

Fig. 1 Expected transfers for the example game and the four scenarios. **a** Scenarios with rel. tax $x \in [0, 1)$. **b** Scenarios with const. tax $y \in [0, 3)$

Theorem 1 (Scenarios with a Relative Tax)
Let $\Gamma = (A, -A)$ be a non-degenerate, fair 2×2 matrix game. For every relative tax rate $x \in [0, 1)$, the following four statements hold.

(i)   *The bimatrix games $\Gamma_{WiT}(x)$ and $\Gamma_{LoT}(x)$ each have exactly one Nash-equilibrium.*
(ii) *$ET(\Gamma_{WiT}(x))$ is either strictly increasing in x or constant.*
(iii) *$ET(\Gamma_{LoT}(x))$ is either strictly increasing in x or constant.*
(iv) *$ET(\Gamma_{WiT}(x)) \geq ET(\Gamma_{LoT}(x)) \geq ET(\Gamma)$.*

Theorem 2 (Scenarios with a Constant Tax)
Let $\Gamma = (A, -A)$ be a non-degenerate, fair 2×2 matrix game and $a := \min\{|a_{11}| + |a_{21}|, |a_{12}| + |a_{22}|, |a_{11}| + |a_{12}|, |a_{21}| + |a_{22}|\}$. For every constant tax $y \in [0, a)$, the following three statements hold.

(i)   *The bimatrix games $\Gamma_{WiT}(y)$ and $\Gamma_{LoT}(y)$ each have exactly one Nash-equilibrium.*
(ii) *$ET(\Gamma_{WiT}(y))$ is either strictly decreasing in y or constant.*
(iii) *$ET(\Gamma_{LoT}(y))$ is either strictly increasing in y or constant.*

Remark 1 The games with mixed optimal strategies and constant expected transfers in Theorems 1 and 2 are always sets of measure 0.

A complete proof of Theorem 1 can be found in [1]. The proof of Theorem 2 works analogously. The limitation of the constant tax to maximally $a := \min\{|a_{11}| + |a_{21}|, |a_{12}| + |a_{22}|, |a_{11}| + |a_{12}|, |a_{21}| + |a_{22}|\}$ substantially means: For a player, the *winner tax* does not convert an originally better option to a worse option if the opponent selects a fixed pure strategy.

4 Conclusion and Open Questions

For small matrix games, the introduction of taxes often results in higher expected transfers. For non-degenerate, fair 2×2 matrix games and the three tax scenarios *relative winner tax*, *relative loser tax*, *constant loser tax* this phenomenon holds in every instance. The scenario *constant winner tax* results in a decreasing expected transfer.

Having results of our model in mind, several interesting questions for real applications occur. Is it possible that after the introduction of taxes, actors speculate and gamble even more aggressively? Would in consequence a tax be much more disturbing than calming for a market? Would on the other hand a tax be a secure source of revenue? If so, what would be a good tax parameter to establish a justifiable balance between disturbance and financial revenue (that in an ideal case could be used sustainably)?

The game theoretic result of increasing expected transfers indeed can not readily be transmitted to real-world applications like a financial transaction tax or taxes on bets or gambling. However, our model shows that one has to be careful in considering taxes as panacea to confine aggressive playing behavior. In case of taxing 2×2 matrix games, often the opposite is true: Taxes encourage higher transfers.

Acknowledgments Special thanks go to Ingo Althöfer, who gave his time for discussions and proofreading. I thank Rahul Savani for his nice online bimatrix solver (http://banach.lse.ac.uk/form.html).

References

1. Althöfer, I., Bärthel, M.: On taxed matrix games and changes in the expected transfer. Game Theory. Hindawi (2014). http://dx.doi.org/10.1155/2014/435092
2. Avis, D., Rosenberg, G.D., Savani, R., von Stengel, B.: Enumeration of Nash equilibria for two-player games. Econ. Theor. **42**(1), 9–37 (2009)
3. Nash, J.: Non-cooperative games. Ann. Math. **54**(2), 286–295 (1951)
4. von Neumann, J.: Zur Theorie der Gesellschaftsspiele. Ann. Math. **100**(1), 295–320 (1928)

Extension and Application of a General Pickup and Delivery Model for Evaluating the Impact of Bundling Goods in an Urban Environment

Stephan Bütikofer and Albert Steiner

Abstract In this work we present results from a research project dealing with the bundling of goods, which was applied to the region of Zurich. The main goal was to develop a model for quantifying the impact of bundling goods on total costs. We focussed on the case of two distributors here and investigated two scenarios, namely no and full cooperation. The scenarios were based on transportation requests from customers and known origin-destination transport flows between postcodes. The costs include salaries, fleet degeneration and amortisation, respectively, and CO_2 emissions. To allow for computing the total costs, a pickup and delivery model from literature was implemented and extended. It could be shown that even for a simple kind of cooperation, savings of around 7 % in total costs are possible.

1 Introduction

Over the past decades, many metropolitan areas were facing a continuing increase of their population leading, amongst others, to a higher demand of transporting goods, both by the industry and suppliers, and by households. In addition, various new types of services were developed to deliver goods. For delivery companies, short delivery times, low costs and high quality are some of the main targets of a shipment, whereas for public authorities and the society, the minimization of green-house gas emissions is of increasing importance.

To address these requirements, various methods for bundling goods were developed. In an applied research project with industrial partners in the region of Zurich, a cooperation platform for several logistic companies, associations and public authorities will be developed. The main goal of the project is to come up with a cooperation platform for the transport of goods with respect to economic and ecological aspects.

S. Bütikofer (✉) · A. Steiner
Institute of Data Analysis and Process Design, Zurich University of Applied
Sciences ZHAW, Rosenstrasse 3, 8401 Winterthur, Switzerland
e-mail: stephan.buetikofer@zhaw.ch

A. Steiner
e-mail: albert.steiner@zhaw.ch

© Springer International Publishing Switzerland 2016
M. Lübbecke et al. (eds.), *Operations Research Proceedings 2014*,
Operations Research Proceedings, DOI 10.1007/978-3-319-28697-6_13

Bundling goods is a key component to achieve this target. The results we present here are part of a pre-study, with the goal of quantifying the impact of bundling goods. For this we considered transportation requests of two distributors with separate depots. In a first scenario each distributor carries out his freight independently. In a second scenario the distributors collaborate by sharing their transportation requests (see Sect. 3 for details).

Therefore, an existing pickup and delivery model from [10] was extended. We refer to [2, 9] and the references there for a good literature survey corresponding to pickup and delivery models. A pickup and delivery model was required especially for the second scenario. Furthermore, in the main project following this pre-study, it will be necessary to handle transportation requests which will be shipped directly between customers without an intermediate depot station. This model used is flexible enough for the extensions developed while at the same time covering all essential requirements (allowing for time windows, splitting orders, etc.). Furthermore, a methodology was developed to allow for meeting multi-criteria objectives (driver and fleet costs, costs due to emissions, etc.).

2   Methodology

In the pickup and delivery problem in [10] a set of routes has to be constructed in order to satisfy transportation requests. A fleet of vehicles is available to serve the routes. Each vehicle has a given capacity, a start location and an end location. Each transportation request specifies the size of the load to be transported, locations where freight has to be picked up (origins) and locations where freight has to be delivered (destinations). Each load has to be transported by one vehicle from its set of origins to its set of destinations without any transshipment at other locations.

Our aim is the integration of conflicting targets (e.g. low driver costs and low CO_2 emissions in one objective function). For this we translate all targets into a common currency (i.e. EUR) and combine them as linear combination in a utility function. In the rest of the paper we make use of the definitions and notations in Table 1.

Optimization
The pickup and delivery model in [10] is defined on a complete, directed and weighted graph (A, F) as an mixed integer problem, where the nodes fulfill $A = (V \cup M^+ \cup M^-)$ with

$$V = \cup_{r \in N}(N_r^+ \cup N_r^-), \quad M^+ = \{k^+ \mid k \in M\}, \quad M^- = \{k^- \mid k \in M\}$$

and c_{ij}^k are the weights of edge $(i, j) \in F$ for vehicle $k \in M$. The graph (A, F) represents the formalized road network in our test system (see Sect. 3). The construction of the graph is discussed in Sect. 3. Instead of restating the formal definitions of this optimization problem, we just describe its feasible set in words, provide the objective function and refer to [10] for details.

Table 1 Notation of sets, variables and parameters together with corresponding units

Classes	Subclass	Description	Notation	Unit
Road network		Set of nodes in road network	$i \in I$	(1)
		Distance between node i and j	d_{ij}	(km)
		Max. speed on edge (i, j)	v_{ij}^0	(km/h)
Fleet vehicle (k)	Characteristics	Set of vehicles	$k \in M$	(1)
		Capacity	Q_k	(kg)
		Emissions of CO_2 on edge (i, j)	E_{ij}^k	(tCO$_2$/km)
	Locations	Average speed on edge (i, j)	v_{ij}^k	(km/h)
		Start node (start depot) of vehicle k	k^+	–
		End node (end depot) of vehicle k	k^-	–
		Time to traverse edge (i, j)	T_{ij}^k	(h)
Order (r)	Locations	Set of orders	$r \in N$	(1)
		Set of pickup nodes	N_r^+	(1)
		Set of delivery nodes	N_r^-	(1)
		Volume or weight	q_r	(m^3) or (kg)
	Handling	Pickup duration at node i	T_i^+	(h)
		Delivery duration at node i	T_i^-	(h)
Costs	Time	Salary of driver of vehicle k	P_k^S	(EUR/h)
		Handling cost (pickup, delivery)	P_k^H	(EUR/h)
	Distance	Cost for vehicle k	P_k^V	(EUR/km)
	Emissions		P^E	(EUR/tCO$_2$)

The feasible set of the pickup and delivery model consists of pickup and delivery plans. A pickup and delivery plan is a set of routes for each vehicle $k \in M$, which form a partition of all nodes in V. For a vehicle $k \in M$ a route heads through a subset of $V_k \subset V$, such that the route starts in k^+ and ends in k^-. Vehicle k visits all locations in V_k exactly once and fulfils all the transportation requests, which are associated to this vehicle. The capacity restriction Q_k is always met. The properties of the routes imply an important fact for constructing the graph. Namely, each node (pickup, delivery, origin and destination depot, respectively) has its own representation even if they are geographically the same nodes (e.g. origin and destination depot are identical).

The objective function $f(x)$ is given by

$$f(x) = \sum_{k \in M} \sum_{(i,j) \in F} x_{ij}^k c_{ij}^k, \tag{1}$$

where x_{ij}^k is a decision variable equal to 1 if vehicle k travels from location i to j and 0 otherwise.

Composition of the Weights c_{ij}^k

The costs c_{ij}^k in (1) for traversing an edge (i, j) of graph (A, F) by vehicle $k \in M$ consist of components depending (i) on time (driving and handling) and (ii) distance, respectively. For the sake of simplicity we don't integrate costs for infrastructure, safety, and environment here.

We start with the salary costs $(c_{ij}^k)^S$ due to the time required by the driver, the costs $(c_{ij}^k)^E$ due emissions and the costs $(c_{ij}^k)^V$ of vehicle $k \in M$ to traverse edge $(i, j) \in F$.

$$(c_{ij}^k)^S = T_{ij}^k P_k^S, \quad (c_{ij}^k)^V = d_{ij} P_k^V, \quad (c_{ij}^k)^E = d_{ij} E_{ij}^k P^E \tag{2}$$

The handling costs $(c_{ij}^k)^H$ are calculated as

$$(c_{ij}^k)^H = \begin{cases} T_j^+ P_k^H, \text{ if } j \in N^+ T_j^- P_k^H, \& \text{ if } j \in N^-; \\ 0, \qquad \text{else.} \end{cases} \tag{3}$$

In a similar manner these costs could also be time dependent (e.g. by replacing T_{ij}^k with $T_{ij}^k(t)$). We omit this here for space reasons. For most vehicles, data regarding their average CO_2 emissions is available (e.g. [5, 8]). From [11], CO_2 emissions can be calculated for a variety of vehicle classes even as a function of speed and other parameters. References [1, 4] provide a very good discussion on the dependency of vehicle emissions from speed, road gradients and weight.

Based on equations in (2) and (3), the resulting overall costs for vehicle $k \in M$ to traverse the edge $(i, j) \in F$ are defined by

$$c_{ij}^k = \alpha_1 (c_{ij}^k)^S + \alpha_2 (c_{ij}^k)^H + \alpha_3 (c_{ij}^k)^V + \alpha_4 (c_{ij}^k)^E$$

where parameters α_1 to α_4 (with $\sum_i \alpha_i = 1$) allow for some weighting of costs, e.g. over-weighting emission costs to optimize for low-emission routes. We have set all parameters to an identical weight of 0.25.

3 Case Study

The road network considered around the city centre of Zurich covers an area of around $1670 \, \text{km}^2$ (width: $62 \, \text{km}$, height: $26.5 \, \text{km}$) and consists of around 29000 nodes and 36000 edges. For each edge (i, j) in the original road network the road type, the maximum speed v_{ij}^0 and the driving directions allowed are known [7]. The average speed v_{ij} was set to $80 \, \%$ of the maximum speed and always below $80 \, \text{km/h}$. For the sake of simplicity, we assumed that turnings are possible between adjacent edges. Furthermore, 242 postcodes were considered, based on data provided by [6].

The number of distributors was two, each one with a fleet consisting of seven vehicles. All vehicles $k \in M$ had an identical payload Q_k of 12 tons. In addition, both distributors had to process 15 orders, where each order $r \in N$ had a weight q_r of 3 tons. We assumed that at the beginning all vehicles and all orders are located at the start depot for each distributor. The end depot was identical to the start depot. For each distributor, the ten postcodes with the highest number of orders (delivery) were determined from survey data on Swiss freight transport for 2008 [6] and scaled down such that the overall number of orders is 15 each (which is necessary for sampling the orders, see section computational procedure below).

For each vehicle $k \in M$ the average salary P_k^S considered for truck drivers is $50 \, \text{EUR/h}$ (average value according to our Swiss distributor partners) and the average time T_i^+ resp. T_i^- required to loading/unloading freight was assumed to be $10 \, \text{min}$. The averages costs P_k^V of a truck delivering general cargo were assumed to be $0.80 \, \text{EUR/km}$ (which is low compared to the $3 \, \text{EUR/km}$ for parcel deliveries where usually larger vehicles are required), and the costs P^E of emitting CO_2 is assumed to be 48 EUR per ton, which is in good agreement with data from [3]. In the long run, a fair price should be even higher.

Model Extensions

We added two additional groups of constraints to the pickup and delivery model from [10], which both led to better performance (i.e. shorter computing times) of the optimization in CPLEX. First we introduced a maximal time duration for a pickup and delivery route of 3 h for each vehicle $k \in M$. Second we add a constraint to break symmetry in the vehicle loadings. A vehicle had to pick up orders in the depot with increasing order number, which excludes a lot of feasible solutions with equal total costs of the pickup and delivery problem.

Computational Procedure

Based on the above mentioned parameters and constraints, we now describe in brief the procedure to calculate the overall costs for the two scenarios:

a. For each edge (i, j) of the road network we determine the costs according to the formulae defined in (2) and (3).
b. For each distributor and for each of its destination postcodes, we sampled randomly the corresponding number of pickup/delivery nodes from all nodes within the postcode area.

c. Calculation of the shortest path with respect to the costs c_{ij}^k for all (i, j) in F (according to b). Only the nodes and edges in (A, F) will be considered later in the pickup and delivery model.

d. Optimization: (i) no collaboration: for each distributor and for each of its orders, the routes are determined by solving the extended pickup and delivery model; (ii) with collaboration (depots and orders stay at the same place but the orders are shared between the distributors): the two results from (i) are used as initial solution for the new optimization run.

4   Results

The results were computed on a laptop with 8 GB RAM and 4 Intel Core 2.4 GHz processors. The model was implemented in GAMS 24.2.1 and solved with CPLEX 12 (up to a relative gap of 10%). With the computational procedure described in Sect. 3 above we produced a typical delivery situation for distributors 1 and 2. Table 2 summarizes the results. For larger instances with more delivery requests (>30) we could not solve the cooperation case within 5 h. In Fig. 1 the computed routes for the

Table 2 Comparison of costs for working individually or in cooperation

Strategy	Distributor 1 (EUR)	Distributor 2 (EUR)	Total (EUR)	Total (%)	CPU (min)
Individual	530.43	446.35	976.81	100.00	6
Cooperation	401.12	504.6	905.72	92.72	120

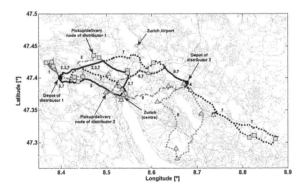

Fig. 1 Cooperation of distributor 1 and 2: *Squares* resp. *Triangles* represent delivery nodes of distributor 1 resp. 2. The numbers 1–7 represent the different tours. Routes 1–3 resp. 4–7 start in the depot of distributor 1 resp. 2. Route 7 starts with distributing requests from depot 2 and then it picks up four requests in depot 1 for which the delivery node is closer to depot 2

case of cooperation are shown. Only route 7 serves nodes of distributor 1 and 2 and is mainly responsible for reduction of 7 % in total costs.

In the large research project following this pre-project, the methodology will be further extended with respect to, amongst others, the general pickup and delivery model and the optimization (speed, solution method, quality of solutions). Moreover, large-scale cases will be investigated considering (i) several distributors collaborating, (ii) a substantially higher number of orders and parameter sensitivity.

Acknowledgments We would like to thank the School of Engineering (ZHAW) for partial support of this research and our colleagues Helene Schmelzer, Merja Hoppe, Andreas Besse, Andreas Christen, Jean-Jacques Keller and Martin Winter (ZHAW).

References

1. Barth, M., Boriboonsomsin, K.: Real-World carbon dioxide impacts of traffic congestion. Transpo. Res. Rec. **2058**, 163–171 (2008)
2. Berbeglia, G., Cordeau, J.F., Gribkovskaia, I., Laporte, G.: Static pickup and delivery problems: a classification scheme and survey. TOP **15**(1), 1–31 (2007)
3. Cai, Y., Judd, K.L., Lontzek, T.S.: The social cost of abrupt climate change. http://www-2.iies.su.se/Nobel2012/Papers/Lontzek_et_al.pdf. Accessed 27 Jun 2014
4. European commission: press release climate action: commission sets out strategy to curb co2 emissions from trucks, buses and coaches, Brussels, 21 May 2014. http://europa.eu/rapid/press-release_IP-14-576_en.pdf. Accessed 21 Jun 2014
5. European environment agency: monitoring of CO2 emissions from passenger cars regulation 443/2009. http://www.eea.europa.eu/data-and-maps/data/co2-cars-emission-6/. Accessed 21 Jun 2014
6. Federal statistical office, Espace de l'Europe 10, 2010 Neuchtel, Switzerland. http://www.bfs.admin.ch/bfs/portal/en/index.html
7. Licenced road data were provided by digital data services GMBH. www.ddsgeo.de
8. Mobitool: environmental data and emission factors (in German and French only). http://www.mobitool.ch/typo/fileadmin/Aktualisierung_v1.1/Umweltdaten_Emissionsfaktoren_mobitool_v1.1.xls. Accessed 28 Jun 2014
9. Parragh, S., Doerner, K., Hartl, R.: A survey on pickup and delivery problems. Journal für Betriebswirtschaft **58**(1), 21–51 (2008) (part I) and **58**(2), 81–117 (2008) (part II)
10. Savelsbergh, M., Sol, M.: The general pickup and delivery problem. Transp. Sci. **29**(1), 17–29 (1995)
11. United States environmental protectio agency: MOVES (Motor Vehicle Emission Simulator). http://www.epa.gov/otaq/models/moves/. Accessed 27 Jun 2014

Optimizing Time Slot Allocation in Single Operator Home Delivery Problems

Marco Casazza, Alberto Ceselli and Lucas Létocart

Abstract Motivated by applications in last-mile home delivery, we tackle the problem of optimizing service time to customers by service providers in an online-realtime scenario. We focus on the particular case where a single operator performs all deliveries. We formalize the problem with combinatorial optimization models. We propose diverse time slot assignment policies and level of service measures. We perform a computational study to evaluate the impact of each policy on the quality of a solution, and to assess the overall effectiveness on the time slot assignment process.

1 Introduction

Home service optimization is becoming a key issue in many industrial sectors. For instance, it is common practice for large technology stores in Europe to offer both the delivery of products at home after purchase, and additional professional services like installation and setup, either for free or at a charge. On one hand, the customer must wait at home for the service to be provided, and is therefore interested in having very well defined, guaranteed, service time slots; he is also interested in choosing as much as possible the placement of the slots that best suits his needs. On the other hand, retailers that must provide such a service are interested in minimizing costs, that mainly consist in limiting the number of operators involved in the home services. Often, the number of operators employed is even fixed in advance by the retailer, who is then interested in offering the best possible level of service to the customer,

M. Casazza · A. Ceselli (✉)
Dipartimento di Informatica, Università Degli Studi di Milano,
Via Bramante 65, 26013 Crema, Italy
e-mail: alberto.ceselli@unimi.it

M. Casazza
e-mail: marco.casazza@unimi.it

L. Létocart
Laboratoire D'Informatique de L'Université Paris Nord, Institut Galilée,
Avenue J.B. Clment, 93430 Villetaneuse, France
e-mail: lucas.letocart@lipn.univ-paris13.fr

© Springer International Publishing Switzerland 2016
M. Lübbecke et al. (eds.), *Operations Research Proceedings 2014*,
Operations Research Proceedings, DOI 10.1007/978-3-319-28697-6_14

taking into account the limited service possibility of her operators. In turn, the skill of matching the agreed service time windows is crucial for a store reputation.

In this context crucial decisions must be taken at different levels, like the tactical definition of time slots and the operational scheduling of the operators; different strategies have been developed, typically trying to trade time slot flexibility with price incentives and discounts [2]. Any approach agrees on a common principle: while a service time slot may be modified with some degree of flexibility, missing a fixed appointment is perceived as a strong disservice by the customer. Later, more focused investigations on the tactical definition of time slots [1], or methods exploiting a-priori knowledge on the set of customers [7] have appeared in the literature. Very recently, customer acceptance policies have been proposed and experimented, allowing the selection among two possible delivery time slots for each customer [5].

None of the works in the literature, however, face at the same time (a) the problem of *designing* service time slots, that is explicitly producing new hard time windows, instead of simply *selecting* a slot in a restricted set of pre-defined ones (b) in an online fashion, that is answering to each customer at his arrival time, without assuming any previous knowledge on future customers, and without the possibility to retract at later time, and (c) with realtime performances, that is with computational methods yielding decision support options in small fractions of seconds.

Hence, in this paper we formalize a time slot allocation problem and we introduce suitable level of service measures (Sect. 2), and we design policies for decision support tools that are able to cope with issues (a), (b) and (c) simultaneously (Sect. 3). We also report on experiments assessing the trade-off that can be achieved between different level of service measures (Sect. 4). We focus on the case where a single operator performs all deliveries, assuming that no a-priori information can be exploited on the distribution of customer's requests.

2   Modeling

Our time slot allocation problem involves three actors: a set of *customers*, that ask for a certain home service at a particular day and time, an *operator* that performs such a service, and a *service provider* that acts as an interface between customers and operators, by assigning service slots and by creating daily schedules. From the point of view of the service provider, we model the service scheduling as the following two-step process.

Time slot assignment. Let I be a sequence of customers. During the day, each customer $i \in I$ appears at the provider's counter in an online fashion, asking for the delivery of a service in a desired time window $[a_i, b_i]$ of a certain day. The provider can either directly accept the customer's request, propose an alternative service time window $[c_i, d_i]$ on the same day, or negate service in the desired day, and defer to alternative service days. Once an agreement is reached, no change in the assigned day and time window is allowed: it is mandatory for the provider to meet the service slot they agreed.

Routing. Then, at the end of the day, a routing for a single operator is computed, in order to service the accepted customers in their assigned time windows. The scarce resource is time: the operator has a limited working shift, that without loss of generality we indicate as $[0, T]$. Moving from the location of a customer i to that of a customer j takes D_{ij} units of time, that is D represents the distance matrix between customers; once at destination, we consider the time needed to perform service at i to be known, and we denote it as w_i. Therefore, computing a feasible routing amounts to find a feasible solution to a Traveling Salesman Problem with Time Windows (TSPTW) [4], considering the operator as a vehicle leaving a depot at time 0, visiting once each of the accepted customers within their assigned time windows, and going back to the depot before time T.

By design, the two service scheduling steps are of radically different nature. The routing problem is an *offline* problem, that can be solved by overnight computations once per day. A provider may assume the traveling cost to be not an issue, as the operator is expected to move in a rather small urban area, or may provide an additional suitable cost function. In any case, once the set of customers and their service time windows is fixed, the problem of finding an optimal schedule turns out to be a traditional TSPTW, for which very efficient exact algorithms are presented in the literature. The time slot assignment, instead, is an *online* problem to be solved with *realtime* efficiency: the sequence of customers is not known in advance, and every time a new customer appears, the provider has to be able to give answers in fractions of seconds. Since at this stage it is crucial not to miss a fixed service, the provider needs a procedure taking in input the desired service day and time window of a new customer, and the set of accepted customers for that day and their assigned time windows, and producing as output either an alternative time window or a 'null' value, indicating that the new customer cannot be serviced in the desired day; in the latter case we say for the sake of simplicity that the customer is *rejected*, although the actual behavior of the provider is to repeat the assignment process on a different day. More formally, let $\bar{I}$ be the set of accepted customers for the desired service day, $[c_i, d_i]$ be the assigned time window for each customer in $\bar{I}$, and $[a, b]$ be the desired time window of the new customer j, a procedure $\sigma(\bar{I}, \bigcup_{i \in \bar{I}}[c_i, d_i], D, j, [a, b]) \rightarrow [c, d]$ is needed in the decision making process, where $[c, d]$ represents an alternative service time window for the new customer, or encodes a 'reject' value $[-\infty, +\infty]$.

Among all possible *feasible* time slot allocations, the service provider may search for ones providing high *level of service* $\mathbb{L}$. There are many ways of defining good plans. In the following we describe three possible measures.

Acceptance rate. First, the provider may be interested in rejecting as few customers as possible, as changing the service delivery day is usually perceived by a customer as the worst level of service. We therefore define the rate of acceptance quality measure as follows: $\mathbb{L}^a = \frac{|\bar{I}|}{|I|}$.

Amount of time shift. If alternative time windows are proposed to a customer, a certain worsening in her perceived level of service is introduced. The most intuitive way of measuring such a worsening is by means of average amount

of shift of the assigned time window with respect to the desired one, that is
$\mathbb{L}^s = \frac{\sum_{i \in \bar{I}} |(a_i + b_i)/2 - (c_i + d_i)/2|}{|\bar{I}|}$.

Amount of window enlargement For services that require the customer to be
at home, a widening of the time window is even more problematic than a shift, as
it forces the customer to take hours off. We therefore define a third level of service
measure as the average amount of time window enlargement $\mathbb{L}^e = \frac{\sum_{i \in \bar{I}} (d_i - c_i) - (b_i - a_i)}{|\bar{I}|}$.

3 Defining and Computing Assignment Policies

We propose online policies to support decisions of the provider during the online
task of assigning time windows to the customers. Formally, to define such an online
decision policy corresponds to provide a definition for the $\sigma()$ procedure introduced
in the previous Section.

We outline four policies. Full details are given in [3]. Each of them is based on
the iterative checking of TSPTW feasibility problems: let $\tau()$ be a procedure taking
in input a set of customers $\bar{I}$, their assigned time windows $[c_i, d_i]$, their pairwise
distances D_{ij} and their service time w_i, for each $i \in \bar{I}$, and giving as output a Boolean
'true' value if the resulting TSPTW problem has a feasible solution, 'false' otherwise.

Fixed. As soon as a new customer arrives, we check if it is possible to service
her and all customers previously accepted in their desired time window. If so, the
customer is accepted without any change in its desired time window. Otherwise, the
customer is rejected.

Shift policy. It aims to accept each new customer, provided that a suitable shift in
his desired time window can be found, allowing the operator to visit him and all the
previously accepted customers. No change in the time window width is performed.

We consider two strategies for performing the time shift: *forward* and *backward*,
in which a certain value s is added (resp. subtracted) to both a_i and b_i, that makes
feasible the TSPTW instance including all accepted customers and the new one;
if no such a value can be found without exceeding the daily deadline T (resp. 0),
the customer is rejected. We also consider the *bidirectional* strategy in which both
forward and backward shifts are computed, and the one requiring minimum shift
is retained. For what concerns the search for a suitable amount of shift s, we take
into account two strategies: in the *coarse* strategy s is chosen as a multiple of a base
constant k, while in the *fine* strategy s can take any value.

Enlarge policy. It aims to accept each new customer by possibly enlarging its
desired time window. Also in this case we consider three strategies (*forward*, *backward*, and *bidirectional*) and two ways of choosing the amount of enlargement e
(*coarse* and *fine*), that are defined as in 'Shift'.

Bucket policy. Finally, we simulated a policy which is often used by industrial
home service providers. We define q time *buckets*, that is a splitting of the daily
working time $[0, T]$ in equal parts $[\ell \cdot T/q, (\ell + 1) \cdot T/q]$ for $\ell = 0 \ldots (q - 1)$.

Then we replace the desired time window of each customer with that of the best
fitting bucket that allows the operator to visit all the accepted customers and the new
one; if no such a bucket can be found, the new customer is rejected. In our case, the

regret of a bucket is computed as the difference between the central instant of the bucket and the central instant of the desired customer time window, being best those buckets having minimum regret.

4 Computational Results

We implemented our algorithms in C, using gcc 4.7 as a compiler, and running a set of experiments on a PC equipped with a 2.7 GHz CPU and 2 GB of RAM, under linux operating system. As a benchmark we considered the set of instances of [6], that were originally drawn from Solomon's dataset. The benchmark consists of 30 feasible TSPTW instances involving up to 44 customers, that include a single depot. Indeed, the size and feature of these instances well represent those of realistic home service delivery problems. In order to check the behavior of our policies as the requests of the customers become more and more tight, we created three scenarii, indicated as datasets A, B, and C in the following, obtained by reducing each original time window $[a'_i, b'_i]$ by 25, 50 and 75 %, that is by setting $\alpha_i = \frac{a'_i + b'_i}{2}$, $\beta_i = b'_i - a'_i$, $a_i = \alpha_i - r \cdot \frac{\beta_i}{2}$, $b_i = \alpha_i + r \cdot \frac{\beta_i}{2}$, for r equal to 0.75, 0.50 and 0.25, respectively. Therefore, our overall testbed consists of 90 instances. Each service time w_i was set to 15; the availability time window of the depot has been left unchanged.

Three main issues arise in the implementation of our policies. First, for efficiently computing the procedure $\tau()$ to solve TSPTW feasibility problems, we embed a dynamic programming algorithm that is adapted from [4]. Second, for computing minima for s and e values to make a new customer reachable, we simply resort to an iterative binary search. Third, for balancing accuracy with speed, for each customer we set the base constant k to half of the width of her desired service time window. In a preliminary round of experiments we found that by fixing a maximum number of labels $\Delta = 2000$ in the dynamic programming procedure gave a good compromise between solution quality and CPU time. As a first observation, we found the average query time to be always less than a tenth of a second, matching our proposal of producing a real-time tool. Our results are given in full details in [3]: in order to highlight the most significant trends, in the following we report only aggregated ones.

In Fig. 1 (top, center-left) a chart for each quality measure is drawn, that reports time windows reduction values on the horizontal axis and average $\mathbb{L}^a$, $\mathbb{L}^s$ and $\mathbb{L}^e$ values on the vertical axes, and includes one line for each policy, indicating the average values over all the instances having a certain time windows reduction, and for all policy variations. In terms of number of accepted customers, the 'bucket' policy performs best, yielding from 6.5 % to almost 20 % improvement with respect to the 'fixed' policy. Policies 'shift' and 'enlarge' perform similarly: they offer a few percentage points improvement with respect to 'fixed' in dataset A, that increases as the desired time windows reduce, reaching more than 10 % on dataset C. The other measures worsen mildly as the time windows reduce. No policy eventually overtakes

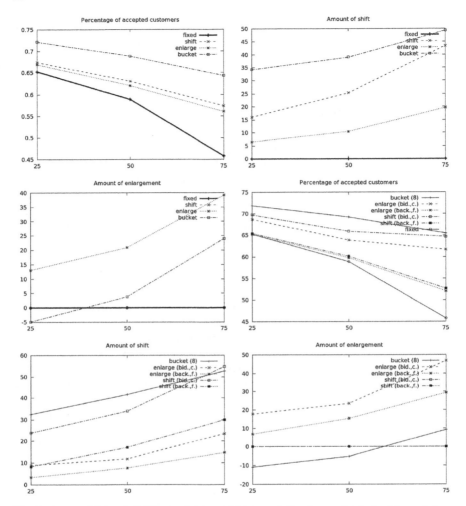

Fig. 1 Average $\mathbb{L}^a$ (*top left*) $\mathbb{L}^s$ (*top right*) and $\mathbb{L}^e$ (*center left*) values as the desired time windows reduce; behavior of the best performing acceptance policies and strategies in terms of average $\mathbb{L}^a$ (*center right*) $\mathbb{L}^s$ (*bottom left*) and $\mathbb{L}^e$ (*bottom right*) values

the others. While 'shift' policy produces by construction solutions having zero $\mathbb{L}^e$ values, the 'enlargement' policy is able to reduce shift amount by enlargement. By using 'buckets' policy, in dataset A and dataset B it is even possible to shrink the desired customer service time windows, at the expense of higher shift values. For what concerns the strategy selection, bidirectional coarse and backward fine showed to be particularly appealing for both 'shift' and 'enlarge'. For the 'bucket' policy, 8 slots seem to provide the best overall behavior, accepting more customers with a modest increase in $\mathbb{L}^s$ and a strong reduction in $\mathbb{L}^e$. We therefore report a synthesis of this comparison in Fig. 1. Bottom and center-right charts have the same structure of the previous ones, and contain one line for each of the following policy-strategy combination: fixed, shift bidirectional coarse, shift backward fine, enlarge bidirectional coarse, enlarge backward fine, bucket 8 slots. No policy dominates the others, but 'bucket' with 8 slots seems to offer the best compromise.

We conclude that our algorithms are suitable for real-time decision support systems, and that the fraction of accepted customers can be substantially increased by allowing for automatic changes in their desired time window.

Acknowledgments The authors wish to thank R. Wolfler Calvo and L. Bussi Roncalini. This work was partially funded by the Italian Ministry of Economic Development under the Industria 2015 contract n.MI01_00173—KITE.IT project.

References

1. Agatz, N., Campbell, A., Fleischmann, M., Savelsbergh, M.: Time slot management in attended home delivery. Transp. Sci. **45**(3), 435–449 (2011)
2. Campbell, A.M., Savelsbergh, M.: Incentive schemes for attended home delivery services. Transpo. Sci. **40**(3), 327–341 (2006)
3. Casazza, M., Ceselli, A., Létocart, L.: Dynamically negotiating time slots in attended home delivery. Univ. degli Studi di Milano—Dipartimento di Informatica, Technical report (2013)
4. Dumas, Y., Desrosiers, J., Gelinas, E., Solomon, M.M.: An optimal algorithm for the traveling salesman problem with time windows. Oper. Res. **43**, 367–371 (1995)
5. Ehmke, J.F., Campbell, A.M.: Customer acceptance mechanisms for home deliveries in metropolitan areas. Eur. J. Oper. Res. **233**(1), 193–207 (2014)
6. Pesant, G., Gendreau, M., Potvin, J.-Y., Rousseau, J.-M.: An exact constraint logic programming algorithm for the traveling salesman problem with time windows. Transp. Sci. **32**(1), 12–29 (1998)
7. Spliet, R., Desaulniers, G.: The discrete time window assignment vehicle routing problem. Technical report, HEC (2012)

Robust Scheduling with Logic-Based Benders Decomposition

Elvin Coban, Aliza Heching, J.N. Hooker and Alan Scheller-Wolf

Abstract We study project scheduling at a large IT services delivery center in which there are unpredictable delays. We apply robust optimization to minimize tardiness while informing the customer of a reasonable worst-case completion time, based on empirically determined uncertainty sets. We introduce a new solution method based on logic-based Benders decomposition. We show that when the uncertainty set is polyhedral, the decomposition simplifies substantially, leading to a model of tractable size. Preliminary computational experience indicates that this approach is superior to a mixed integer programming model solved by state-of-the-art software.

1 Introduction

We analyze a project scheduling problem at a large IT services delivery center in which there are unpredictable delays in start times. This study is motivated by a real problem at a global IT services delivery organization. To design a schedule that is not unduly disrupted by contingencies, we formulate a robust optimization problem. We minimize tardiness cost while informing the customer of a reasonable worst-case completion time.

Due to the impracticality of quantifying joint probability distributions for delays, we apply robust optimization with uncertainty sets rather than probabilistic information [1, 2]. An uncertainty set is an empirically determined space of possible outcomes for which one should realistically plan, without encompassing

E. Coban (✉)
Özyeğin University, Istanbul, Turkey
e-mail: elvin.coban@ozyegin.edu.tr

A. Heching
IBM Thomas J. Watson Research Center, Yorktown Heights, USA
e-mail: ahechi@us.ibm.com

J.N. Hooker · A. Scheller-Wolf
Carnegie Mellon University, Pittsburgh, USA
e-mail: jh38@andrew.cmu.edu

A. Scheller-Wolf
e-mail: awolf@andrew.cmu.edu

© Springer International Publishing Switzerland 2016
M. Lübbecke et al. (eds.), *Operations Research Proceedings 2014*,
Operations Research Proceedings, DOI 10.1007/978-3-319-28697-6_15

99

theoretically worst-case scenarios [5]. To our knowledge, uncertainty sets have not previously been applied to service scheduling. We propose a new solution method based on logic-based Benders decomposition [11, 12]. We show that when the uncertainty set is polyhedral, the problem has convexity properties that result in a simplified decomposition. In addition, the Benders subproblem decouples into many smaller subproblems, each corresponding to an agent or small group of agents.

Robust optimization with uncertainty sets was introduced by [15] and polyhedral uncertainty sets by [3, 4]. Detailed reviews of robust optimization with uncertainty sets can be found in [1, 2, 5].

2 Modeling the Problem

Several hundred agents with different skill sets process thousands of incoming customer projects each year. A project may consist of several ordered tasks, each of which requires certain skills. Late deliveries result primarily from interruptions that require a task to be set aside for an unpredictable period of time. Because processing times are short, we treat an interrupted task as having a delayed start time. We recompute the schedule on a rolling basis as projects arrive. The notation is summarized in Table 1.

Table 1 Notation

Sets	
(J, E)	Precedence graph with task set $J = \{1, 2, ..., n\}$
S_i, S'_j	Skill set of agent i and required for task j
I_α, J_α	α_{th} agent class, jobs assigned to agents in I_α
R	Uncertainty set for release time delays
Parameters	
r_j, d_j	Release time and due date of task j
p_j, c_j	Processing time and unit tardiness cost of task j
$\mathrm{pr}(j, \sigma, y)$	Task performed by agent y_j immediately before task j in sequence σ
$\Delta \bar{r}_j^k$	Task j release time delay in subproblem solution of Benders iteration k
Δr_j^ℓ	Task j release time delay at extreme point ℓ of R
Variables	
$y_j (x_{ij})$	Agent assigned to task j ($x_{ij} = 1$ if $y_j = i$)
s_j	Start time of task j
s_j^k	Task j start time in kth Benders subproblem
s_j^ℓ	Task j start time for extreme point ℓ of R
σ_j	Position of task j in sequence
Δr_j	Task j release time delay in uncertainty subproblem

In the robust model, we require that the tuple of uncertain release time delays $\Delta r = (\Delta r_1, \ldots, \Delta r_n)$ belong to uncertainty set R. We minimize worst-case tardiness cost subject to R as follows, where $\alpha^+ = \max\{0, \alpha\}$:

$$\min_{y, \sigma} \left\{ \max_{\Delta r \in R} \{f(\sigma, y, r + \Delta r, p)\} \,\middle|\, S_j' \subset S_{y_j} \right\} \tag{1}$$

Here $f(\sigma, y, r + \Delta r, p)$ is the cost that results from a greedy schedule in which each agent performs assigned tasks, without overlap, in the order given by σ. Thus we have $f(\sigma, y, r + \Delta r, p) = \sum_j c_j (s_j + p_j - d_j)^+$, where s_j is recursively defined for all $j = 1, \ldots, n$ by

$$s_j = \max \left\{ r_j + \Delta r_j, \, s_{\mathrm{pr}(j, \sigma, y)} + p_{\mathrm{pr}(j, \sigma, y)}, \, \max_{(j', j) \in E} \{s_{j'} + p_{j'}\} \right\} \tag{2}$$

3 Logic-Based Benders Decomposition

Logic-based Benders decomposition (LBBD) is a generalization of Benders decomposition in which the subproblem can in principle be any combinatorial problem, not necessarily a linear or nonlinear programming problem [9, 12, 13]. This approach can reduce solution times by several orders of magnitude relative to conventional methods [6–8, 11, 14].

We apply LBBD as follows. The master problem determines agent assignments y and the task sequence σ:

$$\begin{aligned} &\min \; z \\ &S_j' \subset S_{y_j}, \;\; \text{all } j; \;\; \sigma_j < \sigma_{j'}, \;\; \text{all } (j, j') \in E \\ &\text{Benders cuts} \end{aligned} \tag{3}$$

where each $y_j \in \{1, \ldots, m\}$. The subproblem is

$$\begin{aligned} &\max_{s, \Delta r, \Delta p} \; \sum_j (s_j + p_j - d_j)^+ \\ &s_j = \max \left\{ r_j + \Delta r_j, \, s_{\mathrm{pr}(j, \bar{\sigma}, \bar{y})} + p_{\mathrm{pr}(j, \bar{\sigma}, \bar{y})} \right\}, \;\; \text{all } j; \;\; \Delta r \in R \end{aligned} \tag{4}$$

where $(\bar{\sigma}, \bar{y})$ is the solution of the master problem. It is straightforward to formulate this as an MILP, but the problem becomes very large as Benders cuts accumulate due to the addition of new variables with each cut.

To overcome this, (3) can be solved by a second Benders decomposition scheme in which the subproblem decouples by agent classes, resulting in the three-stage decomposition of Fig. 1a. Agent classes are the smallest sets of agents whose assigned tasks are not coupled by precedence relations with tasks assigned to other sets of agents. When the uncertainty set R is polyhedral, the three-stage decomposition

(a)   **(b)**

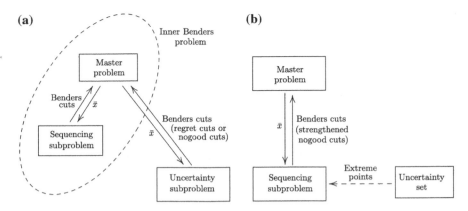

Fig. 1 **a** Three-stage decomposition. **b** Two-stage decomposition

simplifies to the more tractable two-stage decomposition in Fig. 1b. The proof of the following theorem is based on that fact that (4) can be viewed as maximizing a convex function over the polyhedron R.

Theorem 1 *If the uncertainty set R is a bounded polyhedron, then at least one of its extreme points is an optimal solution of the uncertainty subproblem (4).*

We can now minimize worst-case tardiness for a given assignment by finding a sequence that minimizes the maximum tardiness over all extreme points ℓ.

The master problem computes an optimal assignment of tasks to agents:

$$\min z$$
$$S_j' \subset S_i, \quad \text{all } i, j \text{ with } x_{ij} = 1; \quad \sum_i x_{ij} = 1, \quad \text{all } j \qquad (5)$$
$$\text{relaxation and Benders cuts}$$

where $x_{ij} \in \{0, 1\}$ for all i, j. The subproblem minimizes worst-case tardiness T:

$$\min T$$
$$T \geq \sum_j (s_j^\ell + p_j - d_j)^+, \quad \text{all } \ell$$
$$\text{noOverlap}\left(\sigma(\bar{x}, i), s^\ell(\bar{x}, i), p(\bar{x}, i)\right), \quad \text{all } i, \ell \qquad (6)$$
$$s_j^\ell \geq r_j + \Delta r_j^\ell, \quad \text{all } j, \ell; \quad s_j^\ell \geq s_{j'}^\ell + p_{j'} \quad \text{all } (j', j) \in E, \text{ all } \ell$$

The minimum is taken over variables s_j^ℓ and T.

Theorem 2 *If the uncertainty set R is a bounded polyhedron, the three-stage and two-stage decompositions are equivalent optimization problems.*

The number of constraints in the subproblem (6) grows linearly with the number of extreme points, and therefore only linearly with the number of tasks when the polyhedron is a simplex. In addition, (6) decouples by agent class I_α. We can strengthen the master problem (5) with relaxations of the subproblem, which allow the master problem to select more reasonable assignments before many Benders cuts have been accumulated. We use the two relaxations described in [11].

As Benders cuts, we use *strengthened nogood cuts* in the master problem (5) which state that the solution of the subproblem cannot be improved unless certain x's are fixed to different values. If T^* is the optimal value of the subproblem when $x = \bar{x}$, the simplest nogood cut is

$$z \geq \begin{cases} T^* & \text{if } x = \bar{x} \\ -\infty & \text{otherwise} \end{cases} \tag{7}$$

The cut can be strengthened by heuristically removing task assignments and resolving the subproblem until the minimum tardiness is less than T^*.

4   Computational Results

We compared a rudimentary implementation of the two-stage Benders model with an MILP model solved by commercial software (IBM ILOG CPLEX 12.5.0). The MILP model uses a discrete-time formulation, which previous experience shows to be best for MILP [10, 11], and takes advantage of Theorem 1. The time horizon is 5 times the maximum due date in one formulation (MILP1), and 40 plus the maximum due date in a second formulation (MILP2), based on a maximum total tardiness of 40 obtained from the Benders solution. The instances, which have 13 tasks and 2 agents, are based on actual data obtained from an IT services delivery center. The Benders master problem is solved by the CPLEX MILP solver and the subproblems by IBM ILOG CP Optimizer 12.5.0. Table 2 shows that the Benders method is significantly faster than MILP. The advantage is less for MILP2, but MILP2 relies on prior information about the optimal solution.

We increase the number of tasks to 16 from 13 with the same number of agents and generate another 20 instances mimicking the motivating real IT services delivery center. Average computation time of the Benders method increases to 11.2 s (with a maximum of 75.7 s) from 8.4 s (with a maximum of 24.2 s as represented in Table 2) and the Benders method is still significantly faster than MILP.

Table 2 Computational results

Instance	Optimal	Computation times (s)		
	Value	Benders	MILP1	MILP2
1	39	5.4	32.1	7.5
2	20	7.8	122.0	14.5
3	39	5.6	26.4	7.2
4	22	6.1	216.0	29.9
5	33	8.2	68.8	14.7
6	36	8.3	68.8	10.9
7	34	5.6	40.0	10.9
8	25	11.2	61.4	13.7
9	31	5.8	29.5	5.8
10	23	5.8	61.0	14.9
11	30	5.6	54.1	27.5
12	11	8.4	59.9	15.4
13	21	8.1	7.8	6.0
14	33	8.2	79.7	27.9
15	34	6.2	52.1	14.6
16	36	8.9	19.6	17.0
17	38	24.1	78.7	34.1
18	7	5.6	18.5	5.3
19	40	11.0	57.8	13.3
20	40	11.32	27.8	7.2

5 Conclusion

We introduced a novel robust scheduling method for IT services delivery based on logic-based Benders decomposition. We obtained solutions for small but realistic instances in significantly less time that a state-of-the-art mixed integer solver. The advantage of Benders is likely to be much greater as the instances scale up, because the decoupled Benders subproblems remain about the same size as the number of agents increases. In addition, the MILP model grows with the length and granularity of the time horizon, which does not occur in the Benders model. Finally, the Benders model is suitable for distributed computation due to the decoupling of the Benders subproblems.

References

1. Ben-Tal, A., El Ghaoui, L., Nemirovski, A.: Robust Optimization. Princeton University Press (2009)

2. Bertsimas, D., Brown, D.B., Caramanis, C.: Theory and applications of robust optimization. SIAM Rev. **53**(3), 464–501 (2011)
3. Bertsimas, D., Pachamanova, D., Sim, M.: Robust linear optimization under general norms. Oper. Res. Lett. **32**, 510–516 (2004)
4. Bertsimas, D., Sim, M.: The price of robustness. Oper. Res. **52**(1), 35–53 (2004)
5. Bertsimas, D., Thiele, A.: Robust and data-driven optimization: modern decision-making under uncertainty. Tutorials Oper. Res. **4**, 122–195 (2006)
6. Ciré, A.A., Çoban, E., Hooker, J.N.: Mixed integer programming versus logic-based Benders decomposition for planning and scheduling. In: Gomes, C., Sellmann, M., (eds.) CPAIOR 2013, Lecture Notes in Computer Science, vol. 7874, pp. 325–331, Springer (2013)
7. Harjunkoski, I., Grossmann, I.E.: A decomposition approach for the scheduling of a steel plant production. Comput. Chem. Eng. **25**, 1647–1660 (2001)
8. Hooker, J.N.: Integrated Methods for Optimization, 2nd edn. Springer (2012)
9. Hooker, J.N.: Logic-based Benders decomposition. In INFORMS National Meeting (fall), New Orleans (1995)
10. Hooker, J.N.: An integrated method for planning and scheduling to minimize tardiness. Constraints **11**, 139–157 (2006)
11. Hooker, J.N.: Planning and scheduling by logic-based Benders decomposition. Oper. Res. **55**, 588–602 (2007)
12. Hooker, J.N., Ottosson, G.: Logic-based Benders decomposition. Math. Program. **96**, 33–60 (2003)
13. Hooker, J.N., Yan, H.: Logic circuit verification by Benders decomposition. In: Saraswat, V., Van Hentenryck, P. (eds.) Principles and Practice of Constraint Programming: The Newport Papers, pp. 267–288. MIT Press, Cambridge, MA (1995)
14. Jain, V., Grossmann, I.E.: Algorithms for hybrid MILP/CP models for a class of optimization problems. INFORMS J. Comput. **13**, 258–276 (2001)
15. Soyster, A.L.: Convex programming with set-inclusive constraints and applications to inexact linear programming. Oper. Res. **21**(5), 1154–1157 (1973)

Optimal Adaptation Process of Emergency Medical Services Systems in a Changing Environment

Dirk Degel

Abstract Quality of *emergency medical services* (EMS) depends on the EMS infrastructure which is subject to exogenous conditions, like the considered demand area. Demographic changes, increased traffic volume, and structural modifications in the urban infrastructure lead to permanent changes in the demand area. To provide high EMS quality in a strategic perspective, an adaptive planning system to consider these future environmental changes is necessary. An anticipatory adaptation process based on the existing EMS infrastructure and anticipating future developments has to be determined. Assessment criteria of an adaptation process include (1) EMS quality, (2) periodic operating costs, and (3) system adaptation costs. A linear multi-criteria program is developed to support EMS decision makers to dynamically improve an existing EMS infrastructure with respect to multiple requirements during a strategic time horizon.

1 Problem Description and Criteria of an Optimal Adaptation Process

An appropriate EMS infrastructure, including number and positions of EMS facilities, and the configuration of the EMS system (number, positions, and relocations of ambulances) is needed to ensure a high service quality. In particular, the determination of facility locations is a crucial task in the context of planning rescue and emergency medical services. In highly developed areas such as Europe and the USA, environmental changes, especially developments in the urban infrastructure, require anticipatory adaptation of the existing EMS infrastructure to ensure the high service quality. Such developments include modifications of the road network, the development of new residential areas, novel use of former industrial properties, and the incorporation of neighboring cities with a merger of EMS systems. For example, in Munich (Germany) three of the existing ten EMS stations will be relocated and

D. Degel (✉)
Faculty of Management and Economics, Chair of Operations Research
and Accounting, Ruhr University Bochum, Bochum, Germany
e-mail: dirk.degel@rub.de

© Springer International Publishing Switzerland 2016   107
M. Lübbecke et al. (eds.), *Operations Research Proceedings 2014*,
Operations Research Proceedings, DOI 10.1007/978-3-319-28697-6_16

two stations will be newly build until 2030 with costs of about 500 million Euro to counteract the new requirements through expanding cities and an increasing traffic volume [8]. Most existing models, solution approaches, and studies in this research area concentrate on planning without considering the existing EMS infrastructure and dynamic adaptation of the EMS infrastructure. For an overview see [1, 4, 5]. In [2] an approach to reorganize a volunteer fire station network is analyzed but only a dynamic reduction of stations is considered neglecting optional relocations. In contrast, the focus of this paper is a dynamic adaptation process of an existing EMS infrastructure. An anticipatory and future-oriented planning concept that takes merging of cities and new developments into account, is proposed. This allows a smooth EMS infrastructure adaptation and stability. To evaluate an adaptation process, the following assessment criteria are used: (1) A high service quality has to be ensured to guarantee the medical services supply in each time period (2) with respect to adequate periodic operating costs of the EMS system (for the use of existing stations). Furthermore, (3) the costs of the adaptation process (new stations and relocations) have to be as low as possible, corresponding to solution stability, which means robustness and sustainability of the EMS infrastructure with respect to environmental changes. Satisfying the EMS quality criteria in one point in time leads to a solution in this period, which may be completely different from an optimal solution of another period. This instability of solutions results in high adaptation/reconstruction costs of the EMS infrastructure. Due to this trade-off a simultaneous consideration of the three assessment criteria is necessary to reach a stable adaptation process without over-fitting to short-term changes. The remainder of the paper is structured as follows: In Sect. 2 a multi-critera linear programming model is developed and described. The scalarization approach is briefly stated. Results for an illustrative case study are presented in Sect. 3. The paper concludes with a short outlook in Sect. 4.

2 Model and Solution Approach

Due to the trade-off a simultaneous consideration of the three aforementioned assessment criteria is necessary to reach a continuously adjustable but stable adaptation process without over-fitting to short-term environmental changes. For this reason a multi-criteria decision approach is proposed. *Coverage* is one of the most accepted quality criteria in EMS literature [1, 4, 5, 7]. Covering constraints are usually formulated as hard constraints, i. e., (for the notations see next page)

$$\sum_{j \in \mathcal{N}_{it}} y_{jt} \geq 1 \quad \forall i \in \mathcal{I}, \forall t \in \mathcal{T} \tag{1}$$

which means that each demand node i has to be covered by at least one EMS facility within a predefined response time threshold r in period t. In most real world situations using (1) as a hard covering constraint might result in an empty solution space,

i. e., there are uncovered demand nodes at the initial state. Relaxing this constraint and using the constraint in a soft way (see Eq. 5) has the advantage that it allows a much smoother adaptation process, in which small constraint violations are tolerated in sporadic periods. This can be interpreted as a kind of robustness of the adaptation process. To formalize the second assessment criterion, i. e., the periodic operational system costs, the number of stations is minimized. The third assessment criterion, namely solution stability, is incorporated by minimizing the changes in the EMS infrastructure. The three criteria are integrated into a linear multi-criteria programming model, called dynamic set covering problem (dynSCP). The model considers a strategic time horizon of T time periods $t \in \mathscr{T} = \{1, \ldots, T\}$. The subset $\bar{\mathscr{T}} \subset \mathscr{T}$ collects all points in time when EMS infrastructure is expanded. For each time period t, the index set $\mathscr{I}_t$ represents the demand nodes and the sets $\mathscr{J}_t$ indicate potential locations for EMS facilities. The subsets $\mathscr{J}_t^0 \subset \mathscr{J}_t$ $(t \in \bar{\mathscr{T}})$ represent existing facilities of the status quo $(t = 1)$ or at the point in time when cities are merged $(t \in \bar{\mathscr{T}} \setminus \{1\})$. A demand node $i \in \mathscr{I}_t$ is said to be *covered* by an EMS station at node $j \in \mathscr{J}_t$ if and only if the travel time dist_{ijt} is less than or equal to r, where r is a predefined coverage threshold. The sets $\mathscr{N}_{it} := \{j \in \mathscr{J}_t \mid \text{dist}_{ijt} \leq r\}$ characterize the neighborhood sets of a demand node i in period t. $\lambda^-, \lambda^+ \in \mathbb{R}_{\geq 0}$ represent penalty values for one un-/overcovered demand node. $\bar{b}, \underline{b} \in \mathbb{N}_0$ are upper/lower bounds for the amount of newly built/closed facilities in each time period. The decision variable $y_{jt} \in \{0, 1\}$ is equal to 1 if and only if an EMS station is located at node j in period t. The variables $b_t^{\text{close}}, b_t^{\text{open}} \in \mathbb{N}_0$ represent the number of stations built or closed in period t. To capture under-coverage and over-coverage of a demand node i in period t, variables $\Delta_{it}^- \in \{0, 1\}$ respectively $\Delta_{it}^+ \in \mathbb{N}_0$ are used. The dynSCP can be stated as follows:

$$\min \quad \sum_{t \in \mathscr{T}} \sum_{j \in \mathscr{J}_t} y_{jt} \tag{2}$$

$$\min \quad \sum_{t \in \mathscr{T}} \left(b_t^{\text{open}} + b_t^{\text{close}} \right) \tag{3}$$

$$\min \quad \sum_{t \in \mathscr{T}} \sum_{i \in \mathscr{I}_t} \left(\lambda^- \Delta_{it}^- + \lambda^+ \Delta_{it}^+ \right) \tag{4}$$

$$\text{s.t.} \quad \sum_{j \in \mathscr{N}_{it}} y_{jt} + \Delta_{it}^- - \Delta_{it}^+ = 1 \qquad \forall t \in \mathscr{T}, \forall i \in \mathscr{I}_t \tag{5}$$

$$y_{jt} = 1 \qquad \forall t \in \bar{\mathscr{T}}, \forall j \in \mathscr{J}_t^0 \tag{6}$$

$$y_{jt} = 0 \qquad \forall t \in \bar{\mathscr{T}}, \forall j \in \mathscr{J}_t \setminus (\mathscr{J}_{t-1} \cup \mathscr{J}_t^0) \tag{7}$$

$$y_{jt} \geq y_{jt'} \qquad \forall t, t' \in \mathscr{T}, t < t', \forall j \in \bigcup_{\tau=1}^{t} \mathscr{J}_\tau^0 \tag{8}$$

$$y_{jt} \leq y_{jt'} \qquad\qquad\qquad \forall t, t' \in \mathscr{T}, t < t', \forall j \in \mathscr{J}_t \setminus \bigcup_{\tau=1}^{t} \mathscr{J}_\tau^0 \tag{9}$$

$$\sum_{j \in \mathscr{J}_t \setminus \bigcup_{\tau=1}^{t} \mathscr{J}_\tau^0} y_{jt} \leq (t-1) \cdot b_t^{\text{open}} \qquad \forall t \in \mathscr{T} \tag{10}$$

$$\sum_{j \in \bigcup_{\tau=1}^{t} \mathscr{J}_\tau^0} (1 - y_{jt}) \leq (t-1) \cdot b_t^{\text{close}} \qquad \forall t \in \mathscr{T} \tag{11}$$

$$b_t^{\text{open}} \leq \bar{b} \qquad\qquad\qquad \forall t \in \mathscr{T} \tag{12}$$

$$b_t^{\text{close}} \leq \underline{b} \qquad\qquad\qquad \forall t \in \mathscr{T} \tag{13}$$

$$y_{jt} \in \{0, 1\} \qquad\qquad\qquad \forall t \in \mathscr{T}, \forall j \in \mathscr{J}_t \tag{14}$$

$$\Delta_{it}^{+} \in \mathbb{N}_0 \qquad\qquad\qquad \forall t \in \mathscr{T}, \forall i \in \mathscr{I}_t \tag{15}$$

$$\Delta_{it}^{-} \in \{0, 1\} \qquad\qquad\qquad \forall t \in \mathscr{T}, \forall i \in \mathscr{I}_t \tag{16}$$

$$b_t^{\text{close}}, b_t^{\text{open}} \in \mathbb{N}_0 \qquad\qquad\qquad \forall t \in \mathscr{T} \tag{17}$$

The first objective function (2) minimizes the number of EMS stations over the entire time horizon, representing corresponding operating costs of the EMS infrastructure. In contrast, objective (3) minimizes changes in the EMS infrastructure, in order to ensure stability of the EMS system. Finally, objective (4) considers the EMS quality by minimizing uncovered and overcovered demand nodes. Together with constraints (5), the aim is to minimize the number of stations needed to cover all demand points in each period. Constraints (6) and (7) ensure a variable fixing at points $t \in \bar{\mathscr{T}}$ when a given EMS infrastructure has to be incorporated into the model. Equations (8) and (9) stabilize the infrastructure. An existing station that has been closed once cannot be reopened and newly built stations have to be used until the end of the planning horizon. Constraints (10)–(13) state that at least $\bar{b}$ resp. $\underline{b}$ stations can be opened resp. closed in average per period t. Declarations (14)–(17) describe the domain of the decision variables. With definitions $x := (y, \Delta^+, \Delta^-, b^{\text{open}}, b^{\text{close}})$ and $X := \{x \mid x \text{ satisfies } (5) - (13)\}$ the multiobjective program (2)–(17) can be stated in the form

$$\min_{x \in X} \left(f_1(x), \ldots, f_L(x) \right)^T. \tag{18}$$

The scalarization of the weighted sum method is a convex combination of the L objectives ($\mathscr{L} = \{1, \ldots, L\}$) of a multiobjective problem, where the feasible set X is unchanged [9]. The basic formulation of a weighted sum approach is $\min_{x \in X} \sum_{\ell \in \mathscr{L}} u_\ell f_\ell(x)$, with $u = (u_1, \ldots, u_L)$, $\sum_{\ell \in \mathscr{L}} u_\ell = 1$ and $u_\ell > 0$ for all $\ell \in \mathscr{L}$. The magnitudes of the objective functions $f_\ell(x)$ affect how the weights are to be set. Therefore, the following formulation is used [6]:

$$\min_{x \in X} \sum_{\ell \in \mathscr{L}} u_\ell \cdot \frac{f_\ell(x) - f_\ell^\star}{\bar{f}_\ell - f_\ell^\star} \qquad (19)$$

with $x_\ell^\star := \arg\min_{x \in X} f_\ell(x)$, $f_\ell^\star := \min_{x \in X}\{f_\ell(x) + \sum_{\substack{k \in \mathscr{L} \\ k \neq \ell}} \varepsilon f_k(x)\}$ where $\varepsilon > 0$ is a small value to avoid dominated solutions and $\bar{f}_\ell := \max_{k \in \mathscr{L}}\{f_\ell(x_k^\star)\}$.

3　Case Study and Results

In the case study a planning horizon of $T = 20$ periods and a planning area with three adjacent demand areas is considered as shown in Fig. 1: The entire planning area consists of a core city, a development area without an existing infrastructure, and a neighboring city. The EMS infrastructure of the entire area will be planned by anticipating added demand areas and their particular expansion date (development area in $t = 5$, neighboring city in $t = 15$). In the initial state, the existing EMS

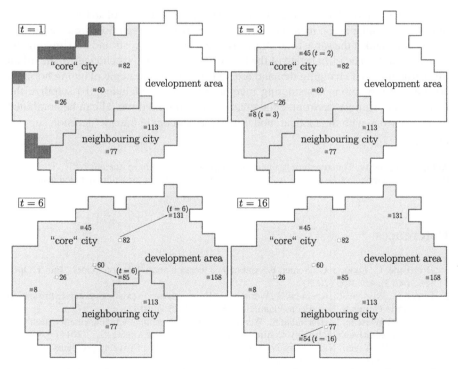

Fig. 1 Three adjacent areas and the existing EMS facilities $\mathscr{I}_1^0 = \{26, 60, 82\}$, $\mathscr{I}_5^0 = \emptyset$, $\mathscr{I}_{15}^0 = \{77, 113\}$ in the beginning. Relocations and new stations at $t = 6$, $t = 16$, and the resulting EMS infrastructure

infrastructure in the core city, described by the set $\mathscr{I}_1^0 = \{26, 60, 82\}$, as well as $|\mathscr{I}_1| = 71$ demand nodes, is infeasible with respect to a hard covering constraint. Nine demand regions cannot be reached within the pre-defined response time (dark gray squares). Thus, in period $t = 3$ one station will be relocated to ensure a full demand area coverage. In period $t = 5$ the development area will be added without an infrastructure present. To ensure high EMS quality and minimal operating costs one existing station will be relocated (from 82 to 131). Additionally, station 60 will be relocated to 85 to cover the development area while anticipating the incorporation of the neighboring city. In total, there are 4 stations less in comparison to the individual solutions for each planning area with hard covering constraints. The solution of the optimization model was determined optimally using Fico Xpress-Optimizer 7.3 on a desktop PC with Intel Core i7 CPU (3.4 GHz) and 8 GB RAM running Windows 7 (64 Bit Version).

4 Conclusion and Outlook

The presented approach identifies an optimal adaptation process of an EMS infrastructure. Optimal adaptation means minimizing the costs of adaptation, i. e., stability of the solution and at the same time minimizing the operating costs and maximizing the service quality. The initial state of the EMS system, exogenous changes in the urban infrastructure, and changing demand are considered in a strategic planning horizon. An inter-city planning considering mergers of cities is included. To stabilize the EMS system against environmental changes the presented model can be combined in further work with the tactical model in [3] in a tactical-strategic decision support process.

Acknowledgments This research is financially supported by *Stiftung Zukunft NRW*. The author thanks staff members of *Feuerwehr und Rettungsdienst Bochum* for providing detailed insights.

References

1. Brotcorne, L., Laporte, G., Semet, F.: Ambulance location and relocation models. Eur. J. Oper. Res. **147**(3), 451–463 (2003)
2. Degel, D., Wiesche, L., Rachuba, S., Werners, B.: Reorganizing an existing volunteer fire station network in Germany. Soc.-Econ. Plann. Sci. **48**(2), 149–157 (2014)
3. Degel, D., Wiesche, L., Rachuba, S., Werners, B.: Time-dependent ambulance allocation considering data-driven empirically required coverage. Health Care Manag. Sci. (2014)
4. Farahani, R., Asgari, N., Heidari, N., Hosseininia, M., Goh, M.: Covering problems in facility location: a review. Comput. Ind. Eng. **62**(1), 368–407 (2012)
5. Li, X., Zhao, Z., Zhu, X., Wyatt, T.: Covering models and optimization techniques for emergency response facility location and planning: A review. Math. Methods Oper. Res. **74**(3), 281–310 (2011)

6. Marler, T.R., Arora, J.S.: The weighted sum method for multi-objective optimization: new insights. Struct. Multi. Optim. **41**(6), 853–862 (2010)
7. Toregas, C., Swain, R., ReVelle, C., Bergman, L.: The location of emergency service facilities. Oper. Res. **19**(6), 1363–1373 (1971)
8. Wimmer, S.: Feuerwehr braucht neue Wachen. http://www.sueddeutsche.de/muenchen/muenchner-stadtrand-feuerwehr-braucht-neue-wachen-1.1796542, 17. Oktober 2013
9. Zadeh, L.: Optimality and non-scalar-valued performance criteria. IEEE Trans. Autom. Control **8**(1), 59–60 (1963)

Strategic Network Planning of Recycling of Photovoltaic Modules

Eva Johanna Degel and Grit Walther

Abstract Due to high subsidies, the installed capacity of photovoltaic (PV) in Germany steeply increased in the last years. Considering an expected lifetime of 20–25 years the related amount of PV waste will increase during the next decades. Thus, PV modules have been integrated into the WEEE directive in 2012. In order to fulfil the WEEE recycling and recovery quotas, it will be sufficient to recover the raw materials with the highest mass, i.e. glass and aluminium. However, with new technologies under development, it will be possible to recover also other rare strategic materials with limited availability in Germany, like silver, copper or indium. Against this background, the aim is to develop a strategic planning approach to analyse the early installation of appropriate collection and recycling infrastructures with the focus on resource criticalities. In order to do so, a multi-period MILP is developed regarding capacity, technology and location decisions for collection and recycling facilities of PV modules. Decisions are evaluated with regard to economic aspects. Additionally, information on resource criticalities derived from criticality indicators are integrated. A case study illustrates the approach and its results.

1 Decision Situation and Model Requirements

The energy production by renewable energy sources is an essential mean to prevent the climate change and its consequences. For this reason the German government has strongly supported the installation of PV modules via different subsidies (e.g. 1,000- and 100,000-roof program) since 1990 [13]. The installed PV capacity reached a power of 35,7 GWp at the end of 2013 [3]. This trend also stimulated the development of PV module technologies, which can be divided into two main groups—crystalline and thin-film modules. According to their type the modules have different resource

E.J. Degel (✉) · G. Walther
School of Business and Economics, Chair of Operations Management,
RWTH Aachen University, Aachen, Germany
e-mail: eva.degel@om.rwth-aachen.de

G. Walther
e-mail: walther@om.rwth-aachen.de

© Springer International Publishing Switzerland 2016
M. Lübbecke et al. (eds.), *Operations Research Proceedings 2014*,
Operations Research Proceedings, DOI 10.1007/978-3-319-28697-6_17

Table 1 Criticality assessment of PV resources [7]

Resource	Aluminium	Silicon	Copper	Tellurium	Silver	Indium
Criticality	II	II	IV	IV	V	V

compositions. Due to the expected life-time of 20–25 years the amount of end-of-life modules will steeply increase in the next decades following the trend of the installed capacity. The focus of this paper is the recycling of PV modules with special consideration of the small portions of rare materials, like silver, copper, tellurium or indium, which are especially used in thin-film modules.

Besides the different module types, there are also different recycling technologies. They can be differentiated by their development stage in technologies, which are in industrial use (e.g. shredders [9]) or in pilot or development stage (e.g. [8, 12]). While the simple technologies focus mostly on the mass parts, glass and aluminium frames, the research activities for the advanced technologies aim to recover the small amounts of scarce materials.

In 2012, the European Union integrated PV modules in the Directive 2012/19/EU on waste electrical and electronic equipment (WEEE), due to the expected amount of waste [6]. To fulfil the specified mass-based recycling and recovery quotas of the WEEE, it is sufficient to recover the low value fractions with the highest mass. But due to limited availability of rare materials, it could be reasonable to focus also on the small portions of scarce resources.

To evaluate the necessary recycling technologies and especially the future strategic potential of different raw materials, the so-called resource criticality indicators are under development. An overview is given in [1, 11]. The indicators and sub-indicators are still under discussion and topic of further research. Table 1 shows one possible criticality assessment of PV resources for Germany in a scale from I (not critical) to VI (very critical) given by [7].

There already exists a lot of planning approaches for strategic recycling networks in the field of Reverse Logistics and Closed-Loop Supply Chain Management (cf. [2, 4, 5, 10]), which could be used for the planning of a strategic PV recycling network. Nevertheless, there is a lack of a systematic integration of resource criticality risks. Hence, this research gap will be filled with the presented approach.

Accordingly, the aim of this paper is to develop a strategic planning approach in order to analyse the early installation of appropriate collection and recycling infrastructures with special focus on resource criticalities. In Sect. 2, a multi-period MILP is developed regarding capacity, technology and location decisions for collection and recycling of PV modules. Decisions are evaluated with regard to economic aspects. Additionally, in Sect. 3, resource price scenarios, derived from criticality indicators, are integrated. By using this information, the model evaluates the effects of resource criticality on economic decisions in recycling network planning. The computational results of a small case study will be shown in Sect. 4. The paper ends with a conclusion and an outlook.

2 Strategic Network Planning Approach for PV Recycling

For PV recycling network planning a linear multi-period mixed-integer problem is developed. The considered recycling network contains existing collection points $m \in M$ and the potential recycling facilities $n \in N$. In each period $t \in T$ and at each side n a recycling facility can be installed. There are different recycling technologies $r \in R$, which can treat different types of PV modules $p \in P$, and recovering different output fractions $h \in H$. As mentioned before, currently recycling technologies focus on the mass parts, glass and aluminium,while advanced technologies, aiming at a recovery of the small amounts of scarce materials, are mostly still in development. For each technology the point in time of availability the market t_r^* is known. The different collection qualities $q \in Q$ influence the applicable recycling technology r. A high collection quality means that the PV modules are separated according to their type p, whereas in a low quality all module types are mixed.

The decision variables of the model can be divided in design, transportation and storing variables. At the collection points the gathered waste quantities in the chosen collection quality can be directly sent to a recycling facility or can be stored at the collection points. The design variable indicates, whether a recycling facility at site n with technology r is build in period t. Here, the point in time of availability at the market of each recycling technology t_r^* has to be respected. Depending on the applied recycling technology r different amounts of the considered output fractions will be recovered. The model structure is shown in Fig. 1.

In order to evaluate the decisions with regard to economic aspects, the net present value (NPV) is used as the objective. It consists of the sum of the discounted payments of all periods $t \in T$. These payments result from investments I_t, transports C_t^T, the operation at the different locations C_t^P and the sales or disposal of the different output fractions E_t^{Out}:

$$\max \; NPV = \sum_{t \in T} \left(-I_t - C_t^T - C_t^P + E_t^{Out} \right) \cdot (1+i)^{-t}. \qquad (1)$$

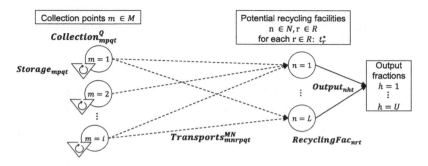

Fig. 1 Model structure

The starting point for the integration of resource criticalities are the payments for the output fractions. In the deterministic model, these are calculated as the price respectively the fees for the disposal multiplied with the mass:

$$E_t^{Out} = \sum_{n \in N} \sum_{h \in H} E_{ht} \cdot x_{nht}^{Out} \qquad \forall s \in S, \forall t \in T. \qquad (2)$$

3 Strategic Network Model Enhanced with Resource Criticalities

GLEICH ET AL. observed a significant correlation between the criticality assessment and the price development of different resources [11]. As a first step, information derived from resource criticality indicators given in [7] are used to deviate resource price scenarios $s \in S$ for each resource $h \in H$. To do so, the payments for the recovered output fractions E_t^{Out} are modelled as being dependent on scenario s:

$$E_{ts}^{Out} = \sum_{n \in N} \sum_{h \in H} E_{hts} \cdot x_{nht}^{Out} \qquad \forall t \in T. \qquad (3)$$

To find an appropriate solution without complete information the approaches of optimisation under uncertainty have to be considered. A robust approach is used as the developed resource criticality indicators include no direct probability information.

In the present situation, the relevant uncertainties appear in the objective function. Hence, the problem is not to find a feasible solution, but to find one with an adequate objective value in all possible scenarios. In such a situation different decision criteria can be applied. In our approach the Hurwicz-Criterion is used, due to the fact that it calculates a linear combination of the best and the worst objective value for each solution (4). This combination reflects the expected opportunity through uncertain price development of the strategic raw materials. The risk attitude of the decision maker is set by a weighting parameter λ. A linear formulation is given in (4)–(8), where z_{max} and z_{min} are auxiliary variables, which represent the best and the worst objective value for one solution considering each scenario s. The variable z_s indicates the scenario s achieving z_{max}.

$$\max \quad \lambda \cdot z_{max} + (1 - \lambda) \cdot z_{min} \qquad (4)$$

$$\text{s. t.} \quad z_{min} \leq NPV_s \qquad \forall s \in S, \qquad (5)$$

$$z_{max} \leq NPV_s + M \cdot (1 - z_s) \qquad \forall s \in S, \qquad (6)$$

$$\sum_{s \in S} z_s = 1, \qquad (7)$$

$$z_s \in \{0, 1\} \qquad \forall s \in S, \qquad (8)$$

+ basic problem constraints.

4 Case Study and Computational Results

The developed model is applied to a case study of North Rhine-Westphalia (NRW). In Germany, data about the installed PV modules, including postal code, location and power, have been published by the transmission network operator since 1990. Thus, the considered data cover a planning horizon of 24 years. With an expected life-time of 25 years the first modules will reach their end of life in 2015. The expected waste quantities are allocated to existing collection points.

For a first integration of the resource criticality, four price scenarios are considered corresponding to the criticality assessment given in Table 1: In each period in (s_1) prices remain stable, in (s_2) prices of criticality class V rise by 5%, in (s_3) prices of criticality class V rise by 10%, in (s_4) prices of criticality class IV rise by 5% and class V by 10%. Three recycling technologies are considered for installation. Two are already available or will be in a short time (shredder, Sunicon). One (Lobbe) is still under development.

The case study is solved with FICO Xpress. For the Hurwicz decision approach the weighting factor is set to $\lambda = 0.8$, which represents a risk-loving decision maker. The chosen technologies in the Hurwicz solution are compared to the optimal solutions in each scenario s in consideration of the installed recycling technologies r (Fig. 2 left) and the reached objective value (Fig. 2 right).

The results show that in the scenarios, where the prices for the critical resources rise (especially s_3, s_4), it is better to wait for the advanced technologies like Lobbe, which can treat every PV module type and recovers also the strategic materials. In the Hurwicz solution, this technology is also installed, but in fewer number. The comparison of the objective values displays that the Hurwicz approach is in an acceptable distance to the scenario optimal solution in every scenario but scenario s_4.

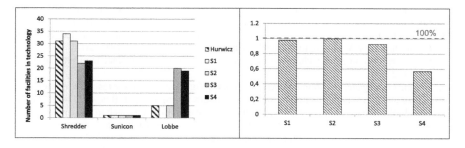

Fig. 2 *Left* Number of installed recycling technologies r; *Right* Hurwicz solution in comparison to the scenario optimal solutions

5 Conclusion and Outlook

In this paper a resource criticality extension of an strategic recycling network planning approach is presented. The approach offers the possibility to include resource price risks within the planning of a recycling network. Thereby it is possible to hedge an economy against scarcity of strategic resources by an early reaction and installation of an appropriate infrastructure. In further research, the methodological derivation of further information from the resource criticality indicators and from experts will be considered to refine the derived and implemented data. Furthermore, other approaches for optimisation under uncertainty will be tested and compared with the currently used approach.

References

1. Achzet, B., Helbig, C.: How to evaluate raw material supply risks–an overview. Resour. Policy **38**(4), 435–447 (2013)
2. Alumur, S., Nickel, S., Saldanha-da-Gama, F., Verter, V.: Multi-period reverse logistics network design. Eur. J. Oper. Res. **220**(1), 67–78 (2012)
3. Bundesverband Solarwirtschaft e. V.: Statistische Zahlen der deutschen Solarbranche (Photovoltaik) (2014)
4. Choi, J., Fthenakis, V.: Economic feasibility of recycling photovoltaic modules. J. Ind. Ecol. **14**(6), 947–964 (2010)
5. Choi, J., Fthenakis, V.: Crystalline silicon photovoltaic recycling planning: macro and micro perspectives. J. Clean. Prod. **66**, 443–449 (2014)
6. Das Europäische Parlamanemt und der Rat der Europäischen Union. Richtlinie 2012/19/EU über Elektro- und Elektronik-Altgeräte: WEEE
7. Erdmann, L., Behrendt, S., Feil, M.: Kritische Rohstoffe für Deutschland: Identifikation aus Sicht deutscher Unternehmen wirtschaftlich bedeutsamer mineralischer Rohstoffe, deren Versorgungslage sich mittel- bis langfristig als kritisch erweisen könnte: Abschlussbericht. Institut für Zukunftsstudien und Technologiebewertung (ITZ) and adelphi, Berlin (2011)
8. Friedrich, B.: Photorec: Rückgewinnung von seltenen strategischen Metallen aus EOL Dünnschicht-PV-Modulen. r3-Kickoff, Posterpräsentation, Freiberg, 18.04.2013
9. Friege, H., Kummer, B.: Photovoltaik-Module: Bedarf und Recycling. UmweltMagazin **3**, 46–49 (2013)
10. Ghiani, G., Laganà, D., Manni, E., Musmanno, R., Vigo, D.: Operations research in solid waste management: a survey of strategic and tactical issues. Comput. Oper. Res. **44**, 22–32 (2014)
11. Gleich, B., Achzet, B., Mayer, H., Rathgeber, A.: An empirical approach to determine specific weights of driving factors for the price of commodities—a contribution to the measurement of the economic scarcity of minerals and metals. Resour. Policy **38**(3), 350–362 (2013)
12. GmbH, Lobbe Industrieservice, Co, K.G.: Entsorgungsunternehmen stellt Weichen für Recycling von Photovoltaik-Modulen. Müll und Abfall **44**(2), 109 (2012)
13. Sander, K., Zangl, S., Reichmuth, M., Schröder, G.: Stoffbezogene Anforderungen an Photovoltaik-Produkte und deren Entsorgung: Umwelt-Forschungs-Plan, FKZ 202 33 304, Endbericht, ökopol, Institut für Energetik und Umwelt, Hamburg/Berlin, 15.01.2004

Designing a Feedback Control System via Mixed-Integer Programming

Lena C. Altherr, Thorsten Ederer, Ulf Lorenz, Peter F. Pelz
and Philipp Pöttgen

Abstract Pure analytical or experimental methods can only find a control strategy for technical systems with a fixed setup. In former contributions we presented an approach that simultaneously finds the optimal topology and the optimal open-loop control of a system via Mixed Integer Linear Programming (MILP). In order to extend this approach by a closed-loop control we present a Mixed Integer Program for a time discretized tank level control. This model is the basis for an extension by combinatorial decisions and thus for the variation of the network topology. Furthermore, one is able to appraise feasible solutions using the global optimality gap.

1 Introduction

Conventional methods can only find an optimal control strategy for technical systems with a fixed setup. Another system topology may enable an even better control strategy. To find the global optimum, it is necessary to find the optimal control strategy for a wide variety of setups, not all of which might be obvious even to an experienced designer. We introduce a Mixed Integer Linear Program (MILP) for a feedback control that can be extended by combinatorial decisions. This approach is the basis for finding the optimal topology of a system and its optimal closed loop control simultaneously.

L.C. Altherr (✉) · U. Lorenz (✉) · P.F. Pelz (✉) · P. Pöttgen (✉)
Chair of Fluid Systems, TU Darmstadt, Darmstadt, Germany
e-mail: lena.altherr@fst.tu-darmstadt.de

U. Lorenz
e-mail: ulf.lorenz@fst.tu-darmstadt.de

P.F. Pelz
e-mail: peter.pelz@fst.tu-darmstadt.de

P. Pöttgen
e-mail: philipp.poettgen@fst.tu-darmstadt.de

T. Ederer (✉)
Discrete Optimization, TU Darmstadt, Darmstadt, Germany
e-mail: ederer@mathematik.tu-darmstadt.de

© Springer International Publishing Switzerland 2016

M. Lübbecke et al. (eds.), *Operations Research Proceedings 2014*,
Operations Research Proceedings, DOI 10.1007/978-3-319-28697-6_18

121

One example for a fluid system with different topological and control strategy options is a tank level control system. Pumps can either be used or not and they can be connected in series or in parallel. To control the water level, the rotational speed of the used pumps has to be adjusted. The goal may either be energy efficiency or a controller property like settling time or stability.

The tank level control system can be represented by a control circuit. To accurately optimize a feedback control system one needs to account for the time-dependent behavior of its components: P (proportional), I (integration) and D (derivation), PT1, PT2 (delay of first or second order) or PTt (dead-time). We discretized the time dependence and obtained a mixed-integer formulation based on a time-expanded flow network. This formulation allows us to appraise feasible solutions using the global optimality gap. Furthermore, we can combine this approach with a variation of the network topology [4].

2 State of the Art

A proportional-integral-derivative (PID) controller is one of the most commonly used closed-loop feedback methods. The control output $u(t)$ is computed from the present error P, the accumulated past errors I, and a prediction of future errors based on the current error derivative D [2]. The corresponding controller parameters k_p, k_i and k_d affect the variation of the error $e(t)$ in time. Adjusting these parameters is called tuning. For instance, with a bad parameter assignment the process variable may oscillate around a given set point, but given a good assignment the controller is fairly robust against small disturbances. However, tuning a controller for specific process requirements is highly nontrivial.

One approach to finding controller parameters is manual tuning. In this case, the proportional parameter k_p is raised until the control variable oscillates. This current value of k_p is then halved to obtain a good assignment. From this starting point, the other controller parameters can be derived [7]. Of course, this approach is rather error-prone and can be improved by computer-aided methods, e.g. by parameter optimization or evolutionary algorithms. However, each of these methods has at least one significant shortcoming: Discrete decisions like topology variations are not possible, or local optima cannot be distinguished from global optima.

MILP [5] is a modeling and solving technique for computer-aided optimization of real-world problems. Its main advantage compared to other optimization methods is that global optimality can be proven. During the last two decades, its application has become increasingly popular in the Operations Research community, e.g. to model problems in logistics, transportation and production planning. Commercial solvers are nowadays able to solve plenty of real-world problems with up to millions of variables and constraints. The field of applying mathematical optimization to physical-technical systems is called Technical Operations Research (TOR) [3].

In previous work we have shown how to optimize a hydrostatic power transmission system via mixed-integer programming [1]. The solution is a system topology with a corresponding open-loop control strategy. A weakness of this approach is that the actuating variable has to be known in advance for each time step. It is not possible to react to the current state of the fluid system. To overcome this, we include a mathematical description of closed-loop controllers into a mixed integer program. The aim of this paper is to illustrate the first step of our current work by presenting a small example.

3 Tank Level Control

We look at the example of a water level control. A tank is filled with water and has an outlet at the bottom out of which a certain amount of water drains, depending on the height of the stored water column. To fill the tank, a pump has to be activated. The filling speed depends on the pressure built up and the volume flow provided by the pump. Figure 1 shows an exemplary setup.

We model this setup by a mathematical graph $G = (V, E)$. The reservoir and the tank are represented by vertices V, pumps and pipes by edges E. The following physical equations and models describe the filling level system and are included into the MILP formulation. First, the continuity equation for incompressible fluids results in the conservation of volume flows Q.

$$\forall t \in T \; \forall v \in V : \quad \sum_{(i,v) \in E} Q_t^{i,v} = \sum_{(v,j) \in E} Q_t^{v,j} \quad (1)$$

The increase or decrease of the tank's filling volume V is approximated in our model by a sequence of static flows over short time intervals Δt.

$$\forall t \in T \setminus \{0\} : \quad V_t = V_{t-1} + \left(\sum_{(i,\text{tank}) \in E} Q_t^{i,\text{tank}} - \sum_{(\text{tank},j) \in E} Q_t^{\text{tank},j} \right) \cdot \Delta t \quad (2)$$

If a pump is activated (indicated by the binary variable $a_t^{i,j}$), the pressure difference between both connectors i and j is fixed to $\Delta p_t^{i,j}$, which is given by the pump

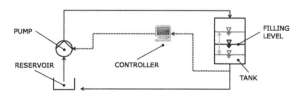

Fig. 1 A tank level control system. A pump conveys water from a reservoir into the tank while water is draining out of it. The rotary speed of the pump is set via controller. The task is to reach and maintain a given filling level in minimum time

characteristic, cf. Eq. (10). Otherwise, the pressure levels of the two connectors are decoupled by means of a big-M formulation.

$$\forall t \in T \,\, \forall (i, j) \in \text{Pumps}: \quad |p_t^j - p_t^i| \le \Delta p_t^{i,j} + M \cdot (1 - a_t^{i,j}) \tag{3}$$

In an ideal pipe and in case of turbulent flow, the pressure loss is described by an origin-rooted parabola. In real systems, an interference due to dynamic effects can be observed. Still, the pressure loss can be well fitted by a general quadratic form.

$$\forall t \in T \,\, \forall (i, j) \in \text{Pipes}: \quad p_t^j - p_t^i = c_2 \cdot Q_t^{i,j,\text{sqr}} + c_1 \cdot Q_t^{i,j} + c_0 \tag{4}$$

We have to introduce an auxiliary variable for the squared flowrate and a piecewise linearization of the square function. For univariate functions, the incremental method was shown to be very efficient [6]. With progress variables $\delta \in [0, 1]$ on linearization intervals D and passing indicators $z \in \{0, 1\}$, this results in:

$$\forall t \in T \,\, \forall (i, j) \in E \,\, \forall (g, h) \in D: \quad z_{t,g}^{i,j} \ge \delta_{t,g,h}^{i,j} \ge z_{t,h}^{i,j} \tag{5}$$

$$\forall t \in T \,\, \forall (i, j) \in E: \quad Q_t^{i,j} = \sum_{(g,h) \in D} (\tilde{Q}_{t,h}^{i,j} - \tilde{Q}_{t,g}^{i,j}) \cdot \delta_{t,g,h}^{i,j} \tag{6}$$

$$\forall t \in T \,\, \forall (i, j) \in E: \quad Q_t^{i,j,\text{sqr}} = \sum_{(g,h) \in D} (\tilde{Q}_{t,h}^{i,j,\text{sqr}} - \tilde{Q}_{t,g}^{i,j,\text{sqr}}) \cdot \delta_{t,g,h}^{i,j} \tag{7}$$

The pump characteristic is a nonlinear relation—caused by the pump geometry—between its rotational speed n, flowrate Q, pressure boost Δp and power input P. This dependence is MILP-representable by a convex combination formulation [6] with weights $\lambda \in [0, 1]$ on nodes K and a selection $\sigma \in \{0, 1\}$ of simplices X.

$$\forall t \in T: \quad \sum_{x \in X} \sigma_{t,x}^{\text{pump}} = \sum_{k \in K} \lambda_{t,k}^{\text{pump}} = a_t^{\text{pump}} \tag{8}$$

$$\forall t \in T \,\, \forall k \in K: \quad \lambda_{t,k}^{\text{pump}} \le \sum_{x \in X(k)} \sigma_{t,x}^{\text{pump}} \tag{9}$$

$$\forall t \in T: \quad n_t^{\text{pump}} = \sum_{k \in K} \lambda_{t,k}^{\text{pump}} \cdot \tilde{n}_k^{\text{pump}}, \quad \Delta p_t^{\text{pump}} = \sum_{k \in K} \lambda_{t,k}^{\text{pump}} \cdot \Delta \tilde{p}_k^{\text{pump}} \tag{10}$$

$$\forall t \in T: \quad Q_t^{\text{pump}} = \sum_{k \in K} \lambda_{t,k}^{\text{pump}} \cdot \tilde{Q}_k^{\text{pump}}, \quad P_t^{\text{pump}} = \sum_{k \in K} \lambda_{t,k}^{\text{pump}} \cdot \tilde{P}_k^{\text{pump}} \tag{11}$$

The pressure at the bottom of the tank is proportional to the water column height, with the tank's area A, the density of water ρ and the gravitational acceleration g.

$$\forall t \in T: \quad p_t^{\text{tank}} = \rho \cdot g \cdot \frac{V_{t-1}^{\text{tank}}}{A} \tag{12}$$

Additional Variables and Constraints for the Control Circuit

We model a PI controller with the equation

$$n^{\text{ref}}(t) = k_p \cdot \Delta V(t) + k_i \cdot \int_0^t \Delta V(\tau)\, d\tau \tag{13}$$

where k_p and k_i are control parameters that determine if and how fast the control deviation converges to zero. Discretizing this relation yields input variables ΔV_t for the proportional and $S_t^V = \sum_0^t V_\tau$ for the integral part.

$$\forall t \in T : \quad n_t^{\text{ref}} = n_t^p + n_t^i, \quad n_t^p = k_p \cdot \Delta V_t, \quad n_t^i = k_i \cdot S_t^V \tag{14}$$

$$\forall t \in T : \quad \Delta V_t = V^{\text{ref}} - V_t^{\text{tank}}, \quad S_t^V = S_{t-1}^V + \Delta V_t \tag{15}$$

The products of continuous variables $k_p \cdot \Delta V_t$ and $k_i \cdot S_t^V$ are linearized by the convex combination method, cf. Eqs. (8) and (10).

If the reference value is larger than the pump's maximum rotational speed or smaller than the pump's minimum rotational speed, it cannot be reached by the actual value. We introduce two auxiliary variables Ω_t and ω_t for each time interval that peak if and only if the reference value is out of bounds.

$$\Omega_t \cdot (n_{\max}^p + n_{\max}^i - n_{\max}) \geq n_t^{\text{ref}} - n_{\max} \geq (1 - \Omega_t) \cdot (n_{\min}^p + n_{\min}^i - n_{\max}) \tag{16}$$

$$\omega_t \cdot (n_{\min}^p + n_{\min}^i - n_{\min}) \leq n_t^{\text{ref}} - n_{\min} \leq (1 - \omega_t) \cdot (n_{\max}^p + n_{\max}^i - n_{\min}) \tag{17}$$

If the out-of-bounds indicators are deactivated, the rotational speed of the pump has to be equal to the output $n_p + n_i$ of the PI controller. If an out-of-bound indicator peaks, the rotational speed has to reach its corresponding bound.

$$\omega_t \cdot (n_{\min}^p + n_{\min}^i - n_{\max}) \leq n_t^{\text{ref}} - n_t \leq \Omega_t \cdot (n_{\max}^p + n_{\max}^i - n_{\min}) \tag{18}$$

$$n_{\max} - n_t \leq (n_{\max} - n_{\min}) \cdot (1 - \Omega_t) \tag{19}$$

$$n_{\min} - n_t \geq (n_{\min} - n_{\max}) \cdot (1 - \omega_t) \tag{20}$$

We want to minimize the time m until the tank's volume stays within a given error bound ε_V. We model this using a big-M formulation and a monotonically increasing binary function $s_t \in \{0, 1\}$ with peak indicator $j_t^s \in \{0, 1\}$.

$$\forall t \in T : \quad |\Delta V_t| \leq \varepsilon_V + M \cdot (1 - s_t), \quad s_t - s_{t-1} = j_t^s \tag{21}$$

$$\sum_{t \in T} j_t^s = 1, \quad \text{minimize} \sum_{t \in T} t \cdot j_t^s \tag{22}$$

Fig. 2 Comparison of optimization result and simulation. The step size is set to 1 s in our model. The simulation result is computed with the DASSL solver with variable order and a time step of 0.02 s

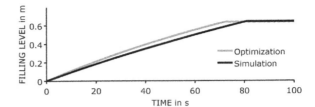

Figure 2 shows the result for a test instance. The MILP model needs 71 s to reach the target volume. In a detailed simulation with the controller parameters obtained from the optimization, the desired set point is reached within approximately 75 s. In contrast to the optimization model, no oscillation of the control variable is observed.

The approximation used in the MILP is comparable to the Euler method, a first-order method with an error proportional to the step size. The computation times are still rather long, ranging from minutes with MILP start to days without. For ongoing research it might be interesting to investigate problem-specific primal heuristics.

4 Conclusion

We motivated a MILP formulation for the design process of a closed-loop control. We started from a model for finding an optimal open-loop control for a filling level control system and extended it to find an optimal closed-loop control. The advantage of this formulation is that the load does not need to be known in advance to adjust the rotational speed of the pump. Instead, the implemented controller is able to compute the actuating variable in realtime based on the current pressure head of the tank. Owing to the mixed-integer formulation one can easily extend our model by integer or binary variables. We aim to include discrete purchase decisions or network topology variations in future projects.

Acknowledgments This work is partially supported by the German Federal Ministry for Economic Affairs and Energy funded project Entwicklung hocheffizienter Pumpensysteme and by the German Research Foundation (DFG) funded SFB 805.

References

1. Dörig, B., Ederer, T., Hedrich, P., Lorenz, U., Pelz, P. F., Pöttgen, P.: Technical operations research (TOR) exemplified by a hydrostatic power transmission system (2014)
2. Kilian, C.: Modern control technology. Thompson delmar learning. Electron. Eng. (2005)
3. Pelz, P., Lorenz, U.: Besser gehts nicht! TOR plant das energetisch optimale Fluidsystem. delta p, Ausgabe 3, ErP Spezial (2013)
4. Pelz, P. F., Lorenz, U., Ederer, T., Lang, S., Ludwig, G.: Designing pump systems by discrete mathematical topology optimization: The artificial fluid systems designer (AFSD) (2012)

5. Schrijver, A.: Theory of Linear and Integer Programming. Wiley (1998)
6. Vielma, J. P., Ahmed, S., Nemhauser, G.: Mixed-integer models for nonseparable piecewise-linear optimization: unifying framework and extensions. Oper. Res. **58**(2), 303–315 (2010)
7. Ziegler, J., Nichols, N.: Optimum settings for automatic controllers. Trans. ASME, **64**(11) (1942)

Fair Cyclic Roster Planning—A Case Study for a Large European Airport

Torsten Fahle and Wolfgang Vermöhlen

Abstract Airport ground staff scheduling has been long known as one of the most challenging and successful application of operations research. In this presentation, we will concentrate on one type of rostering known as cyclic roster. Numerous aspects required in practice have to be taken into account, amongst others crew qualification, work locations and the travel time between each location, government regulations and labor agreements, etc. INFORM's branch-and-price solution approach covers all of these aspects and is in use on many airports world-wide. Cyclic Rosters cover several fairness aspects by construction. In this case study we will discuss why one of our customers wanted to add additional fairness criteria to the roster. We show which new fairness requirements are needed. We present a fast local search post-processing step that transforms a cost optimal shift plan into a fair cyclic shift plan with the same costs. The transformed plans are highly accepted and are in operational use.

1 Introduction

Airport ground staff scheduling has been long known as one of the most challenging and successful application of operations research, in particular column generation method.

Cyclic rosters (equivalently, shift patterns or rotating schedules) represent sequences of shifts designed for a group of employees. One worker starts on each week on the roster, switching cyclically from one week to the next. After finishing one week, each worker switches to the subsequent week. The worker assigned to the last week switches to the shifts planned in week one. All workers therefore rotate over the pattern for a given period of time, e.g. for 4 weeks or for a flight season, see [5].

T. Fahle (✉) · W. Vermöhlen
Aviation Division, INFORM GmbH, Aachen, Germany
e-mail: Torsten.Fahle@inform-software.com

W. Vermöhlen
e-mail: Wolfgang.Vermoehlen@inform-software.com

© Springer International Publishing Switzerland 2016

129

M. Lübbecke et al. (eds.), *Operations Research Proceedings 2014*,
Operations Research Proceedings, DOI 10.1007/978-3-319-28697-6_19

The main goal is to cover all demand of a planning week at low costs. Numerous aspects required in practice have to be taken into account, amongst others crew qualification, work locations and the travel time between each location, government regulations and labor agreements, etc. Our branch-and- price solution approach covers all of these aspects and is in use on many airports world-wide.

The number of weeks in a typical cyclic roster is usually small (4–12) and rarely exceeds the 26 weeks of a season. If the number of worker n assigned to the cyclic roster is larger than the number of weeks w, workers are grouped in teams of size $b \approx n/w$.

Cyclic rosters occur at airlines and airports, but also in call centers, hospitals, emergency services and in public transport, see [1–3, 6].

Cyclic Rosters cover several fairness aspects by construction. E.g. each group of workers has to perform the entire schedule. Thus, all workers perform the same number of day and night shifts in the same order. This is one reason why unions often prefer cyclic rosters over individual shifts.

A drawback of cyclic rosters (especially with large teams) is that the covering quality decreases. E.g. if a department runs a 13-week cyclic roster this means that each week has to be covered by 13 different work lines. If there are 130 workers in the department, each team of 10 workers perform the same line of work. This is efficient only if there is always work for 10 persons. We may avoid over covering by using more individual assignments.

These considerations lead to the idea that rather than using a fully cyclic roster of 13 weeks for teams of 10 persons, we generate a cyclic roster of 130 lines of work for individuals. A worker entering the roster in line i leaves the roster after 13 weeks in line $(i + 13)$ mod 130. Now, the covering of work is good, but the subsequences of weeks per worker may contain different shifts/functions and work is no longer distributed fairly. Thus, external fairness criteria need to be taken into account.

A *shift* defines start and end of work on a day. A *function* here defines what work has to be done. Certain (night) shifts and (leader) functions qualify for premiums, thus shift and functions basically determine the roster costs. A *work stretch (WS)* is a sequence of some days on and some days off. A typical patterns is "6-3"—working 6 days, having 3 days off. Another one covering 3 weeks is "7-3/7-4": working 7 days, 3 days off, again 7 days of work, 4 days off. For an overview on staff rosterring see [4].

2   Customers Fairness Rules

According to our customer, different departments have different definitions of a "fair roster". These definitions often stem from negotiations with unions and reflect ergonomic aspects, monetary fairness or variety of work. There are several detailed rules and exceptions, but the main principles are:

1. Shifts and functions are evenly distributed over the roster (e.g. night shifts or functions that qualify for premiums). If shifts or functions occur in blocks (see below), distribute the blocks equally over the roster.
2. Within a work stretch (WS), different departments follow different rules:

   a. (Dep. A, 6-3 rhythm) a WS should have as much variety of shifts and functions as possible. If a functions occurs more than once, it should not be assigned to consecutive shifts.
   b. (Dep. B, 7-3/7-4 rhythm) The 7 days on are subdivided into blocks of 3 early, 4 late and then 4 early, 3 late shifts. All days in a block get the same function, consecutive blocks must have different functions.

3 Transforming a Roster into a Fair Roster

Unfortunately, these criteria are hard to integrate into the existing branch-and-price approach. In the following we define several measures that calculate the fairness of a given roster R and we define swaps that exchanges parts of a roster such that fairness can be improved.[1]

Let $prio(ST)$ $[prio(F)]$ denote the priority for a certain shift type [function]. Let $pos(i, ST)$ $[pos(i, F)]$ denote the ith occurrence of a shift of shift type ST [of function F] within the entire plan, $pos(1, X) < pos(2, X) < \cdots < pos(n, X)$ for $X \in \{ST, F\}$. We calculate the distance between positions i and its two neighbors $i \pm 1$ and store the larger of these to two:

$$\alpha^{ST}(i) := \max\{pos(i+1, ST) - pos(i, ST), pos(i, ST) - pos(i-1, ST)\} \tag{1}$$

$$\alpha^{F}(i) := \max\{pos(i+1, F) - pos(i, F), pos(i, F) - pos(i-1, F)\} \tag{2}$$

$$obj_1 := \sum_{ST} prio(ST) \cdot \sum_{i} \alpha^{ST}(i), \qquad obj_2 := \sum_{F} prio(F) \cdot \sum_{i} \alpha^{F}(i) \tag{3}$$

Function $count(X, WS)$ counts the appearance of X in work stretch WS. X might be a function or a shift. We penalize counters larger than one.

$$\vartheta_{WS}^{X} := \max\{0, count(X, WS) - 1\}, \quad X \in \{ST, F\} \tag{4}$$

$$obj_3 := \sum_{WS} \left(\sum_{F} \vartheta_{WS}^{F} \right), \qquad obj_4 := \sum_{WS} \left(\sum_{ST} \vartheta_{WS}^{ST} \right) \tag{5}$$

---

[1] We only denote the normal cases here and omit the technical cases where a formula must take into account the wrap-around from the last week in the plan to the first one.

For all days on d in the work plan we use a penalty of one if day $d + 1$ is also a day on and has the same function as day d. Otherwise the penalty is zero. Let $function(d)$ denote the function assigned to the shift on day d:

$$\delta_d^f := \begin{cases} 1, & \text{if } f = function(d) \text{ and } f = function(d + 1) \\ 0, & \text{otherwise} \end{cases} \tag{6}$$

$$obj_5 := \sum_{d,f} \delta_d^f \tag{7}$$

We evaluate the overall fairness of a plan by summing up all partial objective terms. For each term, we use some parameter λ for calibration. In so doing we can put emphasis on different aspects of fairness and adapt the magnitude of the different terms.

$$obj := \sum_{i=1}^{5} \lambda_i \cdot obj_i \tag{8}$$

Next, we need to define some swaps that are capable of improving the fairness of a roster while preserving the desired work stretch structure.

The result of the branch-and-price approach is a valid and cost-optimal rotating roster $R = ((s_{id}, f_{id}), i = 1 \ldots n, d = 1 \ldots 7)$ containing shifts s_{id} and functions f_{id} for each day $d = 1 \ldots 7$ in each week i.

- A *2-swap* swaps shift+functions on day d in week i with those in week j.
- A *3-swap* exchanges shift+functions on day d in weeks i, j, k
- A *work stretch swap* swaps two different work stretches starting on the same day of the week.
- A *crossover swap* exchanges first k days of two different work stretches starting on the same day of the week.

Figure 1 shows these swaps on a roster. Notice, that according to the customers fairness rules (Sect. 2) we have a certain shift substructure in the 7-3/7-4 work stretches. Thus, we can only apply work stretch swaps and crossover swaps with $k = 3, 4$ to these rosters. 2-swap, 3-swap, and cross over swaps with arbitrary k are only valid for department A.

The swaps defined above are embedded in a simple hill climbing algorithm. Each admissible swap for the scenario at hand is called for all possible combinations of days, weeks, work stretches, or parameters k. Each combination is further checked for validity with global rostering rules, e.g. shift start time offset, min time before day off. The parameters that provide the best improvement with respect to objective (8) is applied to the roster at hand. If no parameter set for a swap provides an improvement, the swap is skipped. This procedure is repeated until we cannot improve the current roster anymore.

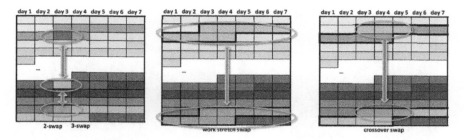

Fig. 1 2-/3-swap, work stretch swap and crossover swap exchange parts of the roster in order to obtain a better fairness value

Notice, that roster costs depend on the selected shifts and functions which are not altered by any swap. Consequently, the roster's cost stays optimal while the roster's fairness improves.

4 Numerical Results

We ran the described hill climbing algorithm on 28 scenarios from different departments of a large European airport. These scenarios contain 1–20 functions and 6–15 shift types. A scenario contains between 9 and 789 persons.[2] Depending on the department's fairness definition, the algorithms applied different swaps and used adapted weights λ in the objective (8). On an up-to-date laptop, 15 scenario took less than 5 s for fairness optimization, 26 scenarios took less than 2 min. In two cases, we applied 3-swaps to increase fairness quality and run time went up to 320 and 600 s (time limit), resp. In all cases, planners at the airport were highly satisfied with the results.

Due to space limitations we present a deeper look into only one example from department A for 189 workers, 12 shift types and 17 functions.[3] Each worker rotates over an 18 weeks subset of this roster. Figure 2 shows the distribution of functions for the scenario before and after fairness optimization. In the initial roster there were 327 days having the same function as the day before (only 7 after fairness). Shifts were not equally distributed, one shift type was assigned 18 times to some worker, other workers had no shifts of this types. After fairness, each worker got 5–9 shifts of this type. Concerning the shift and functions variety, the initial roster only contained 3 work stretch with 5 or 6 different shifts and functions, but 56 work stretches contained

---

[2] One scenario had only 1 function, all other had more. 19 scenarios had between 54 and 252 persons, only 1 had 9 persons, the largest scenarios contained 684, 789 persons, resp.

[3] The roster is a 6-3 roster. It contains $189 \cdot 7 = 1323$ days and $189 \cdot 7/(6+3) = 147$ work stretches. In a 6-3 pattern, $147 \cdot 6 = 882$ shift/functions combinations have to be assigned.

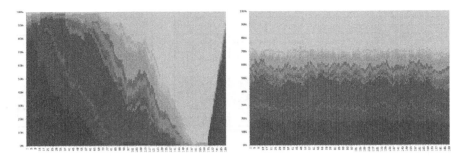

Fig. 2 Function distribution before (l) and after (r) fairness. x-axis is the starting week from 1–189. The column for week w represents the portion of shifts having a certain functions in roster weeks $w \ldots w + 18$. Different *colors* denote the 17 different functions

3 or less different shifts and functions. After fairness, all work stretches contained at least 4 different shifts and 3 different functions. 95 work stretches ($\approx$65 %) were rather diverse, they contained 5 or 6 shifts and 5 or 6 functions.

5   Conclusions

Since 2013, the approach is in operational use and plans several departments with up to 700 persons per cyclic roster. The main gain of the fairness post optimization is that it allows planners to reduce their manual work dramatically when management is in negotiations with unions. Unions typically request several alternative rosters per department for the next season and select one for operations. Planners thus have to prepare several cost optimal and fair rosters per department per season. This manual process easily took 1–2 weeks per roster. As the customer states it:

> A 495 week roster meant assigning 3465 shifts one by one which meant a lot of mouse clicking

Using the fairness post-optimization reduces this process to some minutes run time and some reporting preparation, allowing to react faster and with reproducible quality.

References

1. Bennett, B.T., Potts, R.B.: Rotating roster for a transit system. Trans. Sci. **2**, 14–34 (1968)
2. Çezik, T., Günük, O., Luss, H.: An integer programming model for the weekly tour scheduling problem. Nav. Res. Logist. **48**, 607–624 (2001)
3. Chew, K.L.: Cyclic schedule for apron services. J. Oper. Res. Soc. **42**, 1061–1069 (1991)
4. Ernst, A.T., Jiang, H., Krishnamoorthy, M., Sier, D.: Staff scheduling and rostering: a review of applications, methods and models. EJOR **153**, 3–27 (2004)

5. Herbers, J.: Models and algorithms for ground staff scheduling on airports. Dissertation, RWTH Aachen (2005)
6. Khoong, C.M., Lau, H.C., Chew, L.W.: Automated manpower rostering: techniques and experience. Int. Trans. Oper. Res. **1**, 353–361 (1994)

Why Does a Railway Infrastructure Company Need an Optimized Train Path Assignment for Industrialized Timetabling?

Matthias Feil and Daniel Pöhle

1 Introduction

Today's timetabling process of German rail freight transport is a handicraft make-to-order process. Train paths are only planned when operators apply for specific train services including specific train characteristics such as train weight, train length, accelerating and braking power, etc. What seems customer-friendly, has indeed many disadvantages: frequent mismatch of demand and supply, general lack of service level differentiation and long customer response times for the train operators; inefficient use of network capacity and timetabling resources for the rail infrastructure manager.

Modern and industrialized supply chain concepts do not feature pure make-to-order processes but often prefer a mix of make-to-stock/make-to-order processes, namely assemble-to-order processes ([7, 8]). Products consist of forecast-based made-to-stock modules that are assembled to finished goods only when customers have placed their specific orders. The core idea of assemble-to-order processes is postponement. "Postponement [...] is based on the principle of seeking to design products using common platforms, components or modules but where the final assembly or customization does not take place until the final [...] customer requirement is known" (see [2]). A first advantage of an assemble-to-order supply chain is that the common modules strategy leads to a low total number of modules helping reduce the forecast risk and, hence, reduce inventory stocks. A second advantage is that a company can achieve very high flexibility. This is because it can feature both a high product differentiation by cross-assembling the modules as well as short order lead times since the supply chains order penetration point, or "de-coupling point" respectively, has

M. Feil (✉) · D. Pöhle (✉)
DB Netz AG, Langfristfahrplan Und Fahrwegkapazität,
Theodor-Heuss-Allee 7, D-60486 Frankfurt, Germany
e-mail: matthias.feil@deutschebahn.com

D. Pöhle
e-mail: daniel.poehle@deutschebahn.com; d.poehle@gmx.de

© Springer International Publishing Switzerland 2016   137
M. Lübbecke et al. (eds.), *Operations Research Proceedings 2014*,
Operations Research Proceedings, DOI 10.1007/978-3-319-28697-6_20

Fig. 1 Comparison of both
processes

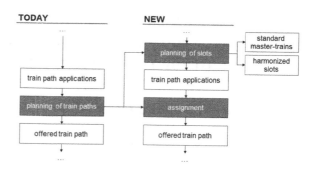

moved backwards (see [8]). "The marketing advantages that such flexibility brings are considerable. It means that in effect the company can cater for the precise needs of multiple customers, and they can offer even higher levels of customization. In todays marketplace where customers seek individuality and where segments or 'niches' are getting ever smaller, a major source of competitive advantage can be gained by linking production flexibility to customers needs for variety." (see [3]). Prominent examples of assemble-to-order products are e.g. electronic devices such as printers,[1] cars[2] or fashion.[3]

German Railways Infrastructure division DB Netz has started to gradually introduce an assemble-to-order process for its rail freight timetabling.[4] Two innovations have been necessary (see Fig. 1): the pre-production, or pre-planning respectively, of standardized train paths ("slots") able to match future demand, and the assignment of operators' train service applications to the pre-planned slots. First analyses of the new timetabling process indicate that the aforementioned advantages from non-transport industries can be carried over to the railway industry: Rail freight operators ("multiple customers") will mainly benefit from a better match of demand and supply ("linking production flexibility to customers' needs for variety"). They will also benefit from much faster response times when applying for train paths ("shorter order lead times"). Finally, they will benefit, as a side-kick, from shorter average O-D-travel times, hence better rolling stock and crew utilization. This is because the pre-planned slots feature better average travel time characteristics than the pure made-to-order train paths. The rail infrastructure manager DB Netz is expected to benefit from increased demand, from a higher network transparency leading to higher network

---

[1]Case study Hewlett Packard: Country-generic printers are combined with country-specific power modules and power cord terminators, see [7], pp. 331ff.

[2]All major OEM: Modules are e.g. body type, color, engine, décor elements, etc.; see e.g. [9].

[3]Case study Benetton: "By developing an innovative process whereby entire knitted garments can be dyed in small batches, they reduced the need to carry inventory of multiple colors, and because of the small batch sizes for dying they greatly enhanced their flexibility", see [3].

[4]Timetabling of passenger traffic will continue to be a make-to-order process. Core reason is the strong customization requirement in passenger traffic leading to timetabling to the minute with lots of definite changing connections at the stations. Both factors eliminate the likelihood that pre-produced and standardized slots would meet future customer demand.

utilization as well as from economies of scale in the pre-planning of slots. Altogether, the rail freight sector is expected to strengthen its competitive position against road and shipping transport. Going forward, this paper will give a basic introduction to the slot assignment innovation. Detailed explanations of the slot assignment's results can be found in [5]. The pre-planning of slots for a long-term timetable is extensively described in [6, 10].

2 Requirements and Constraints of the Slot Assignment Model

The aim of the assignment phase is to find optimal O-D-slot itineraries for all operators' train service applications. A slot is a train path from a node A to a node B at a certain time t. An O-D-slot itinerary p is a string of slots leading from the origin to the destination of a train service r matching the operator's route and departure/arrival time requirements. A pre-processing module verifies which type of pre-planned slots[5] is adequate for the specific train service application. Three different objective functions can be used in the slot assignment model: maximize total revenues from slot fees, maximize fulfillment of demand (equal to minimal rejection of train service applications), or maximize quality of the assigned slot itineraries:

$$\sum_{p,r} \omega_{p,r} \cdot x_{p,r} \rightarrow max$$

$$\forall C_s \in \mathscr{C} : \sum_{p \in C_s} x_{p,r} \leq 1$$

$$\forall H \in \mathscr{H} : \sum_{p \in P(H)} x_{p,r} \leq c_H$$

$$x_{p,r} \in \{0, 1\}$$

$$\omega_{p,r} = \begin{cases} 1 \ , \ max. \ fulfillment \\ u_r \ , \ max. \ revenue \\ (\varrho \cdot \tau(p_r^*) - \tau(p)) \ , \ max. \ quality \end{cases}$$

where $x_{p,r}$ is a binary decision variable indicating whether application r uses itinerary p, c_H is the node's capacity, ϱ is a detour factor and $\tau(p)$ is the travel time for itinerary p. The fulfillment criterion represents slot and node capacities. If the slot or node capacity is insufficient, affected train service applications cannot be fulfilled and must be rejected. Slot capacity means that every slot may only be assigned to at the most one train service. Node capacity means that only a certain number of trains can simultaneously dwell in a node while every node has also a minimum dwelling or

---

[5]The slot types are mainly characterized by maximum train speed. On most lines, a fast (100 km/h) and a slow (80 km/h) slot type is sufficient to cover more than 90% of nowadays train service applications. The balance must continue to be planned in a handicraft make-to-order process.

Fig. 2 Three different levels
of modelling nodes

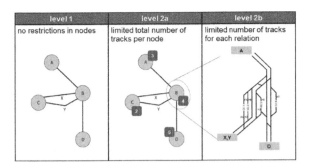

transfer time from an incoming to an outgoing slot. We differentiate three levels of node capacity (see Fig. 2). At level 1, there are no capacity restrictions in the network nodes. All incoming and outgoing slots of a node can be arbitrarily connected, only subject to the node's minimum transfer time. In level 2a, each node is assigned with a global capacity of dwelling positions. This number indicates how many trains can simultaneously dwell in a node waiting for or transferring to the next suitable exit slot. In contrast, at level 2b, the model differentiates all dwelling positions by whether or not they give access to the slots of the desired exit direction and by their length (can the dwelling position cope with the specific train's length?). The revenue criterion represents the infrastructure managers income from slot fees. Every application may have a different fee. The quality criterion quantifies the ratio of assigned travel time and minimum technical travel time.[6] The optimization model is solved by a two-step approach. First, a valid initial solution is generated by a simple heuristic algorithm. The algorithm sorts all train service applications by ascending departure times and assigns them one by one to the slots. Once a slot has been assigned to a train service application, it is no longer available for subsequent applications. Similarly, node dwelling positions are gradually blocked during a train's node dwelling time. This simple heuristic basically represents the strategy a human timetabling engineer would pursue. Second, the initial solution is optimized by the column generation method (see e.g. [1]). A comprehensive explanation of the optimization model is given in [4].

3   Practical Use of the Optimization Model

To be useful in daily business, the optimization model must yield better results in affordable run times than the infrastructure manager's conventional means. Therefore, we tested the optimization model's performance by running the column generation method against the simple "human" heuristic algorithm introduced in Sect. 2. Two different disguised scenarios of the German railway network have been served

---

[6]This number is always greater or equal to 1. A ratio of 1 means assigned travel time and minimum technical travel time are equal.

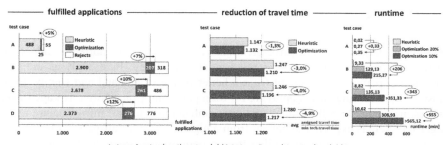

Fig. 3 Performances of the algorithms

as a test environment. In both scenarios, the number of train path applications has been deliberately increased compared to today's numbers. This leads to a certain level of network overload and guarantees some optimization potential for the algorithms. Therefore, no conclusions about real network capacities or bottlenecks may be drawn from the presented results. In the first attempt, we let a group of timetable engineers equipped by today's standard timetabling software compete against the column generation method executed by a standard PC.[7] The attempt was in vain because the engineers quit after a couple of days without a valid solution. They did not fail the individual slot assignment task but couldn't manage to establish an effective administration system of assigned and free slots. So they got lost after some while. As a consequence, in the second attempt, we let the simple "human" heuristic algorithm compete against the column generation method, both executed by a standard PC. Figure 3 display the performances of the two algorithms. For the optimization model, only the objective function is shown that maximizes the assigned slots' quality as it performed best among the three objective functions. One can observe in Fig. 3 that, in all test cases, the optimization model rejects less train service applications than the heuristic algorithm. The number of applications with a valid assignment is raised by 5–12 %. Moreover, the average quality improves, too. The more node constraints are considered, the higher the optimization potential of the optimization model is. In test case D, the optimization models travel time quality is 4.9 % higher than the heuristics one. Run times can be drawn from the right graph in Fig. 3. For the optimization model, run times are those for solutions with a dual gap of less or equal 20 or 10 %. The first solution with a duality gap less than 20 % for test cases B, C and D can be found within two to five hours of optimization. The more node constraints are included (level 2a and 2b) the longer it takes to optimize the assignments. It becomes clear that the optimization model yields much better results than the heuristic at the cost of higher run times. Yet, over-night runs are feasible so that the optimization model will be usable in daily business. Moreover, thanks to the use of column generation, there is always a feasible solution for assignment available independently from how long the optimization takes.

---

[7]CPU 2.5 GHz, RAM 16 GB, OS Debian 64 bit.

4 Conclusion and Outlook

Germanys largest rail infrastructure manager DB Netz has started to adopt an assemble-to-order process for its rail freight timetabling. First analyses indicate that train operators will benefit from a better match of demand and supply, from faster response times when applying for train paths and from shorter average travel times, hence better rolling stock and crew utilization. The rail infrastructure manager DB Netz is expected to benefit from a higher network transparency leading to higher network utilization as well as from economies of scale in the pre-production of slots. Altogether, the rail freight sector is expected to strengthen its competitive position against road transport. For the assembly phase, we have developed a slot assignment optimization model. Different test scenarios have shown that the model is able to generate valid initial solutions and to refine them by a column generation method. The latter one excels the simple initial heuristic algorithm representing today's timetabling logic up to 12 % (fulfillment and average travel time). The overall model's run times allow for over-night runs so that the model is appropriate for daily business. Heading towards day-to-day implementation, we will extend the model's functionality. Development efforts will focus on the extension of the current 1-day model scope to a 365-days scope. They will also focus on improving running times and a more granular representation of the rail network (inclusion of all relevant network nodes instead of only major and mid-sized network nodes). The implementation is accompanied by a major transformation program to enable staff to deal effectively with the new assemble-to-order process.

References

1. Barnhart, C., et al.: Branch-and-price: column generation for solving huge integer programs. Oper. Res. **46**(3), 316–329 (1998)
2. Christopher, M.: The agile supply chain: competing in volatile markets. Ind. Mark. Manag. **29**(1), 37–44 (2000)
3. Martin, C.: Logistics and Supply Chain Management, 4th edn. Pearson Education Limited, Harlow (2011)
4. Nachtigall, K.: Modelling and solving a train path assignment model. In: International Conference on Operations Research, Aachen (2014)
5. Pöhle, D., Feil, M.: What are new insights using optimized train path assignment for the development of railway infrastructure?. In: International Conference on Operations Research, Aachen (2014)
6. Pöhle, D., Weigand, W.: Neue Strategien für die Konstruktion von Systemtrassen in einem industrialisierten Fahrplan. 24. Verkehrswissenschaftliche Tage, Dresden (2014)
7. Simchi-Levi, D., Kaminsky, P., Simchi-Levi, E.: Designing and Managing the Supply Chain: Concepts, Strategies and Case Studies, 3rd edn. McGraw-Hill, New York (2008)
8. Stadtler, H., Kilger, C.: Supply Chain Management and Advanced Planning, 4th edn. Springer, Berlin (2008)
9. Tianjun Feng, T., Zhang, F.: The Impact of Modular Assembly on Supply Chain Efficiency. In: Production and Operations Management (2013)
10. Weigand, W.: Von der Angebotsplanung über den Langfristfahrplan zur Weiterentwicklung der Infrastruktur. In: ETR, No. 7/8, pp. 18–25 (2012)

A Polyhedral Study of the Quadratic Traveling Salesman Problem

Anja Fischer

Abstract In this paper we summarize some of the results of the author's Ph.D.-thesis. We consider an extension of the traveling salesman problem (TSP). Instead of each path of two nodes, an arc, the costs depend on each three nodes that are traversed in succession. As such a path of three nodes, a 2-arc, is present in a tour if the two corresponding arcs are contained in that tour, we speak of a quadratic traveling salesman problem (QTSP). This problem is motivated by an application in biology, special cases are the TSP with reload costs as well as the angular-metric TSP. Linearizing the quadratic objective function, we derive a linear integer programming formulation and present a polyhedral study of the associated polytope. This includes the dimension as well as three groups of facet-defining inequalities. Some are related to the Boolean quadric polytope and some forbid conflicting configurations. Furthermore, we describe approaches to strengthen valid inequalities of TSP in order to get stronger inequalities for QTSP.

1 Introduction

The traveling salesman problem (TSP) is one of the most studied combinatorial optimization problems, see, e.g. [3, 13]. In this paper we consider the *asymmetric quadratic TSP (AQTSP)*, as an extension of the asymmetric TSP (ATSP), where the costs do not depend on two but on three nodes that are traversed in succession by a tour. The QTSP was originally motivated by an application in bioinformatics [12]. Special cases are the *Angular-Metric TSP* [1], which looks for tours minimizing the total angle change for points in the Euclidean space, and the *TSP with reload costs* (relTSP) used in telecommunication and transport networks [2]. In addition to the distance-dependent transportation costs the relTSP takes the costs for changes of transportation means into account.

A. Fischer (✉)
TU Dortmund, Vogelpothsweg 87, 44227 Dortmund, Germany
e-mail: anja.fischer@math.tu-dortmund.de

© Springer International Publishing Switzerland 2016
M. Lübbecke et al. (eds.), *Operations Research Proceedings 2014*,
Operations Research Proceedings, DOI 10.1007/978-3-319-28697-6_21

Because a path of three nodes is contained in a tour if and only if the two associated arcs are present, this leads to a TSP with a quadratic objective function. We derive a linear integer programming formulation by linearizing the objective function by introducing new variables for each path of three nodes. Considering the associated asymmetric quadratic traveling salesman polytope P_{AQTSP}, we determine the dimension of P_{AQTSP} and present three groups of facet-defining inequalities. The inequalities in the first group are related to the Boolean quadric polytope [14]. After this we study which paths of two and three nodes are in pairwise conflict, leading to the *conflicting arcs inequalities*. Third, we exploit the relation of AQTSP to ATSP by strengthening valid inequalities for ATSP in order to get improved valid inequalities for AQTSP. We demonstrate our strengthening approaches on the example of the well-known subtour elimination constraints.

For further details and references as well as for an extensive study of the symmetric QTSP, where the direction of traversal is irrelevant, we refer the reader to [6–8] as well as to the author's Ph.D.-thesis [5].

2 Problem Description

In this section we describe our optimization problem more formally. Let $G = (V, V^{(2)}, V^{(3)}, c)$ be a directed 2-arc-weighted complete graph with node set $V := \{1, \dots, n\}$, $n \geq 3$, set of arcs $V^{(2)} := \{(i, j) : i, j \in V, i \neq j\}$, set of 2-arcs $V^{(3)} := \{(i, j, k) : i, j, k \in V, |\{i, j, k\}| = 3\}$, as well as 2-arc weights $c_a, a \in V^{(3)}$. We often write ij and ijk instead of (i, j) and (i, j, k). A cycle of length k is a set of arcs $C = \{v_1 v_2, v_2 v_3, \dots, v_{k-1} v_k, v_k v_1\}$ with associated set of 2-arcs $C^3 = \{v_1 v_2 v_3, v_2 v_3 v_4, \dots, v_{k-1} v_k v_1, v_k v_1 v_2\}$ with pairwise distinct nodes $v_i \in V$, i.e., $|\{v_1, v_2, \dots, v_k\}| = k$. Then the task is to find a *tour* or *Hamiltonian cycle* C in G, i.e., a cycle of length n, that minimizes the sum of the given weights c_a over all 2-arcs $a \in C^3$. Let $\mathcal{C}_n^a = \{C : C \text{ tour in } G\}$ denote the set of all tours on n nodes. Then the optimization problem reads

$$\min \left\{ c(C) := \sum_{ijk \in C^3} c_{ijk} : C \in \mathcal{C}_n^a \right\}.$$

For a cycle C we define the incidence vector $(x^C, y^C) \in \{0, 1\}^{V^{(2)} \cup V^{(3)}}$ by

$$\forall a \in V^{(2)} : x_a^C = \begin{cases} 1 & \text{if } a \in C, \\ 0 & \text{if } a \notin C, \end{cases} \quad \text{and} \quad \forall a \in V^{(3)} : y_a^C = \begin{cases} 1 & \text{if } a \in C^3, \\ 0 & \text{if } a \notin C^3. \end{cases}$$

Because a 2-arc $ijk \in V^{(3)}$ is part of a tour $C \in \mathcal{C}_n^a$, i.e., $ijk \in C^3$, if the two associated arcs (i, j) and (j, k) are contained in C, our optimization problem can be written as

$$\min \left\{ \sum_{ijk \in V^{(3)}} c_{ijk} x_{ij}^C x_{jk}^C : C \in \mathcal{C}_n^a \right\}. \tag{1}$$

Linearizing, a linear integer programming formulation for (1) is given by

$$\min \sum_{ijk \in V^{(3)}} c_{ijk} y_{ijk}$$

$$\sum_{j \in V \setminus \{i\}} x_{ij} = \sum_{j \in V \setminus \{i\}} x_{ji} = 1, \qquad\qquad i \in V, \tag{2}$$

$$\sum_{ij \in S^{(2)}} x_{ij} \leq |S| - 1, \qquad\qquad S \subset V, 2 \leq |S| \leq n - 2, \tag{3}$$

$$x_{ij} = \sum_{k \in V \setminus \{i,j\}} y_{ijk} = \sum_{k \in V \setminus \{i,j\}} y_{kij}, \qquad\qquad ij \in V^{(2)}, \tag{4}$$

$$x_{ij} \in \{0, 1\}, y_{ijk} \in [0, 1], \qquad\qquad ij \in V^{(2)}, ijk \in V^{(3)}. \tag{5}$$

The *degree constraints* (2) ensure that each node is entered and left exactly once by a tour. Subtours are forbidden via the well-known *subtour elimination constraints* (3), see [4]. Constraints (4) couple the arc- and 2-arc-variables. Indeed, the sum of the in-flow into ij via 2-arcs $kij \in V^{(3)}$ has to be the same as the out-flow out of ij via 2-arcs $ijk \in V^{(3)}$. One can derive these inequalities by multiplying (2) by a variable x_{ki} or x_{ik} and using the integrality of the x-variables as well as the property that subtours of length two are forbidden.

In the following we will present a polyhedral study of the associated *Asymmetric Quadratic Traveling Salesman Polytope* $P_{\mathbf{AQTSP}_n} := \mathrm{conv}\{(x^C, y^C) : C \in \mathcal{C}_n^a\} = \mathrm{conv}\left\{(x, y) \in \mathbb{R}^{V^{(2)} \cup V^{(3)}} : (x, y) \text{ fulfills } (2)–(5)\right\}.$

3 Polyhedral Study of P_{AQTSP_n}

In this section we summarize the main polyhedral results for $P_{\mathbf{AQTSP}_n}$. Before considering its facetial structure we shortly look at its dimension. An upper bound on the dimension of $P_{\mathbf{AQTSP}_n}$ is given by the number of variables reduced by the rank of the constraint matrix, which equals $2n^2 - n - 1$ for $n \geq 4$. One can prove:

Theorem 1 *The dimension of $P_{\mathbf{AQTSP}_n}$ equals $f(n) := n(n-1)^2 - (2n^2 - n - 1) = n^3 - 4n^2 + 2n + 1$ for all $n \geq 8$.*

The proof of this result is constructive. We build $f(n) + 1$ tours whose incidence vectors are affinely independent in a three step approach. First, we determine, independently of n, the rank of some specially structured tours explicitly and take a maximal subset such that all corresponding incidence vectors are affinely independent. In steps two and three we iteratively construct tours in such an order that each of these tours contains at least one 2-arc that is not contained in any previously constructed tour. The same proof structure also allows us to prove the facetness of several classes of valid inequalities, but (partially large) adaptations of the constructions are needed, because we have to ensure that the incidence vectors of the tours define roots of the respective inequality.

The facet-defining inequalities presented next can be divided into three large classes. First, we exploit the relation of P_{AQTSP_n} to the Boolean quadric polytope (BQP) [14]. Then we introduce the *conflicting arcs inequalities*, which forbid certain configuration. Because the asymmetric traveling salesman polytope $P_{\mathrm{ATSP}_n} := \mathrm{conv}\{x^C \in \{0,1\}^{V^{(2)}} : C \in \mathcal{C}_n^a\}$ is a projection of P_{AQTSP_n}, valid inequalities for P_{ATSP_n} remain valid for P_{AQTSP_n} but in most cases a strengthening is possible. Based on the ideas that led to the conflicting arcs inequalities we present approaches to strengthen valid inequalities of P_{ATSP_n} in order to derive stronger inequalities for P_{AQTSP_n}. One strengthening is applicable to constraints with non-negative coefficients, the other one to the so called *clique tree inequalities* [9, 11].

Because we consider a linearization of a quadratic zero-one problem it is natural to ask for connections to the BQP. One can easily show that the inequalities $y_{ijk} \geq 0$, $ijk \in V^{(3)}$, are facet-defining for P_{AQTSP_n}. But the triangle inequalities of the BQP can be improved:

Theorem 2 *For $n \geq 7$ the inequalities*

$$y_{ijk} + y_{kij} \leq x_{ij}, \qquad\qquad ij \in V^{(2)}, \ k \in V \setminus \{i,j\},$$

$$\sum_{ij \in D^{(2)}} x_{ij} - \sum_{ijk \in D^{(3)}} y_{ijk} \leq 1, \qquad\qquad D \subset V, \ |D| = 3,$$

define facets of P_{AQTSP_n}.

The conflicting arcs inequalities exploit the fact that short subtours or T-structures are not allowed. In their most general form they can be written as

$$x_{ij} + x_{ji} + \sum_{k \in S \cup S_1} y_{ikj} + \sum_{k \in S \cup S_2} y_{jki} + \sum_{\substack{k \in S_1, \\ l \in S_2}} y_{kil} + \sum_{\substack{k \in S_1, \\ l \in T}} y_{kil} + \sum_{\substack{k \in T, \\ l \in S_2}} y_{kil} + \sum_{\substack{k,l \in T, \\ k \neq l}} y_{kil} \leq 1 \quad (6)$$

with $V = \{i,j\} \dot\cup S \dot\cup T \dot\cup S_1 \dot\cup S_2$, $i \neq j$, being a partition of V. One can prove the following result:

Theorem 3 *Inequalities (6) define facets of P_{AQTSP_n} if*

1. $n \geq 7$ and $T = S_1 = S_2 = \emptyset$,
2. $n \geq 8$ and $|S| \geq 3, |T| \geq 3, S_1 = S_2 = \emptyset$,
3. $n \geq 6$ and $|S_1| \geq 2, |S_2| \geq 2, S = T = \emptyset$,
4. $n \geq 6$ and $S_1, S_2 \neq \emptyset$ as well as $((|S_1| \geq 2, |S_2| \geq 2, |T| = 1), \ (|S_1| \geq 3, |S_2| \geq 3)$ or $(|T| \geq 2))$.

Furthermore, let $V = \{i, j\} \dot\cup S \dot\cup T$ with $T = \{k, l\}, k \neq l$, then inequalities

$$x_{ij} + x_{ji} + y_{kil} + y_{kjl} + y_{lik} + y_{ljk} + \sum_{m \in S} y_{imj} + \sum_{m \in S} y_{jmi} \leq 1 \tag{7}$$

define facets of P_{AQTSP_n}, $n \geq 7$.

Some groups of these inequalities have only polynomial size and can therefore be separated by enumeration. If we consider the two cases $(|S| \geq 3, |T| \geq 3, S_1 = S_2 = \emptyset)$ and $(|S_1| \geq 2, |S_2| \geq 2, S = T = \emptyset)$, then there are exponentially many inequalities (6) of both types. However, in the first case the separation problem is solvable in polynomial time and in the second case the separation problem is NP-hard even if the tested vector $(\bar{x}, \bar{y})$ fulfills (2)–(4) as well as $(\bar{x}, \bar{y}) \in [0, 1]^{V^{(2)} \cup V^{(3)}}$.

As mentioned above, valid inequalities of P_{ATSP_n} are also valid for P_{AQTSP_n}, but in many cases they can be improved. Our strengthening approaches are motivated by the conflicting arcs inequalities (6). Let $i, j \in V, i \neq j$, be fixed. Then the 2-arcs $imj, jmi \in V^{(3)}$ almost act like the two arcs ij, ji themselves expressing that the two nodes i, j are close in a tour. Based on this observation we derive:

Theorem 4 *Let $a^T x \leq b$ be a valid inequality of P_{ATSP_n} with $a \geq 0$. Define $V' := \{i \in V : \exists j \in V$ with $a_{ij} + a_{ji} > 0\}$. Then the inequalities*

$$a^T x + \sum_{\substack{ikj \in V^{(3)}: \\ a_{ik} = a_{kj} = 0}} a_{ij} y_{ikj} \leq b \quad \text{if } V' < \frac{n}{2}, \quad \text{and} \quad a^T x + \sum_{\substack{ikj \in V^{(3)}: \, k \neq \bar{i}, \\ a_{ik} = a_{kj} = 0}} a_{ij} y_{ikj} \leq b$$

for a node $\bar{i} \in V \setminus V'$ are valid for P_{AQTSP_n}.

Applying this general strengthening to the subtour elimination constraints (3) we even derive facet-defining inequalities of P_{AQTSP_n}.

Theorem 5 *For $n \geq 7$ the inequalities*

$$\sum_{ij \in S^2} x_{ij} + \sum_{\substack{ikj \in V^{(3)}: \\ i,j \in S, k \in V \setminus S}} y_{ikj} \leq |S| - 1 \quad \Leftrightarrow \quad \sum_{\substack{ijk \in V^{(3)}: \\ i \in S, j, k \in V \setminus S}} y_{ijk} \geq 1 \tag{8}$$

define facets of P_{AQTSP_n} for all $S \subset V, 2 \leq |S| < \frac{n}{2}$. Furthermore, the inequalities

$$\sum_{ij \in S^2} x_{ij} + \sum_{\substack{ikj \in V^{(3)}: \\ i,j \in S, \\ k \in V \setminus (S \cup \{\bar{i}\})}} y_{ikj} \leq |S| - 1 \quad \Leftrightarrow \quad \sum_{\substack{ijk \in V^{(3)}: \\ i \in S, \\ j, k \in V \setminus S}} y_{ijk} + \sum_{\substack{\bar{i}ij \in V^{(3)}: \\ i,j \in S}} y_{\bar{i}ij} \geq 1$$

define facets of P_{AQTSP_n}, $n \geq 11$, for all $S \subset V, \frac{n}{2} \leq |S| \leq n - 5, \bar{i} \in V \setminus S$.

It is well-known that the separation problem for the subtour elimination constraints (3) can be solved in polynomial time. But, it is NP-hard to determine a maximally violated inequality of type (8) for points $(\bar{x}, \bar{y}) \in [0, 1]^{V^{(2)} \cup V^{(3)}}$ satisfying (2) and (4).

Unfortunately, the strengthening approach does not leads to facets of P_{AQTSP_n} in general if we apply it to facet-defining inequalities of P_{ATSP_n}. Examples are further lifted versions of some of the so called D_k^+- and D_k^--inequalities [10].

Our second lifting approach can only be applied to the clique tree inequalities [9, 11], which are a large class of valid inequalities of ATSP including the subtour elimination constraints and comb inequalities. It combines the idea of "replacing" 2-arcs of the last approach with "outer" 2-arcs, see the 2-arcs with nodes i or j between two nodes of T (or S_1, S_2) [7]. Strengthened, even facet-defining, versions of (3) read, e.g.,

$$\sum_{ij \in I^{(2)}} x_{ij} + \sum_{\substack{ikj \in V^{(3)}: \\ i,j \in I, k \in S}} y_{ikj} + \sum_{\substack{kil \in V^{(3)}: \\ i \in I \setminus \{\bar{\imath}\}, k, l \in T}} y_{kil} \leq |I| - 1$$

for appropriate sets I, S, T with $V = I \dot\cup S \dot\cup T$, $|I| \geq 2$, $\bar{\imath} \in I$ and $n \geq 9$. There also exist non-coefficient-symmetric strengthened variants of (3) similar to (6).

Using some of the newly derived cutting planes in a branch-and-cut framework allowed us to solve real-world instances from biology resp. bioinformatics surprisingly well. Instances with up to 100 nodes could be solved in less than 700 s improving the results in the literature by several orders of magnitude. Without the new cutting planes the root gaps as well as the running times were much higher for these instances. On most of the randomly generated instances additional separators reduced the root gaps and the numbers of nodes in the branch-and-cut tree significantly, often even the running times.

Acknowledgments This work was partially supported by the European Union and the Free State of Saxony funding the cluster eniPROD at Chemnitz University of Technology.

References

1. Aggarwal, A., Coppersmith, D., Khanna, S., Motwani, R., Schieber, B.: The angular-metric traveling salesman problem. SIAM J. Comput. **29**, 697–711 (1999)
2. Amaldi, E., Galbiati, G., Maffioli, F.: On minimum reload cost paths, tours, and flows. Networks **57**, 254–260 (2011)
3. Applegate, D.L., Bixby, R.E., Chvatal, V., Cook, W.J.: The Traveling Salesman Problem: A Computational Study (Princeton Series in Applied Mathematics). Princeton University Press, Princeton (2007)
4. Dantzig, G., Fulkerson, R., Johnson, S.: Solution of a large-scale traveling-salesman problem. Oper. Res. **2**, 393–410 (1954)
5. Fischer, A.: A polyhedral study of quadratic traveling salesman problems. Ph.D. thesis, Chemnitz University of Technology, Germany (2013)
6. Fischer, A.: An analysis of the asymmetric quadratic traveling salesman polytope. SIAM J. Discrete Math. **28**(1), 240–276 (2014)
7. Fischer, A., Fischer, F.: An extended approach for lifting clique tree inequalities. J. Comb. Optim. **30**(3), 489–519 (2015). doi:10.1007/s10878-013-9647-3

8. Fischer, A., Helmberg, C.: The symmetric quadratic traveling salesman problem. Math. Prog. **142**(1–2), 205–254 (2013)
9. Fischetti, M.: Clique tree inequalities define facets of the asymmetric traveling salesman polytope. Discrete Appl. Math. **56**(1), 9–18 (1995)
10. Grötschel, M., Padberg, M.W.: Lineare Charakterisierungen von Travelling Salesman Problemen. Zeitschrift für Operations Research, Series A **21**(1), 33–64 (1977)
11. Grötschel, M., Pulleyblank, W.R.: Clique tree inequalities and the symmetric travelling salesman problem. Math. Oper. Res. **11**(4), 537–569 (1986)
12. Jäger, G., Molitor, P.: Algorithms and experimental study for the traveling salesman problem of second order. LNCS **5165**, 211–224 (2008)
13. Lawler, E.L., Lenstra, J.K., Kan, A.H.G.R., Shmoys, D.B. (eds.): The Traveling Salesman Problem. A Guided Tour of Combinatorial Optimization. Wiley, Chichester (1985)
14. Padberg, M.: The Boolean quadric polytope: some characteristics, facets and relatives. Math. Prog. **45**, 139–172 (1989)

New Inequalities for 1D Relaxations of the 2D Rectangular Strip Packing Problem

Isabel Friedow and Guntram Scheithauer

Abstract We investigate a heuristic for the two-dimensional rectangular strip packing problem that constructs a feasible two-dimensional packing by placing one-dimensional cutting patterns obtained by solving the horizontal one-dimensional bar relaxation. To represent a solution of the strip packing problem, a solution of a horizontal bar relaxation has to satisfy, among others, the vertical contiguous condition. To strengthen the one-dimensional horizontal bar relaxation with respect to that vertical contiguity new inequalities are formulated. Some computational results are also reported.

1 Introduction

Given a set of rectangles $I := \{1, \ldots, n\}$ of width w_i and height h_i, $i \in I$, the objective of the rectangular two-dimensional strip packing problem (2D-SPP) is to pack the set into a strip of width W without overlap and minimal needed height H. The dimensions of the rectangles and the strip are integers and rotation of rectangles is not allowed. The problem has several industrial applications and is known to be NP-hard. Numerous heuristic algorithms have been proposed in literature. For a survey of some of the most common see [7]. In [2] the iterative heuristic SVC(SubKP) is presented that utilizes 1D knapsack problems and the sequential value correction method. The results achieved improve those of SPGAL [3] and GRASP [1].

We investigate a heuristic based on solutions of 1D bar relaxations [8] or more precisely on packing the 1D cutting patterns obtained. A brief description of the constructive heuristic approach is given in Sect. 2. The strip is filled from bottom

I. Friedow (✉) · G. Scheithauer (✉)
Institute of Numerical Mathematics, Technical University of Dresden,
Dresden, Germany
e-mail: isabel.friedow@tu-dresden.de

G. Scheithauer
e-mail: guntram.scheithauer@tu-dresden.de

© Springer International Publishing Switzerland 2016   151
M. Lübbecke et al. (eds.), *Operations Research Proceedings 2014*,
Operations Research Proceedings, DOI 10.1007/978-3-319-28697-6_22

to top. We define bottom patterns (Sect. 3) and formulate inequalities that maintain the bottom-up filling by continuing the bottom patterns. In Sect. 4 we show how to carry on the concept of continuing for non-bottom patterns.

2   LP-Based Constructive Algorithm

Our algorithm bases on two linear problems, the one-dimensional cutting stock problem (1D-CSP) and the one-dimensional multiple length cutting stock problem (1D-MCSP) [8]. The 1D-CSP consists in cutting stock bins of size W into smaller items of size w_i to meet order demands h_i, $i = 1, \ldots, n$, while minimizing the number of stock bins used. A cutting pattern $a \in Z_+^n$ describes which items are obtained by cutting a stock bin. That means the ith element is the number of items of size w_i in a. If stock bins of different sizes W_k, $k \in K = \{1, \ldots, q\}$, are applicable for cutting, the considered problem is the 1D-MCSP. Stock bins of size W_k, $k = 1, \ldots, p < q$ can be used u_k-times. The supply of the other stock bins is not restricted. The goal is again to meet all demands while minimizing the number of stock bins. Let J_k, $k \in K$, describe the set of feasible cutting patterns a^j for stock size W_k. Thus $j \in J_k$, if $w^T a^j \le W_k$. In the case of 1D-CSP $j \in J$, if a^j satisfies $w^T a^j \le W$. How often a cutting pattern a^j is used is represented by $x_j \in Z_+$, $j \in J$, and matrix A consists of columns representing the cutting patterns.

2.1   1D-CSP and 1D-MCSP Adapted to 2D-SPP

We assume that a rectangle $i \in I$ is represented by an item with size w_i and demand h_i. So we need to cut exactly h_i items of size w_i, $i = 1, \ldots, n$, and we consider binary patterns $a \in \{0, 1\}^n$. When the strip is empty, we have only one stock bin size W. It results the one-dimensional binary horizontal bar relaxation (1DHBRb):

$$\sum_{j \in J} x_j \to \min \quad \sum_{j \in J} a_i^j x_j = h_i \ \forall i \in I, \quad x_j \in Z_+ \ \forall j \in J \tag{1}$$

Now, imagine there are rectangles that already have been packed into the strip and I is the set of unpacked rectangles. We describe the resulting free space by stock bins of size W_k and supply u_k with the help of the packing skyline [6]. Let $(s^1, \ldots, s^{p+1})$ be the vector of line segments of pairwise different, in ascending order sorted, y-coordinates s_y^k. Furthermore let $(v_1, \ldots, v_{p+1})$ be the vector of the lengths of line segments. The free space between line segment s^k and s^{k+1} is represented by a stock bin of size $W_k = W - \sum_{i=k+1}^{p+1} v_i$ and available number $u_k = s_y^{k+1} - s_y^k$ for $k = 1, \ldots, p$ (see Fig. 1). We get exactly one stock bin of unrestricted supply $(p + 1 = q)$ and size $W =: W_q$ which represents the space above the highest line

Fig. 1 Free space
represented by stock bins of
size $W_1, \ldots, W_3$ and supply
$u_1, \ldots, u_3$

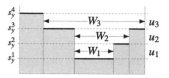

segment. Because the strip height needed is at least s_y^{p+1} we minimize only the usage
of patterns for stock size W. The objective function becomes $\sum_{j \in J_q} x_j \to$ min. Furthermore we demand that $\sum_{j \in J_k} x_j = u_k$ for all $k = 1, \ldots, p$ to ensure best usage
of the free space below the highest line segment. With $J := \bigcup_{k \in K} J_k$ we get again
$\sum_{j \in J} a_i^j x_j = h_i$ for all $i \in I$ and $x_j \in Z_+$ for all $j \in J$. We refer to this model with
1DHBRb-ML.

The solution of 1DHBRb or 1DHBRb-ML only represents a solution of 2D-SPP
if there exists such an ordering of cutting patterns that all items representing one
rectangle are in consecutive patterns (vertical contiguous condition) and all items
have the same position in each stock bin [2]. The objective value of 1DHBRb respectively 1DHBRb-ML is a lower bound $LB := \sum_{j \in J} x_j$ for the 2D-SPP.

2.2 Algorithm

The integer linear problems 1DHBRb and 1DHBRb-ML are also NP-hard problems [8]. To obtain 1D cutting patterns that represent sets of rectangles that can
be placed in the strip side by side in x-direction we solve the relaxations ($x_j \geq 0$)
referred to as HBR respectively HBR-ML. To ensure the vertical contiguous condition and the consistency of x-positions we generate and pack the cutting patterns
iteratively. The algorithm works as follows:

At first HBR is solved and the 1D cutting patterns are obtained in the form of the
columns of A. Now we consider the patterns a^j that are part of the solution, which
means $x_j > 0$. They are called candidates. To evaluate the quality of a candidate
we examine the packing that would result after placing the rectangles in a^j. The
arising skyline and the updated data of unpacked rectangles define the input data for
the next linear problem needed to solve, HBR or HBR-ML. The solution of HBR
respectively HBR-ML delivers on the one hand the major quality criterion, the lower
bound LB for the candidate, and on the other hand the cutting patterns for the next
construction step. The candidate of best quality is chosen and the corresponding
rectangles are really packed. If HBR-ML was solved for that candidate only cutting
patterns a^j, $j \in J_1$, have to be considered next otherwise all cutting patterns in A are
possible candidates. Again the quality of the new candidates is evaluated and so on.
The algorithm ends when all rectangles are packed. To get a more detailed insight
see [4].

3 Constraints for Bottom Patterns

As described in Sect. 2.2 in the later steps of the algorithm the possible candidates are taken from set J_1 that represents the stock bin with the smallest size $W_1 < W$. The number of candidates is limited. In the initial step every pattern a^j with $x_j > 0$ can be chosen for placing at bottom. With increasing n the number of candidates increases but differences between the resulting lower bounds LB are only minor or do not exist. Thus the lower bounds are not helpful in order to decide which candidate is chosen for packing. Because we want to minimize the height of the packing it seems reasonable to pack high rectangles as early as possible. Furthermore packing long rectangles in the end may cause a lot of waste. Thus we would choose a candidate that contains for example especially high or long rectangles. For these reasons we define in an appropriate way a set of rectangles I_0 that can be packed on the strip bottom and require that a *bottom pattern* only contains rectangles $i \in I_0$. Let be $J^0 := \{j \in J : a_i^j = 0 \ \forall i \in I \setminus I_0\}$ the set of bottom patterns and $u_0 := \min_{i \in I_0} h_i$ the height of the lowest rectangle in I_0. The linear problem HBR is extended to HBR-BP (bottom pattern) by the following inequality

$$\sum_{j \in J^0} x_j \geq u_0 \tag{2}$$

With that constraint we can reduce the number of bottom patterns and we ensure the existence of patterns with somehow suitable properties.

Rectangles $i \in I(a^j) := \{i \in I : a_i^j = 1\}$ are placed at the bottom of the strip if $j \in J^0$. Because $y_i = 0$ for all $i \in I(a^j)$, $j \in J^0$, we know their positions relative to each other in y-direction. Let us consider a bottom pattern a^j, $j \in J^0$, that contains rectangles with $t(j) \leq |I(a^j)|$ pairwise different heights $\tilde{h}_1 < \cdots < \tilde{h}_{t(j)}$. When placing them at the bottom a skyline $(s^1, \ldots, s^{t(j)})$ arises where $s_y^i = \tilde{h}_i$, $i = 1, \ldots, t(j)$. For every $r \in C_j := \{1, \ldots, t(j) - 1\}$ there is a horizontal bar between line segment s^r and s^{r+1} that intersects with every rectangle of height $h_i > s_y^r$ but not with the others. That properties are represented by $c^{j_r} \in \{-1, 0, 1\}^n$, $r = 1, \ldots, t(j) - 1$, with

$$c_i^{j_r} = \begin{cases} 1 & \text{if } i \in I(a^j) \wedge h_i > s_y^r, \\ -1 & \text{if } i \in I(a^j) \wedge h_i \leq s_y^r \\ 0 & \text{else} \end{cases}$$

called *combinations* induced by bottom pattern a^j, $j \in J^0$ (see Fig. 2).

A pattern a^k *fulfills* a combination c^{j_r} if $a_i^k = 0$ for all $i \in I^+(c^{j_r}) := \{i \in I : c_i^{j_r} = 1\}$ and $a_i^k = 0$ for all $i \in I^-(c^{j_r}) := \{i \in I : c_i^{j_r} = -1\}$. So a *strip pattern* a^k, $k \in J^S := J \setminus J^0$, continues the bottom pattern a^j if $(a^k)^\top c^{j_r} = n(c^{j_r})$ for a $r \in C_j$, where $n(c^{j_r}) := |I^+(c^{j_r})|$.

Fig. 2 *Horizontal bars*
represented by combinations
$c^{j_1}, \ldots, c^{j_3}$ induced by a^j,
$j \in J^0$

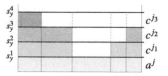

Remember that HBR, and so it's extension (2), does not ensure the vertical contiguous condition (Sect. 2.2). Thus there is no guarantee for the existence of strip patterns $j \in J^S$ that continue bottom patterns. Now we formulate inequalities that ensure the following: If a bottom pattern a^j, $j \in J^0$, is used in the solution of HBR with usage $x_j > 0$ and a^j induces at least one combination, then there exists a strip pattern a^k with usage $x_k > 0$ that continues a^j.

The usage x_j of a bottom pattern a^j is restricted by $x_j \leq b_0^j := \min_{i \in I(a^j)} h_i$. The height of the horizontal bar represented by c^{j_r} is $s_y^{r+1} - s_y^r$. Thus the usage of patterns that fulfills combination c^{j_r} is restricted by $b_r^j := s_y^{r+1} - s_y^r$, $r \in C_j$.

Let $J(c^{j_r}) := \{k \in J \setminus J^0 : (a^k)^\top c^{j_r} = n(c^{j_r})\}$ be the index set of patterns that fulfills combination c^{j_r}, $r \in C_j$, induced by bottom pattern a^j, $j \in J^0$. The linear problem HBR-BP is extended to HBR-cBP (continuous bottom pattern) by the following linear inequalities

$$\frac{\sum_{k \in J(c^{j_r})} x_k}{b_r^j} \geq \frac{x_j}{b_0^j} \quad j \in J^0, r = 1, \ldots, t(j) - 1. \tag{3}$$

4 Higher Level Continuing

The presented concept of continuing bottom patterns can be applied to strip patterns if they continue a bottom pattern or an other strip pattern. Let us consider a bottom pattern a^j, $j \in J^0$, and a strip pattern a^k, $k \in J^S$, that fulfills the first-level combination $c^j := c^{j_1}$. Let $I(a^k) = \{i_1, \ldots, i_p\}$ and $I_c := \{i \in I(a^k) : c_i^j = 1\}$ be the set of items that continue the rectangles of the bottom pattern and $I_{new} := \{i \in I(a^k) : c_i^j = 0\}$. If a^j and a^k are chosen for packing then $y_i = b_0^j$ for all $i \in I_{new}$. Because of that the argumentation described in Sect. 3 can be applied. The difference is that we have to consider the reduced heights $\widetilde{h}_i := h_i - b_0^j$ for $i \in I_c$ for continuing. With $\widetilde{h}_i := h_i$ for all $i \in I_{new}$ the maximum usage of a^k is $b_0^k = \min\{\widetilde{h}_i : i \in I(a^k)\}$. Remember that rectangles $i \in I^-(c^j)$ are not allowed in a^k and so naturally not in any pattern continuing a^k. The induced first-level combination is

$$c_i^k = \begin{cases} 1 & \text{if } i \in I(a^k) \wedge \widetilde{h}_i > b_0^k \\ -1 & \text{if } i \in I(a^k) \wedge \widetilde{h}_i \leq b_0^k \quad \vee \quad i \in I^-(c^j) \\ 0 & \text{else} \end{cases}$$

The usage of patterns that fulfills combination c^k is restricted by $b_1^k := \min\{\tilde{h}_i - b_0^k :$
$i \in I(a^k), \tilde{h}_i - b_0^k > 0\}$. With $J(c^k) := \{l \in J \setminus J^0 : (a^l)^\top c^k = n(c^k)\}$ we get the
inequality

$$\frac{\sum_{l \in J(c^k)} x_l}{b_1^k} \geq \frac{x_k}{b_0^k} .$$

The introduced continuing strategy leads to a solution matrix A with a sub matrix
that has the consecutive ones property. Note that the presented approaches can also
be applied to HBR-ML where the lowest line segment of the skyline represents the
bottom and $I_0 \subseteq \{i \in I : w_i \leq W_1\}$.

5   Numerical Experiments and Conclusions

We tested our algorithm, among others, for the instances of the waste-free classes
T1–T5 from Hopper [5] because optimal strip height is known. Each of the 5 classes
contains 5 instances with strip width $W = 200$ and optimal strip height $H_{opt} = 200$.
The average number of items per instance is n. Table 1 contains the results of the
algorithm (Sect. 2.2) using the linear problems HBR, HBR-BP, HBR-cBP and HBR-
cBP combined with HBR-ML with continuing bottom patterns (ML-cBP). For better
comparison we pack all rectangles of bottom patterns in decreasing order of heights.
Column HBR-BP* contains the results obtained with a more specialized packing
rule. Column SVC contains the results of the iterative heuristic SVC(SubKP) of [2].
For SVC and HBR-BP* the percentage deviation from H_{opt} is given in column gap.

The introduced concept of continuing bottom patterns enables our algorithm to
start at a somehow suitable initial point but the obtained forecast is still a short one.
With the higher level continuing the vertical contiguous condition can be fulfilled
but that does not ensure a feasible two-dimensional packing. Thus, further work will
focus on the realization of constant location of items in one-dimensional cutting
patterns.

Table 1 Results obtained by the constructive algorithm with and without bottom patterns

Class	n	H_{opt}	HBR	HBR-BP	HBR-cBP	ML-cBP	HBR-BP*	gap	SVC	gap
T1	17	200	200,4	200,0	200,0	200,0	200,0	0,0	201,8	0,9
T2	25	200	207,4	206,6	205,4	201,4	201,6	0,8	207,0	3,5
T3	29	200	207,2	205,2	206,4	205,6	203,6	1,8	206,6	3,3
T4	49	200	206,8	206,4	206,2	206,2	205,8	2,9	205,0	2,5
T5	73	200	206,0	205,6	205,6	205,2	205,0	2,5	204,2	2,1
		gap_{av}						1,6		2,5

References

1. Alvarez-Valdez, R., Parreño, F., Tamarit, J.M.: Reactive grasp for the strip-packing problem. Comput. Oper. Res. **35**, 1065–1083 (2008)
2. Belov, G., Scheithauer, G., Mukhacheva, E.A.: One-dimensional heuristic adapted for two-dimensional rectangular strip packing. J. Oper. Soc. **59**, 823–832 (2008)
3. Bortfeld, A.: A genetic algorithm for the two-dimensional strip packing problem with rectangular pieces. Eur. J. Oper. Res. **172**, 814–837 (2006)
4. Friedow, I.: LP-basierte Heuristiken zur Lösung des Streifenpackproblems. Diploma thesis Technical University of Dresden (2012)
5. Hopper, E.: Two-dimensional packing utilising evolutionary algorithms and other meta-heuristic methods. Ph.D. thesis University of Wales, Cardiff School of Engineering (2000)
6. Lim, A., Oon, W., Wei, L., Zhu, W.: A skyline heuristic for the 2D rectangular packing and strip packing problems. Eur. J. Oper. Res. **215**, 337–346 (2011)
7. Lodi, A., Martello, S., Monaci, M.: Two-dimensional packing problems—a survey. Eur. J. Oper. Res. **141**, 241–252 (2002)
8. Scheithauer, G.: LP-based bounds for the container and multi-container loading problem. Int. Trans. Oper. Res. **6**, 199–213 (1999)

Representing Production Scheduling with Constraint Answer Set Programming

Gerhard Friedrich, Melanie Frühstück, Vera Mersheeva,
Anna Ryabokon, Maria Sander, Andreas Starzacher and Erich Teppan

Abstract Answer Set Programming and Constraint Programming constitute declarative programming approaches with different strengths which have already been shown to be highly effective for many hard combinatorial problems. In this article we discuss two hybrid Constraint Answer Set Programming approaches with regard to their suitability for encoding production scheduling problems. Our exemplifications are done on the basis of a production scheduling problem of Infineon Technologies Austria AG.

1 Introduction

Scheduling is one of the most important and also hard problems in industrial production planning. Various methods such as SAT-Solving, Dynamic Programming or state-based search have been applied to scheduling since the late 1940s. In handling

The authors are funded by FFG (grant 840242) and listed in alphabetical order.

G. Friedrich · M. Frühstück · V. Mersheeva (✉) · A. Ryabokon · E. Teppan (✉)
Alpen-Adria Universität Klagenfurt, Klagenfurt, Austria
e-mail: vera.mersheeva@aau.at

E. Teppan
e-mail: erich.teppan@aau.at

G. Friedrich
e-mail: gerhard.friedrich@aau.at

M. Frühstück
e-mail: melanie.fruhstuck@aau.at

A. Ryabokon
e-mail: anna.ryabokon@aau.at

M. Sander · A. Starzacher (✉)
Infineon Technologies Austria AG, Villach, Austria
e-mail: andreas.starzacher@infineon.com

M. Sander
e-mail: maria.sander@infineon.com

© Springer International Publishing Switzerland 2016
M. Lübbecke et al. (eds.), *Operations Research Proceedings 2014*,
Operations Research Proceedings, DOI 10.1007/978-3-319-28697-6_23

159

the diversity and dynamics of different scheduling domains, declarative approaches have proven to be highly effective. Those approaches have a long history in Artificial Intelligence in general, and in automatic planning and scheduling [5] in particular. In this circumstance, Constraint Programming (CP) is a very successful approach which offers good performance for numerical calculations and a rich set of specialized global constraints. Another approach is Answer Set Programming (ASP) which constitutes a decidable subset of first-order logic. The big advantage of ASP is its high-level knowledge representation features. Different strengths of these approaches—the representation abilities of ASP and the numerical calculation features of CP—led to the development of a hybrid approach called Constraint ASP (CASP).

This article introduces two CASP approaches with respect to production scheduling. Our exemplifications are based on a simplified semiconductor manufacturing scheduling problem of Infineon Technologies Austria AG. Infineon is one of the biggest semiconductor manufacturers world-wide and offers system solutions in the fields of automotive-, industry-, smartcard- and security electronics. A semiconductor chip is an integrated circuit consisting of thousands of components being assembled in highly complex workflows. Consequently, the Infineon case constitutes a representative real-world case.

2   Lot Scheduling Problem

Input contains a set of lots (silicon wafer entities) that have to be scheduled starting from the given current time. Each lot belongs to an order and consists of several wafers that are of a certain product type. Every order includes wafers of only one type. A lot should be finished by the given due date, otherwise, it is late. The time by which a lot is late is called tardiness. Every lot has to be processed in accordance with its product's workflow that is a sequence of tasks. A workflow can have time coupling intervals. A time coupling is the maximal time interval between the end of one task and the beginning of a defined subsequent task. Available machines are listed with their current setup, i.e. task–product pair, and possible tasks of products that they can perform. For every possible task a processing time per wafer is given. Machines can process only one lot at a time. A machine might not be available in one or several time periods when it cannot process lots. Such periods are defined as start and end time points and can be either planned (service) or unplanned (break down). In the latter case, a rescheduling procedure is triggered. Another parameter is the changeover. The notion of changeover expresses the time which is needed for setting up a machine to perform another type of task. During a changeover no lots can be processed on the affected machine. At the beginning of the schedule some lots might have already started their workflow. Such lots are given in a set together with information about their current status. If a lot is being processed by a machine, its entry includes a machine, a sequence number of the current task and its remaining time. If a lot is waiting for the next task to start, it is listed with a sequence number of the last finished task. A legacy schedule might also be provided as an input. A *solution* of the problem is a set of admissible timed assignments of lots to the existing machines. Currently we consider a decision problem based on summed-up total tardiness which is not allowed to exceed a predefined maximal value. In this case, optimization can be performed by an iterative approach. Additionally, it is possible to add any further optimization criteria such as minimization of finish time of each lot or number of changeovers.

3 Constraint Answer Set Programming

ASP is a declarative programming approach having its roots in deductive databases and logic programming. Problem solutions correspond to answer sets which are models of ASP programs under the stable model semantics [2] and its extensions, e.g. [6]. An ASP program contains logic rules possibly including variables starting with capital letters. During the process of finding a solution the variables are substituted by constant symbols. This process is called *grounding*. Constant symbols start with lower letters. Generally, if the body of a rule is satisfied, its head can be derived. A special form of an ASP rule are *choice rules*, e.g. *1 {st(J,S):time(S)} 1 :- job(J),in_sched(J)*. The meaning of this rule is that exactly one starting time is assigned to every job which must be in the schedule.

A Constraint Satisfaction Problem (CSP) is a triple (V, D, C) where V is a set of variables associated with a domain D and a set of constraints C. A solution of a CSP is an assignment of domain values to variables such that no constraint is violated. CASP approaches have been developed for defining CSPs within answer set programs in order to combine the high-level representation abilities of ASP and the computation facilities of state-of-the-art CP solvers. In general, CASP allows to improve ASP grounding and solving performance, since it gives a possibility to represent constraints over large (infinite) domains and pass them to specialized constraint programming systems. Several CASP systems have been developed [3]. In our paper we discuss two of them, *clingcon* [4] and *ezcsp* [1], and outline their characteristics.

The *clingcon* system [4] is part of the Potassco collection.[1] It is a hybrid solver which combines the high performance Boolean solving capacities of ASP with techniques for using non-Boolean constraints from the area of constraint programming. The system differentiates between regular ASP atoms and constraint atoms. As shown in Fig. 1 for *clingcon*, the input file is written in an extended input language for *gringo*. It includes specific constraint programming operators marked with a preceding \$ symbol. This involves arithmetic constraints ($+$, $*$, etc.), the global constraints *count* and *distinct* as well as the optimization statements *minimize* and *maximize*. After grounding, a partial Boolean assignment of regular and constraint atoms (interpretation) is initialized. The ASP and CP solvers (*clasp* and *gecode*) extend this Boolean assignment until a full assignment of the atoms is reached. In case of a conflict, the

---

[1]*clingcon*, *clasp* and *gringo* are available on http://sf.net/projects/potassco/files/.

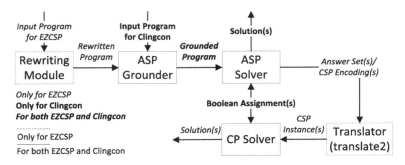

Fig. 1 General architecture of *clingcon* and *ezcsp* systems

conflict is analysed and backjumping is done. Eventually, the CP solver fixes the values for the CSP variables such that the CSP constraints are not violated. Such interpretation including the assignments of domain values to variables constitutes an answer set.

In [1], *ezcsp* [2] has been introduced. Its language comprises statements for declaration of variables, their domains as well as constraints over these variables. The atom $cspdomain(D)$ specifies a domain type D of all CSP variables occurring in a program. The system *ezcsp* supports three types of domains: fd—finite domain, r—real numbers and q—rational numbers. The atom $cspvar(X, L, U)$ declares CSP variables where a term X stands for the CSP variable name, whereas L and U stand for the lower and upper values from the selected domain. Finally, the atom $required(\gamma)$ defines constraints on CSP variables and any solution of a CSP, i.e. assignment of values to CSP variables, must satisfy all constraints. The constraint atoms are treated as regular atoms by an ASP grounder, e.g. *gringo*. A grounded program is then solved by an ASP solver, e.g. *clasp*. If a returned answer set includes constraint atoms, they are mapped by *ezcsp* into a CSP which is processed by a CP solver, e.g. *bprolog*, as shown in Fig. 1. The performance of a CP solver might be improved by using global constraints, variable and value orderings.

4 Modeling Production Scheduling

To keep the model more general, notions of lots, products and workflows are transformed to jobs, where a job is a single task of a lot. For example, *lot1* with three tasks will be transformed to three jobs: *lot1_1*, *lot1_2* and *lot1_3*, where the last numbers stand for sequential indexes of the tasks. All data related to the mentioned notions is transformed as well.

The input is defined using the following atoms. Job information is given in *job(Job)* and *deadline(Job,Time)*. The last atom defines the due time by which a job

---

[2]http://www.mbal.tk/ezcsp/index.html.

should be accomplished. Order of tasks belonging to one lot is expressed by an atom *prec(Job1,Job2)* meaning that *Job2* cannot start before its preceding *Job1* is finished. The time that the job takes on a machine is given by *job_len(Job,Machine,Time)*. An atom *cur_t(Time)* defines the time when the new schedule will start, whereas *max_t(Time)* indicates the maximal possible time value. Finally, *max_p(Value)* defines the maximal possible value of the optimization criteria (total tardiness).

A solution is expressed using atoms *st(Job,Time)* and *on_inst(Job,Machine)* in ASP. They specify the time when a job will start and a machine that will perform it. These atoms are defined only for the jobs that have not been finished yet. Such jobs are listed by an atom *in_sched(Job)*. In CASP, the representation is slightly different, e.g. *st(Job)=Time*. Tardiness of a job is defined using an atom *td(Job,Value)*, whereas the total tardiness of all jobs is expressed as an atom *total_td(Value)*.

Below the encodings of several essential rules are provided for both *clingcon* and *ezcsp*.[3] First of all, solution generation is performed in *clingcon* as follows:

```
st(J) $>= 0 $and st(J) $<= MT :- job(J),max_t(MT),in_sched(J).
1 {on_inst(J,M) : job_len(J,M,_)} 1 :- job(J),in_sched(J).
```

The same rules in *ezcsp* are expressed slightly different:

```
cspvar(st(J),0,MT) :- job(J),max_t(MT),in_sched(J).
cspvar(on_inst(J),1,N) :- job(J),in_sched(J),nMachines(N).
job_m(J,M) :- job(J),in_sched(J),job_len(J,M,_).
required(on_inst(J)  != M) :- machine(M),in_sched(J),not job_m(J,M).
```

One example of common constraints is that no job that must be scheduled can start before the current time. This constraint looks in *clingcon* as follows:

```
st(J) $>= CT :- cur_t(CT),in_sched(J).          .
```

The same constraint can be rewritten in *ezcsp* in the following way:

```
required(st(J) >= CT) :- cur_t(CT),in_sched(J).
```

The rule for tardiness that cannot be greater than the maximum value is expressed in *clingcon* as:

```
td(J) $>= 0 $and td(J) $<= MT :- job(J),max_t(MT),in_sched(J).
td(J) $== $max{0,st(J) $+ L $- D} :- job(J),deadline(J,D),
    on_inst(J,M),job_len(J,M,L),in_sched(J).
$sum{td(J) : job(J) : in_sched(J)} $== total_td.
total_td $<= MP :- max_p(MP).
```

---

[3]In our paper only parts of the models are provided due to the space limit. The full encodings as well as input and output formats can be found at http://isbi.aau.at/hint/problems.

Table 1 Maximal (average) time and grounding size measured during the evaluation of (C)ASP approaches

	ASP (pure)		clingcon		ezcsp	
	Grounding	Solving	Grounding	Solving	Grounding	Solving
Success rate	0%	0%	100%	90%	100%	0%
time, s			1.1 (0.8)	597.7 (107.3)	2.4 (2.0)	
size, # lines			94301 (69752)		64853 (56216)	

Its analog in *ezcsp* looks like the following set of rules:

```
cspvar(td(J),0,MT)  :- job(J),in_sched(J),max_t(MT).
required( (td(J)  == max(0, st(J) + L - D)) \/ on_inst(J) != M)  :-
    job(J),in_sched(J),deadline(J,D),job_len(J,M,L).
cspvar(total_td,0,MP) :- max_p(MP).
required(sum([td/1],==,total_td)).
```

5   Results

We have evaluated the following solvers within 900 seconds: ASP (pure)—grounder GRINGO 3.0.3 and solver CLASP 2.0.4; CLINGCON 2.0.3 and EZSCP 1.6.20B30.[4] The evaluation was conducted on a set of real-world instances[5] which include 107 machines each. The results for a set of 20 mid-size instances with up to 101 jobs are presented in Table 1. In contrast to pure ASP, CASP approaches could ground the program and required at most 2.4 s. Generally *clingcon* was faster but *ezcsp* generated smaller grounded programs. However, only *clingcon* could solve all but two instances.

Generally, finding (optimal) schedules remains a challenge for realistic problem cases. The principal difficulties for ASP-based approaches are the explosion of grounding and the completeness of search. Solutions for the two instances, i.e. with 84 and 101 jobs, could not be computed, although realistic schedules might include even higher number of jobs. Due to computational complexity, the described approaches seem to be inapplicable for large practical scenarios without incorporation of heuristics. Therefore, the design of methods which are able to (automatically) create heuristics for high-level knowledge representation languages such as ASP is a promising direction for future work.

---

[4]The experiments were performed on a system with Intel i7-3930K CPU (3.20 GHz), 64 GB of RAM, running Ubuntu.

[5]Instances are available at http://isbi.aau.at/hint/images/lsp/benchmarks.zip.

References

1. Balduccini, M.: Representing constraint satisfaction problems in answer set programming. In: ASPOCP (2009)
2. Gelfond, M., Lifschitz, V.: The stable model semantics for logic programming. In: ICLP/SLP, pp. 1070–1080 (1988)
3. Lierler, Y.: Relating constraint answer set programming languages and algorithms. Artif. Intell. **207**, 1–22 (2014)
4. Ostrowski, M., Schaub, T.: ASP modulo CSP: the clingcon system. TPLP **12**(4–5), 485–503 (2012)
5. Shaw, M.J.P. Whinston, A.B.: Automatic planning and flexible scheduling: a knowledge-based approach. In: ICRA, pp. 890–894 (1985)
6. Simons, P., Niemelä, I., Soininen, T.: Extending and implementing the stable model semantics. Artif. Intell. **138**(1), 181–234 (2002)

Day-Ahead Versus Intraday Valuation of Flexibility for Photovoltaic and Wind Power Systems

Ernesto Garnier and Reinhard Madlener

Abstract This paper takes the perspective of a photovoltaic (PV) or wind power plant operator who wants to optimally allocate demand-side flexibility to maximize realizable production value. We compare two allocation alternatives: (1) use of flexible loads to maximize relative day-ahead market value by shifting the portfolio balance in view of day-ahead prices; (2) use of flexible loads in intraday operations to minimize the costs incurred when balancing forecast errors. We argue that the second alternative yields a greater average value than the first in continuous-trade intraday markets. The argument is backed by a market data analysis for Germany in 2013.

1 Introduction and Background

Two effects decrease the competitiveness of non-dispatchable renewable power production in energy markets: a *low relative market value* and *balancing costs*.

The former is a direct result of non-dispatchability. Since the output of PV or wind power systems is dictated by weather conditions, operators cannot freely choose their output levels. Further, weather conditions are similar within fairly large geographical regions. Consequently, wind or PV power plant operators within the same market area face strongly correlated production patterns. This implies relatively low (high) market prices in times of strong (weak) production. The issue is amplified by the expansion of PV and wind power capacities in energy markets [5].

Balancing costs result from PV and wind forecast errors, i.e. deviations between day-ahead projections and actual power production. Shortages and surpluses need to

E. Garnier (✉)
RWTH Aachen University, Templergraben 55, 52056 Aachen, Germany
e-mail: ernesto.garnier@rwth-aachen.de

R. Madlener (✉)
Institute for Future Energy Consumer Needs and Behavior (FCN),
School of Business and Economics/E.ON Energy Research Center,
RWTH Aachen University, Mathieustrasse 10, 52074 Aachen, Germany
e-mail: RMadlener@eonerc.rwth-aachen.de

© Springer International Publishing Switzerland 2016
M. Lübbecke et al. (eds.), *Operations Research Proceedings 2014*,
Operations Research Proceedings, DOI 10.1007/978-3-319-28697-6_24

be balanced within portfolios or in intraday markets. All remaining deviations must be compensated in the imbalance market, often at substantial surcharges. Considering both opportunity costs from foregone day-ahead sales and actual balancing costs, the impacts of production forecast errors can easily destroy up to 10 % of day-ahead sales value [8].

In this paper, we investigate temporal wise flexibility (storage, flexible loads) as one of the most promising resources to both increase relative market value and minimize balancing costs. Relative market value could be improved by shifting *expected* supply surpluses and resulting (day-ahead) sales into delivery slots with high prices. Balancing costs could be minimized by using flexibility to shift *unexpected* short or long positions arising from forecast errors. We address temporal flexibility by assessing the value created by demand response (DR)—i.e. flexible, responsive loads—when applied to either day-ahead marketing or intraday balancing of forecast errors for PV or wind power assets. Our interest for DR is based on its vast potential [6] and on the gap we suspect regarding science-based support for operators who face DR allocation decisions. Previous contributions have focused on either day-ahead or intraday use [1, 6], or modeled DR application as a unit commitment problem without explicitly trading off between markets [7]. Given the challenges that power markets face in light of the expansion of PV and wind power, a better understanding of the relative value of flexibility in real-world power markets is paramount.

Section 2 suggests a simple valuation logic for day-ahead and intraday DR applications. Section 3 compares both application strategies from an analytical and from a real data perspective for Germany. Section 4 provides an outlook on upcoming research steps and discusses limitations.

2   Value Formulation

We consider two strategies for DR: (1) a day-ahead application (denoted by DA) to increase sales revenue, and (2) an intraday application (denoted by ID) to minimize forecast error balancing costs. DR is modeled as L^t units of flexible load, to be understood as a share of (perfectly predictable) demand D^t for any time slot that can be shifted by r slots within a predetermined range $R^t \rightarrow t - r, \ldots, t + r$.[1] D^t is set below the day-ahead supply forecast Y^t_{DA} at all times.[2]

In strategy (1), the operator has different amounts of excess power available for delivery or sale across time slots within R^t. Value is derived from flexibility by shifting demand into times of low market prices, such that supply volumes are freed for selling at times of high prices. Assuming that flexible loads can be shifted from

---

[1] The amount of slots within R^t depends on the time span by which demand can be shifted and on the granularity by which time is dissected into slots (e.g., 15 min versus one hour).

[2] Note that these requirements are given in order to focus on the aspect under investigation—relative market value optimization for PV or wind. Relaxing the constraints is easily possible; however, it would introduce other effects, i.e. short day-ahead portfolios and demand volatility.

their initial consumption slot at time t into the least-cost available slot within the range R^t, $M \rightarrow \min P_{DA}^t \{R^t\}$, the value of units shifted day-ahead is defined by

$$V\left(L^t\right)_{DA} = L^t \times \left[P_{DA}^t - P_{DA}^M + \frac{c_{s\,DA}^M - c_{s\,DA}^t}{2} - \left(C_{DA}^O + C_{DA}^A\right) \right]. \quad (1)$$

It shows that the value of shifting loads away from t is determined by differences in price P and bid-ask spread (BAS) c_s between the current slot and the slot M. The costs of shifting loads arise from both opportunity costs of customers C_{DA}^O and variable activation costs C_{DA}^A. Evidently, the value of flexibility at slot t increases with the difference in day-ahead prices between t and M.

For strategy (2), valuation not only depends on prices, but also on forecast error dynamics. Depending on the forecast error sign, an operator is either short, and has to buy intraday for the current slot t, or he is long and can sell volumes intraday for the current slot at time t. In both settings, there are two options for the operator:

(a) to shift loads from the current short (long) position slot into another slot with a short (long) position within range R^t that offers a lower price, or
(b) to net (enhance) an open position, by moving loads from the current short (long) position slot into a slot with a long (short) position within R^t. This is only possible if R^t is "mixed", i.e. if there are slots with both long and short positions.

The valuation of (a) is similar to the day-ahead valuation. For long positions, the formula is identical to Eq. (1), except that intraday variables replace day-ahead variables. At short position slots, the only difference occurs regarding the BAS terms: instead of $(c_{s\,ID}^M - c_{s\,ID}^t)$, we have $(c_{s\,ID}^t - c_{s\,ID}^M)$. This is because the short position at time t is reduced, leading to BAS savings. Meanwhile, the short position at M is enhanced, leading to more BAS incurred at M.

The valuation of (b) calls for further differentiation. In the case of a *short position*, the current shortage is *netted* with a long position within R^t. To this end, the long position with the lowest corresponding market price is chosen: $N \rightarrow \min P_{ID}^t \{R^t | Y_{DA}^t < Y_{ID}^t\}$. We define the value of flexible load in this setting as

$$V\left(L^t\right)_{ID}^{Net} = L^t \times \left[\left(P_{ID}^t + \frac{c_{s\,ID}^t}{2} + TC \right) - \left(P_{ID}^N - \frac{c_{s\,ID}^N}{2} - TC \right) - \left(C_{ID}^O + C_{ID}^A\right) \right]. \quad (2)$$

The first term within the squared brackets refers to the value created by shifting loads away from the current short slot. Here, purchasing volumes at price P_{ID}^t with the corresponding (half) spread and transaction costs (TC) is avoided. The second term defines the value created by shifting these loads into the lowest-priced long slot within R^t. While revenues are foregone by reducing the volumes sold at price P_{ID}^N, a half spread and transaction costs for the sale are avoided.

A comparison between options (a) and (b) shows that, for short positions, it is beneficial to opt for option (b) whenever $[(c_{s\,ID}^N + c_{s\,ID}^M)/2 + 2TC + P_{ID}^M > P_{ID}^N]$. Just from the balance of variables, this seems to hold in the majority of cases. Indeed, market mechanisms make it likely that, on average, $P_{ID}^M > P_{ID}^N$, as long as the portfolio of the operator is sufficiently correlated with the portfolio of the other PV or wind power plant operators in the market. This is because the aggregate balance of market actors is reflected in the market prices. Hence, with correlation, prices are likely higher for slots with short positions than for slots with long positions.

In contrast, if the operator decides to shift flexible loads away from a long position towards a short position, he does not net but rather *enhance* positions. Since demand is reduced at time t, the supply surplus is even larger. Similarly, the shortage at the short slot to which demand is shifted, $E \to \min P_{ID}^t \{R^t | Y_{DA}^t > Y_{ID}^t\}$, is increased. Consequently, higher transaction costs are incurred at both the long and short position slots. We thus have the (already simplified) term

$$V\left(L^t\right)_{ID}^{\text{Enhance}} = L^t \times \left[P_{ID}^t - P_{ID}^E - \frac{c_{s\,ID}^t + c_{s\,ID}^E}{2} - 2TC - \left(C_{ID}^O + C_{ID}^A\right)\right]. \quad (3)$$

Applying a similar logic as before, we can show that applying option (b) to long position slots by shifting demand into a short position slot E only increases value if $\left[P_{ID}^E + (c_{s\,ID}^E + c_{s\,ID}^M)/2 + 2TC < P_{ID}^M\right]$. Market mechanics further imply that, on average, $P_{ID}^E > P_{ID}^M$. Thus, (b) will not be executed at many long position slots.

3 Value Comparison

Ignoring differences in prices, BAS, and (transaction) costs between day-ahead and intraday markets for now, we can derive analytically that the application of flexible demand creates greater value intraday. This can be shown by setting the day-ahead value equation equal to the intraday valuations for the different scenarios (see [3] for a detailed analysis). In summary, intraday application yields additional value due to the possibility to mitigate short positions and thus BAS and transaction costs. For long intraday positions, the most likely outcome is that value is equal to day-ahead application value. Consequently, the share of short position slots intraday drives the value advantage of intraday DR application. That share amounted to a remarkable 61 % for the dominating German operators in 2013.[3] Table 1 summarizes the results from setting DA = ID and eliminating variables where possible.

Our analytical comparison treats corresponding day-ahead and intraday parameters as equal. In practice, they differ, with the effect that intraday application of flexible demand will prove even more advantageous. Recalling Equations (1)–(3), we find four value determinants: prices, BAS, and other transaction costs increase the DR value, whereas DR costs lower it. Focusing on the value-increasing elements,

---

[3]In Germany, the bulk of PV and wind power production is currently integrated into the market by the four transmission system operators (TSOs). The 61 % short positions refer to actual production deviations from forecasts, at an aggreagate level (sum of TSOs) for 15-min delivery slots.

Table 1 Setting valuations DA = ID and reducing to differentiating variables

Slot positions in R^t	Current slot t: long	Current slot t: short
Homogeneous[a]	DA = ID → 0	$c_s{}^M = c_s{}^t$
(Option (a))		
Mixed[a]	$-P_{DA}^M + c_{sDA}^M/2 =$	$-P_{DA}^M + c_{sDA}^M/2 =$
(Option (b))	$-P_{ID}^E - c_{sID}^E/2 - 2TC$	$-P_{ID}^N + c_s{}^t + c_{sID}^N/2 + 2TC$

[a]Homogeneous ranges only include either short or long positions. Mixed ranges include both

prices have the largest impact. BAS constitutes a fraction of price; other transaction costs for trading amount to only a fraction of the BAS and can thus be neglected.

When investigating market data (EPEX) for Germany in 2013, we find numerous indications that intraday price dynamics increase the value of DR more than day-ahead dynamics. First, absolute prices of hourly deliveries average higher intraday (38.5 €/MWh versus 37.9 €/MWh), while variance is much higher (+20%). For an exemplary range size of $R^t = 5$,[4] intraday variance exceeds variance day-ahead for the same range regarding the 2013 average value (68.9 versus 59.5, +16%). Further, the spread between the highest and the lowest price intraday exceeds the day-ahead spread by more than 9% (16.0 versus 14.6). All of these measures support the notion that intraday price volatility is higher, and that absolute price movements are larger as well. With respect to our valuations, this implies greater DR value intraday than day-ahead. Another very important aspect should be mentioned here: while day-ahead trading is possible for hourly delivery and longer slots, intraday balancing of forecast errors is also commonly conducted at 15-min granularity. When we leave the time distance by which demand can be shifted unchanged, but assume 15-min granularity, we have a range of $R^t = 17$.[5]Volatility then significantly exceeds the values for hourly slots and $R^t = 5$, with variance amounting to 292.2 (versus 43.9 for day-ahead hours). The difference between the highest and lowest price averages 51.1 (versus 14.6 day-ahead). All in all, using DR when balancing forecast errors at 15-min granularity appears much more valuable than any other option.

While BAS dynamics are hard to quantify and less relevant for DR valuation than prices, we can expect them to also boost the value of intraday DR application. The BAS is much larger intraday than day-ahead. [4] find an intraday BAS of 3 €/MWh versus 0.25 €/MWh day-ahead. Given the higher absolute BAS values, and the higher volatility of prices intraday, it seems plausible to assume BAS volatility to be higher intraday as well. Another factor is that, derived from our analytical comparison, DR value is enhanced more through BAS effects intraday than day-ahead (i.e., avoidance of BAS through netting).

---

[4]This means that demand can be brought forward or postponed by up to two hours.

[5]One hour covers four 15-min intervals. Considering the two hours prior and after the current delivery hour, we get a total of 17 slots, including the current slot.

4 Conclusion

This paper addressed the economic benefits of access to flexible demand for operators of PV and wind power plants. Two alternative allocation strategies were investigated: (1) the use of DR to improve market value in day-ahead sales, and (2) the use of DR to optimize the balancing of forecast errors intraday. Both from an analytical and from a data perspective, we find evidence for the advantages of using flexible demand in the latter setting (2). In light of these findings, we advocate a more explicit consideration of intraday dynamics in future research on DR decision support and resource allocation. As a contribution to this, in a more detailed paper than the present one [3], we aim at integrating DR allocation strategies into a previously developed intraday bidding strategy [2] to holistically address PV or wind forecast errors under uncertainty. A caveat of our study is the lack of an analysis of DR activation costs. While this is on our research agenda, it cannot be addressed analytically and requires another, yet to be developed approach.

References

1. Feuerriegel, S., Neumann, D.: Measuring the financial impact of demand response for electricity retailers. Energy Policy **65**, 359–368 (2014). doi:10.1016/j.enpol.2013.10.012
2. Garnier, E., Madlener, R.: Balancing forecast errors in continuous-trade intraday markets. FCN Working Paper No. 2/2014, RWTH Aachen University (2014). doi:10.2139/ssrn.2463199
3. Garnier, E., Madlener, R.: Day-ahead versus intraday valuation of demand-side flexibility for photovoltaic and wind power systems. FCN Working Paper No. 17/2014, RWTH Aachen University (2014). doi:10.2139/ssrn.2556210
4. Hagemann, S, Weber, C.: An empirical analysis of liquidity and its determinants in the German intraday market for electricity. EWL Working Papers 1317, University of Duisburg-Essen (2013). http://ideas.repec.org/p/dui/wpaper/1317.html
5. Hirth, L.: The market value of variable renewables: the effect of solar wind power variability on their relative price. Energy Economics **38**, 218–236 (2013)
6. Klobasa, M.: Analysis of demand response and wind integration in Germany's electricity market. IET Renew. Power Gener. **4**(1), 55–63 (2010)
7. Madaeni, S.H., Sioshansi, R.: The impacts of stochastic programming and demand response on wind integration. Energy Syst. **4**(2), 109–124 (2013). doi:10.1007/s12667-012-0068-7
8. Von Roon, S.: Empirische Analyse über die Kosten des Ausgleichs von Prognosefehlern der Wind- und PV-Stromerzeugung. Wien, 7. Internationale Energiewirtschaftstagung an der TU Wien, 16–18 Feb 2011

A Real Options Model for the Disinvestment in Conventional Power Plants

Barbara Glensk, Christiane Rosen and Reinhard Madlener

Abstract The liberalization of the energy market and the promotion of renewables lead to difficulties in the profitable operation even of many modern conventional power plants. Although such state-of-the-art plants are highly energy-efficient, they are often underutilized or even mothballed. Decisions about further operation or shut-down of these conventional power plants are in most cases characterized as being irreversible, implying uncertainty about future rewards, and being flexible in timing. A useful approach for evaluating (dis-)investment projects with uncertainties is the real options approach (ROA) [2, 14]. This valuation technique is based on option pricing methods used in finance that have been developed by Black, Scholes, and Merton [1, 11]. In the last two decades, real options models have been widely applied to analyze investment decisions under dynamic market conditions. In recent years, however, also the analysis of disinvestment decisions considering market uncertainties has gained in importance (e.g. in studies on the agricultural and dairy sector). Moreover, ignoring disinvestment options in decision-making processes can lead to incorrect valuations of investment strategies at the firm level. In this paper, we develop a real options model for the disinvestment in conventional power plants, with the aim of determining the optimal timing for the shut-down of unprofitable power plants.

B. Glensk · C. Rosen (✉) · R. Madlener
Institute for Future Energy Consumer Needs and Behavior (FCN),
School of Business and Economics / E.ON Energy Research Center,
RWTH Aachen University, Mathieustrasse 10,
52074 Aachen, Germany
e-mail: CRosen@eonerc.rwth-aachen.de

B. Glensk
e-mail: BGlensk@eonerc.rwth-aachen.de

R. Madlener
e-mail: RMadlener@eonerc.rwth-aachen.de

© Springer International Publishing Switzerland 2016
M. Lübbecke et al. (eds.), *Operations Research Proceedings 2014*,
Operations Research Proceedings, DOI 10.1007/978-3-319-28697-6_25

1 Introduction

The liberalization of energy markets has increased the sources of uncertainty. In particular, electricity producers today face market risks (regarding future demand and supply and thus also prices) and also regulatory risks related to uncertainty about the future legal environment impacting electricity generation activities.

The traditional school of economic thought proposes the net present value (NPV) criterion to evaluate investment decisions under uncertain future market conditions. However, such a static approach does not provide sufficient guidance regarding the estimation of the expected streams of profits, inflation, or discount rates. In particular, the NPV criterion does not properly capture any existing managerial flexibility to adapt decisions dynamically to unexpected market developments. Moreover, dissatisfaction with the NPV criterion by academics as well as corporate practitioners has been an important motivation for the development of new project valuation methods.

A relatively new approach for evaluating (dis-)investment projects under uncertainty is the real options approach (ROA) [2, 14]. Hereby, the use of continuous time stochastic processes enables modeling of uncertain cash flows, prices, as well as returns or asset values. Methodologically, the approach builds upon option pricing theory (options in financial securities) by applying it to non-financial, i.e. physical (or "real") assets viewed as investment options. Several kinds of options can be implemented, such as simple timing options with the option to invest or abandon the project, compound timing options, and switching or learning options (for more details see, e.g. [7, 12]). Regarding the investment decisions in the energy sector, the option to invest is one of the most popular ones, whereas the options to delay, expand or abandon a project support the definition of an optimal policy decision. Fleten and Näsäkkälä [6] use a real options model to analyze whether it is sensible for an energy company to build a gas-fired power plant (vis-a-vis a biofuel plant). On the other hand, sequential modular investments in the energy sector are discussed in Jain et al. [9]. Unfortunately, real options models for disinvestments have so far only been discussed in a few applied articles, especially in the fields of agriculture [13], dairy [5], and production planning [4]. From a dynamic perspective, the research question addressed here is at what point in time an energy company should shutdown an existing power plant that does not, or no longer, generate any profits. This disinvestment decision is related to the market situation and, therefore, the capacity factor, i.e. the achieved output divided by the potential output, for a specified period of time (in our case full-load hours in one year).

2 Model Specification

In general, a real options model is based on three factors, namely the existence of uncertainty about future cash flows, investment irreversibility, and flexible timing regarding project initiation or (as in our case) project abandonment. In our approach, we discretize the problem and set up a discrete-valued lattice, for which we apply a dynamic programming model, developed to numerically value the investment decision, that is solved by backward induction. The uncertain capacity factor (number of full-load hours) of power plants serves as the stochastic variable (the underlying risky asset). Furthermore, it is assumed that the capacity factor is normally distributed and then approximated by a binomial distribution, allowing for the use of the standard binomial lattice approach. The binomial lattice is just one of several real options solution approaches (such as closed-form solutions, partial-differential equations, finite-differences, or simulations). Due to its better tractability, it has been more widely accepted by the industry than others. This also means that when using this approach for the real options analysis, the procedures and results can be more easily explained to and accepted by executives. The binomial approach specifies how the underlying assets change over time. It means that from the current state only two future states are possible, so-called "up" and "down" movements corresponding to good and bad market (situation) development. The "up" and "down" movements as well as risk-neutral probabilities can be calculated using the distribution parameters from the underlying asset [9, 12]. We further anticipate that at the beginning of the investigated period the expected discounted future cash flows are significantly higher than the residual value of the power plant. This difference should then decrease over time. The proposed method consists of the following steps:

1. Based on the assumed normal distribution of the capacity factor, the "up" and "down" movements are determined as follows:

$$up = e^{(\sigma\sqrt{\Delta t})} \text{ and } down = e^{(-\sigma\sqrt{\Delta t})} \tag{1}$$

where σ is the associated volatility, and Δt is the time step. The "up" and "down" movements are subsequently used to set up the binomial tree.

2. The future cash-flow values of the existing project (the power plant) in each period are calculated for different values of the capacity factor obtained in step 1. In order to account for the stochastic character of some of the cash-flow elements, a Monte Carlo simulation is employed.

3. The optimal project value as a function of the capacity factor, $PV_{i,t}(CF_{i,t})$, is given by

$$PV_{i,t}(CF_{i,t}) = max \begin{cases} RV_t \\ PCF_{i,t} + \frac{\alpha \cdot PV_{i,t+1}(CF_{i,t+1}) + (1-\alpha) \cdot PV_{i+1,t+1}(CF_{i+1,t+1})}{1+r_f} \end{cases} \tag{2}$$

where RV_t denotes the residual value, $PCF_{i,t}$ the project cash flow for the ith "down" move at current time period t, α defines the probability for an "up" movement, $CF_{i,t}$ denotes the capacity factor for the ith "down" move at time t, r_f the risk-free rate, and i is the number of "down" movements ($i = 1 \ldots T - 1$). The risk-neutral probability α is calculated according to the formula[1]: $\alpha = \frac{K-down}{up-down}$, where $K = E[CF] - (E[R_M] - r_f)\beta$, $E[CF]$ is the expected

---

[1] Note that the underlying asset is not a price of a traded asset (for more information see [7]).

proportional change in the state variable, $E[R_M]$ is the expected return on the market portfolio, and β is the beta coefficient. The optimal project value is then calculated using recursive dynamic programming. As soon as $PV_{i,t}$ is equal to the residual value (RV_t), the power plant should be shut down. In contrast, if the optimal project value is equal to the second part of Eq. (2), the power plant should be kept in operation.

3   Case Study and Results

The case study presented here considers one of the highly efficient and recently built gas-fired power plants in Germany. While such power plants are environmentally friendly in terms of their efficiency, their CO_2 emissions, and their ability to operate flexibly (and thereby to counter the fluctuating generation of renewables), they are economically not viable. This is due to two reasons: On the one hand, the gas price has developed unfavorably in the past years and on the other hand the electricity prices at the wholesale market have decreased significantly. In combination with the increasing share of renewable energy sources, such as wind and solar power, with extremely low marginal costs, this leads to the so-called merit order effect [10]. This effect describes the fact that power plants with larger marginal costs, such as gas-fired power plants, are not dispatched any longer because they are pushed to the right on the merit order curve. Their operation becomes unprofitable and often the remaining option is to liquidate the plant altogether. The important question hereby is at what point in time the operation of the power plant should be given up.

The maximum time period during which the decision should be made is assumed to be six years. Note that the capacity factor and its stochastic values influence the plant's output and thus also the current value of the power plant, thus playing a crucial role for the analysis. In the model, both the capacity factor and the electricity, gas, and CO_2 prices (the latter obtained from EEX databases, the former from [3]) are represented as stochastic variables with corresponding probability distributions.

The power plant analyzed was commissioned in year 2010, with a net installed capacity of 845 MW. Its net thermal efficiency is 59.7%, and the total investment volume 400 million Euros [3]. Further assumptions considering technical characteristics as well as economic parameter values are based on expert interviews and a thorough literature review (more detailed information can be obtained from the authors upon request). The analytical procedure used to compute the optimal project values is based on the methodology presented in the previous section.

Based on the existing literature, and available online data regarding full-load hours of gas-fired power plants, the impact of different values (distributions) of the capacity factor on the optimal project value and the final decision are analyzed. The results of three different values for the capacity factor and emerging decisions are presented in Table 1.

From the results presented here, one can see that the higher the capacity factor value is, the longer is the time period during which the power plant can remain in

Table 1 Value of capacity factor and corresponding decision (C—continue or S—stop operation)

Period	Capacity factor	Decision	Capacity factor	Decision	Capacity factor	Decision
1	0.3000	C	0.1100[a]	S	0.1500	C
2	0.3154	C	0.1133	S	0.1546	C
2	0.2854	C	0.1067	S	0.1456	C
3	0.3316	C	0.1168	S	0.1593	C
3	0.3000	C	0.1100	S	0.1500	C
3	0.2715	C	0.1036	S	0.1413	C
4	0.3486	C	0.1204	S	0.1641	C
4	0.3154	C	0.1133	S	0.1546	C
4	0.2854	C	0.1067	S	0.1456	C
4	0.2582	C	0.1005	S	0.1371	S
5	0.3664	C	0.1240	S	0.1691	C
5	0.3316	C	0.1168	S	0.1593	C
5	0.3000	C	0.1100	S	0.1500	C
5	0.2715	C	0.1036	S	0.1413	S
5	0.2456	C	0.0976	S	0.1330	S
6	0.3852	C	0.1278	S	0.1743	C
6	0.3486	C	0.1204	S	0.1641	C
6	0.3154	C	0.1133	S	0.1546	C
6	0.2854	C	0.1067	S	0.1456	C
6	0.2582	C	0.1005	S	0.1371	S
6	0.2336	C	0.0947	S	0.1291	S
7	0.4050[a]	S	0.1317	S	0.1796[a]	S
7	0.3664	S	0.1240	S	0.1691	S
7	0.3316	S	0.1168	S	0.1593	S
7	0.3000	S	0.1100	S	0.1500	S
7	0.2715	S	0.1036	S	0.1413	S
7	0.2456	S	0.0976	S	0.1330	S
7	0.2222	S	0.0919	S	0.1253	S

[a] Stopping values (more information on this can be found in [8])

service. Nevertheless, it should be mentioned that the residual value defined in Eq. (2) also impacts the final decision regarding how long the power plant's operation can be continued. In this case study, the residual value is assumed to be constant, computed by accounting for linear depreciation; a more sophisticated calculation of the *RV* variable will be an important next step in the further development of the model.

4 Conclusion

The increased use of technologies for renewable electricity production has a significant impact on the merit order of power plant dispatch, and leads to difficulties in the profitable operation of many highly energy-efficient conventional power plants. Due to these circumstances, numerous power plant operators today are forced to revise their strategy and decide about the continued operation, mothballing, or shut-down of their conventional power plants.

In this short paper, we presented a real options model which can support such a decision-making process. We find that it is highly dependent on the initial capacity factor and its subsequent development. This also means that the final decision is path-dependent. For this reason, the results suggest the stopping of the plant's operation at very different values of the capacity factor. For a starting value of 0.300, operation should be ceased at a value of 0.405, whereas for a starting value of 0.110, operation should be stopped immediately. In contrast, for a more moderate starting value of 0.150, operation should be stopped at a value of about 0.180. The higher threshold values for stopping seem paradoxical, but result from the upward movements in the binomial tree, which nonetheless have unfavorable prospects.

We plan to further investigate the residual value of the power plant, including a thorough analysis aimed at determining its exact value. Further sensitivity testing of the other parameters is also planned to check the robustness of the model. Future research will be dedicated to evaluate whether a temporary shut-down instead of a complete disinvestment of the power plant is economically reasonable.

References

1. Black, F., Scholes, M.: The pricing of options and corporate liabilities. J. Polit. Econ. **81**(3), 637–654 (1973)
2. Dixit, A.K., Pindyck, R.S.: Investment under Uncertainty. Princeton University Press, Princeton (1994)
3. E. ON, http://www.eon.com/de/ueber-uns/struktur/asset-finder/irsching.html. Accessed 10 July 2014
4. Fontes, D.B.M.M.: Fixed versus flexible production systems: a real options analysis. Eur. J. Oper. Res. **188**(1), 169–184 (2008)
5. Feil, J-H., Musshoff, O.: Investment, disinvestment and policy impact analysis in the dairy sector: a real options approach. SiAg-Working Paper 16, DFG-Forschergruppe 986, Humboldt-Universität zu Berlin (2013)
6. Fleten, S.-E., Näsäkkälä, E.: Gas-fired power plants: investment timing, operating flexibility and CO_2 capture. Energy Econ. **32**(4), 805–816 (2010)
7. Guthrie, G.: Real Options in Theory and Practice. Oxford University Press, New York (2009)
8. Glensk, B., Rosen, C., Madlener, R.: A Real Options Model for the Disinvestment in Conventional Power Plants. FCN Working Paper (in prep.), RWTH Aachen University (2015)
9. Jain, S., Roelofs, F., Oosterlee, C.W.: Valuing modular nuclear power plants in finite time decision horizon. Energy Econ. **36**, 625–636 (2013)

10. Lohwasser, R., Madlener, R.: Simulation of the European electricity market and CCS development with the HECTOR model. FCN Working Paper No. 6/2009, RWTH Aachen University, November (2000)
11. Merton, R.C.: Theory of rational option pricing. Bell J. Econ. Manag. Sci. **4**(1), 141–183 (1973)
12. Mun, J.: Real Options Analysis: Tools and Techniques for Valuing Strategic Investment and Decisions. Wiley, Hoboken (2006)
13. Musshoff, O., Odening, M., Schade, C., Maart-Noelck, S.C., Sandri, S.: Inertia in disinvestment decisions: experimental evidence. Eur. Rev. Agric. Econ. **40**(3), 463–485 (2012)
14. Schwartz, E.S., Trigeorgis, L.: Real Options and Investment under Uncertainty: Classical Readings and Recent Contributions. The MIT Press, Cambridge (2001)

Political Districting for Elections to the German Bundestag: An Optimization-Based Multi-stage Heuristic Respecting Administrative Boundaries

Sebastian Goderbauer

Abstract According to the legal requirements for Elections to the German Bundestag the problem of partitioning Germany into electoral districts can be formulated as a multi-criteria graph partition problem. To solve this regularly current problem, an optimization-based heuristic is introduced and successfully applied to German population data.

1 Electoral Districts in Elections to the German Bundestag

In general, the election to the German federal parliament, the Bundestag, takes place every four years. The 299 electoral districts play an important role in those elections. In fact, the voters of each district elect one representative into parliament ensuring that each part of the country is represented. These elected representatives make up half of the members of the Bundestag. The allocation of electoral districts needs regular updates due to an ever-changing population distribution and is subject to a variety of legal requirements as listed in the following.

In order to comply with the principle of electoral equality as anchored in the German constitution, the differences in population between the districts have to be preferably small. The law defines a tolerance limit, saying that the amount of deviation from the average district population should not exceed 15 %. Moreover, an amount of deviation beyond 25 % is illegal. Every district should be a contiguous area and it is prefered that its allocation aligns with existing administrative boundaries. In addition to that, the law demands that the districts strictly comply with the borders of the German federal states. The law specifies that the Sainte-Laguë method [7] has to be used to distribute the 299 districts among the 16 states. The electoral districts ought to be visually compact counteracting the suspicion of applying Gerrymandering [6, 9, 11]. In the context of setting electoral districts, Gerrymandering is a practice

S. Goderbauer (✉)
RWTH Aachen University, Operations Research, Kackertstrasse 7,
52072 Aachen, Germany
e-mail: goderbauer@or.rwth-aachen.de; sebastian.goderbauer@rwth-aachen.de

© Springer International Publishing Switzerland 2016
M. Lübbecke et al. (eds.), *Operations Research Proceedings 2014*,
Operations Research Proceedings, DOI 10.1007/978-3-319-28697-6_26

that attempts to create an advantage or disadvantage for a certain political party or candidate by manipulating district boundaries.

In this contribution, which is an extended abstract of the author's master's thesis, the problem of dividing a country into electoral districts is defined as a multi-criteria graph partition problem. To solve this regularly current practical problem, an optimization-based multi-stage heuristic is introduced and successfully applied to population data of the latest German census [12]. The computed results show that the presented algorithm allocates electoral districts, which are not only in accordance with the law, but also fulfill the tolerances mentioned in the law more closely than the current districting.

2   The Political Districting Problem

The Political Districting Problem is defined on the basis of a so-called population graph. In a population graph

$$G = (V, E)$$

a node $i \in V$ represents an geographical area, e.g., the area of a municipality, and is weighted with its population p_i. An undirected edge $(i, j) \in E$ with nodes $i, j \in V$ exists, iff the corresponding areas share a border. The Political Districting Problem is an optimization problem in which a node-weighted population graph has to be partitioned into a given number of connected, weight-restricted subgraphs.

More precisely, let S be the set of all 16 German states. Given the total number of electoral districts $d \in \mathbb{N}$, which has to be set, the number of districts $d(s) \in \mathbb{N}$ for each state $s \in S$ is computable with the mentioned Sainte-Laguë method [7]. Of course, $\sum_{s \in S} d(s) = d$ holds. Furthermore let $\varnothing_p := \frac{1}{d} \sum_{i \in V} p_i$ be the average population of an electoral district. Finally, a partition

$$D_k \subseteq V, k = 1, \ldots, d \quad \text{with} \quad D_l \cap D_m = \emptyset, \ l \neq m \quad \text{and} \quad \cup_k D_k = V$$

is called a feasible solution (districting) for the Political Districting Problem, if the following holds:

$$\forall 1 \leq k \leq d \quad \forall i, j \in D_k : i \text{ and } j \text{ are in same state,} \tag{1}$$

$$\forall s \in S : |\{D_k : \text{state } s \text{ contains district } D_k\}| = d(s), \tag{2}$$

$$\forall 1 \leq k \leq d : G[D_k] \text{ connected,} \tag{3}$$

$$\forall 1 \leq k \leq d : 0.75\varnothing_p \leq \sum_{i \in D_k} p_i \leq 1.25\varnothing_p. \tag{4}$$

Fig. 1 The population graph of North Rhine-Westphalia (NRW) consists of 396 nodes and 1 084 edges at municipality level. All in all, there are 11 339 municipalities in Germany, thus NRW's population graph is one of the smaller ones

To complete the definition and as a result of analyzing the legal requirements, the multi-criteria objective is as follows:

$$\mathbf{max}\ |\{D_k\ :\ 1 \le k \le d \text{ and } 0.85\varnothing_p \le \sum_{i \in D_k} p_i \le 1.15\varnothing_p\}|, \tag{5}$$

$$\mathbf{min}\ \text{amount of deviations between district population} \sum_{i \in D_k} p_i \text{ and } \varnothing_p, \tag{6}$$

max match between district and existing administrative boundaries, (7)

max geographical and visual compactness of the districts. (8)

To obtain the population graph of Germany and thus the required graphs for each German state, data of the latest German census [12] were combined with geoinformation [8] (cf. Fig. 1).

The Political Districting Problem includes incomparable and conflicting objective criteria. Analyzing the complexity of the subproblems, in which only one objective is considered, Altman [1] concludes the complexity of the Political Districting Problem.

Theorem 1 *The Political Districting Problem is NP-hard.*

To gain a profound understanding of the complexity, it is possible to analyze the graph partition problems on which the Political Districting Problem is based [2, 4, 5]. It is possible to compute feasible districtings in polynomial or even linear time on paths and special trees. Suitable partitions with minimal differences in population between the components can be found in polynomial time as well. Considering general trees a feasible districting is computable in polynomial time, but the problem gets NP-hard by adding the objective of minimizing the differences in population.

3 Optimization-Based Multi-stage Heuristic

Since the 1960 s the Political Districting Problem has been discussed and approached by many authors in operations research and social science. In 1961 Vickrey [9] provided a multi-kernel growth procedure. Although his proposal was rather informal and rudimentary it marked the start of a large variety of work on this topic. Hess et al. [3] considered the problem of setting electoral districts as a modified facility location problem in 1965. The technique of column generation was applied by Mehrotra et al. [6] in 1998. In 2009 Yamada [10] formulated the problem as a spanning forest problem and presented a local search based heuristic.

Most heuristics and exact methods in the literature implement the requirements and objectives of the German Political Districting Problem only partially, e.g., support for matching district boundaries with existing administrative bounderies is disregarded. Beyond that, the numbers of nodes and edges in the population graphs of most German states outnumber all graph orders and sizes considered in the literature. The optimization-based heuristic described hereafter was developed to overcome these shortcomings. The multi-stage algorithm uses the existing hierarchical administrative divisions in Germany (cf. Fig. 2) and iteratively divides the Political Districting Problem into smaller subproblems. This has two advantages: The goal of aligning electoral district boundaries with existing administrative bounderies is realizable in an adequate way and in addition to that, the graphs of the subproblems will be of manageable size.

1st stage: states As mentioned above, it is required by law to align the districts with the boundaries of the German states. Therefore, a union of electoral districts for the individual states is a solution of the Political Districting Problem for Germany.

2nd stage: governmental districts The four most highly populated states are composed of so called governmental districts. For those states, the number of electoral districts of a state is distributed over the governmental districts by reapplying the Sainte-Laguë method [7]. With regard to the final solution, this is mostly a good and valid choice. If it is not valid, two neighbouring governmental districts are merged and seen as one in the application of the Sainte-Laguë method [7].

3rd stage: rural districs, urban districts Subsequently, the population graphs on rural and urban district level of a state (or governmental district) are considered.

Fig. 2 The multi-stage algorithm is based on the structure of the hierarchical administrative divisions of Germany

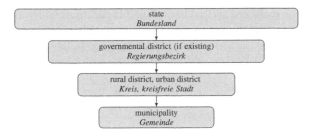

(a) **(b)**

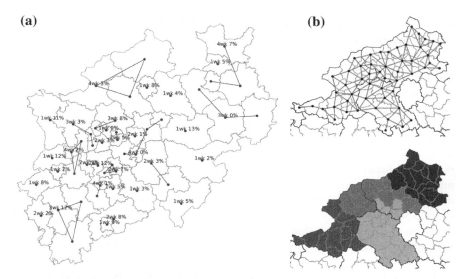

Fig. 3 **a** A solution of the set partitioning problem at the third stage for North Rhine-Westphalia. **b** The population graph on municipality level of the rural districts Borken, Coesfeld, and Steinfurt (*top*) and the allocation of four electoral districts after applying the heuristic (*bottom*)

Intuitively, each node represents a rural or urban district. On this population graph a modified set partitioning problem is solved: A graph partition into connected subgraphs is computed and a number of electoral districts is assigned to each subgraph. The sum of those numbers has to be equal to the number of districts assigned to that state (or governmental district) (cf. Fig. 3a). As the major part of the objective in this set partitioning problem the resulting average differences in district population are minimized. Components and thus subproblems with exactly one electoral district are solved, because the electoral district is already set.

4th stage: municipalities Subproblems which are still open after the third stage are solved at the municipality level. For setting the remaining electoral districts almost all algorithms from the literature can be used, because the problem size is at this point mostly manageable. In this work a simple heuristic was implemented (cf. Fig. 3b).

The districtings computed by the optimization-base multi-stage heuristic are useable for the elections to the German Bundestag in general, because latest population data was used and the heuristic respects all legal requirements. It is worth to note that the developed algorithm supports matches between existing administrative boundaries and electoral district boundaries. This aspect is hard to implement in a compact formulation of the problem. Beside this, the computed districts follow the mentioned objectives transferred from the law more closely than the current districting applied in elections to the Bundestag in 2013.

4   In Search of an Optimal Number of German Electoral Districts

Currently, Germany is divided into 299 districts for elections to the Bundestag. The question arises if that is a well chosen number of electoral districts. In the following, one option of approaching this issue is outlined.

The Sainte-Laguë method [7] distributes the districts between the states on the basis of the state's population. However, the population of a state is usually not an integer multiple of the average electoral district population, thus differences in population between the districts are unavoidable. A question is which realizable number of German electoral districts causes the lowest maximal amount of average deviation in district population in a state. Considering the distribution of 299 electoral districts with the Sainte-Laguë method [7], the state Bremen has the highest average deviation of 17.2 %. This value is greater than the tolerance limit of 15 % specified in the law. However, when Germany is divided into 242 districts, the maximal average deviation is reduced to a mere 5.2 % (Bremen again). This value is the minimum in the range between 1 and 376 distributed districts. It can therefore be concluded, that 242 districts would observe the rules of the law more closely than the current choice of 299 districts. In other words, 242 districts embody the German population distribution between the 16 states better than 299. Interestingly, when the number of districts is increased to 319 no legal districting can be found as the maximum deviation limit of 25 % would always be exceeded in Bremen.

References

1. Altman, M.: Is automation the answer? The computational complexity of automated redistricting. Rutgers Comput. Law Technol. J. **23**(1), 81–142 (1997)
2. De Simone, C., Lucertini, M., Pallottino, S., Simeone, B.: Fair dissections of spiders, worms, and caterpillars. Networks **20**(3), 323–344 (1990)
3. Hess, S.W., Weaver, J.B., Siegfeldt, H.J., Whelan, J.N., Zitlau, P.A.: Nonpartisan political redistricting by computer. Oper. Res. **13**(6), 998–1006 (1965)
4. Ito, T., Zhou, X., Nishizeki, T.: Partitioning a graph of bounded tree-width to connected subgraphs of almost uniform size. J. Discret. Algorithm. **4**(1), 142–154 (2006)
5. Lucertini, M., Perl, Y., Simeone, B.: Most uniform path partitioning and its use in image processing. Discret. Appl. Math. **42**(2–3), 227–256 (1993)
6. Mehrotra, A., Johnson, E.L., Nemhauser, G.L.: An optimization based heuristic for political districting. Manag. Science. **44**(8), 1100–1114 (1998)
7. Sainte-Laguë, A.: La représentation proportionnelle et la méthode des moindres carrés. Annales scientifiques de l'École Normale Supérieure. **27**, 529–542 (1910)
8. Shapefile der Verwaltungsgrenzen: Bundesamt für Kartographie und Geodäsie. (2013)—http://www.zensus2011.de, http://www.bkg.bund.de. Cited 30 Jul 2014
9. Vickrey, W.S.: On the prevention of gerrymandering. Pol. Sci. Quar. **76**(1), 105–110 (1961)

10. Yamada, T.: A mini-max spanning forest approach to the political districting problem. Int. J. Syst. Sci. **40**(5), 471–477 (2009)
11. Young, H.P.: Measuring the compactness of legislative districts. Legis. Stud. Q. **13**(1), 105–115 (1988)
12. Zensus 2011: Bevölkerung—Ergebnisse des Zensus am 9. Mai 2011. Statistisches Bundesamt, Wiesbaden. (2014)—http://www.zensus2011.de. Cited 30 Jul 2014

Duality for Multiobjective Semidefinite Optimization Problems

Sorin-Mihai Grad

Abstract In this note we introduce a new multiobjective dual problem for a given multiobjective optimization problem consisting in the vector minimization with respect to the corresponding positive semidefinite cone of a matrix function subject to both geometric and semidefinite inequality constraints.

1 Introduction and Preliminaries

Matrix functions play an important role in optimization especially in connection to the cone of symmetric positive semidefinite matrices which induces the Löwner partial ordering on the corresponding space of symmetric matrices. Such functions were used mainly in scalar optimization as constraint or penalty functions, but one can find contributions to vector optimization involving them, for instance [8], where convex multiobjective optimization problems subject to semidefinite constraints were considered, or [5, 6], where multiobjective optimization problems consisting in vector minimizing matrix functions with respect to the corresponding cone of the symmetric positive semidefinite matrices under semidefinite constraints were investigated. Motivated by them and by the vector dual problems inspired by [7] we assigned in [1, 2, 4] to linear vector optimization problems, we propose in this note a new multiobjective dual for multiobjective optimization problems similar to the ones from [5, 6] mentioned above.

We denote the set of the *symmetric* $k \times k$ real matrices by $\mathcal{S}^k$. The cone of the *positive semidefinite* symmetric $k \times k$ matrices is $\mathcal{S}^k_+$, while its interior, the set of the *positive definite* symmetric $k \times k$ matrices is $\hat{\mathcal{S}}^k_+$. The *entries* of a matrix $A \in \mathbb{R}^{k \times k}$ will be denoted by A_{ij}, $i, j = 1, \ldots, k$, while its *trace* by $\mathrm{Tr}\, A$. The *Löwner partial ordering* induced by $\mathcal{S}^k_+$ on $\mathcal{S}^k$ is "$\leqq$", defined by $A \leqq B \Leftrightarrow B - A \in \mathcal{S}^k_+$, where $A, B \in \mathcal{S}^k$. When $A \leqq B$ and $A \neq B$ we write "$A \precneqq B$". Recall that the cone $\mathcal{S}^k_+$ is

S.-M. Grad (✉)
Faculty of Mathematics, Chemnitz University of Technology,
D-09107 Chemnitz, Germany
e-mail: grad@mathematik.tu-chemnitz.de

© Springer International Publishing Switzerland 2016   189
M. Lübbecke et al. (eds.), *Operations Research Proceedings 2014*,
Operations Research Proceedings, DOI 10.1007/978-3-319-28697-6_27

self-dual and the *Frobenius inner product* of two matrices $A, B \in \mathcal{S}^k$ is defined as $\langle A, B \rangle = \mathrm{Tr}(A^\top B)$.

Given a set $U \subseteq \mathbb{R}^k$, $\mathrm{ri}\,U$ denotes its *relative interior* of U, while δ_U is its *indicator function*. For a function $f : \mathbb{R}^n \to \overline{\mathbb{R}}$ we use the classical notations for its *domain* dom $f = \{x \in X : f(x) < +\infty\}$ and *epigraph* epi $f = \{(x, r) \in X \times \mathbb{R} : f(x) \leq r\}$. The *conjugate function* of f is $f^* : \mathbb{R}^n \to \overline{\mathbb{R}} = \mathbb{R} \cup \{\pm\infty\}$, $f^*(y) = \sup\{y^\top x - f(x) : x \in \mathbb{R}^n\}$. Between a function and its conjugate there is the *Young-Fenchel inequality* $f^*(y) + f(x) \geq y^\top x$ for all $x, y \in \mathbb{R}^n$. A matrix function $H : \mathbb{R}^n \to \mathcal{S}^k$ is said to be $\mathcal{S}_+^k$-convex if $H(tx + (1-t)y) \leq tH(x) + (1-t)H(y)$ for all $x, y \in \mathbb{R}^n$ and all $t \in [0, 1]$.

2 Duality for Multiobjective Semidefinite Optimization Problems

Let the nonempty set $S \subseteq \mathbb{R}^n$ and the matrix functions $F : \mathbb{R}^n \to \mathcal{S}^k$ and $H : \mathbb{R}^n \to \mathcal{S}^m$. For $i, j \in \{1, \ldots, k\}$, denote by $f_{ij} : \mathbb{R}^n \to \mathbb{R}$ the function defined as $f_{ij}(x) = (F(x))_{ij}$. The primal *multiobjective semidefinite optimization problem* we consider is

(PVS)
$$\mathrm{Min}_{x \in \mathcal{A}}\, F(x),$$

where

$$\mathcal{A} = \left\{x \in S : H(x) \in -\mathcal{S}_+^m\right\},$$

where the vector minimization is done with respect to the cone $\mathcal{S}_+^k$.

An element $\bar{x} \in \mathcal{A}$ is said to be an *efficient solution* to (PVS) if there exists no $x \in \mathcal{A}$ such that $F(x) \preceq F(\bar{x})$, and the set of all the efficient solutions to (PVS) is denoted by $\mathcal{E}(PVS)$. An element $\bar{x} \in \mathcal{A}$ is said to be a *properly efficient solution* to (PVS) (in the sense of linear scalarization) if there exists a $\Lambda \in \hat{\mathcal{S}}_+^k$ such that $\mathrm{Tr}(\Lambda^\top F(\bar{x})) \leq \mathrm{Tr}(\Lambda^\top F(x))$ for all $x \in \mathcal{A}$, and the set of all the properly efficient solutions to (PVS) (in the sense of linear scalarization) is denoted by $\mathcal{PE}_{LS}(PVS)$. A properly efficient solution $\bar{x}$ to (PVS) is also efficient to (PVS), but the opposite implication fails to hold in general.

Remark 1 Similar vector optimization problems were considered, for instance, in [5, 6], with all the involved functions taken cone-convex and differentiable, without the geometric constraint $x \in S$ and by considering finitely many similar semidefinite inequality constraints. Besides delivering optimality conditions regarding the ideal efficient points to considered vector optimization problems, some investigations via duality were performed for them, too, Lagrange and Wolfe dual problems being assigned to the attached scalarized problems. Moreover, a Lagrange type vector dual for a multiobjective semidefinite optimization problem was considered in [5], but with a different construction than the one proposed below.

The vector dual problem we assign to (PVS) is inspired by the ones proposed in [1–4] for primal linear vector optimization problems and by the ones considered in [7, 8] for vector optimization problems whose image spaces were partially ordered by the corresponding nonnegative orthants, being

(DVS) $$\operatorname*{Max}_{(\Lambda,Q,P,V)\in\mathcal{B}_S} H_S(\Lambda, Q, P, V),$$

where

$$\mathcal{B}_S = \Big\{ (\Lambda, Q, P, V) \in \hat{\mathcal{S}}_+^k \times \mathcal{S}_+^m \times (\mathbb{R}^n)^{k\times k} \times \mathbb{R}^{k\times k} : P = (p_{ij})_{i,j=1,\dots,k},$$
$$p_{ij} \in \operatorname{dom} f_{ij}^* \ \forall i,j \in \{1, \dots, k\} \text{ s.t. } \Lambda_{ij} \neq 0,$$
$$-\sum_{i,j=1}^k \Lambda_{ij} p_{ij} \in \operatorname{dom} (QH)_S^*, \operatorname{Tr}(\Lambda^\top V) = 0 \Big\}$$

and, for $i, j = 1, \dots, k$,

$$(H_S(\Lambda, Q, P, V))_{ij} = V_{ij} - \begin{cases} f_{ij}^*(p_{ij}) + \frac{1}{z(\Lambda)\Lambda_{ij}}(QH)_S^*\Big(-\sum_{i,j=1}^k \Lambda_{ij} p_{ij}\Big), & \text{if } \Lambda_{ij} \neq 0, \\ 0, & \text{otherwise,} \end{cases}$$

where $z(\Lambda)$ denotes the *number of nonzero entries* of the matrix Λ.

Remark 2 One can replace in $\mathcal{B}_S$ the constraint equality $\operatorname{Tr}(\Lambda^\top V) = 0$ by $\operatorname{Tr}(\Lambda^\top V) \leq 0$, obtaining thus another vector dual problem to (PVS) with a larger feasible set and, consequently, image set, than (DVS). We will not treat it here separately because the duality investigations regarding it follow analogously and its efficient solutions coincide with the ones of (DVS).

Remark 3 If $(\Lambda, Q, P, V) \in \mathcal{B}_S$, one can easily note that $V \notin (\mathcal{S}_+^k \cup (-\mathcal{S}_+^k))\backslash\{0\}$.

Now let us formulate the weak duality statement for (PVS) and (DVS).

Theorem 1 *There exist no $x \in \mathcal{A}$ and $(\Lambda, Q, P, V) \in \mathcal{B}_S$ such that $F(x) \preceq H_S$ (Λ, Q, P, V).*

Proof Assume the existence of $x \in \mathcal{A}$ and $(\Lambda, Q, P, V) \in \mathcal{B}_S$ such that $F(x) \preceq H_S(\Lambda, Q, P, V)$. Then $0 > \operatorname{Tr}\big(\Lambda^\top(F(x) - H_S(\Lambda, Q, P, V))\big) = \sum_{i,j=1}^k \Lambda_{ij}(f_{ij}(x) + f_{ij}^*(p_{ij})) + (QH)_S^*\big(-\sum_{i,j=1}^k \Lambda_{ij} p_{ij}\big) \geq (\sum_{i,j=1}^k \Lambda_{ij} p_{ij}^\top x - \operatorname{Tr}(Q^\top H(x)) - \delta_S(x) - (\sum_{i,j=1}^k \Lambda_{ij} p_{ij})^\top x \geq 0$ because $x \in \mathcal{A}$. As this cannot happen, the assumption we made is false. $\qquad\square$

In order to prove strong duality for the primal-dual pair of multiobjective optimization problems $(PVS) - (DVS)$ one needs additional hypotheses. The *regularity conditions* we consider are inspired by the ones used in [3, 4] and they are a *generalized Slater* type one

$$(RCV_1^S)\Big| \; \exists x' \in S \text{ such that } H(x') \in -\hat{\mathcal{S}}_+^m,$$

an *interiority type* one

$$(RCV_2^S)\Big| \; 0 \in \mathrm{ri}(H(S) - C),$$

and, respectively, a *closedness type* one

$$(RCV_3^S)\Big| \; S \text{ is closed and } \mathrm{epi}(\Lambda F)^* + \bigcup_{Q \in \mathcal{S}_+^m} \mathrm{epi}(QH)_S^* \text{ is closed for any } \Lambda \in \hat{\mathcal{S}}_+^k.$$

Theorem 2 *If S is a convex set, f_{ij}, $i, j = 1, \ldots, k$, are convex functions, H is $\mathcal{S}_+^m$-convex, $\bar{x} \in \mathcal{PE}_{LS}(PVS)$ and one of the regularity conditions (RCV_i^S), $i \in \{1, 2, 3\}$, is fulfilled, there exists $(\bar{\Lambda}, \bar{Q}, \bar{P}, \bar{V}) \in \mathcal{E}(DVS)$ such that $F(\bar{x}) = H_S(\bar{\Lambda}, \bar{Q}, \bar{P}, \bar{V})$.*

Proof Since $\bar{x}$ is properly efficient to (PVS), there exists a $\bar{\Lambda} \in \hat{\mathcal{S}}_+^k$ such that $\mathrm{Tr}(\Lambda^\top F(\bar{x})) \leq \mathrm{Tr}(\Lambda^\top F(x))$ for all $x \in \mathcal{A}$. The fulfillment of any of the considered regularity conditions yields (cf. [3, Sect. 3.2]) strong duality for the scalarized optimization problem attached to (PVS)

$$\inf_{x \in \mathcal{A}} \mathrm{Tr}(\Lambda^\top F(x))$$

and its Fenchel-Lagrange dual

$$\sup_{\substack{Q \in \mathcal{S}_+^m, \\ T \in \mathbb{R}^n}} \left\{ -(\Lambda F)^*(T) - (QH)_S^*(-T) \right\},$$

thus the latter has the optimal solutions $\bar{Q}$ and $\bar{T}$ that fulfill

$$\mathrm{Tr}(\Lambda^\top F(\bar{x})) = -(\bar{\Lambda}F)^*(\bar{T}) - (\bar{Q}H)_S^*(-\bar{T}).$$

Because f_{ij}, $i, j = 1, \ldots, k$, are convex functions defined on $\mathbb{R}^n$ with full domain they are continuous, too, consequently there exist $\tilde{p}_{ij} \in \mathbb{R}^n$, $i, j = 1, \ldots, k$, taken $\tilde{p}_{ij} = 0$ if $\bar{\Lambda}_{ij} = 0$, such that

$$(\bar{\Lambda}F)^*(\bar{T}) = \sum_{\substack{i,j=1, \\ \bar{\Lambda}_{ij} \neq 0}}^{k} (\bar{\Lambda}_{ij} f_{ij})^*(\tilde{p}_{ij}) = \sum_{\substack{i,j=1, \\ \bar{\Lambda}_{ij} \neq 0}}^{k} \bar{\Lambda}_{ij} f_{ij}^* \left(\frac{\tilde{p}_{ij}}{\bar{\Lambda}_{ij}} \right)$$

and $\sum_{i,j=1}^{k} \tilde{p}_{ij} = \bar{T}$. For $i, j \in \{1, \ldots, k\}$ take $\bar{p}_{ij} = \tilde{p}_{ij}/\bar{\Lambda}_{ij}$ and

$$\bar{V}_{ij} = f_{ij}(\bar{x}) + f_{ij}^*(\bar{p}_{ij}) + \frac{1}{z(\bar{\Lambda})\bar{\Lambda}_{ij}}(\bar{Q}H)_S^* \left(-\sum_{i,j=1}^{k} \tilde{p}_{ij} \right)$$

if $\bar{\Lambda}_{ij} \neq 0$ and $\bar{p}_{ij} = \tilde{p}_{ij}$ and $\bar{V}_{ij} = f_{ij}(\bar{x})$ otherwise. Then

$$\mathrm{Tr}(\bar{\Lambda}^\top \bar{V}) = \sum_{\substack{i,j=1,\\ \bar{\Lambda}_{ij} \neq 0}}^{k} \bar{\Lambda}_{ij} \left(f_{ij}(\bar{x}) + f_{ij}^*(\bar{p}_{ij}) + \left(\frac{1}{z(\bar{\Lambda})\bar{\Lambda}_{ij}} \right) (\bar{Q}H)_S^* \left(- \sum_{i,j=1}^{k} \bar{\Lambda}_{ij} \bar{p}_{ij} \right) \right) = 0.$$

Consequently, after denoting $\bar{P} = (\bar{p}_{ij})_{i,j=1,\ldots,k}$, one notices that $(\bar{\Lambda}, \bar{Q}, \bar{P}, \bar{V}) \in \mathcal{B}_S$. Assuming that $(\bar{\Lambda}, \bar{Q}, \bar{P}, \bar{V}) \notin \mathcal{E}(DVS)$, Theorem 1 yields a contradiction, therefore $(\bar{\Lambda}, \bar{Q}, \bar{P}, \bar{V}) \in \mathcal{E}(DVS)$. $\qquad\square$

Last but not least, let us give necessary and sufficient optimality conditions for the primal-dual pair of multiobjective optimization problems $(PVS) - (DVS)$.

Theorem 3 (a) *If S is a convex set, f_{ij}, $i, j = 1, \ldots, k$, are convex functions, H is $\mathcal{S}_+^m$-convex, $\bar{x} \in \mathcal{PE}_{LS}(PVS)$ and one of the regularity conditions (RCV_i^S), $i \in \{1, 2, 3\}$, is fulfilled, there exists $(\bar{\Lambda}, \bar{Q}, \bar{P}, \bar{V}) \in \mathcal{E}(DVS)$ such that*

(i) $F(\bar{x}) = H_S(\bar{\Lambda}, \bar{Q}, \bar{P}, \bar{V})$;
(ii) $f_{ij}(\bar{x}) + f_{ij}^*(\bar{p}_{ij}) = \bar{p}_{ij}^\top \bar{x}$ *when* $\bar{\Lambda}_{ij} \neq 0$;
(iii) $(\bar{Q}H)_S^* \left(- \sum_{i,j=1}^{k} \bar{\Lambda}_{ij} \bar{p}_{ij} \right) = - \left(\sum_{i,j=1}^{k} \bar{\Lambda}_{ij} \bar{p}_{ij} \right)^\top \bar{x}$;
(iv) $\mathrm{Tr}(\bar{Q}^\top H(\bar{x})) = 0$;
(v) $\mathrm{Tr}(\bar{\Lambda}^\top \bar{V}) = 0$;

(b) *Assume that $\bar{x} \in \mathcal{A}$ and $(\bar{\Lambda}, \bar{Q}, \bar{P}, \bar{V}) \in \hat{\mathcal{S}}_+^k \times \mathcal{S}_+^m \times (\mathbb{R}^n)^{k \times k} \times \mathbb{R}^{k \times k}$ fulfill the relations $(i) - (v)$, where $\bar{P} = (\bar{p}_{ij})_{i,j=1,\ldots,k}$. Then $\bar{x} \in \mathcal{PE}_{LS}(PVS)$ and $(\bar{\Lambda}, \bar{Q}, \bar{P}, \bar{V}) \in \mathcal{E}(DVS)$.*

Proof (a) The existence of a $(\bar{\Lambda}, \bar{Q}, \bar{P}, \bar{V}) \in \mathcal{E}(DVS)$, where $\bar{P} = (\bar{p}_{ij})_{i,j=1,\ldots,k}$, such that $F(\bar{x}) = H_S(\bar{\Lambda}, \bar{Q}, \bar{P}, \bar{V})$ is guaranteed by Theorem 2. The relations (i) and (v) are thus satisfied. Moreover,

$$\mathrm{Tr}(\bar{\Lambda}^\top \bar{V}) = \sum_{\substack{i,j=1,\\ \bar{\Lambda}_{ij} \neq 0}}^{k} \bar{\Lambda}_{ij} \left(f_{ij}(\bar{x}) + f_{ij}^*(\bar{p}_{ij}) + \left(\frac{1}{z(\bar{\Lambda})\bar{\Lambda}_{ij}} \right) (\bar{Q}H)_S^* \left(- \sum_{i,j=1}^{k} \bar{\Lambda}_{ij} \bar{p}_{ij} \right) \right),$$

and this is actually equal to 0. On the other hand, the Young-Fenchel inequality yields $f_{ij}(\bar{x}) + f_{ij}^*(\bar{p}_{ij}) \geq \bar{p}_{ij}^\top \bar{x}$, for all $i, j \in \{1, \ldots, k\}$ such that $\bar{\Lambda}_{ij} \neq 0$, and

$$(\bar{Q}H)_S^* \left(- \sum_{i,j=1}^{k} \bar{\Lambda}_{ij} \bar{p}_{ij} \right) + \mathrm{Tr}\big(\bar{Q}^\top H(\bar{x})\big) \geq \left(- \sum_{i,j=1}^{k} \bar{\Lambda}_{ij} \bar{p}_{ij} \right)^\top \bar{x},$$

which, taking into consideration the equality from above, imply $\mathrm{Tr}(\bar{Q}^\top H(\bar{x})) \geq 0$. But $\mathrm{Tr}(\bar{Q}^\top H(\bar{x})) \leq 0$ because $Q \in \mathcal{S}_+^m$ and $H(\bar{x}) \in -\mathcal{S}_+^m$, consequently the Young-Fenchel inequalities given above are fulfilled as equalities and $\text{Tr}(\bar{Q}^\top H(\bar{x})) = 0$, hence relations $(ii) - (iv)$ are fulfilled, too.

(b) From (ii) it follows that $\bar{p}_{ij} \in \text{dom } f_{ij}^*$ for all $i,j \in \{1, \ldots, k\}$ such that $\bar{\Lambda}_{ij} \neq 0$, while (iii) yields $-\sum_{i,j=1}^{k} \bar{\Lambda}_{ij} \bar{p}_{ij} \in \text{dom } (QH)_S^*$. Because of (v), it follows that $(\bar{\Lambda}, \bar{Q}, \bar{P}, \bar{V}) \in \mathcal{B}_S$.

Multiplying (ii) for f_{ij} with $\bar{\Lambda}_{ij}$ and summing up these relations and also $(iii) - (iv)$ one obtains

$$\sum_{\substack{i,j=1,\\ \bar{\Lambda}_{ij} \neq 0}}^{k} \bar{\Lambda}_{ij} \left(f_{ij}(\bar{x}) + f_{ij}^*(\bar{p}_{ij}) + \left(\frac{1}{z(\bar{\Lambda})\bar{\Lambda}_{ij}}\right) (\bar{Q}H)_S^* \left(-\sum_{i,j=1}^{k} \bar{\Lambda}_{ij} \bar{p}_{ij} \right) \right) = 0,$$

that yields because of the strong duality for the scalarized optimization problem attached to (PVS)

$$\inf_{x \in \mathcal{A}} \text{Tr}\left(\Lambda^\top F(x) \right)$$

and its Fenchel-Lagrange dual that $\bar{x} \in \mathcal{PE}_{LS}(PVS)$. The efficiency of $(\bar{\Lambda}, \bar{Q}, \bar{P}, \bar{V})$ to (DVS) follows immediately by (i) and Theorem 1. $\square$

Remark 4 Of interest and subject to further research are a converse duality statement regarding the primal-dual pair of multiobjective optimization problems $(PVS) - (DVS)$ (possibly inspired by corresponding assertions from [1–4], but no direct consequence of any of them) and investigations on how to derive numerical methods based on the provided duality results and optimality conditions for concretely solving such problems.

Acknowledgments Research partially supported by DFG (German Research Foundation), project WA 922/8-1. The author is indebted to Y. Ledyaev and L.M. Graña Drummond for the useful discussions which led to this note.

References

1. Boţ, R.I., Grad, S.-M.: On linear vector optimization duality in infinite-dimensional spaces. Numer. Algebra Control Optim. **1**, 407–415 (2011)
2. Boţ, R.I., Grad, S.-M., Wanka, G.: Classical linear vector optimization duality revisited. Optim. Lett. **6**, 199–210 (2012)
3. Boţ, R.I., Grad, S.-M., Wanka, G.: Duality in Vector Optimization. Springer, Berlin (2009)
4. Grad, S.-M.: Vector Optimization and Monotone Operators via Convex Duality. Springer, Berlin (2015)
5. Graña Drummond, L.M., Iusem, A.N.: First order conditions for ideal minimization of matrix-valued problems. J. Convex Anal. **10**, 1–19 (2003)
6. Graña Drummond, L.M., Iusem, A.N., Svaiter, B.F.: On first order optimality conditions for vector optimization. Acta Math. Appl. Sin. Engl. Ser. **19**, 371–386 (2003)

7. Wanka, G., Boţ, R.I.: A new duality approach for multiobjective convex optimization problems. J. Nonlinear Convex Anal. **3**, 41–57 (2002)
8. Wanka, G., Boţ, R.I., Grad, S.-M.: Multiobjective duality for convex semidefinite programming problems. Z. Anal. Anwend. **22**, 711–728 (2003)

Algorithms for Controlling Palletizers

Frank Gurski, Jochen Rethmann and Egon Wanke

Abstract Palletizers are widely used in delivery industry. We consider a large palletizer where each stacker crane grabs a bin from one of k conveyors and position it onto a pallet located at one of p stack-up places. All bins have the same size. Each pallet is destined for one customer. A completely stacked pallet will be removed automatically and a new empty pallet is placed at the palletizer. The FIFO STACK-UP problem is to decide whether the bins can be palletized by using at most p stack-up places. We introduce a digraph and a linear programming model for the problem. Since the FIFO STACK-UP problem is computational intractable and is defined on inputs of various informations, we study the parameterized complexity. Based on our models we give xp-algorithms and fpt-algorithms for various parameters, and approximation results for the problem.

1 Introduction

We consider the combinatorial problem of stacking up bins from a set of conveyor belts onto pallets. A detailed description of the practical background of this work is given in [1, 9]. The bins that have to be stacked-up onto pallets reach the palletizer on a conveyor and enter a *cyclic storage conveyor*, see Fig. 1. From the storage conveyor the bins are pushed out to *buffer conveyors*, where they are queued. The equal-sized bins are picked-up by stacker cranes from the end of a buffer conveyor and moved

F. Gurski (✉) · E. Wanke
Institute of Computer Science, University of Düsseldorf,
40225 Düsseldorf, Germany
e-mail: frank.gurski@hhu.de

E. Wanke
e-mail: e.wanke@hhu.de

J. Rethmann
Faculty of Electrical Engineering and Computer Science,
Niederrhein University of Applied Sciences, 47805 Krefeld, Germany
e-mail: jochen.rethmann@hs-niederrhein.de

© Springer International Publishing Switzerland 2016
M. Lübbecke et al. (eds.), *Operations Research Proceedings 2014*,
Operations Research Proceedings, DOI 10.1007/978-3-319-28697-6_28

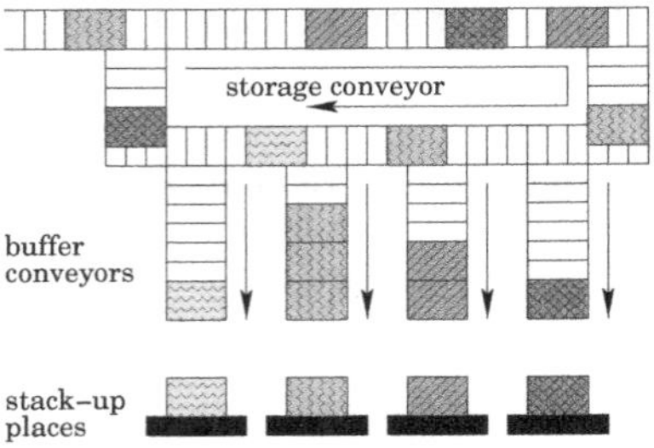

Fig. 1 A real stack-up
system

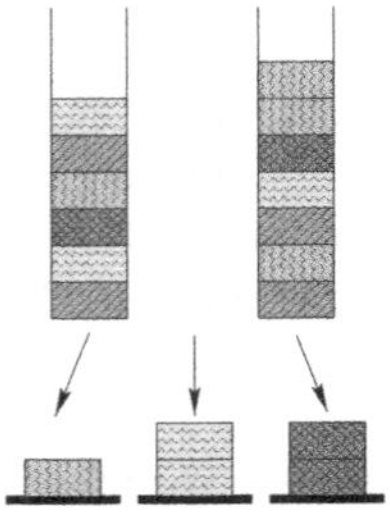

Fig. 2 A FIFO stack-up
system

onto pallets, which are located at some *stack-up places*. Full pallets are carried to
trucks by automated guided vehicles (AGVs).

The cyclic storage conveyor enables a smooth stack-up process irrespective of
the real speed the cranes and conveyors are moving. Such details are unnecessary
to compute an order in which the bins can be palletized with respect to the given
number of stack-up places. For the sake of simplicity, we disregard the cyclic storage
conveyor. Figure 2 shows a sketch of a simplified stack-up system with 2 buffer
conveyors and 3 stack-up places.

From a theoretical point of view, we are given k sequences $q_1, \ldots, q_k$ of bins and a
positive integer p. Each bin is destined for exactly one pallet. The FIFO STACK-UP
problem is to decide whether one can remove iteratively the bins of the k sequences
such that in each step only the first bin of one of the sequences will be removed and
after each step at most p pallets are open. A pallet t is called open, if at least one
but not all bins for pallet t has already been stacked-up. If a bin b is removed from a
sequence then all bins located behind b are moved-up one position to the front.

Our model is the second attempt to capture important parameters necessary for an
efficient and provable good algorithmic controlling of stack-up systems. Many facts
are known on a stack-up system model that uses a random access storage instead
of buffer queues, see [9, 10]. We think that buffer queues model the real stack-up
system more realistic than a random access storage.

The FIFO STACK-UP problem is NP-complete even if the number of bins per
pallet is bounded [3]. In this paper we give parameterized algorithms for the FIFO
STACK-UP problem based on a digraph model and a linear programming model.

2   Preliminaries

We consider *sequences* $q_1 = (b_1, \ldots, b_{n_1}), \ldots, q_k = (b_{n_{k-1}+1}, \ldots, b_{n_k})$ of pairwise
distinct *bins*. These sequences represent the buffer queues (handled by the buffer
conveyors) in real stack-up systems. Each bin b is labeled with a *pallet symbol plt(b)*.
We say bin b is destined for pallet $plt(b)$. The labels of the pallets can be chosen
arbitrarily, because we only need to know whether two bins are destined for the same
pallet or for different pallets. The set of all pallets of the bins in some sequence q_i
is denoted by $plts(q_i) = \{plt(b) \mid b \in q_i\}$. For a list of sequences $Q = (q_1, \ldots, q_k)$
we denote $plts(Q) = plts(q_1) \cup \cdots \cup plts(q_k)$. For some sequence $q = (b_1, \ldots, b_n)$
we say bin b_i is *on the left of* bin b_j in sequence q if $i < j$. A sequence $q' =
(b_j, b_{j+1}, \ldots, b_n)$, $j \geq 1$, is called a *subsequence* of sequence $q = (b_1, \ldots, b_n)$,
and we write $q - q' = (b_1, \ldots, b_{j-1})$.

Let $Q = (q_1, \ldots, q_k)$ and $Q' = (q'_1, \ldots, q'_k)$ be two lists of sequences of bins,
such that each sequence q'_j, $1 \leq j \leq k$, is a subsequence of sequence q_j. Each such
pair (Q, Q') is called a *configuration*. In every configuration (Q, Q') the first entry Q
is the initial list of sequences of bins and the second entry Q' is the list of sequences of
bins that remain to be processed. A pallet t is called *open* in configuration (Q, Q'), if a
bin of pallet t is contained in some $q'_i \in Q'$ and if another bin of pallet t is contained in
some $q_j - q'_j$ for $q_j \in Q$, $q'_j \in Q'$. The *set of open pallets* in configuration (Q, Q')
is denoted by $open(Q, Q')$. A pallet $t \in plts(Q)$ is called *closed* in configuration
(Q, Q'), if $t \notin plts(Q')$, i.e. no sequence of Q' contains a bin for pallet t. Initially all
pallets are *unprocessed*. From the time when the first bin of a pallet t is processed,
pallet t is either open or closed.

The FIFO Stack-Up Problem Let (Q, Q') be a configuration. The removal of the
first bin from one subsequence $q' \in Q'$ is called a *transformation step*. A sequence
of transformation steps that transforms the list Q of k sequences from the initial
configuration (Q, Q) into the final configuration (Q, Q'), where $Q' = (\emptyset, \ldots, \emptyset)$
containing k empty subsequences is called a *processing* of Q.

We define the FIFO STACK- UP problem as follows. Given a list $Q = (q_1, \ldots, q_k)$
of sequences and a positive integer p. Is there a processing of Q, such that in each
configuration during this processing at most p pallets are open?

In the analysis of our algorithms we use the following variables: k denotes the
number of sequences, and p stands for the number of stack up places. Furthermore,
m represents the number of pallets in $plts(Q)$, and n denotes the total number of bins
in all sequences. Finally, $N = \max\{|q_1|, \ldots, |q_k|\}$ is the maximum sequence length.
In view of the practical background, it holds $p < m$, $k < m$, $m < n$, and $N < n$.

Obviously, the order in which the bins are removed from the sequences determines
the number of stack-up places necessary to process the input. Consider a processing
of a list Q of sequences. Let $B = (b_{\pi(1)}, \ldots, b_{\pi(n)})$ be the order in which the bins are
removed during the processing of Q, and let $T = (t_1, \ldots, t_m)$ be the order in which
the pallets are opened during the processing of Q. Then B is called a *bin solution*
of Q, and T is called a *pallet solution* of Q. Examples can be found in [3]. Let s_i

denote the sequence q_{s_i} from which in the i-th decision configuration (Q, Q_i) a bin for pallet t_i will be removed. Then $S = (s_1, \ldots, s_m)$ is called a *sequence solution*.

During a processing of a list Q of sequences there are often configurations (Q, Q') for which it is easy to find a bin b that can be removed from Q' such that a further processing with p stack-up places is still possible. This is the case, if bin b is destined for an already open pallet [3]. A configuration (Q, Q') is called a *decision configuration*, if the first bin of each sequence $q' \in Q'$ is destined for a non-open pallet. We can restrict FIFO stack-up algorithms to deal with such decision configurations, in all other configurations the algorithms automatically remove a bin for some already open pallet.

For some positive integer n, let $[n] = \{1, \ldots, n\}$ be the set of all positive integers between 1 and n. The following theorems are easy to show.

Theorem 1 *For some pallet order* $(t_1, \ldots, t_m)$, $t_i \in plts(Q)$ *and* $t_i \neq t_j$ *for* $i \neq j$, *we can verify in time* $\mathcal{O}(n \cdot k) \subseteq \mathcal{O}(n \cdot m) \subseteq \mathcal{O}(n^2)$, *whether it represents a sequence of transformation steps to process* Q *with* p *stack-up places.*

Theorem 2 *Likewise, for a sequence order* $(s_1, \ldots, s_m)$, $s_i \in [k]$, *it can be verified in time* $\mathcal{O}(n^2)$ *whether it represents a processing of* Q.

The sequence graph Next we consider a useful relation between an instance of the FIFO STACK- UP problem and the directed pathwidth of a directed graph model. The notion of directed pathwidth was introduced by Reed, Seymour, and Thomas around 1995 and relates to directed treewidth introduced by Johnson, Robertson, Seymour, and Thomas in [5]. The problem of determining the directed pathwidth of a digraph is NP-complete [8].

The *sequence graph* $G_Q = (V, E)$ for an instance $Q = (q_1, \ldots, q_k)$ of the FIFO STACK- UP problem is defined by vertex set $V = plts(Q)$ and the following set of arcs. There is an arc $(u, v) \in E$ if and only if there is a sequence $q_i = (b_{n_{i-1}+1}, \ldots, b_{n_i})$ with two bins b_{j_1}, b_{j_2} such that (1) $j_1 < j_2$, (2) b_{j_1} is destined for pallet u, (3) b_{j_2} is destined for pallet v, and (4) $u \neq v$.

If $G_Q = (V, E)$ has an arc $(u, v) \in E$ then $u \neq v$ and for every processing of Q, pallet u is opened before pallet v is closed. Digraph $G_Q = (V, E)$ can be computed in time $\mathcal{O}(n + k \cdot |E|) \subseteq \mathcal{O}(n + k \cdot m^2)$, see [3].

Theorem 3 ([3]) *Let* $Q = (q_1, \ldots, q_k)$ *then digraph* $G_Q = (V, E)$ *has directed pathwidth at most* $p - 1$ *if and only if* Q *can be processed with at most* p *stack-up places.*

3   Algorithms for the FIFO Stack-Up Problem

Exponential time Since the directed pathwidth of a digraph $D = (V, E)$ can be computed in time $\mathcal{O}(1.89^{|V|})$ by [8], the FIFO STACK- UP problem can be solved using G_Q by Theorem 3 in time $\mathcal{O}(1.89^m + n + k \cdot m^2)$.

In the following paragraphs we use standard definitions for parameterized algorithms from the textbook [2]. A *parameterized problem* is a pair (Π, κ), where Π is a decision problem, $\mathscr{I}$ the set of all instances of Π and $\kappa : \mathscr{I} \to \mathbb{N}$ a so-called *parameterization*. $\kappa(I)$ is expected to be small for all inputs $I \in \mathscr{I}$.

An algorithm A is an *fpt-algorithm with respect to* κ, if there is a computable function $f : \mathbb{N} \to \mathbb{N}$ and a constant $c \in \mathbb{N}$ such that for every instance $I \in \mathscr{I}$ the running time of A on I is at most $f(\kappa(I)) \cdot |I|^c$. If there is an fpt-algorithm with respect to κ that decides Π then Π is called *fixed-parameter tractable*. For example, vertex cover can be solved in time $\mathscr{O}(2^k \cdot |V|^2)$ relating to parameter k.

An algorithm A is an *xp-algorithm with respect to* κ, if there are two computable functions $f, g : \mathbb{N} \to \mathbb{N}$ such that for every instance $I \in \mathscr{I}$ the running time of A on I is at most $f(\kappa(I)) \cdot |I|^{g(\kappa(I))}$. If there is an xp-algorithm with respect to κ which decides Π then Π is called *slicewise polynomial*.

XP-Algorithms In [4] we have shown that we can compute a solution for the FIFO STACK- UP problem in time $\mathscr{O}((N + 1)^{2k})$, which is an xp-algorithm with respect to the parameter k.

Since the directed pathwidth of a digraph $D = (V, E)$ can be computed in time $\mathscr{O}(|E| \cdot |V|^{\mathrm{d\text{-}pw}(D)+1}) \subseteq \mathscr{O}(|V|^{\mathrm{d\text{-}pw}(D)+3})$ by [11] and G_Q can be computed in time $\mathscr{O}(n + k \cdot m^2) \subseteq \mathscr{O}(n + m^3)$ the FIFO STACK- UP problem can be solved by Theorem 3 in time $\mathscr{O}(n + m^{\mathrm{d\text{-}pw}(G_Q)+3})$, which is an xp-algorithm with respect to the parameter p.

FPT-Algorithms Since there are at most $m!$ different pallet orders $(t_1, \ldots, t_m)$, $t_i \in plts(Q)$, $t_i \neq t_j$ for $i \neq j$, and every of these can be verified in time $\mathscr{O}(n^2)$ by Theorem 1, the FIFO STACK- UP problem can be solved in time $\mathscr{O}(n^2 \cdot m!)$, which is an fpt-algorithm with respect to parameter m.

Since there are at most k^m different sequence orders $(s_1, \ldots, s_m)$, $s_i \in [k]$, and every of these can be verified in time $\mathscr{O}(n^2)$ by Theorem 2, we can solve the FIFO STACK- UP problem in time $\mathscr{O}(n^2 \cdot k^m)$, which is an fpt-algorithm with respect to the combined parameter k^m. In practice k is much smaller than m, since there are much fewer buffer conveyors than pallets, thus this solution is better than $\mathscr{O}(n^2 \cdot m!)$.

Linear Programming To realize a bijection $\pi : [n] \to [n]$ we define n^2 binary variables $x_i^j \in \{0, 1\}$, $i, j \in [n]$, such that x_i^j is equal to 1, if and only if $\pi(i) = j$. In order to ensure π to be surjective and injective, we use the conditions

$$\sum_{i=1}^n x_i^j = 1 \text{ for every } j \in [n] \quad \text{and} \quad \sum_{j=1}^n x_i^j = 1 \text{ for every } i \in [n].$$

Further we have to ensure that all variables x_i^j, $i, j \in [n]$, are in $\{0, 1\}$. We will denote the previous $n^2 + 2n$ conditions by n-PERMUTATION(x_i^j).

By Theorem 3 the minimum number of stack-up places can be computed by the directed pathwidth of sequence graph G_Q plus one. In the following we use the fact that the directed pathwidth equals the directed vertex separation number [12].

For a graph $G = (V, E)$, we denote by $\Pi(G)$ the set of all bijections $\pi : [|V|] \rightarrow [|V|]$ of its vertex set. Given a bijection $\pi \in \Pi(G)$ we define for $1 \leq i \leq |V|$ the vertex sets $L(i, \pi, G) = \{u \in V \mid \pi(u) \leq i\}$ and $R(i, \pi, G) = \{u \in V \mid \pi(u) > i\}$. Every position $1 \leq i \leq |V|$ is called a *cut*. This allows us to define the directed vertex separation number for a digraph $G = (V, E)$ as follows.

$$\text{d-vsn}(G) = \min_{\pi \in \Pi(G)} \max_{1 \leq i \leq |V|} |\{u \in L(i, \pi, G) \mid \exists v \in R(i, \pi, G) : (v, u) \in E\}|$$

An integer program for computing the vertex separation number for some given sequence graph $G_Q = (V, E)$ on m vertices is as follows. To realize permutation π we define m^2 binary variables $x_i^j \in \{0, 1\}, i, j \in [m]$. Additionally we use an integer valued variable p in order to count the vertices on the left which are adjacent to vertices of the right side of the cuts.

For every list Q of sequences we can compute the minimum number of stack-up places p in a processing of Q by minimizing p subject to m-PERMUTATION(x_i^j) and

$$\sum_{j=1}^{c} Y(j, c) \leq p \text{ for every } c \in [m - 1]$$

where

$$Y(j, c) = \bigvee_{j' \in \{c+1,\dots,m\} i, i' \in [m], (v_{i'}, v_i) \in E} (x_i^j \wedge x_{i'}^{j'}).$$

For the correctness note that subexpression $Y(j, c)$ is equal to one if and only if there exists an arc from a vertex on a position $j' > c$ to a vertex on position j. The shown integer program has $m^2 + 1$ variables and a polynomial number of constraints.

By [6] integer linear programming is fixed-parameter tractable for the parameter number of variables, thus the FIFO STACK- UP problem is fixed-parameter tractable for the parameter m. Since $m \leq n$ the problem is also fixed-parameter tractable for the parameter n.

Approximation Since the directed pathwidth of a digraph $D = (V, E)$ can be approximated by a factor $\mathcal{O}(\log^{1.5} |V|)$ by [7], the FIFO STACK- UP problem can be approximated using the sequence graph G_Q by Theorem 3 by a factor $\mathcal{O}(\log^{1.5} m)$.

References

1. de Koster, R.: Performance approximation of pick-to-belt orderpicking systems. Eur. J. Oper. Res. **92**, 558–573 (1994)
2. Flum, J., Grohe, M.: Parameterized Complexity Theory. Springer, Berlin (2006)
3. Gurski, F., Rethmann, J., Wanke, E.: Complexity of the FIFO stack-up problem. ACM Comput. Res. Repos. (2013). arXiv:1307.1915

4. Gurski, F., Rethmann, J., Wanke, E.: Moving bins from conveyor belts onto pallets using FIFO queues. In: Operations Research Proceedings 2013, Selected Papers, pp. 185–191. Springer-Verlag (2014)
5. Johnson, T., Robertson, N., Seymour, P.D., Thomas, R.: Directed tree-width. J. Comb. Theory Ser. B **82**, 138–155 (2001)
6. Kannan, R.: Minkowski's convex body theorem and integer programming. Math. Oper. Res. **12**, 415–440 (1987)
7. Kintali, S., Kothari, N., Kumar, A.: Approximation algorithms for directed width parameters (2013). arXiv:1107.4824v2
8. Kitsunai, K., Kobayashi, Y., Komuro, K., Tamaki, H., Tano, T.: Computing directed pathwidth in $O(1.89^n)$ time. In: Proceedings of International Workshop on Parameterized and Exact Computation. LNCS, vol. 7535, pp. 182–193. Springer-Verlag (2012)
9. Rethmann, J., Wanke, E.: Storage controlled pile-up systems, theoretical foundations. Eur. J. Oper. Res. **103**(3), 515–530 (1997)
10. Rethmann, J., Wanke, E.: Stack-up algorithms for palletizing at delivery industry. Eur. J. Oper. Res. **128**(1), 74–97 (2001)
11. Tamaki, H.: A polynomial time algorithm for bounded directed pathwidth. In: Proceedings of Graph-Theoretical Concepts in Computer Science. LNCS, vol. 6986, pp. 331–342. Springer-Verlag (2011)
12. Yang, B., Cao, Y.: Digraph searching, directed vertex separation and directed pathwidth. Discrete Appl. Math. **156**(10), 1822–1837 (2008)

Capital Budgeting Problems: A Parameterized Point of View

Frank Gurski, Jochen Rethmann and Eda Yilmaz

Abstract A fundamental financial problem is budgeting. A firm is given a set of financial instruments $X = \{x_1, \ldots, x_n\}$ over a number of time periods T. Every instrument x_i has a return of r_i and for time period $t = 1, \ldots, T$ a price of $p_{t,i}$. Further for every time period t there is budget b_t. The task is to choose a portfolio X' from X such that for every time period $t = 1, \ldots, T$ the prices of the portfolio do not exceed the budget b_t and the return of the portfolio is maximized. We study the fixed-parameter tractability of the problem. For a lot of small parameter values we obtain efficient solutions for the capital budgeting problem. We also consider the connection to pseudo-polynomial algorithms.

1 Introduction

Capital budgeting can be regarded as a tool for maximizing a companys profit since most companies are able to manage only a limited number of projects at the same time. See [9] for a survey on capital budgeting problems.

From a computational point of view the capital budgeting problem is intractable. Since the problem is defined on inputs of various informations, in this paper we consider the fixed-parameter tractability for several parameterized versions of the problem. The idea behind fixed-parameter tractability is to split the complexity into two parts—one part that depends purely on the size of the input, and one part that depends on some *parameter* of the problem that tends to be small in practice. For

F. Gurski (✉) · E. Yilmaz
Institute of Computer Science, Algorithmics for Hard Problems Group,
University of Düsseldorf, 40225 Düsseldorf, Germany
e-mail: frank.gurski@hhu.de

E. Yilmaz
e-mail: eda.yilmaz@hhu.de

J. Rethmann
Faculty of Electrical Engineering and Computer Science,
Niederrhein University of Applied Sciences, 47805 Krefeld, Germany
e-mail: jochen.rethmann@hs-niederrhein.de

© Springer International Publishing Switzerland 2016
M. Lübbecke et al. (eds.), *Operations Research Proceedings 2014*,
Operations Research Proceedings, DOI 10.1007/978-3-319-28697-6_29

example, vertex cover can be solved in time $\mathcal{O}(2^k \cdot n^2)$, which is exponential, but usable in practice for small values of k. We also address the connection between these problems and pseudo-polynomial algorithms.

The capital budgeting problem can be solved in time $\mathcal{O}(T \cdot n \cdot 2^n)$ by enumerating all possible subsets of the n financial instruments. We present better algorithms relating to parameters. In this paper we use standard definitions for parameterized algorithms and pseudo-polynomial algorithms from the textbooks [1, 3, 4].

2 Single-Period Capital Budgeting Problem

The simplest of all capital budgeting models has just one time period ($T = 1$). It has drawn a lot of attention in the literature, see for example [8, 9]. For some positive integer n, let $[n] = \{1, \ldots, n\}$ be the set of all positive integers between 1 and n.

Name MAX SINGLE- PERIOD CAPITAL BUDGETING (or MAX SPCB for short)
Instance A set $X = \{x_1, \ldots, x_n\}$ of n financial instruments, for every instrument $x_i, i \in [n]$, there is a return of r_i and a price of p_i and there is a budget b.
Task Find a subset $X' \subseteq X$ such that the prices of the portfolio X' do not exceed the budget b and the return of the portfolio X' is maximized.

Parameters n, r_i, p_i, and b are assumed to be positive integers. Let $r_{\max} = \max_{1 \leq i \leq n} r_i$ and $p_{\max} = \max_{1 \leq i \leq n} p_i$. The problem is also known as the 0/1-knapsack problem (KP), see [7]. For some instance I its size $|I|$ can be bounded by $\mathcal{O}(n + n \cdot \log_2(r_{\max}) + n \cdot \log_2(p_{\max}) + \log_2(b))$. We can bound the number n and prices of the instruments of an instance I as follows.

Theorem 1 *Every instance of* MAX SPCB *can be transformed into an equivalent instance, such that* $n \in \mathcal{O}(b \cdot \log(b))$.

Proof First we can assume, that there is no instrument in X, whose price is larger than the budget b, i.e. $p_i \leq b$ for every $1 \leq i \leq n$. In general, for every $1 \leq a \leq b$ and $\lfloor \frac{b}{a+1} \rfloor + 1 \leq p \leq \lfloor \frac{b}{a} \rfloor$ there are at most $n_p = a$ instruments of price p in X. Since for every $1 \leq a \leq b$ the number of integer valued prices in interval $\lfloor \frac{b}{a+1} \rfloor + 1 \leq p \leq \lfloor \frac{b}{a} \rfloor$ is at most $a \cdot (\lfloor \frac{b}{a} \rfloor - (\lfloor \frac{b}{a+1} \rfloor + 1) + 1) = a \cdot (\lfloor \frac{b}{a} \rfloor - \lfloor \frac{b}{a+1} \rfloor) \in \mathcal{O}(\frac{b}{a+1})$, and by the harmonic series, we always can bound the number n of instruments in X by $\mathcal{O}(\frac{b}{2} + \frac{b}{3} + \frac{b}{4} + \frac{b}{5} + \frac{b}{6} + \cdots + \frac{b}{b+1}) = \mathcal{O}(b \cdot \log(b))$.

If we have given an instance for MAX SPCB with more than the mentioned number n_p of instruments of price p, we remove all of them except the n_p instruments of the highest return. $\qquad \square$

By choosing a boolean variable y_i for every instrument $x_i \in X$, indicating whether or not instrument x_i is chosen into the portfolio, a binary integer programming (BIP) version of this problem is as follows.

$$\max \sum_{i=1}^{n} r_i y_i \quad \text{s.t.} \quad \sum_{i=1}^{n} p_i y_i \le b \text{ and } y_i \in \{0, 1\} \text{ for } i \in [n] \qquad (1)$$

Dynamic programming solutions for MAX SPCB can be found in [7, 10].

Theorem 2 MAX SPCB *can be solved in time* $\mathscr{O}(n \cdot b)$.

Theorem 3 MAX SPCB *can be solved in time* $\mathscr{O}(n \cdot \sum_{i=1}^{n} r_i) \subseteq \mathscr{O}(n^2 \cdot r_{\max})$.

The algorithm used in the proof of Theorem 3 is pseudo-polynomial by its running time $\mathscr{O}(n^2 \cdot r_{\max})$. Therefore the following hold:

Theorem 4 *There is a pseudo-polynomial algorithm that solves* MAX SPCB *in time* $\mathscr{O}(n^2 \cdot r_{\max})$.

Parameterized Algorithms By adding a threshold value k for the return to the instance and choosing a parameter $\kappa(I)$ from the instance I, we define the following parameterized problem.

Name $\kappa(I)$-SINGLE- PERIOD CAPITAL BUDGETING (or $\kappa(I)$-SPCB for short)
Instance An instance of MAX SPCB and a positive integer k.
Parameter $\kappa(I)$
Question Is there a subset $X' \subseteq X$ such that the prices of the portfolio X' do not exceed the budget b and the return of the portfolio X' is at least k?

By Theorem 3.3.2.1 of [5] and Theorem 4 we conclude that $\kappa(I)$-SPCB is fixed-parameter tractable with respect to parameter $\kappa(I) = $ "maximum length of the binary encoding of all numbers within I".

Parameterization by number of instruments n A brute force solution is to check all 2^n possible subsets of X within BIP (1), which leads to an algorithm of time complexity $\mathscr{O}(n \cdot 2^n)$.

Theorem 5 *There is an fpt-algorithm that solves* n-SPCB *in time* $\mathscr{O}(n \cdot 2^n)$.

Alternatively one can use the result of [6], which implies that integer linear programming is fixed-parameter tractable for the parameter number of variables. Thus using BIP (1) the n-SPCB problem is fixed-parameter tractable.

Parameterization by standard parameter k When choosing the threshold value of the return as our parameter, i.e. $\kappa(I) = k$ we obtain the so-called *standard parameterization* of the problem.

Theorem 6 *There is an fpt-algorithm that solves* k-SPCB *in time* $\mathscr{O}(n^2 \cdot k)$.

Proof The k-SPCB problem is a special case of the 0/1-knapsack problem which allows an FPTAS of running time $\mathscr{O}(n^2 \cdot \frac{1}{\epsilon})$, see [7]. By Theorem 3.2 of [2] a polynomial fpt-algorithm that solves k-SPCB in time $\mathscr{O}(n^2 \cdot k)$ follows. $\qquad\square$

Parameterization by budget b From a practical point of view choosing k as a parameter is not useful, since a large return of the portfolio X' violates the aim that a good parameterization is small for every input. So we suggest it is better to choose the budget as parameter, i.e. $\kappa(I) = b$.

Theorem 7 *There is an fpt-algorithm that solves* b-SPCB *in time* $\mathcal{O}(n \cdot b)$.

Proof The running time of dynamic programming algorithm in the proof of Theorem 2 is in $\mathcal{O}(n \cdot b)$.

3 Multi-period Capital Budgeting Problem

Name MAX MULTI- PERIOD CAPITAL BUDGETING (or MAX MPCB for short)
Instance A set $X = \{x_1, \ldots, x_n\}$ of n financial instruments, a number of time periods T, for every x_i, $i \in [n]$, there is a return of r_i, for period $t \in [T]$ instrument x_i has a price of $p_{t,i}$, and for every period $t \in [T]$ there is a budget b_t.
Task Find a subset $X' \subseteq X$ such that for every period $t \in [T]$ the prices of portfolio X' do not exceed the budget b_t and the return of X' is maximized.

Parameters n, r_i, $p_{t,i}$, and b_t and are assumed to be positive integers. Let $r_{\max} = \max_{1 \le i \le n} r_i$, $p_{\max} = \max_{1 \le i \le n, 1 \le t \le T} p_{t,i}$, and $b_{\max} = \max_{1 \le i \le T} b_i$. The problem is also known as the multi-dimensional 0/1-knapsack problem (MKP), see [7]. For some instance I its size $|I|$ can be bounded by $\mathcal{O}(n \cdot T + n \cdot \log_2(r_{\max}) + n \cdot T \cdot \log_2(p_{\max}) + T \cdot \log_2(b_{\max}))$.

Theorem 8 *Every instance of* MAX MPCB *can be transformed into an equivalent instance, such that* $n \in \mathcal{O}(\sum_{t=1}^{T} b_t \cdot \log(b_t))$.

Proof By the proof of Theorem 1 for every $1 \le t \le T$ there are $\mathcal{O}(b_t \cdot \log(b_t))$ instruments, the union of all these instruments leads to $n \in \mathcal{O}(\sum_{t=1}^{T} b_t \cdot \log(b_t))$. $\square$

By choosing a boolean variable y_i for every instrument $x_i \in X$, a binary integer programming (BIP) version of this problem is as follows.

$$\max \sum_{i=1}^{n} r_i y_i \quad \text{s.t.} \quad \sum_{i=1}^{n} p_{t,i} y_i \le b_t \text{ for } t \in [T] \text{ and } y_i \in \{0, 1\} \text{ for } i \in [n] \quad (2)$$

Dynamic programming solutions for MAX MPCB can be found in [7, 10].

Theorem 9 MAX MPCB *can be solved in time* $\mathcal{O}(n \cdot T \cdot (b_{\max})^T)$.

Pseudo-polynomial Algorithms The existence of pseudo-polynomial algorithms for MAX MPCB depends on the assumption whether the number of time periods T is given in the input or is fixed.

Theorem 10 MAX MPCB *is not pseudo-polynomial.*

Proof Every NP-hard problem for which every instance I only contains numbers x, such that the value of x is polynomial bounded in $|I|$ is strongly NP-hard and thus not pseudo-polynomial (cf. Theorem 3.18 in [1]). Thus we can use a pseudo-polynomial reduction from MAX INDEPENDENT SET to show that our problem is not pseudo-polynomial in general. The problem is given a graph $G = (V, E)$ and the task is to find a subset $V' \subseteq V$ such that no two vertices of V' are adjacent and V' has a maximum size.

Let graph $G = (V, E)$ be an input for the MAX INDEPENDENT SET problem. For every vertex v_i of G we define an instrument x_i and for every edge e_i of G we define a time period within an instance I_G for MAX MPCB. The return of every instrument is equal to 1 and the budget for every time period is equal to 1, too. The price of an instrument within a time period is equal to 1, if the vertex corresponding to the instrument is involved in the edge corresponding to the time period, otherwise the price is 0.   □

Theorem 11 *For every fixed T there is a pseudo-polynomial algorithm that solves* MAX MPCB *in time* $\mathcal{O}(n \cdot T \cdot (b_{\max})^T)$.

Proof The algorithm in the proof of Theorem 9 given in [7, 10] has running time in $\mathcal{O}(n \cdot T \cdot (b_{\max})^T)$. If T is fixed, $n \cdot T$ is polynomial in the input size and $(b_{\max})^T$ is polynomial in the value of the largest occurring number in every input.   □

Parameterized Algorithms By adding a threshold value k for the return to the instance and choosing a parameter $\kappa(I)$ from the instance I, we define the following parameterized problem.

Name $\kappa(I)$-MULTI- PERIOD CAPITAL BUDGETING (or $\kappa(I)$-MPCB for short)
Instance An instance of MAX MPCB and a positive integer k.
Parameter $\kappa(I)$
Question Is there a subset $X' \subseteq X$ such that for every time period $t \in [T]$ the prices of portfolio X' do not exceed the budget b_t and the return of X' is at least k?

By Theorem 3.3.2.1 of [5] and Theorem 11 we conclude that for every fixed T problem $\kappa(I)$-MPCB is fixed-parameter tractable with respect to parameter $\kappa(I) =$ "maximum length of the binary encoding of all numbers within I".

Parameterization by number of instruments n A brute force solution is to check all 2^n possible subsets of X each in time $\mathcal{O}(T \cdot n)$ by BIP (2).

Theorem 12 *There is an fpt-algorithm that solves n-MPCB in time* $\mathcal{O}(T \cdot n \cdot 2^n)$.

One also can use the result of [6], which implies that integer linear programming is fixed-parameter tractable for the parameter number of variables. Thus by BIP (2) problem $\kappa(I)$-MPCB is fixed-parameter tractable for the parameter $\kappa(I) = n$.

Parameterization by standard parameter k When choosing $\kappa(I) = k$ we obtain the standard parameterization of the problem.

Theorem 13 k-MPCB *is* $W[1]$-*hard.*

Proof The proof of Theorem 10 describes a parameterized reduction from the k-INDEPENDENT SET problem, which is $W[1]$-hard, see [3]. □

Theorem 14 *There is an xp-algorithm that solves k-MPCB in time $\mathcal{O}(T \cdot n^{k+1})$.*

Proof Let I be an instance of k-MPCB and X' be a solution which complies the budgets of every time period and the return of X' is at least k. Whenever there are at least $k + 1$ instruments in X' we can remove one of the instruments of smallest return r' and obtain a solution X'' which still complies the budgets of every time period. Further since all returns are positive integers, the return of X'' is at least $r' \cdot (k + 1) - r' = r' \cdot k \geq k$. Thus we can assume that every solution X' of k-MPCB has at most k instruments which allows us to check at most n^k possible solutions within BIP (2), which implies an xp-algorithm w.r.t. parameter k. □

Parameterization by the budgets $b_1, \ldots, b_T$ Again, from a practical point of view choosing $\kappa(I) = k$ is not useful. So we suggest it is better to choose the sum of the budgets $\kappa(I) = \sum_{t=1}^{T} b_t \cdot \log(b_t)$ as a parameter.

Theorem 15 *There is an fpt-algorithm that solves $(\sum_{t=1}^{T} b_t \cdot \log(b_t))$-MPCB in time $\mathcal{O}(T \cdot n \cdot 2^{\mathcal{O}(\sum_{t=1}^{T} b_t \cdot \log(b_t))})$.*

Proof By Theorem 8 we have to check at most $2^{\mathcal{O}(\sum_{t=1}^{T} b_t \cdot \log(b_t))}$ possible portfolios $X' \subseteq X$ each in time $\mathcal{O}(T \cdot n)$ by BIP (2). □

Parameterization by number of time periods T When choosing $\kappa(I) = T$ the parameterized problem is at least $W[1]$-hard, since an fpt-algorithm with respect to T would imply a polynomial time algorithm for every fixed T. But even for $T = 1$ the problem is NP-hard.

Theorem 16 T-MPCB *is at least* $W[1]$-*hard, unless* $P = NP$.

For the same reasons there is no xp-algorithm for T-MPCB.

Theorem 17 *There is no xp-algorithm that solves T-MPCB, unless* $P = NP$.

References

1. Ausiello, G., Crescenzi, P., Gambosi, G., Kann, V., Marchetti-Spaccamela, A., Protasi, M.: Complexity and Approximation: Combinatorial Optimization Problems and Their Approximability Properties. Springer, Berlin (1999)
2. Cai, L., Chen, J.: On fixed-parameter tractability and approximability of NP optimization problems. J. Comput. Syst. Sci. **54**, 465–474 (1997)
3. Downey, R.G., Fellows, M.R.: Fundamentals of Parameterized Complexity. Springer, New York (2013)
4. Flum, J., Grohe, M.: Parameterized Complexity Theory. Springer, Berlin (2006)

5. Hromkovic, J.: Algorithmics for Hard Problems: Introduction to Combinatorial Optimization, Randomization, Approximation, and Heuristics. Springer, Berlin (2004)
6. Kannan, R.: Minkowski's convex body theorem and integer programming. Math. Oper. Res. **12**, 415–440 (1987)
7. Kellerer, H., Pferschy, U., Pisinger, D.: Knapsack Problems. Springer, Berlin (2010)
8. Šedová, J., Šeda, M.: A comparison of exact and heuristic approaches to capital budgeting. In: Proceedings of World Congress on Science, Engineering and Technology WCSET 2008, vol. 35, pp. 187–191 (2008)
9. Weingartner, H.M.: Capital budgeting of interrelated projects: survey and synthesis. Manage. Sci. **12**(7), 485–516 (1966)
10. Weingartner, H.M., Ness, D.N.: Methods for the solution of the multidimensional 0/1-knapsack problem. Oper. Res. **15**(1), 83–103 (1967)

How to Increase Robustness of Capable-to-Promise

A Numerical Analysis of Preventive Measures

Ralf Gössinger and Sonja Kalkowski

Abstract Reliable delivery date promises offer supply chains a chance to differentiate from competitors. Order planning therefore not only aims at short-term profit maximization, but also at robustness. For planning purposes different preventive measures are proposed to enhance robustness if order- and resource-related uncertainty is present. With regard to profit and robustness this paper analyzes the interaction of preventive measures applied in capable-to-promise approaches.

1 Problem

In *capable-to-promise* (CTP) approaches proposed for answering to customer order inquiries [1] the order- and resource-related uncertainty is not yet adequately taken into account. As a result, promised delivery dates often cannot be met. To enhance reliability, order planning can be done robust against fluctuating planning data in two respects [2]: *Solution robustness* is given if the objective function value fluctuates only within a tolerable range [3]. The term *planning robustness* is used if the extent of adjustments required to restore an optimal plan is tolerable [4]. Preventive measures applied for generating robustness are: (a) *Capacity nesting* (CN): The basic idea is to split up capacity into multiple segments and to define segment-specific utilization costs [5]. Hence, more lucrative orders have access to a bigger share of capacity than less lucrative orders. The risks of having to reject lucrative orders and of not being able to meet delivery dates of accepted orders due to accepting less lucrative orders are reduced. This measure is adopted to deal with order-related uncertainty [1, 6]. (b) *Providing safety capacity* (SC): In order to cover resource-related uncertainty a

R. Gössinger (✉) · S. Kalkowski (✉)
Department of Business Administration, Production Management and Logistics,
University of Dortmund, Martin-Schmeißer-Weg 12, 44227 Dortmund, Germany
e-mail: ralf.goessinger@udo.edu

S. Kalkowski
e-mail: sonja.kalkowski@tu-dortmund.de

© Springer International Publishing Switzerland 2016
M. Lübbecke et al. (eds.), *Operations Research Proceedings 2014*,
Operations Research Proceedings, DOI 10.1007/978-3-319-28697-6_30

213

risk-averse estimation of available capacity is performed in such a way that the planned capacity utilization meets constraints with an acceptable probability [6]. Thus, the available capacity usually exceeds the estimated values and consequently delayed deliveries occur less often. However, if the available capacity is exceeded, currently processed orders need to be rescheduled. Therefore, the probability parameter has to balance lost sales and the costs of delayed order fulfillment due to scarce capacity. In situations where these cost components cannot be measured with acceptable effort, the parameter value is normally set according to customary service level standards. (c) *Interactive order promising* (IP): Desired, but unrealistic delivery dates cannot be confirmed without rescheduling currently processed orders. The option of proposing delivery dates that deviate from the desired ones delegates the order acceptance decision to the customer. For order planning customers response to proposed deviating delivery dates needs to be anticipated with the probability of winning an order in dependence of the deviation [7, 8]. In the CTP context this response function is taken into consideration for model tests [6] and as a central part of the planning approach [9].

Up until now, the effectiveness of these measures has been proven in isolated analyses. Although they are directed to different sources of uncertainty, it cannot be assumed that the observed impacts unfold independently. Hence, in the present paper the *impacts of a joint measure application* on profit and solution resp. planning robustness are numerically analyzed based on batch CTP models proposed in [9].

2 Numerical Testbed

The CTP approach focuses on order-related planning of the processes manufacturing, intermediate storing and delivery of ordered quantities in a supply chain. These processes are initiated by customers order inquiries and controlled by decisions on order acceptance, delivery dates and production quantities. To capture *order-related uncertainty* real order data on the seven best-selling product configurations of a manufacturer of customized products from a period of 3 months are taken as the basis (order scenario 1). Additionally, four order scenarios are generated which are statistically equal to the observed order situation (Table 1). *Resource-related uncertainty* is modelled by a random variable that considers fluctuating capacity availability and follows period-specific triangular distributions. While scenarios I and II represent a certain capacity situation, uncertainty is present in scenarios III and IV. For each uncertain scenario four streams of random variables are generated. The parameters of CN and SC are varied systematically in the situations with and without IP. *CN* differentiates between standard (SCap) and premium capacity (PCap). The cost rates k^P for utilizing PCap are set to 500, 1000, 1500, 2000, 2500, 3000 and 6000 and the share of PCap is varied with the values 1/3, 1/2 and 2/3. *SC* is varied with the values $1 \cdot \sigma^c$ (scenario a) and $2 \cdot \sigma^c$ (scenario b). In each test 13 batch runs are performed with and without IP so that 51480 runs result (99.98% solved to optimality). In order to take the *interaction with the customer* into account,

Table 1 Order and capacity data

Product configuration		1	2	3	4	5	6	7
Order quantity	$\mu^q|\sigma^q$	4.7|3.3	4.9|2.5	7.8|3.9	4.7|6.5	7.7|0.6	4.8|3.3	6.4|2.6
Interarrival time	$\mu^t|\sigma^t$	13.0|5.9	11.4|12.7	9.1|4.9	22.8|16.8	30.3|22.5	22.8|31.5	11.4|7.4
Profit margin			5120.07	3555.07	3523.5	2958.81	2641.35	2388.28	2077.56
Capacity scenario	μ^c	σ^c (Periods within batch interval)					σ^c (Periods after batch interval)	
		1	2	3	4	5	6 ...T	
I	2.75	0	0	0	0	0	0	
II	2.5	0	0	0	0	0	0	
III	2.75	0	0.0204	0.0408	0.0612	0.0816	0.1021	
IV	2.5	0	0.0408	0.0816	0.1225	0.1633	0.2041	

a two-stage planning approach is applied. At the non-interactive stage only those order inquiries become accepted orders that can be fulfilled in time and maximize the profit. The remaining order inquiries are preliminarily rejected and integrated at the interactive stage. This process allows for interaction with customers whose order inquiries could be accepted with a deviating delivery date. For this purpose a profit-enhancing delivery date is determined with respect to anticipated customer response. In case of its rejection, the order is finally rejected, otherwise the modified order is accepted. Customer response is empirically estimated as a discrete cumulative distribution function of the deviation-dependent acceptance probability $\beta(V)$ (with the values V; $\beta(V)$): (>25; $1 \cdot 10^{-10}$), (>10; 0.2), (>0; 0.6), ($\leq$0; 1).

3 Numerical Results

The combination of parameter values results in six scenarios which are characterized by increasing SC (e.g. scenario I, IIIa, IIIb) and/or increasing resource-related uncertainty (e.g. scenario IIIa, IVa) (Fig. 1). In all scenarios the CTP approach generates a positive *average profit* for each parameter constellation. If *IP* is applied, on average 9 % more orders than without IP are accepted if the costs for PCap are between 500 and 3000. For PCap costs of 6000 the percentage is much higher (15 %). This increases profits by 5 % or 40 %, respectively. In all scenarios parameter constellations exist in which *CN* raises profits (dashed shading). In particular, costs for PCap between 2000 and 3000 combined with a high share of PCap are advantageous. The advantage range grows with increasing SC and increasing capacity uncertainty as well with applying IP. This can be attributed to the increasing scarcity of capacity and to the increasing profits going along with the growing number of accepted orders induced by IP. However, by applying CN the number of accepted orders is reduced with an increasing share of PCap and increasing costs for its utilization. *SC* results in a shortage of capacity accessible for planning so that less orders are accepted and lower profits are generated the greater the safety level and the capacity

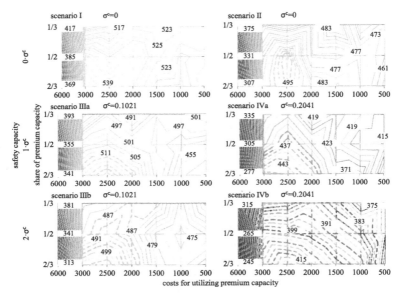

Fig. 1 Generated average profit in case of applying all preventive measures

uncertainty are. Simultaneously costs of delayed order fulfillment are reduced, but the reduction is overcompensated by lost profits. Thus, orientating towards customary service level standards does not necessarily enhance profits.

The preventive measures are not able to completely cover uncertainty. Therefore, a rescheduling of currently processed orders is necessary which might cause penalty costs due to delays if the present production situation lies outside the covered range. In terms of *solution robustness*, it is determined to which extent the planning approach compensates the uncertainty of planning data with regard to the objective value. Table 2 visualizes the uncertainty by coefficients of variation (CV). Despite the three sources of uncertainty, profits have very low CV if the costs for utilizing PCap are between 500 and 3000. This indicates a high level of solution robustness. The CV rise with increasing uncertainty and increasing SC, whereas IP leads to their reduction. On the other hand, the CV have considerably higher values if the cost rate for utilizing PCap is very high (6000) and the influence tendencies of parameters are slightly different: SC partially compensates increasing uncertainty and the impact of IP on CV is non-monotonic.

Since the extent of plan adjustments required to cope with uncertainty is an indicator for *planning robustness*, the penalty costs (PC) as the average weighted deviation between planned and achieved delivery dates can be used for evaluation [10]. In Fig. 2 the evolution of PC is exemplary illustrated for the representative scenario IV. In case of *CN* PC decrease with an increasing share of PCap and increasing costs for its utilization. It has the strongest impact on planning robustness, but the influence direction is dependent on the parameter setting and the application of the other

Table 2 Coefficients of variation of planning data and objective value

Planning data					Generated profits							
Order data (quantity, interarrival time)			Capacity data		$k^P \in \{500, \ldots, 3000\}$				$k^P = 6000$			
					Non-interactive		Interactive		Non-interactive		Interactive	
1	2	3	I	II	I	II	I	II	I	II	I	II
(0.69,0.45)	(0.52,1.11)	(0.50,0.54)	0	0	I 0.06	II 0.08	I 0.06	II 0.07	I 0.13	II 0.12	I 0.16	II 0.11
4	5	6	III		IIIa 0.07	IIIb 0.08	IIIa 0.06	IIIb 0.07	IIIa 0.12	IIIb 0.13	IIIa 0.10	IIIb 0.15
(0.75,0.74)	(0.08,0.74)	(0.69,1.38)	0.04									
7			IV		IVa 0.09	IVb 0.10	IVa 0.07	IVb 0.08	IVa 0.12	IVb 0.18	IVa 0.14	IVb 0.16
(0.40,0.65)			0.08									

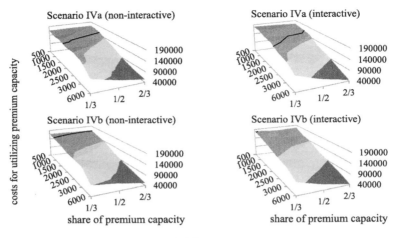

Fig. 2 Evolution of penalty costs in scenario IV

measures. The black line marks the PC level for a PCap of zero. That is, planning robustness can be improved by CN if costs of PCap are high enough. *IP* always leads to a moderate reduction of PC and the advantage area of CN increases with increasing SC. For a high level of capacity uncertainty (IV) PC have the strongest reduction if a high level of SC is utilized. The evolution of PC induced by CN and SC is leveraged by *IP* which results mostly in a PC increase.

4 Conclusions

In the present paper the impacts of applying the preventive measures capacity nesting, safety capacity and interactive order promising on the assessment criteria profit as well as planning and solution robustness of CTP approaches are analyzed regarding the interactions of the measures. The numerical analysis reveals that the simultaneous application of analyzed measures can increase both, profits and solution robustness, if the parameter settings are coordinated. In the numerical analysis this is given in combinations with a high share of premium capacity, medium utilization costs as well as low safety capacity and interactive order promising. Compared to the case without capacity nesting planning robustness is also increased in this area. Outside this area firstly a trade-off between planning robustness, solution robustness as well as generated profits exists and then, in extreme cases, the measure application reduces the values of the three assessment criteria simultaneously. Regardless of the other measures interactive order promising has to be highlighted as a suitable measure to increase profits and solution robustness. Providing safety capacity according to an externally defined service level enhances planning robustness but goes along with the risk of an unbalance between costs of delayed order fulfillment and lost sales.

Capacity nesting has the strongest increasing/reducing impact on solution and planning robustness and the strength of effects depends on the extent of safety capacity (planning robustness enhancing resp. solution robustness reducing) and the application of interactive order promising (effect-strengthening). For constant order-related uncertainty an area of advantage has been identified whose surface area depends on the application of the two other measures and the resource uncertainty. However, the behavior at different levels of order-related uncertainty has to be analyzed in future studies.

References

1. Chen, C.-Y., Zhao, Z., Ball, M.O.: Quantity and due date quoting available to promise. Inf. Syst. Front. **3**, 477–488 (2001)
2. Roy, B.: Robustness in operational research and decision aiding: a multi-faceted issue. Eur. J. Oper. Res. **200**, 629–638 (2010)
3. Mulvey, J.M., Vanderbei, R.J., Zenios, S.A.: Robust optimization of large-scale systems. Oper. Res. **43**, 264–281 (1995)
4. Kimms, A.: Stability measures for rolling schedules with applications to capacity expansion planning, master production scheduling and lot sizing. Omega **26**, 355–366 (1998)
5. Harris, F.H.deB., Pinder, J.P.: A revenue management approach to demand management and order booking in assemble-to-order manufacturing. J. Oper. Manage. **13**, 299–309 (1995)
6. Pibernik, R., Yadav, P.: Dynamic capacity reservation and due date quoting in a make-to-order system. Naval Res. Logist. **55**, 593–611 (2008)
7. Simmonds, K.: Competitive bidding: deciding the best combination of non-price features. Oper. Res. Q. **19**, 5–14 (1968)
8. Kingsman, B.G., Mercer, A.: Strike rate matrices for integrating marketing and production during the tendering process in make-to-order subcontractors. Int. Trans. Oper. Res. **4**, 251–257 (1997)
9. Gössinger, R., Kalkowski, S.: Planung robuster Lieferterminzusagen unter Berücksichtigung des Kundenverhaltens. In: Gössinger, R., Zäpfel, G. (eds.) Management integrativer Leistungserstellung, pp. 429–454. Duncker & Humblot, Berlin (2014)
10. Sridharan, S.V., Berry, W.L., Udayabhanu, V.: Measuring master production schedule stability under rolling planning horizons. Decis. Sci. **19**, 147–166 (1988)

An Iterated Local Search
for a Re-entrant Flow Shop Scheduling
Problem

Richard Hinze and Dirk Sackmann

Abstract This paper discusses a re-entrant permutation flow shop scheduling problem with missing operations. The two considered objective functions are makespan and total flow time. Re-entrant flows are characterized by a multiple processing of jobs on more than one machine. We propose a heuristic for solving the problem. Since there have been promising approaches in literature on other scheduling problems, we chose the iterated local search (ILS). This meta-heuristic framework combines the advantages of local search algorithm and still tries to avoid being stuck in local optima by a so called shaking step. The initial solution for the ILS is obtained by a dispatching rule. Various rules have been tested, e.g., total job processing time and total processing time of job levels. A hill climbing algorithm has been implemented as the integrated local search method of the ILS. The ILS is compared to a MIP formulation from literature. The results show, that the ILS can deliver better results.

1 Introduction

The jobs in many real world production systems need to be processed more than once on at least one manufacturing facility within a production system. This re-entrant property can be determined either by technological reasons, e.g., in semiconductor wafer fabrication, paint shops, software development and aircraft manufacturing or by rework operations due to quality issues. The re-entrant feature in scheduling was first described by [3] in a flow shop problem. Later a variety of re-entrant scheduling problems have been considered in literature. An overview on research in this field can be found in [2, 9, 10]. Chamnanlor et al. [1] examine a re-entrant flow shop problem for a hard-disk manufacturing system with time windows. In [6] and [7] a

R. Hinze (✉) · D. Sackmann (✉)
University of Applied Sciences Merseburg,
Eberhard-Leibnitz-Str. 2, 06217 Merseburg, Germany
e-mail: richard.hinze@hs-merseburg.de

D. Sackmann
e-mail: dirk.sackmann@hs-merseburg.de

© Springer International Publishing Switzerland 2016   221
M. Lübbecke et al. (eds.), *Operations Research Proceedings 2014*,
Operations Research Proceedings, DOI 10.1007/978-3-319-28697-6_31

genetic algorithm with an integrated analytic hierarchy process (AHP) is applied to a flexible re-entrant flow shop. The proposed combination outperformed a pure genetic algorithm. For a two stage flexible re-entrant flow shop problem a hybrid genetic algorithm and a random key genetic algorithm have been compared to modified versions of shortest processing time, longest processing time and NEH algorithms by [4]. The hybrid genetic algorithm obtained the best results for makespan minimization.

This paper considers a re-entrant permutation flow shop problem (RPFSP). The detailed problem assumptions are shown in Sect. 2. Different dispatching rules are suggested to initialize an iterated local search (ILS). The ILS is described in Sect. 3. Section 4 presents the computational results.

2   Problem Description

The RPFSP considers the problem of scheduling N jobs on M machines, with at least one job visiting one or more machines multiple times. The number of entries of each job into the production system is counted in levels L [8]. The machine sequence for each job is identical, as well as the sequence of job levels on all machines. As the number of the jobs' operations may differ from machine to machine, we consider a problem with missing operations. Missing operations occur, when a job re-enters the production system on a other machine than the first one. That can have technological reasons as well as quality reasons, which lead to rework operations.

Each job level needs to be assigned to a position in the processing sequence in order to optimize a selected objective function. Therefore the processing sequence has $B = N \cdot L$ positions. Makespan and total flow time are the examined objective functions.

The processing times of job i in a level l on machine k are represented by a parameter p_{lk}^i. The following example shows the processing times for a job with $L = 2$, that re-entrers the system on machine $k = 2$. This leads to a missing operation on machine $k = 1$.

$$p_{lk}^i = \begin{pmatrix} 4\ 6\ 2\ 7\ 4\ 5 \\ 0\ 2\ 3\ 4\ 2\ 3 \end{pmatrix}$$

The re-entries caused by rework lead to a necessity to consider missing operations and to shuffle levels of different jobs. The processing times need to be adjusted in case of rework:

$$p_{lk}^i = \begin{pmatrix} 4\ 6\ 2\ 7\ 4\ 5 \\ 0\ 2\ 3\ 0\ 0\ 0 \\ 0\ 0\ 1\ 4\ 2\ 3 \end{pmatrix}$$

A repeated processing on machine $k = 3$ is necessary, if a mistake after operating the job on machine $k = 3$ in level $l = 2$ is recognized. The rework operation on $k = 3$

is lead to an additional job level. The rework time on the third machine is $p_{33}^i = 1$. Zero processing times represent the operations, that need to be skipped before the rework is done. The processing of the job continues with the normal processing times after correcting the quality problems. This deterministic, static representation of rework is used in Sect. 2, to deliver an estimation of the lower bound of the problem. The importance of a proper dealing with zero processing times for makespan and throughput time minimization has been shown in [5].

3 Proposed Iterated Local Search

We propose an algorithm to the RPFSP with greedy and random moves within an interchange neighborhood. Figure 1 shows the elements of the ILS framework. In the following the different steps are explained in detail.

3.1 Initial Solution

Different dispatching rules have been compared for calculating an initial solution.

The maximum sum of processing times rule (MaxSP rule) uses each job's sum of processing times. The first level of the job with the maximum sum of processing times is put on sequence position one. The first level of the job with the second highest sum of processing times follows on the second position and so on. The levels $l = 2$ follow in the same job sequence, after all levels $l = 1$ have been assigned to

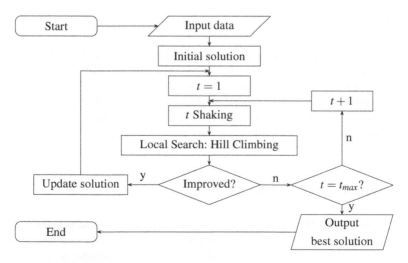

Fig. 1 Iterated Local Search

their positions. Assigning the sequence positions in that manner results into blocks of the size N within the sequence.

The minimum sum of processing times rule (MinSP rule) is similar to the first procedure, but prefers the lowest sum of processing times for first sequence position.

Similar to the method using the sum of all processing times of a job as criteria the sum of processing times of each level are used as a criteria (MaxLSP rule and MinLSP rule). These methods avoid to assign a complete block of sequence positions with job levels of the same rank l. A job level l is allowed to be assigned to a sequence position, if the predecessor level $l - 1$ of the same job has a lower sequence position and has completed its last operation.

3.2 Shaking

The shaking phase provides the possibility to leave local optima in order to reach a better solution than the current best in an additional local search phase. The performed operations are pairwise swaps of level sequence positions in the current best solution. That means, that a level of a job swaps the sequence position with a level of another job. The two jobs i with level l and ii with level ll are selected randomly. The conditions for a valid swap are: (1) the two selected jobs are not identical, $i \neq ii$, (2) the new sequence positions of the levels are between their predecessors and successors. The number of pairwise swaps within one iteration of the ILS is t. t is increased by one, if the local search phase did not find a better solution than the current best. The maximum number of pairwise swaps is t_{max}.

3.3 Local Search

A hill climbing algorithm as local search method is applied after the shaking phase the current best solution. It searches for the best pairwise swap of sequence positions in the current solution, iterating until there is no improvement in this neighborhood possible anymore. The following shaking phase continues with $t = 1$, if there is an improvement of the current best solution. Otherwise the shaking follows with $t + 1$ follows, if $t < t_{max}$. The algorithm quits in case of $t = t_{max}$.

4 Computational Experiments

Two versions of the ILS have been compared to a MIP model of [8] (P&C model) on an Intel Core2 Duo CPU E8500 3.16 GHz, 3.44 GB memory system. The two versions of the ILS use different methods of initialization (MinSP or MaxSP rule). In preliminary tests the MinSP rule reached better results for the total flow time in

Table 1 Computational results, N—number of jobs, L—number of levels, L^*—number of levels including rework, M—number of machines

Problem size				Makespan		Total flow time	
N	L	L^*	M	ILS	P&C model	ILS	P&C model
5	5	7	10	759	739	2870	3560
		10	20	1638	1609	5660	8045
		11	30	2411	2613	8880	12813
		13	40	3600	3685	11105	17589
		13	50	3612	4236	17100	20967
	10	14	10	1576	1532	6525	7559
		18	20	3153	3232	12735	16037
		23	30	5690	5582	17595	27750
		25	40	6616	7327	28600	36635
		27	50	8023	9103	36930	45169
10	5	8	10	983	962	6500	9391
		10	20	1997	1987	10350	19457
		13	30	3362	3283	19950	32190
		14	40	4077	4453	26100	43768
		18	50	5896	6125	37900	60992
	10	16	10	2014	1951	15890	18793
		19	20	3928	3981	31940	39169
		22	30	6236	6185	47060	60522
		27	40	8909	9145	65900	89478
		32	50	11842	12150	82170	120608

a majority of problems. The MaxSP rule outperformed the other dispatching rules, considering makespan as the objective. The ILS is coded in C++, the MIP is solved with CPLEX 12.4. The results are shown in Table 1. The initial test processing times are randomly generated and equal distributed between 1 and 20. The test problems consider missing operations (see example shown in Sect. 2). L is the number of re-entries per job without rework, L^* includes rework operations. N is the number of jobs to be scheduled on M machines. The performance on makespan and total flow time minimization has been tested in separate test runs. The ILS finds better solutions than the MIP of Pan and Chen for 14 of the 20 tested problem instances, if the objective is makespan. Total flow time reached by the ILS is in every case lower than the flow time calculated by the MIP. This improvement is possible because of a proper dealing with processing times equal to zero. The average improvement of makespan is 3.09 and 27.71 % for total flow time.

5 Conclusions

Different dispatching rules have been tested to initialize the ILS. In most cases the MinSP rule leads to the best results regarding makespan. The MaxSP rule obtains better total flow time. The ILS reaches mostly lower values of makespan as well as total flow time than the model of [8]. The algorithm is able to find initial solutions for large problem sizes, e.g., $N = 50$, $L^* = 69$, $M = 50$, within several seconds. Further research is necessary to improve the results regarding objective values and computation time. The hill climbing algorithm needs relatively long to identify a local minimum. An improvement can be obtained by applying different methods to generate an initial solution as well as implementing an alternative local search algorithm.

References

1. Chamnanlor, C., Sethanan, K., Chien, C., Gen, M.: Reentrant flow-shop scheduling with time windows for hard-disk manufacturing by hybrid genetic algorithms. In: Asia Pacific Industrial Engineering and Management Systems Conference, pp. 896–907 (2012)
2. Danping, L., Lee, C.K.: A review of the research methodology for the re-entrant scheduling problem. Int. J. Prod. Res. **49**, 2221–2242 (2011)
3. Graves, S.C.: A review of production scheduling. Oper. Res. Inf. **29**, 646–675 (1981)
4. Hekmatfar, M., Fatemi Ghomi, S.M.T., Karimi, B.: Two stage reentrant hybrid flow shop with setup times and the criterion of minimizing makespan. Appl. Soft. Comput. **11**, 4530–4539 (2011)
5. Hinze, R., Sackmann, D., Buscher, U., Aust, G.: A contribution to the reentrant flow-shop scheduling problem. In: Proceedings of IFAC Conference Manufacturing Modelling, Management, and Control, pp. 718–723. IFAC, St. Petersburg (2013)
6. Lin, D., Lee, C., Wu, Z.: Integrated GA and AHP for re-entrant flow shop scheduling problem. In: 2011 IEEE International Conference on Quality and Reliability (ICQR), pp. 496–500. IEEE (2011)
7. Lin, D., Lee, C., Wu, Z.: Integrating analytical hierarchy process to genetic algorithm for re-entrant flow shop scheduling problem. Int. J. Prod. Res. **50**, 1813–1824 (2012)
8. Pan, J., Chen, J.: Minimizing makespan in re-entrant permutation flow-shops. J. Oper. Res. Soc. **54**, 642–653 (2003)
9. Uzsoy, R., Lee, C.Y., Martin-Vega, L.A.: A review of production planning and scheduling models in the semiconductor industry. Part I: System characteristics, performance evaluation and production planning. IIE Trans. **24**, 47–60 (1992)
10. Uzsoy, R., Lee, C.Y., Martin-Vega, L.A.: A review of production planning and scheduling models in the semiconductor industry. Part II: Shop-floor control. IIE Trans. **26**, 44–55 (1994)

Selecting Delivery Patterns for Grocery Chains

Andreas Holzapfel, Michael Sternbeck and Alexander Hübner

Abstract On a tactical level retailers face the problem of determining on which weekdays stores should be delivered, and of setting a frame for short-term vehicle routing. We therefore propose a binary program that considers the decision-relevant costs and capacities at the distribution center, in transportation and instore. To resolve the trade-off between the different cost components aligned to the delivery pattern decision, we propose a sequential solution approach. A numerical example illustrates first results.

1 Problem Description

Especially in the grocery retail trade, repetitive weekly delivery patterns are used to increase planning stability for the stores and to balance picking workload at the distribution center (DC). A delivery pattern is defined as a store-specific combination of weekdays on which a delivery takes place. Assuming that there must be at least one delivery every week, and six weekdays, there is a total of 63 potential delivery patterns for each store.

As several processes in the logistics subsystems DC, transportation and instore logistics of the retail supply chain are influenced by the delivery pattern decision, an integrated approach is necessary to solve the problem. Existing approaches in literature, however, neglect important areas influenced by the delivery patterns and often focus on short-term vehicle routing. E.g., Ronen and Goodhart [4] propose

A. Holzapfel (✉) · M. Sternbeck · A. Hübner
Department of Operations, Catholic University Eichstätt-Ingolstadt,
Auf der Schanz 49, 85049 Ingolstadt, Germany
e-mail: andreas.holzapfel@ku.de

M. Sternbeck
e-mail: michael.sternbeck@ku.de

A. Hübner
e-mail: alexander.huebner@ku.de

© Springer International Publishing Switzerland 2016   227
M. Lübbecke et al. (eds.), *Operations Research Proceedings 2014*,
Operations Research Proceedings, DOI 10.1007/978-3-319-28697-6_32

a hierarchical approach in a cluster-first, route-second manner. They first cluster stores with similar characteristics and preselect patterns. Then they assign patterns to stores using a MIP, and afterwards build transportation routes applying a periodic vehicle routing problem (PVRP). Cardos and Garcia Sabater [1] design a decision support system to determine delivery schemes for stores focusing on inventory and transportation costs. They preselect feasible solutions in the first step. Then they evaluate minimal cost paths for each delivery frequency for each store and afterwards solve a PVRP for each delivery frequency combination. A major disadvantage of both approaches is that they do not consider instore handling costs, despite the fact that these determine the main cost block of operational logistics costs according to several empirical studies (see [3]). Finally Sternbeck and Kuhn [5] try to act on some of the limitations of the papers previously named. They present an integrative approach taking into account relevant cost components and capacity restrictions in three subsystems of the retail supply chain (DC, transportation and the stores). A limitation is that they do not model bundling effects across stores.

The model we propose in the following section is based on the approach of Sternbeck and Kuhn [5], but avoids preselection of patterns and incorporates bundling issues, thus creating a more exact decision tool for retailers trying to resolve the problem of selecting optimal delivery patterns for their stores.

Our focus within this paper is on the three major subsystems of the retail supply chain that are affected by the selection of delivery patterns, namely the DC, transportation and the stores (see [5]). We concentrate on one DC supplying its affiliated stores with a dry goods assortment. We focus on operational costs that are directly dependent on the selection of delivery patterns. Basing our model on the findings of Sternbeck and Kuhn [5], we include the following cost terms: picking costs at the DC, transportation costs, instore inventory holding costs and instore handling costs that can be separated into ordering and receiving costs, initial shelf restocking costs and costs of restocking shelves from the backroom.

There is a trade-off between different cost components that are aligned to the delivery pattern decision. While most cost components increase with higher delivery frequency, instore inventory holding costs and shelf restocking costs have an inverse relationship. Picking costs, transportation costs, ordering and receiving costs as well as initial shelf restocking costs increase with higher delivery frequency as they consist of a fixed portion for each delivery and a variable component depending on the quantities delivered. As a consequence, less frequent deliveries lead to a reduction in fixed costs and therefore decreasing total costs per week. As delivery frequency goes down, however, the quantities delivered exceed daily demand and shelf space capacity to a greater degree, leading to higher inventory holding costs and more shelf restocking activities instore as well as higher costs.

Without considering capacity limits at the DC and in transportation, each store could realize its own optimal delivery pattern. Retailers, however, try to balance the workload at the DC to be able to use existing capacity efficiently. We therefore assume minimum and maximum picking capacity as well as limited transportation capacity on each day. Limited space in a store's backroom can limit the potential delivery frequencies for the store. We therefore include store-specific receiving capacity in our decision model. The mathematical model reflects the relevant cost components as well as the restricting capacities, and models a twofold decision:

- Selection of delivery patterns: Which stores are supplied on which days?
- Construction of base tours: Which stores build a delivery cluster?

As our decision is a tactical one, we assume deterministic but dynamic store- and article-specific demand. Furthermore, each store has its specific shelf layout with specific shelf space capacity as well as different backroom capacities. We assume a homogeneous transportation fleet, but this assumption can easily be avoided by slightly adapting the model proposed in the following section. Each delivery includes the demand of all articles until the next delivery day. In contrast to the model proposed by Sternbeck and Kuhn [5], we explicitly model bundling effects in transportation across stores and determine the base tours that build the basis for short-term vehicle routing.

2 Model Formulation

Considering the setting and assumptions above, we propose the following binary program to model the joint delivery pattern and base tour decision:

$$Minimize \sum_{f \in F} \sum_{r \in R} x_{f,r} \cdot \hat{c}_{f,r} + \sum_{k \in K} \sum_{t \in T} y_{k,t} \cdot c_k^{transtour} + \sum_{f \in F} \sum_{k \in K} \sum_{r \in R} x_{f,k,r} \cdot c_{f,k,r}^{transstop} \tag{1}$$

s.t.

$$\sum_{r \in R} x_{f,r} = 1 \qquad\qquad \forall f \in F \tag{2}$$

$$\sum_{k \in K} x_{f,k,r} = x_{f,r} \qquad\qquad \forall f \in F, r \in R \tag{3}$$

$$\sum_{f \in F} \sum_{r \in R} x_{f,k,r} \cdot pall_{f,r,t} \leq y_{k,t} \cdot cap^{trans} \qquad\qquad \forall k \in K, t \in T \tag{4}$$

$$\sum_{f \in F} \sum_{r \in R} x_{f,r} \cdot pick_{f,r,t} \leq maxcap_t^{pick} \qquad\qquad \forall t \in T \tag{5}$$

$$\sum_{f \in F} \sum_{r \in R} x_{f,r} \cdot pick_{f,r,t} \geq mincap_t^{pick} \qquad\qquad \forall t \in T \tag{6}$$

$$\sum_{r \in R} x_{f,r} \cdot pall_{f,r,t} \leq cap_f^{rec} \qquad\qquad \forall f \in F, t \in T \tag{7}$$

$$x_{f,r} \in \{0; 1\} \qquad\qquad \forall f \in F, r \in R \tag{8}$$

$$x_{f,k,r} \in \{0; 1\} \qquad\qquad \forall f \in F, k \in K, r \in R \tag{9}$$

$$y_{k,t} \in \{0; 1\} \qquad\qquad \forall k \in K, t \in T \tag{10}$$

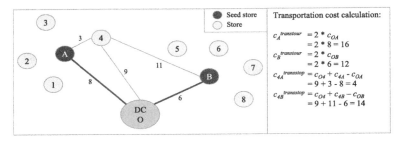

Fig. 1 Determination of transportation costs: schematic illustration and calculation example

The objective function (1) minimizes the sum of total operational costs. These consist of the costs independent of the delivery tour settings $\hat{c}_{f,r}$ and tour-specific costs. While costs independent of the tour settings solely depend on which pattern r is assigned to store f, reflected in the binary decision variable $x_{f,r}$, tour-specific costs also depend on the assignment of a store f to a delivery cluster k, reflected in the two other types of decision variables $x_{f,k,r}$ and $y_{k,t}$, which decides if a cluster k is supplied on day t. The associated tour-specific costs are costs reflecting tour length and stops, $c_k^{transtour}$ and $c_{f,k,r}^{transstop}$. The underlying idea of the determination of tour-specific transportation costs is the logic developed by Fisher and Jaikumar [2]. They approximate tour costs on a tactical level by assigning stores to seed points. They thereby construct clusters that build so-called base tours for which the exact sequence is not known, but transportation costs can be approximated quite accurately. Figure 1 illustrates the determination of transportation costs.

Besides the objective function, the model consists of several constraints (2)–(10). The first two conditions (2) and (3) ensure that each store f gets exactly one delivery pattern r, and is assigned to exactly one cluster k. Constraint (4) ensures that transportation capacity is met and a delivery to cluster k on day t only takes place if at least one of the stores assigned to cluster k is served on this day. Conditions (5) and (6) ensure that the picking effort at the DC is within the range $[mincap_t^{pick}; maxcap_t^{pick}]$ on each day. Constraint (7) ensures that the receiving capacity restriction holds. The decision variables are defined in conditions (8)–(10).

3   Solution Approach

As computational efforts can be extensive when applying the MIP proposed with its simultaneous clustering decision and pattern selection, we propose a sequential approach. The methodology is a cluster-first, assign-second procedure. In a first step, stores are assigned to clusters. This assignment can be done on the basis of predefined delivery areas of service providers or via a two-step algorithm. First, seed points are set (e.g., by applying a clustering algorithm), and then an adapted algorithm on the basis of Fisher and Jaikumar [2] can be used to assign stores to clusters. The clustering step results in an assignment parameter $z_{f,k}$ that assigns store f to cluster k. After determining the clusters a model for assigning delivery patterns to stores can be applied that is analogous to the model presented in Sect. 2, with the main difference being that no store-cluster assignment decision is incorporated. The decision variable $x_{f,k,r}$ is replaced by the combination of decision variable $x_{f,r}$ and parameter $z_{f,k}$.

4 Numerical Example

To illustrate the results of the model and approach proposed, we use real-life data from a major European grocery chain. The sample taken for this example consists of 35 small- and medium-sized stores attached to one central DC. The product ranges are store-specific and amount to a maximum of 618 different articles. Shelf space capacities and demand are also store-specific. For supplying the 35 stores, five trucks are available with capacity constraints according to the share of the products focused on the total assortment. To limit the analysis, a sequential approach is used, clustering the stores into 10 clusters according to geographical and sales issues in a first step. Applying the data obtained, CPLEX solves the model within a few seconds.

The results show a distribution of operational costs to the subsystems of the retail supply chain that is analogous to findings presented in literature: DC 27%, transportation 30%, and instore 43%. The stores are served 1.86 times per week on average, with a maximum of three and a minimum of one delivery per week. As the sample consists of merely small- and medium-sized stores, this result is reasonable. The capacity utilization of the trucks is 87% on average. There are 2.24 stops per tour on average. This means that the logic of Fisher and Jaikumar can serve as a good approximation of transportation costs because of a manageable amount of stops per tour on average. Post-calculations show that the transportation cost approximation

Cluster 1

Store	Mon	Tue	Wed	Thu	Fri	Sat
1	X	0	X	0	0	0
2	X	0	0	X	0	0
3	0	0	X	0	0	0
4	X	0	0	X	0	0

Cluster 2

Store	Mon	Tue	Wed	Thu	Fri	Sat
5	0	X	0	0	X	0
6	0	X	0	0	0	X
7	0	0	X	0	0	X
8	0	0	X	0	0	X

X: delivery

Fig. 2 Numerical results: delivery patterns of stores of two clusters

only differs by less than 1 % from the real transportation costs that apply to the resulting tours.

A main criticism of the existing approach of Sternbeck and Kuhn [5] was the missing incorporation of bundling effects in transportation across stores. Figure 2 illustrates the delivery patterns of the stores of two clusters and shows the synchronization that is obtained with our new model and solution approach. The stores of cluster 1 are served by specific tours on Monday, Wednesday and Thursday. The stores of cluster 2 are mainly served on Tuesday, Wednesday and Saturday. An extra delivery for store 5 is needed on Friday due to transportation capacity limits.

5 Summary

We proposed a binary programming model that considers the decision-relevant costs and capacities at the DC, in transportation and instore to solve the problem of determining optimal store delivery patterns for grocery chains. To resolve the trade-off between the different cost components aligned to the delivery pattern decision, we proposed a sequential solution approach. We revealed significant bundling potential by applying the model and approach proposed using a case from a major European grocery retailer. The logistics levers identified can be applied to further optimize delivery patterns in retail distribution.

References

1. Cardos, M., Garcia-Sabater, J.P.: Designing a consumer products retail chain inventory replenishment policy with the consideration of transportation costs. Int. J. Prod. Econ. **104**(2), 525–535 (2006)
2. Fisher, M.L., Jaikumar, R.: A generalized assignment heuristic for vehicle routing. Networks **11**(1981), 109–124 (1981)
3. Kuhn, H., Sternbeck, M.G.: Integrative retail logistics—an exploratory study. Oper. Manage. Res. **6**(1–2), 2–18 (2013)
4. Ronen, D., Goodhart, C.A.: Tactical store delivery planning. J. Oper. Res. Soc. **59**(8), 1047–1054 (2008)
5. Sternbeck, M.G., Kuhn, H.: An integrative approach to determine store delivery patterns in grocery retailing. Transp. Res. Part E: Logist. Transp. Rev. **70**(1), 205–224 (2014)

Towards a Customer-Oriented Queuing in Service Incident Management

Peter Hottum and Melanie Reuter-Oppermann

Abstract The provision of services hinges considerably on the contribution of the provider and the customer and—if present—on their involved networks. In this paper we focus on incident management—a service domain that is highly relevant for all kinds of industries and is described from a provider internal perspective in the ITIL documentation.

1 Introduction

The provision of services hinges considerably on the contribution of the provider and the customer and—if present—on their involved networks. In this paper we focus on incident management—a service domain that is highly relevant for all kinds of industries and is described from a provider internal perspective in the ITIL documentation [1].

By understanding the influence of a customer's contribution to a service, the provider should be able to improve the interaction quality in general. Furthermore, the provider should be able to determine and control his effort based on the expected customer's contribution.

In incident management, tickets can arrive by call, email or web interface. For this research we just assume tickets to arrive by web interface as done by many big companies. This has two implications: On the one hand, tickets have a predefined structure, such as a predefined content in general, and on the other hand, the interactions between the customer and the provider are asynchronous—therefore it is possible to collect tickets for some time and then assign them using the knowledge about the other tickets in the queue. This results in an online problem with lookahead or in the extreme case even in an offline problem if we collect all tickets that arrive

P. Hottum (✉) · M. Reuter-Oppermann (✉)
KSRI, Karlsruhe Institute of Technology (KIT), Karlsruhe, Germany
e-mail: peter.hottum@kit.edu

M. Reuter-Oppermann
e-mail: melanie.reuter@kit.edu

© Springer International Publishing Switzerland 2016
M. Lübbecke et al. (eds.), *Operations Research Proceedings 2014*,
Operations Research Proceedings, DOI 10.1007/978-3-319-28697-6_33

within a certain period of time (e.g., a day or a week) and schedule them the next period (e.g., next day or week). It also means that the content of the tickets can be analyzed and the tickets can therefore be categorized. In contrast, in a regular call center tickets often have to be assigned right away. In addition, no incident ticket would quit the queue for new tickets before scheduling, whereas waiting customers would do, if their processing lasts too long.

In previously conducted studies, we have derived result influencing factor classes and instantiated a framework based on qualitatively and textual analyzed service incident tickets from a worldwide operating IT service provider. We have proven the customer induced contribution to the service generation and aggregated a customer contribution factor (ccf). By complementing these provider-centric service processes with the ccf, we are able to use information about the customer's ability to contribute, which was not able to process before, in order to classify the tickets in more detail.

The aim is to build a decision support tool that assigns tickets to servers (agents in service incident management) based on a set of rules depending on the underlying objectives and including ticket characteristics as well as the ccf.

In the paper at hand, we address the question: How can the customer's potential to contribute be used to organize the queuing in service incident management in a customer-oriented way? We present and solve a mathematical formulation for assigning tickets to servers and discuss first results of a discrete event simulation. We use this simulation to compare basic assignment rules based on the ticket complexity and the servers' level of experience to the optimal solution. We also study the impact of the ccf in a small example. In addition, run time experiments for the formulation are presented.

2 Problem Formulation and Solution Approach

Service providers in incident management have to handle different topics on several levels of complexity. For each incident ticket providers have to determine topic and expertise level needed to solve the incident. In the model each incoming ticket is given a set of attributes and each service agent has a specific skill level for each topic he or she is working on.

Figure 1 visualizes the problem of assigning tickets to agents and the setup that is formalized in this short paper. If new incident tickets are reported to the incident management web interface, it has to be decided which agent should work on it. This depends on different aspects: which agent is currently available and fits best to the present topic and necessary expert level? Agents are allowed to work on an incident with a complexity equal or less compared to their expertise level, but not higher.

For the deterministic formulation we assume a set of incoming service incident tickets C including $|C|$ tickets that are known but that cannot be handled before their release date, i.e., arrival time, e_c for all $c \in C$. This is needed to compare the results to the ones of the simulation. Each ticket $c \in C$ has a processing time d_c. The set of topics is represented by T. As described, tickets that arrive have a different level of

Fig. 1 Examined scenario of ticket scheduling

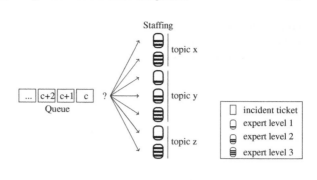

complexity, which we represent be the set of levels L. The binary parameter g_{ctl} is equal to 1 if ticket $c \in C$ has topic $t \in T$ with complexity level $l \in L$ whereas for each ticket $c \in C$ only one parameter is equal to 1. The tickets are handled by a set of agents S. Each agent can only serve a defined subset of topics $t \in T$ and for each topic he has a certain knowledge level that matches with the levels of complexity $l \in L$. f_{stl} is equal to 1 if agent $s \in S$ can solve a ticket with topic $t \in T$ at level $l \in L$ and 0 else. Due to work regulations and as agents are the most valuable resource, their workload should not exceed α percent of the daily working time W. By P we denote the number of consecutive days we are looking at, i.e., the length of the considered period. We assume that it is possible to schedule all tickets within the planning horizon and that each agent is only able to work on one ticket at a time. By M we denote a sufficiently large number. For the formulation, we need three artificial ticket sets C_0 that includes 0 as a starting point, C_1 that includes $|C| + 1$ as an ending point and C_{01} that includes both 0 and $|C| + 1$. In addition, we introduce the following decision variables:

$$x_{bcs} = \begin{cases} 1 & \text{if agent } s \in S \text{ solves ticket } c \in C_1 \text{ after ticket } b \in C_0 \\ 0 & \text{else} \end{cases}$$

$$y_{cs} = \begin{cases} 1 & \text{if agent } s \in S \text{ solves ticket } c \in C \\ 0 & \text{else} \end{cases}$$

$z_{cs} \geq 0$ starting time for solving ticket $c \in C_{01}$ by server $s \in S$.

The objective function (1) minimizes the workload for the agents with the highest skill levels. Constraints (2) assure that the service of a ticket cannot start before the release date, constraints (3)–(5) set the starting times for serving the tickets. By (6) all tickets must be finished within the planning horizon. Of course an agent can only serve a ticket with the right topic and level that he is able to solve (7). Constraints (8)–(11) make sure that we start and end a schedule for each agent once, that each ticket is served and that the same agent starts and ends serving a ticket. We assume that each agent has to solve at least one ticket. Agents shall not work more than α % of the daily working hours in average throughout the considered period as expressed in (12). Constraints (13) and (14) connect the decision variables. (15)–(17) are the domain constraints. Note that if the formulation shall be used in practice, constraints (2) are not needed, of course.

The formulation looks as follows:

$$\min \quad \sum_{b,c\in C, b\neq c}\sum_{s:f_{st|L|}=1} x_{bcs}d_c \tag{1}$$

$$\text{s.t.} \qquad z_{cs} \geq e_c \cdot y_{cs} \qquad\qquad \forall c \in C, s \in S \tag{2}$$

$$z_{cs} \geq z_{bs} + d_b - M\,(1 - x_{bcs}) \qquad \forall b, c \in C, s \in S \tag{3}$$

$$z_{cs} \geq z_{0s} - M\,(1 - x_{0cs}) \qquad\qquad \forall c \in C, s \in S \tag{4}$$

$$z_{|C|+1,s} \geq z_{cs} + d_c - M\,\big(1 - x_{c,|C|+1,s}\big) \;\; \forall c \in C, s \in S \tag{5}$$

$$z_{|C|+1,s} \leq P \cdot W \qquad\qquad \forall s \in S \tag{6}$$

$$y_{cs} \leq \sum_{t\in T}\sum_{l\in L}\,(g_{ctl}\,f_{stl}) \qquad \forall c \in C, s \in S \tag{7}$$

$$\sum_{s\in S} y_{cs} = 1 \qquad\qquad \forall c \in C \tag{8}$$

$$\sum_{c\in C} x_{0cs} = 1 \qquad\qquad \forall s \in S \tag{9}$$

$$\sum_{c\in C} x_{c,|C|+1,s} = 1 \qquad\qquad \forall s \in S \tag{10}$$

$$\sum_{a\in C_0, b\neq a} x_{abs} - \sum_{c\in C_1, b\neq c} x_{bcs} = 0 \quad \forall b \in C, s \in S \tag{11}$$

$$\sum_{c\in C} y_{cs}d_c \leq \alpha \cdot P \cdot W \qquad\qquad \forall s \in S \tag{12}$$

$$y_{cs} \geq \sum_{b\in C_0} x_{bcs} \qquad\qquad \forall c \in C_1, s \in S \tag{13}$$

$$z_{cs} \leq M \cdot y_{cs} \qquad\qquad \forall c \in C, s \in S \tag{14}$$

$$x_{bcs} \in \{0, 1\} \qquad\qquad \forall b, c \in C, s \in S \tag{15}$$

$$y_{cs} \in \{0, 1\} \qquad\qquad \forall c \in C, s \in S \tag{16}$$

$$z_{cs} \geq 0 \qquad\qquad \forall c \in C, s \in S \tag{17}$$

Based on already examined studies in that domain [2–5] we assume the following conditions for an example scenario that we want to study in a discrete event simulation as well as use as an input for the formulation:

We examine the incident management of a medium-sized company. Seven employees with different levels of expertise are working on their day-to-day operations and additionally have to solve incidents that are reported by customers via the company's incident management web interface. We assume an equal distribution of daily business and incident management. Furthermore, we assume an average availability of each expert of less than 70 % of the working time (a so called "shrinkage" with over 30 %), which results in a maximum workload of 35 % per expert for incident management tasks in general. Each expert could gain a level of expertise from low (1) to medium (2) to high (3) for each topic. In our model there are tickets in the domains of 3 different topics (topic x, topic y, and topic z). Each ticket has a complexity of low (1) or medium (2) or high (3). The agents work on the tickets on maximum five days per week for 8 h. The incidents, reported via the incident management web interface, are Poisson distributed with an arrival rate of 1/50 min. The customer contribution is rate-able for each ticket as ccf from low (0) to medium (1) to high (2).

The discussed approach is based on the assumption that a high customer contribution has at least two effects on incident management—incidents could better be identified and solved and incidents are less complicated as there is more information given to solve them.

The first aspect has an influence on the processing time (time to resolve) that is calculated in the model by $20\,min + max(complexity - ccf; 0) \cdot t$ where t is normally distributed with a mean of 60 min and a standard deviation of 10 min.

The second aspect has an influence on the requirements of the agent's expert level, that is represented in the model as follows $complexity^* = complexity - \lfloor ccf \cdot s + 0.5 \rfloor$ where s as the provider's sensitivity for customer contribution is distributed according to the Beta distribution and the product of ccf and s is rounded-to-nearest. By skewing the Beta distribution to the left or right, the sensitivity of the provider's setup and applied scheduling methods could be represented. In this short paper a balanced Beta distribution with $p = q = 2$ was chosen.

An incident ticket always has to be scheduled to the available agent with the lowest expert level. This is important to give the highly educated (and therefore higher paid) experts more time for solving issues in their day-to-day operations. Every incident ticket in the queue is scheduled based on $complexity^*$ by the first-come-first-serve principle.

3   Computational Results

Based on the above described model, we simulated the scheduling of tickets and the utilization of corresponding agents with AnyLogic to also study the impact of the percentage of tickets with a high ccf. Therefore we used ten base seeds each to reduce variations for different shares of tickets with a high customer contribution— from 0 to 1 in steps by 1 %. The remaining share of tickets with a low and a medium customer contribution have been divided equally.

In addition, we solved the formulation on an Intel Core i7-4600U CPU with CPLEX 12.6 using the same data as in the simulation for the instances from 0 to 1 in steps by 10 %.

In Fig. 2a the effects of different shares of tickets with a high ccf on the utilization of the agents is presented. Given a maximum utilization limit of 35 % per agent, it is obvious that with an increase of ccf the utilization of agents with higher expert levels reduces. Additionally, a minimum utilization of those agents with higher expert levels can be determined by solving the presented formulation. It shows the potential of collecting the tickets first and then scheduling them offline instead of online while they are arriving. In Fig. 2b the runtime of CPLEX for different settings of tickets and agents as well as variations of the model is described. It can be seen that the model is calculable, with the restriction that the processing time grows exponential.

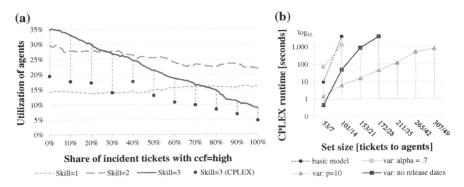

Fig. 2 **a** Utilization of service agents relatively to the customer contribution. **b** Runtime for solving
the scenarios with different variations

As we designed the scenario motivated by practice, determining an optimal solution
for a whole week (53 tickets to 7 agents) or even for larger problem instances could
be done with reasonable effort. By varying restrictions the processing time could be
reduced considerably.

4 Conclusions and Recommendations for Further Research

In this short paper it could be shown that the customer contribution factor ccf can
help to reduce the unbalanced utilization of service agents by assigning tickets to
agents that are able to handle them properly. By applying information about their
customers, providers could be able to save resources and time internally and—at the
same time—serve their customers more individual, faster and with no more effort.

Within this short paper we were not able to apply our approach to real world
data, where we took our motivation and initial set up from. As the exact cause effect
relationships of the ccf are estimated in the starting model, the next step in our
research is to prove these effects with the real interaction data, captured with our
application partner. From a mathematical point of view, we will use queuing theory
to further study waiting times, business of agents and the time a ticket stays in the
system. In addition, further tests with real data are necessary to determine running
times for an optimal solution if daily runs and schedules are desired.

References

1. Steinberg, R.A.: ITIL Service Operation—2011 Edition, 2nd edn. TSO, Belfast (2011)
2. Giurgiu, I., Bogojeska, J., Nikolaiev, S., Stark, G., Wiesmann, D.: Analysis of labor efforts
   and their impact factors to solve server incidents in datacenters. In: Proceedings of the 14th

IEEE/ACM International Symposium on Cluster, Cloud and Grid Computing, pp. 424–433 (2014)
3. Reynolds, P.: Call center metrics—Fundamentals of call center staffing and technologies. NAQC Issue Paper. Phoenix (2010)
4. Mazzuchi, T.A., Wallace, R.B.: Analyzing skill-based routing call centers using discrete-event simulation and design experiment. In: Proceedings of the 2004 Winter Simulation Conference, pp. 1812–1820 (2004)
5. Mehrotra, V., Fama, J.: Call center simulation modeling—Methods, challenges, and opportunities. In: Proceedings of the 2003 Winter Simulation Conference, pp. 135–143 (2003)

A Lagrangian Relaxation Algorithm for Modularity Maximization Problem

Kotohumi Inaba, Yoichi Izunaga and Yoshitsugu Yamamoto

Abstract Modularity proposed by Newman and Girvan is one of the most common measure when the nodes of a graph are grouped into communities consisting of tightly connected nodes. We formulate the modularity maximization problem as a set partitioning problem, and propose an algorithm based on the Lagrangian relaxation. To alleviate the computational burden, we use the column generation technique.

1 Introduction

As social network services grow, clustering on graphs has been attracting more attention. Since Newman and Girvan [5] proposed the modularity as a graph clustering measure, modularity maximization problem became one of the central subjects of research. Most of the solution methods proposed so far are heuristic algorithms due to its *NP*-hardness, which was shown by Brandes et al. [2], while few exact algorithms have been proposed.

Aloise et al. [1] formulated the problem as a set partitioning problem, which has to take into account all, exponentially many, nonempty subsets of the node set. Therefore one cannot secure the computational resource to hold the problem when the number of nodes is large. Their algorithm is based on the linear programming relaxation, and uses the column generation technique. Although it provides a tight

K. Inaba · Y. Izunaga (✉) · Y. Yamamoto
University of Tsukuba, Ibaraki 305-8573, Japan
e-mail: s1130131@sk.tsukuba.ac.jp

K. Inaba
e-mail: s1320486@sk.tsukuba.ac.jp

Y. Yamamoto
e-mail: yamamoto@sk.tsukuba.ac.jp

© Springer International Publishing Switzerland 2016   241
M. Lübbecke et al. (eds.), *Operations Research Proceedings 2014*,
Operations Research Proceedings, DOI 10.1007/978-3-319-28697-6_34

upper bound of the optimal value, it can suffer a high degeneracy due to the set partitioning constraints.

In this paper, based on the set partitioning formulation, we propose a Lagrangian relaxation algorithm, and apply the column generation technique in order to alleviate the computational burden. We also report on some computational experiments.

2   Modularity Maximization Problem

Let $G = (V, E)$ be an undirected graph with the set $V = \{1, 2, \ldots, n\}$ of n nodes and the set $E = \{1, 2, \ldots, m\}$ of m edges. We say that $\Pi = \{C_1, C_2, \ldots, C_k\}$ is a *partition* of V if $V = \bigcup_{p=1}^{k} C_p$, $C_p \cap C_q = \emptyset$ for any distinct p and q, and $C_p \neq \emptyset$ for any p. Each member C_p of a partition is called a *community*. For $i, j \in V$ let e_{ij} be the (i, j) element of the adjacency matrix of graph G, and d_i be the degree of node i, and $\pi(i)$ be the index of community which node i belongs to, i.e., $\pi(i) = p$ means $i \in C_p$. Then *modularity*, denoted by $Q(\Pi)$, of a partition Π is defined as

$$Q(\Pi) = \frac{1}{2m} \sum_{i \in V} \sum_{j \in V} \left(e_{ij} - \frac{d_i d_j}{2m} \right) \delta(\pi(i), \pi(j)),$$

where δ is the Kronecker delta. *Modularity maximization problem*, *MM* for short, is the problem of finding a partition of V that maximizes the modularity $Q(\Pi)$.

Let $\mathscr{P}$ denote the family of all nonempty subsets of V. Note that $\mathscr{P}$ is composed of $2^n - 1$ subsets of V. Introducing a binary variable z_C for each $C \in \mathscr{P}$, a partition Π is represented by the $(2^n - 1)$-dimensional binary vector $z = (z_C)_{C \in \mathscr{P}}$ defined as

$$z_C = \begin{cases} 1 & \text{when } C \in \Pi \\ 0 & \text{otherwise.} \end{cases}$$

For each $i \in V$ and $C \in \mathscr{P}$ we define a constant a_{iC} to describe whether node i belongs to C, i.e., $a_{iC} = 1$ when $i \in C$ and $a_{iC} = 0$ otherwise. The column $a_C = (a_{1C}, \ldots, a_{nC})^{\top}$ is called the incidence vector of community C, i.e., $C = \{i \in V \mid a_{iC} = 1\}$. For each $C \in \mathscr{P}$, let f_C be

$$f_C = \frac{1}{2m} \sum_{i \in V} \sum_{j \in V} w_{ij} a_{iC} a_{jC},$$

where $w_{ij} = (e_{ij} - d_i d_j / 2m)$. The constant f_C represents the contribution of community C to the objective function when C is selected as a member of the partition Π. Thus *MM* is formulated as the following integer programming (P):

$$(P) \quad \begin{array}{l} \text{maximize } \sum_{C \in \mathscr{P}} f_C z_C \\[2mm] \text{subject to } \sum_{C \in \mathscr{P}} a_{iC} z_C = 1 \ (\forall i \in V) \\[2mm] \qquad\qquad z_C \in \{0, 1\} \qquad (\forall C \in \mathscr{P}) \end{array}$$

We call the first set of constraints *set partitioning constraints*.

3 Lagrangian Relaxation and Lagrangian Dual Problem

The problem (P) is a difficult problem due to both its integrality and the set partitioning constraints. The well-known technique in order to obtain the useful information about the solution of (P) is *Linear Programming relaxation*, LP relaxation for short. Although LP relaxation provides a tight upper bound of the optimal value of (P), it usually suffers the high degeneracy due to the set partitioning constraints. To overcome this degeneracy, several techniques have been proposed in the literature, for example [3, 4]. In this paper we employ the *Lagrangian relaxation* instead of LP relaxation. Now we will give a brief review of Lagrangian relaxation and Lagrangian dual problem.

We relax the set partitioning constraints and add them to the objective function as a penalty with Lagrangian multiplier vector $\lambda = (\lambda_1, \ldots, \lambda_n)^\top$, and obtain the following Lagrangian relaxation problem $(LR(\lambda))$ with only the binary variable constraints:

$$(LR(\lambda)) \quad \begin{array}{l} \text{maximize } \sum_{C \in \mathscr{P}} f_C z_C + \sum_{i \in V} \lambda_i (1 - \sum_{C \in \mathscr{P}} a_{iC} z_C) \\[2mm] \text{subject to } z_C \in \{0, 1\} \qquad (\forall C \in \mathscr{P}). \end{array}$$

Let $\gamma_C(\lambda) = f_C - \sum_{i \in V} \lambda_i a_{iC}$, then the objective function of $(LR(\lambda))$ is written as

$$L(z, \lambda) = \sum_{C \in \mathscr{P}} \gamma_C(\lambda) z_C + \sum_{i \in V} \lambda_i.$$

For a given multiplier vector λ, we can obtain an optimal solution $z(\lambda)$ of $(LR(\lambda))$ by simply setting $z_C(\lambda) = 1$ if $\gamma_C(\lambda) > 0$, and $z_C(\lambda) = 0$ otherwise. We denote the optimal value of $(LR(\lambda))$ by $\omega(LR(\lambda))$, then $\omega(LR(\lambda))$ provides an upper bound of $\omega(P)$ for any λ. The problem of finding the best upper bound of $\omega(P)$ is called the Lagrangian dual problem (LRD), which is given as:

$$(LRD) \quad \begin{array}{l} \text{minimize } \omega(LR(\lambda)) \\ \text{subject to } \lambda \in \mathbb{R}^n. \end{array}$$

One of the most commonly used method for this problem is the subgradient method. This method uses the subgradient $d(\lambda) = (d_i(\lambda))_{i \in V}$ at λ, defined by $d_i(\lambda) = 1 - \sum_{C \in \mathcal{P}} a_{iC} z_C(\lambda)$ for $i \in V$, and updates the Lagrangian multiplier vector to the direction of $d(\lambda)$ with a step size μ. We employ the well-known Armijo rule to determine the step-size μ.

4 Proposed Algorithm

As we discussed in the previous section, the optimal solution $z(\lambda)$ can be obtained by checking the sign of $\gamma_C(\lambda)$. However it is hard to compute all of $\gamma_C(\lambda)$ owing to the huge number of variables. The number of variables which are positive at an optimal solution of (P) is at most the number of nodes, hence we need only a small number of variables. Therefore we use the column generation technique in order to alleviate the computation burden. Namely, we start the algorithm with a small number of variables and gradually add variables as the computation goes on.

We consider a small subfamily $\mathcal{C}$ of $\mathcal{P}$ and deal with the following subproblem $(P(\mathcal{C}))$:

$$(P(\mathcal{C})) \quad \left|
\begin{array}{l}
\text{maximize} \displaystyle\sum_{C \in \mathcal{C}} f_C z_C \\[2ex]
\text{subject to} \displaystyle\sum_{C \in \mathcal{C}} a_{iC} z_C = 1 \ (\forall i \in V) \\[2ex]
z_C \in \{0, 1\} \qquad (\forall C \in \mathcal{C}).
\end{array}
\right.$$

We denote the the Lagrangian relaxation problem and the Lagrangian dual problem corresponding to $(P(\mathcal{C}))$ by $(LR(\mathcal{C}, \lambda))$ and $(LRD(\mathcal{C}))$, respectively. Let $\lambda(\mathcal{C})$ be an optimal solution of $(LRD(\mathcal{C}))$. Since the variables z_C for $C \in \mathcal{P} \setminus \mathcal{C}$ are not considered in the problem $(LR(\mathcal{C}, \lambda(\mathcal{C})))$, an optimal solution $z(\lambda(\mathcal{C}))$ is not necessarily optimal to $(LR(\lambda(\mathcal{C})))$. When $\gamma_C(\lambda(\mathcal{C})) \leq 0$ for all $C \in \mathcal{P} \setminus \mathcal{C}, z(\lambda(\mathcal{C}))$ is an optimal solution of $(LR(\mathcal{C}, \lambda(\mathcal{C})))$. On the other hand $\gamma_C(\lambda(\mathcal{C})) > 0$ holds for some $C \in \mathcal{P} \setminus \mathcal{C}$, adding this C to $\mathcal{C}$ can lead to an improvement of the optimal value of $(LR(\mathcal{C}, \lambda(\mathcal{C})))$, i.e., $\omega(LR(\mathcal{C}', \lambda(\mathcal{C}))) > \omega(LR(\mathcal{C}, \lambda(\mathcal{C})))$ where $\mathcal{C}' = \mathcal{C} \cup \{C\}$. Note that $\lambda(\mathcal{C})$ is not necessarily an optimal solution of $(LRD(\mathcal{C}'))$, hence we solve the problem $(LRD(\mathcal{C}'))$ again to obtain an optimal Lagrangian multiplier $\lambda(\mathcal{C}')$ by the subgradient method.

According to the formulation of Xu et al. [6], the problem of finding C that maximizes $\gamma_C(\lambda)$ is formulated as the problem $(AP(\lambda))$ with a quadratic concave objective function:

$$(AP(\lambda)) \quad \begin{vmatrix} \text{maximize } \dfrac{1}{m} \sum_{r=1}^{m} x_r - \dfrac{1}{4m^2} \left(\sum_{i \in V} d_i y_i \right)^2 - \sum_{i \in V} \lambda_i y_i \\ \text{subject to } x_r \le y_i \quad (\forall r = \{i, j\} \in E) \\ \qquad\qquad\; x_r \le y_j \quad (\forall r = \{i, j\} \in E) \\ \qquad\qquad\; x_r \in \{0, 1\} \quad (\forall r \in E) \\ \qquad\qquad\; y_i \in \{0, 1\} \quad (\forall i \in V). \end{vmatrix}$$

For each edge $r = \{i, j\} \in E$, a binary variable x_r is equal to 1 when both end nodes i, j of edge r belong to the community that maximizes $\gamma_C(\lambda)$, and for each $i \in V$ a variables y_i is equal to 1 when node i belongs to the community and 0 otherwise.

From the above discussion, our proposed algorithm is given as follows.

Algorithm LCG

Step 1 : Let $\mathscr{C}$ and λ be an initial family of nonempty subsets of V and an initial multiplier vector, respectively.

Step 2 : Solve $(LRD(\mathscr{C}))$ to obtain a near optimal solution λ and the objective value $\omega(LRD(\mathscr{C}))$ by the subgradient method.

Step 3 : Solve $(AP(\lambda))$ and set y^* be an optimal solution.

Step 4 : If $\omega(AP(\lambda)) \le 0$, then set $\mathscr{C}^* := \mathscr{C}$ and $\omega^* := \omega(LRD(\mathscr{C}))$. Output $\mathscr{C}^*$ and ω^*, and terminate.

Step 5 : Otherwise set $C := \{i \in V \mid y_i^* = 1\}$ and increment $\mathscr{C} := \mathscr{C} \cup \{C\}$. Return to Step 2.

When this algorithm terminates, we construct the problem $(P(\mathscr{C}^*))$ from the obtained $\mathscr{C}^*$, and solve $(P(\mathscr{C}^*))$ by an IP solver.

The following proposition shows that we can obtain an upper bound of $\omega(P)$ at each iteration of the algorithm.

Proposition 1 *Let t be an upper bound of the number of communities at an optimal solution of (P). Then $\sum_{i \in V} \lambda_i + t \cdot \omega(AP(\lambda))$ is an upper bound of $\omega(P)$ for any $\lambda \in \mathbb{R}^n$.*

If the difference between the upper bound and $\omega(LRD(\mathscr{C}))$ is small, we can stop the algorithm even if $\omega(AP(\lambda)) \le 0$ does not hold.

5 Computational Experiments

We report the computational experiment with Algorithm LCG. The experiment was performed on a PC with an Intel Core i7, 3.20 GHz processor and 12.0 GB of memory. We implemented the algorithm in Python 2.7, and used Gurobi 5.6.2 as the IP solver. We solved the benchmark instances provided by DIMACS. The size and the known optimal value of each instance is given in Table 1.

Table 1 Instances

Name	n	m	$\omega(P)$
Karate	34	78	0.4198
Dolphins	62	159	0.5285
Football	115	613	0.6046

Table 2 Computational results of Algorithm LCG

Instance	$|\mathscr{C}^*|$	ω^*	$\omega(P(\mathscr{C}^*))$	Gap (%)	Time (s)
Karate	62	0.4198	0.4198	0.000	7
Dolphins	112	0.5302	0.5222	1.192	37
Football	192	0.6054	0.6043	0.049	34509

We set $\mathscr{C}$ initially to the family of all singletons, i.e., $\mathscr{C} = \{\{1\}, \ldots, \{n\}\}$, and set an initial multiplier vector $\lambda = 0$. Table 2 shows the results of the proposed algorithm for each instance. The columns $|\mathscr{C}^*|$ and $\omega(P(\mathscr{C}^*))$ represent the cardinality of the final family of $\mathscr{C}^*$ and the optimal value of $(P(\mathscr{C}^*))$, respectively. The column Gap indicates relative gap defined by

$$\text{Gap} = \left(\frac{\omega(P) - \omega(P(\mathscr{C}^*))}{\omega(P)} \right) \times 100.$$

The column Time indicates the computation time in seconds.

From Table 2, we observe that Algorithm LCG solves Karate to optimality and fails to solve the others, but the Gap is less than 2 %. Moreover the number of $|\mathscr{C}^*|$ is quite small.

Figure 1 shows $\omega(LRD(\mathscr{C}))$ and the upper bound in Proposition 1 at each iteration of the algorithm for the instance Dolphins. We set t to the optimal number of

Fig. 1 $\omega(LRD(\mathscr{C}))$ versus iterations for Dolphins

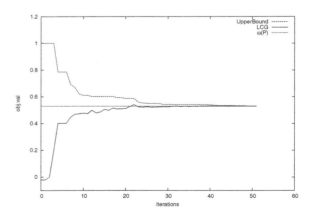

communities in calculating an upper bound of $\omega(P)$. $\omega(LRD(\mathscr{C}))$ rapidly increases at an early stage, and increases slowly as the algorithm goes on. Since we observed the similar results in other instances, we omitted the figures of the others.

References

1. Aloise, D., Cafieri, S., Caporossi, G., Hansen, P., Liberti, L., Pellon, S.: Column generation algorithms for exact modularity maximization in networks. Phys. Rev. E **82** (2012)
2. Brandes, U., Delling, D., Gaertler, M., Görke, R., Hoefer, M., Nikoloski, Z., Wagner, D.: On modularity clustering. IEEE Trans. Knowl. Data Eng. **20**, 172–188 (2008)
3. Boschetti, M., Minggozzi, A., Ricciardelli, S.: A dual ascent procedure for the set partitioning problem. Discrete Optim. **5**, 735–747 (2008)
4. du Merle, O., Villeneuve, D., Desrosiers, J., Hansen, P.: Stabilized column generation. Discrete Math. **194**, 229–237 (1999)
5. Newman, M.E.J., Girvan, M.: Finding and evaluating community structure in networks. Phys. Rev. E **69** (2004)
6. Xu, G., Tsoka, S., Papageorgiou, L.: Finding community structures in complex networks using mixed integer optimization. Eur. Phys. J. B **50**, 231–239 (2007)

Row and Column Generation Algorithm for Maximization of Minimum Margin for Ranking Problems

Yoichi Izunaga, Keisuke Sato, Keiji Tatsumi
and Yoshitsugu Yamamoto

Abstract We consider the ranking problem of learning a ranking function from the data set of objects each of which is endowed with an attribute vector and a ranking label chosen from the ordered set of labels. We propose two different formulations: primal problem, primal problem with dual representation of normal vector, and then propose to apply the kernel technique to the latter formulation. We also propose algorithms based on the row and column generation in order to mitigate the computational burden due to the large number of objects.

1 Introduction

This paper is concerned with a multi-class classification problem of n objects, each of which is endowed with an m-dimensional *attribute vector* $x^i = (x_1^i, x_2^i, \ldots, x_m^i)^\top \in \mathbb{R}^m$ and a *label* ℓ_i. The underlying statistical model assumes that object i receives label k, i.e., $\ell_i = k$, when the latent variable y_i determined by $y_i = w^\top x^i + \varepsilon_i = \sum_{j=1}^m w_j x_j^i + \varepsilon_i$ falls between two thresholds p_k and p_{k+1}, where ε_i represents a random noise whose probabilistic property is not known. Namely, attribute vectors of objects are loosely separated by hyperplanes $H(w, p_k) = \{ x \in \mathbb{R}^m \mid w^\top x = p_k \}$ for $k = 1, 2, \ldots, l$ which share a common normal vector w, then each object is given a label according to the layer it is located in. Note that neither y_i's, w_j's nor p_k's are

Y. Izunaga (✉) · Y. Yamamoto
University of Tsukuba, Ibaraki 305-8573, Japan
e-mail: s1130131@sk.tsukuba.ac.jp

Y. Yamamoto
e-mail: yamamoto@sk.tsukuba.ac.jp

K. Sato
Railway Technical Research Institute,Tokyo 185-8540, Japan
e-mail: sato.keisuke.49@rtri.or.jp

K. Tatsumi
Osaka University, Osaka 565-0871, Japan
e-mail: tatsumi@eei.osaka-u.ac.jp

© Springer International Publishing Switzerland 2016
M. Lübbecke et al. (eds.), *Operations Research Proceedings 2014*,
Operations Research Proceedings, DOI 10.1007/978-3-319-28697-6_35

observable. Our problem is to find the normal vector $w \in \mathbb{R}^m$ as well as the thresholds $p_1, p_2, \ldots, p_l$ that best fit the input data $\{ (x^i, \ell_i) \mid i = 1, 2, \ldots, n \}$.

This problem is known as the *ranking problem* and frequently arises in social sciences and operations research. See, for instance [2–5, 7]. It is a variation of the multi-class classification problem, for which several learning algorithms of the *support vector machine* (*SVM* for short) have been proposed. We refer the reader to [1, 8, 9]. What distinguishes the problem from other multi-class classification problems is that the identical normal vector should be shared by all the separating hyperplanes. In this paper based on the formulation *fixed margin strategy* by Shashua and Levin [5], we propose a row and column generation algorithm to maximize the minimum margin for the ranking problems.

Throughout the paper $N = \{1, 2, \ldots, i, \ldots, n\}$ denotes the set of n objects and $x^i = (x^i_1, x^i_2, \ldots, x^i_m)^\top \in \mathbb{R}^m$ denotes the attribute vector of object i. The predetermined set of labels is $L = \{0, 1, \ldots, k, \ldots, l\}$ and the label assigned to object i is denoted by ℓ_i. Let $N(k) = \{ i \in N \mid \ell_i = k \}$ be the set of objects with label $k \in L$, and for notational convenience we write $n(k) = |N(k)|$ for $k \in L$. For succinct notation we define $X = [x^i]_{i \in N} \in \mathbb{R}^{m \times n}$, $X_W = [x^i]_{i \in W} \in \mathbb{R}^{m \times |W|}$ for $W \subseteq N$, and the corresponding Gram matrices $K = X^\top X \in \mathbb{R}^{n \times n}$, $K_W = X_W^\top X_W \in \mathbb{R}^{|W| \times |W|}$. We denote the k-dimensional zero vector and vector of 1's by $\mathbf{0}_k$ and $\mathbf{1}_k$, respectively.

2 Hard Margin Problem for Separable Case

Henceforth we assume that $N(k) \neq \emptyset$ for all $k \in L$ for the sake of simplicity, and adopt the notational convention that $p_0 = -\infty$ and $p_{l+1} = +\infty$. We say that an instance $\{ (x^i, \ell_i) \mid i \in N \}$ is *separable* if there exist $w \in \mathbb{R}^m$ and $p = (p_1, p_2, \ldots, p_l)^\top \in \mathbb{R}^l$ such that $p_{\ell_i} < w^\top x^i < p_{\ell_i+1}$ for any $i \in N$. Clearly an instance is separable if and only if there are w and p such that $p_{\ell_i} + 1 \leq w^\top x^i \leq p_{\ell_i+1} - 1$ for any $i \in N$.

Then the margin between $\{ x^i \mid i \in N(k - 1) \}$ and $\{ x^j \mid j \in N(k) \}$ is at least $2/\|w\|$. Hence the maximization of the minimum margin is formulated as the quadratic programming

$$(H) \quad \left| \begin{array}{l} \text{minimize } \|w\|^2 \\ \text{subject to } p_{\ell_i} + 1 \leq (x^i)^\top w \leq p_{\ell_i+1} - 1 \quad \text{for } i \in N. \end{array} \right.$$

The constraints therein are called *hard margin* constraints.

A close look at the primal problem (H) shows that the following property holds for an optimum solution w^*. See, for example [1, 5, 6].

Lemma 1 *Let $(w^*, p^*) \in \mathbb{R}^{m+l}$ be an optimum solution of (H). Then $w^* \in \mathbb{R}^m$ lies in the range space of X, i.e., $w^* = X\lambda$ for some $\lambda \in \mathbb{R}^n$.*

The representation $w = X\lambda$ is called the *dual representation*. Substituting $X\lambda$ for w yields another primal hard margin problem $(\bar{H})$:

$$(\bar{H}) \quad \begin{vmatrix} \text{minimize } \lambda^\top K \lambda \\ \text{subject to } p_{\ell_i} + 1 \le (k^i)^\top \lambda \le p_{\ell_i+1} - 1 \quad \text{for } i \in N, \end{vmatrix}$$

where $(k^i)^\top = ((x^i)^\top x^1, (x^i)^\top x^2, \ldots, (x^i)^\top x^n)$ is the ith row of the matrix K. Since n is typically by far larger than m, problem $(\bar{H})$ might be less interesting than problem (H). However, the dimension m of the attribute vector is usually much smaller than the number of objects, hence we need a small number of attribute vectors for the dual representation, and it is likely that most of the constraints are redundant at an optimal solution. Then we propose to start the algorithm with a small number of attribute vectors as W and then increment it as the computation goes on. Moreover the fact that this formulation only requires the matrix K will enable the application of the kernel technique to the problem. The sub-problem to solve is

$$(\bar{H}(W)) \quad \begin{vmatrix} \text{minimize } \lambda_W^\top K_W \lambda_W \\ \text{subject to } p_{\ell_i} + 1 \le (k_W^i)^\top \lambda_W \le p_{\ell_i+1} - 1 \text{ for } i \in W, \end{vmatrix}$$

where $(k_W^i)^\top$ is the row vector consisting of $(x^i)^\top x^j$ for $j \in W$. Note that the dimension of λ_W varies when the size of W changes as the computation goes on.

Algorithm RC$\bar{H}$ (Row and Column Generation Algorithm for $(\bar{H})$)

Step 1 : Let W^0 be an initial working set, and let $\nu = 0$.
Step 2 : Solve $(\bar{H}(W^\nu))$ to obtain λ_W^ν and p^ν.
Step 3 : Let $\Delta = \{ i \in N \backslash W^\nu \mid (\lambda_W^\nu, p^\nu) \text{ violates } p_{\ell_i} + 1 \le (k_W^i)^\top \lambda_W \le p_{\ell_i+1} - 1 \}$.
Step 4 : If $\Delta = \emptyset$, terminate.
Step 5 : Otherwise choose $\Delta^\nu \subseteq \Delta$, let $W^{\nu+1} = W^\nu \cup \Delta^\nu$, increment ν by 1 and go to Step 2.

The following lemma shows that Algorithm RC$\bar{H}$ solves problem $(\bar{H})$ upon termination.

Lemma 2 *Let $(\hat{\lambda}_W, \hat{p}) \in \mathbb{R}^{|W|+l}$ be an optimum solution of $(\bar{H}(W))$. If*

$$\hat{p}_{\ell_i} + 1 \le (k_W^i)^\top \hat{\lambda}_W \le \hat{p}_{\ell_i+1} - 1 \quad \text{for all } i \in N \backslash W,$$

then $(\hat{\lambda}_W, 0_{N \backslash W}) \in \mathbb{R}^n$ together with $\hat{p}$ forms an optimum solution of $(\bar{H})$.

The validity of the algorithm follows from the above lemma.

Theorem 1 *The Algorithm RC$\bar{H}$ solves problem $(\bar{H})$.*

3 Kernel Technique for Hard Margin Problem

The matrix K in the primal hard margin problem $(\bar{H})$ is composed of the inner products $(x^i)^\top x^j$ for $i, j \in N$. This enables us to apply the *kernel technique* simply by replacing them by $\kappa(x^i, x^j)$ for some appropriate kernel function κ.

Let $\phi\colon \mathbb{R}^m \to \mathbb{F}$ be a function, possibly unknown, from $\mathbb{R}^m$ to some higher dimensional inner product space $\mathbb{F}$, so-called the *feature space* such that $\kappa(x, y) = \langle \phi(x), \phi(y) \rangle$ holds for $x, y \in \mathbb{R}^m$, where $\langle \cdot, \cdot \rangle$ is the inner product defined on $\mathbb{F}$. In the sequel we denote $\tilde{x} = \phi(x)$. The kernel technique considers the vectors $\tilde{x}^i \in \mathbb{F}$ instead of $x^i \in \mathbb{R}^m$, and finds a normal vector $\tilde{w} \in \mathbb{F}$ and thresholds $p_1, \ldots, p_l$. Therefore the matrices X and K should be replaced by $\tilde{X}$ composed of vectors $\tilde{x}^i$ and $\tilde{K} = \left[\langle \tilde{x}^i, \tilde{x}^j \rangle \right]_{i,j \in N}$, respectively. Note that the latter matrix is given as $\tilde{K} = [\kappa(x^i, x^j)]_{i,j \in N}$ by the kernel function κ. Denote the i-th row of $\tilde{K}$ by $(k^i)^\top$, then the problem to solve is

$$(\tilde{H}) \quad \left| \begin{array}{l} \text{minimize } \lambda^\top \tilde{K} \lambda \\ \text{subject to } p_{\ell_i} + 1 \le (\tilde{k}^i)^\top \lambda \le p_{\ell_i+1} - 1 \quad \text{for } i \in N. \end{array} \right.$$

In the same way as for the hard margin problem $(\bar{H})$ we consider the sub-problem

$$(\tilde{H}(W)) \quad \left| \begin{array}{l} \text{minimize } \lambda_W^\top \tilde{K}_W \lambda_W \\ \text{subject to } p_{\ell_i} + 1 \le (\tilde{k}_W^i)^\top \lambda_W \le p_{\ell_i+1} - 1 \text{ for } i \in W, \end{array} \right.$$

where $\tilde{K}_W$ is the sub-matrix consisting of the rows and columns of $\tilde{K}$ with indices in W, and $(\tilde{k}_W^i)^\top$ is the row vector of $\kappa(x^i, x^j)$ for $j \in W$.

Algorithm RC$\tilde{H}$ (Row and Column Generation Algorithm for $(\tilde{H})$)

Step 1 : Let W^0 be an initial working set, and let $\nu = 0$.
Step 2 : Solve $(\tilde{H}(W^\nu))$ to obtain λ_W^ν and p^ν.
Step 3 : Let $\Delta = \{i \in N \backslash W^\nu \mid (\lambda_W^\nu, p^\nu) \text{violates} p_{\ell_i} + 1 \le (\tilde{k}_W^i)^\top \lambda_W \le p_{\ell_i+1} - 1\}$.
Step 4 : If $\Delta = \emptyset$, terminate.
Step 5 : Otherwise choose $\Delta^\nu \subseteq \Delta$, let $W^{\nu+1} = W^\nu \cup \Delta^\nu$, increment ν by 1 and go to Step 2.

Theorem 2 *The Algorithm RC$\tilde{H}$ solves problem $(\tilde{H})$.*

4   Soft Margin Problems for Non-separable Case

Introducing nonnegative slack variables ξ_{-i} and ξ_{+i} for $i \in N$ relaxes the hard margin constraints to *soft margin* constraints:

$$p_{\ell_i} + 1 - \xi_{-i} \le w^\top x^i \le p_{\ell_i+1} - 1 + \xi_{+i} \quad \text{for } i \in N.$$

Positive values of variables ξ_{-i} and ξ_{+i} mean misclassification, hence they should be as small as possible. We penalize positive ξ_{-i} and ξ_{+i} by adding $\delta(\xi_-)$ and $\delta(\xi_+)$ to the objective function via a penalty function δ, where $\xi_- = (\xi_{-i})_{i \in N}$ and $\xi_+ = (\xi_{+i})_{i \in N}$. Then we have the following *primal soft margin problem.*

$$(S) \quad \left| \begin{array}{l} \text{minimize } \|w\|^2 + c \ (\delta(\xi_-) + \delta(\xi_+)) \\ \text{subject to } p_{\ell_i} + 1 - \xi_{-i} \le (x^i)^\top w \le p_{\ell_i+1} - 1 + \xi_{+i} \text{ for } i \in N \\ \xi_-, \xi_+ \ge 0_n, \end{array} \right.$$

where c is a penalty parameter. When 1-norm function (resp., 2-norm function) is employed as the function δ, we call the above problem *soft margin problem with 1-norm penalty* (resp., *2-norm penalty*). As we discussed in the previous section, we can replace $\|w\|^2$ and $(x^i)^\top w$ in the problem (S) by $\lambda^\top K \lambda$ and $(k^i)^\top \lambda$ to obtain the primal problem with the dual representation of the normal vector. Then we have

$$(\bar{S}) \quad \left| \begin{array}{l} \text{minimize } \lambda^\top K \lambda + c \ (\delta(\xi_-) + \delta(\xi_+)) \\ \text{subject to } p_{\ell_i} + 1 - \xi_{-i} \le (k^i)^\top \lambda \le p_{\ell_i+1} - 1 + \xi_{+i} \text{ for } i \in N \\ \xi_-, \xi_+ \ge 0_n. \end{array} \right.$$

The sub-problem $(\bar{S}(W))$ for the working set W will be

$$(\bar{S}(W)) \quad \left| \begin{array}{l} \text{minimize } \lambda_W^\top K_W \lambda_W + c \ (\delta(\xi_{-W}) + \delta(\xi_{+W})) \\ \text{subject to } p_{\ell_i} + 1 - \xi_{-i} \le (k_W^i)^\top \lambda \le p_{\ell_i+1} - 1 + \xi_{+i} \text{ for } i \in W \\ \xi_{-W}, \xi_{+W} \ge 0_{|W|}, \end{array} \right.$$

where $\xi_{-W} = (\xi_{-i})_{i \in W}$ and $\xi_{+W} = (\xi_{+i})_{i \in W}$.

Algorithm RC$\bar{S}$ (Row and Column Generation Algorithm for $(\bar{S})$)

Step 1 : Let W^0 be an initial working set, and let $\nu = 0$.
Step 2 : Solve $(\bar{S}(W^\nu))$ to obtain $(\lambda_W^\nu, p^\nu, \xi_{-W}^\nu, \xi_{+W}^\nu)$.
Step 3 : Let $\Delta = \{ i \in N \backslash W^\nu \mid (\lambda_W^\nu, p^\nu) \text{ violates } p_{\ell_i} + 1 \le (k_W^i)^\top \lambda_W \le p_{\ell_i+1} - 1 \}$.
Step 4 : If $\Delta = \emptyset$, terminate.
Step 5 : Otherwise choose $\Delta^\nu \subseteq \Delta$, let $W^{\nu+1} = W^\nu \cup \Delta^\nu$, increment ν by 1 and go to Step 2.

Lemma 3 *Let* $(\hat{\lambda}_W, \hat{p}, \hat{\xi}_{-W}, \hat{\xi}_{+W})$ *be an optimum solution of* $(\bar{S}(W))$. *If*

$$\hat{p}_{\ell_i} + 1 \le (k_W^i)^\top \hat{\lambda}_W \le \hat{p}_{\ell_i+1} - 1 \quad \text{for all } i \in N \backslash W,$$

then $((\hat{\lambda}_W, 0_{N \backslash W}), \hat{p}, (\hat{\xi}_{-W}, 0_{N \backslash W}), (\hat{\xi}_{+W}, 0_{N \backslash W}))$ *is an optimum solution of* $(\bar{S})$.

Theorem 3 *The Algorithm RC$\bar{S}$ solves problem* $(\bar{S})$.

Since the kernel technique can apply to the soft margin problem in the same way as discussed in Sect. 3, we omit the kernel version of soft margin problem.

5   Illustrative Example and Conclusion

We show with a small instance how different models result in different classifications. The instance is the grades in calculus of 44 undergraduates. Each student is given one of the four possible grades A, B, C, and D according to his/her total score of mid-term exam, end-of-term exam and a number of in-class quizzes. We take the scores of mid-term and end-of-term exams to form the attribute vector, and his/her grade as a label.

Since the score of quizzes is not considered as an attribute, the instance is not separable, hence the hard margin problem (H) is infeasible. The solution of the soft margin problem (S) with 1-norm penalty is given in Fig. 1. We set the parameter c to 15.

Using the Gaussian kernel defined as $\kappa(x, y) = \exp(-\|x - y\|^2/2\,\sigma^2)$, we solved $(\tilde{H})$. The result with $\sigma = 7$ is given in Fig. 2, where one can observe that the problem $(\tilde{H})$ is exposed to the risk of over-fitting. Other kernel functions with a combination of various parameter values should be tested.

In this paper, we considered the ranking problem and proposed a row and column generation algorithm to alleviate the computational burden. Furthermore we proved the validity of the algorithm.

Fig. 1 Classification by (S) with 1-norm penalty

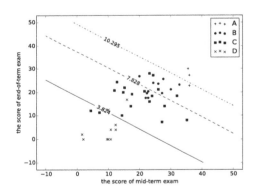

Fig. 2 Classification by $(\tilde{H})$

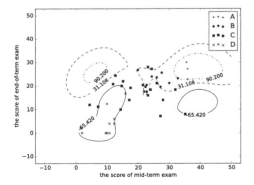

References

1. Bishop, C.M.: Pattern Recognition and Machine Learning. Springer, New York (2006)
2. Crammer, K., Singer, Y.: Pranking with ranking. In: Dietterich, T.G., Becker, S., Ghahramani, Z. (eds.) Advances in Neural Information Processing Systems 14, pp. 641–647. MIT Press, Cambridge (2002)
3. Herbrich, R., Graepel, T., Obermayer, K.: Large margin rank boundaries for ordinal regression. In: Smola, A.J., Bartlette, P., Schölkopt, B., Schuurmans, D. (eds.) Advances in Large Margin Classifiers, pp. 115–132. MIT Press, Cambridge (2000)
4. Liu, T.-Y.: Learning to Rank for Information Retrieval. Springer, Heidelberg (2011)
5. Shashua, A., Levin, A.: Ranking with large margin principles: two approaches. In: Adv. Neural Inf. Process. Syst. **15** (NIPS 2002), 937–944 (2003)
6. Schölkopf, B., Herbrich, R., Smola, A.J.: A generalized representer theorem. In: Helmbold, D., Williamson, B. (eds.) Computational Learning Theory, Lecture Notes in Computer Science, vol. 2111, pp. 416–426 (2001)
7. Shawe-Taylor, J., Cristianini, N.: Kernel Methods for Pattern Analysis. Cambridge University Press, Cambridge (2004)
8. Tatsumi, K., Hayashida, K., Kawachi, R., Tanino, T.: Multiobjective multiclass support vector machines maximizing geometric margins. Pac. J. Optim. **6**, 115–140 (2010)
9. Vapnik, V.N.: Statistical Learning Theory. Wiley, New York (1998)

Dual Sourcing Inventory Model with Uncertain Supply and Advance Capacity Information

Marko Jakšič

Abstract We model a periodic review, single stage dual sourcing inventory system with non-stationary stochastic demand, where replenishment can occur either through a regular stochastic capacitated supply channel with immediate delivery and/or an alternative uncapacitated supply channel with a longer fixed lead time. We focus on describing a situation in which upfront information on capacity availability of an unreliable supply channel is available, denoted as *advance capacity information* (ACI), to the customer. We derive the optimal dynamic programming formulation and we show some of the properties of the optimal policy by carrying out a numerical analysis. Additionally, our numerical results on the benefits of dual sourcing and the value of sharing ACI reveal several managerial insights.

1 Introduction

In the age of agile supply chains the two main determinants of the customer service level are the speed of replenishment and its reliability. To guarantee the customer satisfaction the companies are seeking for a supply base that would enable them to pursue these two goals. It is often the case, that a supplier might offer fast delivery while its reliability will suffer occasionally. This has forced the companies to search for alternative supply channels, through which they would improve the supply process reliability, where often more reliable supply comes with the price, either in higher purchasing costs per unit of a product or longer replenishment lead time.

In this paper we model the problem of a company that primarily relies on a regular supplier, which offers fast replenishment but the order might not be delivered in full due to the limited on-hand stock availability for instance. When the decision maker within the company anticipates the supply shortage, he can rely on an alternative

M. Jakšič (✉)
Faculty of Economics, Department of Management and Organization,
University of Ljubljana, Kardeljeva ploščad 17, Ljubljana, Slovenia
e-mail: marko.jaksic@ef.uni-lj.si

© Springer International Publishing Switzerland 2016
M. Lübbecke et al. (eds.), *Operations Research Proceedings 2014*,
Operations Research Proceedings, DOI 10.1007/978-3-319-28697-6_36

supplier, whose lead time is longer, but he is able to deliver the entire order with certainty. While most of the multiple supplier research explores the trade-off between purchasing costs and indirect costs of holding safety inventory to cover against demand and supply variability, we study the effect of capacity and lead time on supply reliability and the order allocation decision to suppliers. That is the decision between unreliable capacitated supplier with short lead time and reliable infinite capacity supplier with longer lead time. In addition, we study the effect of ACI on the capacity availability, and how this information if revealed by the supplier, can help the company to reduce the negative effects of unreliability of the faster supply channel.

The way we model the supply availability of a regular supplier is in line with the work of [1–4], where the random supply/production capacity determines a random upper bound on the supply availability in each period. For a finite horizon stationary inventory model they show that the optimal policy is of order-up-to type, where the optimal base-stock level is increased to account for possible, albeit uncertain, capacity shortfalls in future periods.

For a general review of the multiple supplier inventory models we refer the interested reader to [5]. A more focused review on multiple sourcing inventory models when supply components are uncertain by [6] reveals that most of these models consider uncertainty in supply lead time, supply yield, or supplier availability. In a deterministic lead time setting, several papers discuss the setting in which the lead times of the two suppliers differ by a fixed number of periods [7, 8], where they all assume infinite supply capacity or at most a fixed capacity limit on one or both suppliers. However, when there is uncertainty in the supplier capacity, diversification through multiple sourcing has received very little attention. The exception to this are [9, 10], where they study a single period problem with multiple capacitated suppliers and develop the optimal policy to assign orders to each supplier. Also, all of the capacitated multiple sourcing papers cited above assume identical lead time suppliers. Our paper makes a contribution to the literature by introducing a dual sourcing inventory model with a capacitated unreliable supplier and a reliable supplier with longer lead time. In addition, we study a situation in which the unreliable supplier provides an upfront information on exact capacity availability, a situation that was studied by [4] in a single supplier setting.

The remainder of the paper is organized as follows. We present the model formulation in Sect. 2. In Sect. 3 we provide the characteristics of the optimal policy based on the numerical results. In Sect. 4, we present the results of a numerical study to quantify the benefits of ACI, and in Sect. 5 we summarize our findings.

2 Model Formulation

In this section, we give the notation and the model description. A regular, zero lead time, supply channel is stochastically capacitated, where the supply capacity is exogenous to the customer and the actual capacity realization is only revealed upon replenishment. However, when ACI is available, the supply capacity availability is known prior to placing the order to the regular supplier, either as ACI on current period's capacity or on capacity of one period in the future. An alternative supply channel is modeled as an uncapacitated with a fixed one period lead time. The end-customer demand and supply capacity of a reliable supply channel are assumed to be stochastic non-stationary with known distributions in each time period, however, independent from period to period. In each period the customer places the order either to a regular, or to an alternative supplier, or both.

Presuming that unmet demand is fully backordered, the goal is to find an optimal policy that would minimize the inventory holding costs and backorder costs over a finite planning horizon T. We intentionally do not consider any product unit price difference and fixed ordering costs as we are primarily interested into the trade-off between capacity uncertainty associated with regular ordering and the delay in the replenishment of an alternative order. We assume the following sequence of events:

(1) At the start of the period, the manager reviews inventory position before ordering x_t, where $x_t = \hat{x}_t + v_{t-1}$ is a sum of the on-hand stock $\hat{x}_t$ and the order to the alternative supplier from the previous period v_{t-1}.
(2) The order to the regular supplier z_t and the order to the alternative supplier v_t are placed and the inventory position is raised to inventory position after ordering y_t, $y_t = x_t + z_t + v_t$. In the case when ACI is available the regular order is placed only up to the available supply capacity q_t.
(3) The order from the alternative supplier from the previous period and the current period's regular order are replenished, and the inventory position is corrected according to the realized capacity availability of the regular supplier $y_t - [z_t - q_t]^+ = x_t + min(z_t, q_t) + v_t$ in the case when ACI is not available.
(4) At the end of the period, demand d_t is observed and satisfied through on-hand inventory; otherwise it is backordered. The system's costs consist of inventory holding c_h and backorder c_b costs charged on end-of-period on-hand inventory, $\hat{x}_{t+1} = y_t - [z_t - q_t]^+ - v_t - d_t$. Correspondingly, the expected single period cost function $C_t(y_t, z_t) = \alpha E_{Q_t, D_t} \tilde{C}_t(\hat{x}_{t+1})$, where $\tilde{C}_t(\hat{x}_{t+1}) = c_h[\hat{x}_{t+1}]^+ + c_b[\hat{x}_{t+1}]^-$ is the regular loss function.

Correspondingly, for the case without ACI, the minimal discounted expected cost function that optimizes the cost over a finite planning horizon T from period t onward, starting in the initial state x_t, can be written as:

$$f_t(x_t) = \min_{z_t \geq 0, v_t \geq 0} \{C_t(y_t - [z_t - Q_t]^+ - v_t - D_t)$$
$$+ \alpha E_{Q_t, D_t} f_{t+1}(y_t - [z_t - Q_t]^+ - D_t)\}], \text{ for } 1 \leq t \leq T \quad (1)$$

and the ending condition is defined as $f_{T+1}(\cdot) \equiv 0$.

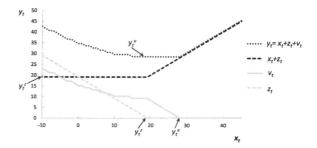

Fig. 1 The optimal inventory position after ordering and the optimal regular and alternative order sizes

3 Characterization of the Optimal Policy

To study the structure of the optimal policy, we depict the optimal order sizes for orders made at a regular unreliable supplier and an alternative reliable supplier depending on the initial inventory position x_t in Fig. 1. Looking at the $x_t + z_t$ line, we observe that the ordering to an unreliable supplier takes place in the manner of the base stock policy. For $x_t \le y_t^z$ we order a regular order z_t up to a base stock level y_t^z, while for $x_t > y_t^z$ no regular order is placed. Similarly, looking at the alternative order v_t to a reliable supplier, it is placed only if $x_t \le y_t^v$, and on $y_t^z \le x_t < y_t^v$, where no regular order is placed, the inventory position x_t is increased to a base stock level y_t^v. For $x_t < y_t^z$ size of an alternative order also depends on the anticipated supply shortage of a regular order (either through supply capacity probability distribution in the case without ACI or upfront information on exact regular order realization in the case with ACI), thus the optimal base stock level y_t^v is state dependent, depending on the size of the regular order. The effect the shortage anticipation has on the size of an alternative order is what makes the optimal policy in our model different from the optimality results of the corresponding uncapacitated model, where the optimal policy instructs that with each of the orders the inventory position is increased to a corresponding constant optimal base stock level [11].

4 The Value of Advance Capacity Information

In this section we present the results of the numerical analysis, which was carried out to determine the value of sharing ACI of the regular supply channel. Numerical calculations were done by solving the dynamic programming formulation given in (1).

In Fig. 2 we present the cost comparison between the base setting without ACI and the settings in which ACI is available for the current and for the future period, for different utilizations and levels of supply capacity variability of the regular supply channel. The cost curve depicted as $Util = 0$ represents the scenario with the lowest costs, where ordering is done solely to a regular supplier with infinite capacity. The worst case scenario is the case where regular supplier has no capacity and all orders

Fig. 2 System costs under ACI for different unreliable supply channel utilizations

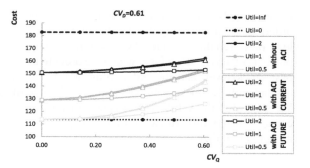

are placed to an alternative supplier, and is depicted with a $Util = Inf$ cost curve. Our interest lies in studying the intermediate scenarios, where we are predominantly interested in the effect the supply capacity variability (denoted as supply capacity coefficient of variation CV_Q) has on the costs under both, the situation without and with ACI.

Observe, when $CV_Q = 0$ there is no supply capacity uncertainty at a regular supplier, therefore the costs of both situations are equal. It is expected that the reduction in costs through ACI is increasing when CV_Q is increasing, as ACI effectively reduces the uncertainty of the supply capacity availability. The value of ACI, defined as the relative difference in costs between a situation without and with ACI, is up to 1.5 % in the case of ACI on current period's capacity. For the case of future ACI availability the cost savings become considerable, approaching 20 %. These largest savings are attained at settings with high CV_Q, no matter what the utilization of the regular supplier is. For lower CV_Q the value of ACI decreases, but still remains relatively high particularly when regular supplier is highly utilized. While the supply capacity variability of the regular supplier has the predominant effect on the benefits of ACI, the demand uncertainty also influences the possible savings attained through ACI. In the case of a highly utilized regular supplier the alternative supplier is used more heavily. While ACI helps the decision maker to come up with a better ordering decision to an alternative supplier, the delayed replenishment together with demand uncertainty results in high costs due to demand and supply mismatches, thus the value of ACI diminishes. This effect is lower for the case of low utilization of a regular supplier. Here ACI is helpful in covering the mismatches mainly through a regular supplier, therefore the exposure to the demand uncertainty risk faced by a supply through an alternative supplier is lower.

5 Conclusions

In this paper we establish the optimal inventory control policies for a finite horizon stochastically capacitated dual sourcing inventory system in which the regular supply is done through a fast albeit unreliable supply channel. The upfront information on regular supply capacity availability is available to a decision maker to improve his ordering decision to a regular supplier and/or decides to utilize an alternative, longer lead time, reliable supplier. We show that the structure of the optimal inventory policy is of a base-stock type, where the order to a regular supplier is up to a constant base stock level, and the order to an alternative supplier is placed up to a state dependent base stock level, depending on the size of the outstanding regular order. We show that the value of ACI is considerable in the case of high regular supply capacity uncertainty, already for the settings in which regular supply channel is only moderately utilized.

References

1. Ciarallo, F.W., Akella, R., Morton, T.E.: A periodic review, production planning model with uncertain capacity and uncertain demand—optimality of extended myopic policies. Manag. Sci. **40**, 320–332 (1994)
2. Khang, D.B., Fujiwara, O.: Optimality of myopic ordering policies for inventory model with stochastic supply. Oper. Res. **48**, 181–184 (2000)
3. Iida, T.: A non-stationary periodic review production-inventory model with uncertain production capacity and uncertain demand. Eur. J. Oper. Res. **140**, 670–683 (2002)
4. Jakšič, M., Fransoo, J., Tan, T., de Kok, A.G., Rusjan, B.: Inventory management with advance capacity information. Naval Res. Logistics **58**, 355–369 (2011)
5. Minner, S.: Multiple-supplier inventory models in supply chain management: a review. Int. J. Prod. Econ. **81**, 265–279 (2003)
6. Tajbakhsh, M.M., Zolfaghari, S., Lee, C.: Supply uncertainty and diversification: a review. In: Jung, H., Chen, F., Seong, B. (eds.) Trends in Supply Chain Design and Management: Technologies and Methodologies. Springer Ltd., London (2007)
7. Fukuda, Y.: Optimal policies for the inventory problem with negotiable leadtime. Manag. Sci. **10**(4), 690–708 (1964)
8. Bulinskaya, E.: Some results concerning optimal inventory policies. Theory Prob. Appl. **9**, 502–507 (1964)
9. Dada, M., Petruzzi, N.C., Schwarz, L.B.: A newsvendors procurement problem when suppliers are unreliable. Manuf. Serv. Oper. Manag. **9**(1), 9–32 (2007)
10. Federgruen, A., Yang, N.: Optimal supply diversification under general supply risks. J. Oper. Res. **57**(6), 909–925 (2009)
11. Veeraraghavan, S., Sheller-Wolf, A.: Now or later a simple policy for effective dual sourcing in capacitated systems. Oper. Res. **56**(4), 850–864 (2008)

Integrated Line Planning and Passenger Routing: Connectivity and Transfers

Marika Karbstein

Abstract The integrated line planning and passenger routing problem is an important planning problem in service design of public transport. A major challenge is the treatment of transfers. In this paper we show that analysing the connectivity aspect of a line plan gives a new idea how to integrate a transfer handling.

1 Introduction

The *integrated line planning and passenger routing problem* is an important planning problem in service design of public transport. The infrastructure of the public transport system can be represented by a graph where the edges correspond to streets and tracks and the nodes correspond to stations/stops. We are further given the number of passengers that want to travel from one point in the network to another point. A line is a path in the network, visiting a set of stops/stations in a predefined order. Passengers can travel along these lines and they can change from one line to another line in a stop/station if these lines intersect. Bringing capacities into play, the task is to find paths in the infrastructure network for lines and passengers such that the capacities of the lines suffice to transport all passengers. There are two main objectives for a line plan, namely, minimization of line operation costs and minimization of passenger discomfort measured in, e.g., travel times and number of transfers.

In general, the computed line system should be connected, i.e., one can travel from one station to any other station along the lines. Associating cost with the lines and searching for a cost minimum set of lines that connects all stations gives rise to a combinatorial optimization problem which we denote the *Steiner connectivity problem*. Its solution gives a lower bound on the costs of a line plan.

In this paper we present some results for the Steiner connectivity problem and show that they can be used to handle the transfer aspect for the line planning problem. In Sect. 2, the Steiner connectivity problem is introduced in more detail. We focus on the special case to connect two nodes via a set of paths. In Sect. 3, we propose

M. Karbstein (✉)
Zuse Institute Berlin, Takustr. 7, 14195 Berlin, Germany
e-mail: karbstein@zib.de

© Springer International Publishing Switzerland 2016   263
M. Lübbecke et al. (eds.), *Operations Research Proceedings 2014*,
Operations Research Proceedings, DOI 10.1007/978-3-319-28697-6_37

a new model for the integrated line planning and passenger routing problem that favors direct connections and involves a version of the 2-terminal Steiner connectivity problem as pricing problem. We briefly discuss computational results and the optimized line plan for ViP Potsdam of the year 2010 in Sect. 4.

2 Steiner Connectivity Problem

The Steiner connectivity problem is a generalization of the well-known *Steiner tree problem*. Given a graph with costs on the edges, the Steiner tree problem is to find a cost minimum set of edges that connects a subset of nodes. The Steiner connectivity problem is to choose a set of paths instead of edges. Steiner trees are fundamental for network design in transportation and telecommunication; see Dell'Amico, Maffioli, and Martello [1] for an overview. In fact, the Steiner tree problem can be seen as the prototype of all problems where nodes are connected by installing capacities on individual edges or arcs. In the same way, the Steiner connectivity problem can be seen as the prototype of all problems where nodes are connected by installing capacities on *paths* which is exactly the case in line planning. Hence, the significance of the Steiner connectivity problem for line planning is similar to the significance of the Steiner tree problem for telecommunication network design.

A formal description of the Steiner connectivity problem (SCP) is as follows. We are given an undirected graph $G = (V, E)$, a set of *terminal nodes* $T \subseteq V$, and a set of elementary *paths* $\mathcal{P}$ in G. The paths have nonnegative costs $c \in \mathbb{R}^{\mathcal{P}}_{\geq 0}$. The problem is to find a subset of paths $\mathcal{P}' \subseteq \mathcal{P}$ of minimal cost $\sum_{p \in \mathcal{P}'} c_p$ that *connect the terminals*, i.e., such that for each pair of distinct terminal nodes $t_1, t_2 \in T$ there exists a path q from t_1 to t_2 in G such that each edge of q is covered by at least one path of $\mathcal{P}'$. We can assume w.l.o.g. that every edge is covered by a path, i.e., for every $e \in E$ there is a $p \in \mathcal{P}$ such that $e \in p$; in particular, G has no loops. Figure 1 gives an example of a Steiner connectivity problem and a feasible solution.

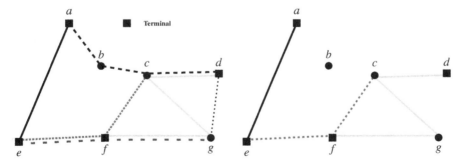

Fig. 1 Example of a Steiner connectivity problem. *Left* A graph with four terminal nodes ($T = \{a, d, e, f\}$) and six paths ($\mathcal{P} = \{p_1 = (ab, bc, cd), p_2 = (ef, fg), p_3 = (ae), p_4 = (ef, fc), p_5 = (gd), p_6 = (fg, gc, cd)\}$). *Right* A feasible solution with three paths ($\mathcal{P}' = \{p_3, p_4, p_6\}$)

Main results about complexity, approximation, integer programming formulations, and polyhedra can be generalized from the Steiner tree problem to the Steiner connectivity problem, see [5, 7]. In the following we want to consider the two-terminal case of the Steiner connectivity problem, see also [3]. The problem is to find a minimum set of paths connecting two given nodes s and t.

We call a set of paths $\mathcal{P}' \subseteq \mathcal{P}$ an st-*connecting set* if s and t are connected in the subgraph $H = (V, E(\mathcal{P}'))$, i.e., $\mathcal{P}'$ is a solution for the Steiner connectivity problem with $T = \{s, t\}$. A set $\mathcal{P}' \subseteq \mathcal{P}$ is st-*disconnecting* if $\mathcal{P} \setminus \mathcal{P}'$ is not an st-connecting set.

Algorithm 1 computes the cost of a minimum st-connecting set. It generalizes Dijkstra's algorithm to our setting. The distances from node s are stored in node labels $d(v)$. The algorithm can be extended such that it also determines the minimum st-connecting set $\mathcal{P}'$.

A cut formulation for the SCP with $T = \{s, t\}$ is:

$$\text{(MCS)} \quad \min \quad \sum_{p \in \mathcal{P}} c_p \, x_p$$

$$\text{s.t.} \quad \sum_{p \in \mathcal{P}_{\delta(W)}} x_p \geq 1 \qquad \forall \, s \in W \subseteq V \setminus \{t\}$$

$$x_p \in \{0, 1\} \qquad \forall \, p \in \mathcal{P}.$$

Here, x_p is a 0/1-variable that indicates whether path p is chosen ($x_p = 1$) or not ($x_p = 0$). Furthermore, $\mathcal{P}_{\delta(W)} := \{p \in \mathcal{P} : \delta(W) \cap p \neq \emptyset\}$ is the set of all paths that cross the cut $\delta(W) = \{\{u, v\} \in E : |\{u, v\} \cap W| = 1\}$ at least one time.

Algorithm 1: Primal-dual minimum st-connecting set algorithm.

Input : A connected graph $G = (V, E)$, a set of paths $\mathcal{P}$ with costs $c \in \, {}_{\geq 0}^{\mathcal{P}}$ that covers all edges E, $s, t \in V$.

Output: The value $d(t)$ of a minimum cost st-connecting set.

1   $d(s) := 0$, $d(v) := \infty \, \forall \, v \in V \setminus \{s\}$; all nodes are unmarked
2   **while** t is unmarked **do**
3   Choose v with $v = \operatorname{argmin} \{d(u) : u \text{ unmarked}\}$
4   **for all** $p \in \mathcal{P}$ with $v \in p$ **do**
5   **for** unmarked w with $w \in p$ **do**
6   **if** $d(w) > d(v) + c_p$ **then**
7           $d(w) := d(v) + c_p$
8   **end**
9   **end**
10   **end**
11   mark v
12 **end**

Theorem 1 *The inequality system of (MCS) is TDI.*

This can be shown by extending Algorithm 1 to a primal-dual algorithm that defines integer solutions for (MCS) and its dual program, compare with [3].

Setting $c \equiv 1$ in Algorithm 1 and interpreting the set of paths $\mathcal{P}$ as lines and s and t as origin and destination stations, then the algorithm computes the minimum number of lines that are necessary to connect s and t. This number corresponds to the minimum number of transfers minus 1 that are necessary to travel from s to t. The calculation of the minimum number of transfers is the basic idea of the model introduced in the next section.

3   A Transfer Model for Line Planning

In this section we want to propose a model for line planning and passenger routing that accounts for the number of *unavoidable transfers*. Each passenger path is associated with its number of minimum transfers with respect to the given set of all possible lines. More precisely, considering a certain passenger path, it may not be possible to cover this path by a single line or even by two lines, i.e., in any definition of a line plan, passengers on the path under consideration have to transfer at least once or twice, respectively. We call such transfers *unavoidable*.

We use the following notation. Consider a public transportation network as a graph $N = (V, E)$, whose nodes and edges correspond to stations and connections between these stations, respectively. Denote by $\mathcal{L}$ the line pool, i.e., a set of paths in N that represent all valid lines and by $\mathcal{F} \subseteq$ the set of possible frequencies at which these lines can be operated. If line ℓ is operated with frequency f, $\kappa_{\ell,f} \in$ $_+$ denotes the capacity and $c_{\ell,f} \in$ $_+$ the operation cost of this line. Let further $(d_{st}) \in$ $_+^{V \times V}$ be an origin-destination (OD) matrix that gives the travel demand between pairs of nodes, and denote by $D = \{(s, t) \in V \times V : d_{st} > 0\}$ the set of all OD-pairs with positive demand. Derive a directed passenger routing graph $\bar{N} = (V, A)$ from N by replacing each edge $e \in E$ with two antiparallel arcs $a(e)$ and $\bar{a}(e)$. Denote by $\mathcal{P}_{(s,t)}$ the set of all possible directed (s, t)-paths in $\bar{N}$ for $(s, t) \in D$, and by $\mathcal{P} = \bigcup_{(s,t) \in D} \mathcal{P}_{(s,t)}$ the set of all such paths; these represent travel routes of passengers. Associated with each arc $a \in A$ and path $p \in \mathcal{P}$ are travel times $\tau_a \in$ $_+$ and $\tau_p = \sum_{a \in p} \tau_a$, respectively, and with each transfer a (uniform) penalty $\sigma \in$ $_+$. Let k_p be the minimum number of transfers that passengers must do on path p if all lines in $\mathcal{L}$ would be built. A path $p \in \mathcal{P}$ with k_p unavoidable transfers has travel and transfer time $\tau_{p,k} = \tau_p + k_p \sigma$. Let $e(a)$ be the undirected edge corresponding to $a \in A$, and let us interpret a(n undirected) line in N in such a way that passengers can travel on this line in both directions in $\bar{N}$. The unavoidable transfer model is then

$$(\text{UT}) \min \lambda \sum_{\ell \in \mathscr{L}} \sum_{f \in \mathscr{F}} c_{\ell,f} \, x_{\ell,f} + (1 - \lambda)\left(\sum_{p \in \mathscr{P}} \tau_{p,k_p} \, y_{p,k_p}\right)$$

$$\sum_{p \in \mathscr{P}_{st}} y_{p,k_p} = d_{st} \qquad\qquad \forall (s,t) \in D \qquad\qquad (1)$$

$$\sum_{p \in \mathscr{P}:a \in p} y_{p,k_p} \leq \sum_{\ell \in \mathscr{L}:e(a) \in \ell} \sum_{f \in \mathscr{F}} \kappa_{\ell,f} \, x_{\ell,f} \qquad \forall a \in A \qquad\qquad (2)$$

$$\sum_{f \in \mathscr{F}} x_{\ell,f} \leq 1 \qquad\qquad \forall \ell \in \mathscr{L} \qquad\qquad (3)$$

$$x_{\ell,f} \in \{0,1\} \qquad\qquad \forall \ell \in \mathscr{L}, \forall f \in \mathscr{F} \qquad (4)$$

$$y_{p,k_p} \geq 0 \qquad\qquad \forall p \in \mathscr{P}. \qquad\qquad (5)$$

Model (UT) minimizes a weighted sum of line operating costs and passenger travel times. We use binary variables $x_{\ell,f}$ for the operation of line $\ell \in \mathscr{L}$ at frequency $f \in \mathscr{F}$. The continuous variables y_{p,k_p} account for the number of passengers that travel on path $p \in \mathscr{P}$ doing *at least* k_p transfers. Equations (1) enforce the passenger flow. Inequalities (2) guarantee sufficient total transportation capacity on each arc. Inequalities (3) ensure that a line is operated at one frequency at most.

Algorithm 1 can be extended such that it computes a travel-time minimal path from a given node $s \in V$ to all other nodes including a uniform transfer penalty $\sigma \in _+$ for each transfer w.r.t. a given set of lines $\mathscr{L}$. More precisely, replacing c_p by the travel time on line ℓ from v to w in lines 6 and 7 of the algorithm and adding a σ for $v \neq s$ in the same lines, yields the following proposition.

Proposition 1 *The pricing problem for the passenger path variables in model (UT) can be solved in polynomial time.*

The number k_p accounts for the minimum number of transfers w.r.t. all lines $\mathscr{L}$. In a final line plan usually only a small subset of lines $\mathscr{L}' \subseteq \mathscr{L}$ is established, i.e., the number of necessary transfers on a path p might be much larger. Since offering direct connections is a major goal in line planning, we included constraints to ensure enough capacities for passenger paths considered as direct connections. Let $\mathscr{L}_{st}$ be the number of lines supporting a direct connection from s to t, $\mathscr{L}_{st}(a) = \{\ell \in \mathscr{L}_{st} : a \in \ell\}$ be the direct connection lines for (s,t) containing arc a, and $\mathscr{P}_{st}^0 = \{p \in \mathscr{P}_{st} : k_p = 0\}$ be the set of all passenger paths from s to t with 0 unavoidable transfers. Then we can define *direct connection constraints* for each arc and each OD pair

$$\sum_{(u,v) \in D} \sum_{\substack{p \in \mathscr{P}_{uv}^0 : a \in p, \\ \mathscr{L}_{uv}(a) \subseteq \mathscr{L}_{st}(a)}} y_{p,0} \leq \sum_{\ell \in \mathscr{L}_{st}(a)} \sum_{f \in \mathscr{F}} \kappa_{\ell,f} \, x_{\ell,f} \qquad \forall a \in A, (s,t) \in D. \qquad (6)$$

These constraints are a combinatorial subset of the so-called *dcmetric inequalities* [4] that enforce sufficient transportation capacity to route all st-paths with 0 transfers via arc a. For each path $p \in \mathscr{P}^0 = \cup_{(s,t) \in D} \mathscr{P}_{st}$ we then have an additional variable

$y_{p,1}$ which come into play if the associated direct connection constraints for $y_{p,0}$ are not satisfied. Then the path can still be chosen in the optimization model but it is associated with at least one transfer and incurs one transfer penalty.

4   Computational Results and Potsdam Line Plan 2010

We made computations for several instances, e.g., a SiouxFalls instance from the Transportation Network Test Problems Library of Bar-Gera, a Dutch instance for the train network introduced by Bussieck in the context of line planning [6], an artificial China instance based on the 2009 high speed train network and some real world instances provided by our cooperation partner ViP Verkehrsbetriebe Potsdam GmbH (ViP), the public transport company of Potsdam.

For the SiouxFalls, Dutch, and China instances it turned out that it already suffice to distinguish passenger paths on direct connections and passenger path with one transfer and to consider the direct connection constraints. Indeed, evaluating the computed line plans shows that each passenger path of these instances is either a direct connection path or involves exactly one transfer [4]. Since the Potsdam instances are real multi-modal public transportation networks, there exist several passenger paths containing two or more transfers. However, modeling transfers between different transportation modes via transfer arcs (including a transfer penalty) and distinguishing direct connection paths from paths with at least one transfer for paths of one transportation mode via the direct connection constraints (6) yields a tractable model also for the Potsdam instances that estimates the travel times and transfers quite accurately [4].

A study to optimize the 2010 line plan for Potsdam was organized within the project "Service Design in Public Transport" of the DFG Research Center MATHEON *Mathematics for key technologies* together with ViP. A reorganization of the line plan in Potsdam became necessary when ViP took over six additional bus lines that were formerly operated by Havelbus Verkehrsgesellschaft mbH. The new line plan should minimize the travel time at a same cost level, and ViP emphasized the importance of a minimal number of transfers.

Our mathematically optimized solution for the Potsdam line plan 2010 minimizes the total number of transfers by around 5 % in comparison to a "hand made" plan on the basis of experience, see [2]. It further reduces the cost by around 4 % and the perceived travel time by around 6 %. ViP also certified that this line plan was indeed practicable and established a slightly modified version of our optimized solution.

Acknowledgments The work of Marika Karbstein was supported by the DFG Research Center MATHEON "Mathematics for key technologies"

References

1. Balakrishnan, A., Mangnanti, T.L., Mirchandani, P.: Network design. In: Dell'Amico, M., Maffioli, F., Martello, S. (eds.) Annotated Bibliographies in Combinatorial Optimization, chapter 18, pp. 311–334. Wiley, Chichester (1997)
2. Borndörfer, R., Friedow, I., Karbstein, M.: Optimierung des Linienplans 2010 in Potsdam. Der Nahverkehr **30**(4), 34–39 (2012)
3. Borndörfer, R., Hoàng, N.D., Karbstein, M., Koch, T., Martin, A.: How many Steiner terminals can you connect in 20 years? In: Jünger, M., Reinelt, G. (eds.) Facets of Combinatorial Optimization; Festschrift for Martin Grötschel, pp. 215–244. Springer (2013)
4. Borndörfer, R., Karbstein, M.: Metric inequalities for routings on direct connections with application to line planning. Discrete Optimization, **18**, pp. 56–73 (2015)
5. Borndörfer, R., Karbstein, M., Pfetsch, M.E.: The Steiner connectivity problem. Math. Program. **142**(1), 133–167 (2013)
6. Bussieck, M.R.: Optimal lines in public rail transport. Ph.D. thesis, TU Braunschweig (1997)
7. Karbstein, M.: Line planning and connectivity. Ph.D. thesis, TU Berlin (2013)

Robust Discrete Optimization Problems with the WOWA Criterion

Adam Kasperski and Paweł Zieliński

Abstract In this paper a class of combinatorial optimization problems with uncertain costs is discussed. The uncertainty is modeled by specifying a discrete scenario set containing all possible vectors of the costs which may occur. In order to choose a solution the Weighted Ordered Weighted Averaging aggregation operator (WOWA) is used. The WOWA operator allows decision makers to take both their attitude towards the risk and subjective probabilities for scenarios into account. The complexity of the problem is described and an approximation algorithm with some guaranteed worst case ratio is constructed.

1 Problem Formulation and Motivation

Let $E = \{e_1, \ldots, e_n\}$ be a finite ground set and let $\Phi \subseteq 2^E$ be a set of feasible solutions. In a deterministic case, each element $e_i \in E$ has a nonnegative cost c_i and we seek a feasible solution $X \in \Phi$, which minimizes the total cost $F(X) = \sum_{e_i \in X} c_i$. We denote such a deterministic combinatorial optimization problem by $\mathcal{P}$. This formulation encompasses a wide class of problems. We obtain, for example, a class of network problems by identifying E with edges of a graph G and Φ with some objects in G such as paths, spanning trees, or matchings. Usually, $\mathcal{P}$ is represented as a 0-1 programming problem whose constraints describe Φ in compact form.

In many practical situations, the element costs are not known in advance and their values depend on a state of the world, which is possible to occur. We model this uncertainty by specifying a scenario set $\Gamma = \{c_1, \ldots, c_K\}$, where $c_j = (c_{j1}, \ldots, c_{jn})$ is a *cost scenario* corresponding to the jth state of the world, $j \in [K]$. The cost of

A. Kasperski (✉)
Department of Operations Research, Wrocław University of Technology,
Wrocław, Poland
e-mail: adam.kasperski@pwr.edu.pl; adam.kasperski@pwr.wroc.pl

P. Zieliński (✉)
Department of Computer Science (W11/K2), Wrocław University of Technology,
Wrocław, Poland
e-mail: pawel.zielinski@pwr.edu.pl; pawel.zielinski@pwr.wroc.pl

© Springer International Publishing Switzerland 2016
M. Lübbecke et al. (eds.), *Operations Research Proceedings 2014*,
Operations Research Proceedings, DOI 10.1007/978-3-319-28697-6_38

271

solution X depends on scenario $c_j \in \Gamma$ and we will denote it by $F(X, c_j) = \sum_{e_i \in X} c_{ji}$. If no additional information with Γ is provided, then the maximum criterion is typically used to aggregate the solution costs and to choose a solution. Namely, we seek a solution $X \in \Phi$ which minimizes the value of $\max_{j \in [K]} F(X, c_j)$, which leads to the robust MIN- MAX $\mathscr{P}$ problem. There are, however, several known drawbacks of this approach. The maximum criterion is very conservative and is appropriate for very risk-averse decision makers. There are examples of decision problems, for which the maximum criterion gives unreasonable solutions (see, e.g. [8]). Moreover, in many situations decision makers have some knowledge about which scenarios are more likely to occur. This knowledge can be modeled by subjective probabilities, which can be derived from the individual preferences of decision makers (see, e.g. [10]). Hence, there is a need of criterion which takes into account both the probabilities for scenarios and various attitudes of decision makers towards a risk.

In [11] Torra proposed an aggregation criterion, called the *Weighted Ordered Weighted Averaging* operator (WOWA), defined as follows. Let $v = (v_1, \ldots, v_K)$ and $p = (p_1, \ldots, p_K)$ be two weight vectors such that $v_j, p_j \in [0, 1]$ for each $j \in [K]$, $\sum_{i \in [K]} v_j = 1$, and $\sum_{j \in [K]} p_j = 1$. Given a vector of reals $a = (a_1, \ldots, a_K)$, let σ be a sequence of $[K]$ such that $a_{\sigma(1)} \geq \cdots \geq a_{\sigma(K)}$. Then

$$\text{wowa}_{(v,p)}(a) = \sum_{j \in [K]} \omega_j a_{\sigma(j)},$$

where

$$\omega_j = w^*(\sum_{i \leq j} p_{\sigma(i)}) - w^*(\sum_{i < j} p_{\sigma(i)}),$$

and w^* is a nondecreasing function that interpolates the points $(0, 0)$ and $(j/K, \sum_{i \leq j} v_i)$ for $j \in [K]$. The function w^* is required to be a straight line when the points can be interpolated in this way. In this paper, we will assume that $v_1 \geq v_2 \geq \cdots \geq v_K$ and the function w^* is linear between the points $(0, 0)$, $(j/K, \sum_{i \leq j} v_i)$, $j \in [K]$. Under these assumptions, w^* is a concave and piecewise linear function.

Figure 1 shows three sample functions w^* with two boundary cases, where $v^1 = (1, 0, \ldots, 0)$ and $v^2 = (1/K, \ldots, 1/K)$. The vector v^2 models the weighted mean, i.e. in this case we get $\text{wowa}_{(v^2,p)}(a) = \sum_{j \in [K]} p_j a_j$. The vector v^1 models the weighted maximum, which in the case of uniform $p = (1/K, \ldots, 1/K)$ is the usual maximum operator. In general, for arbitrary v and uniform $p = (1/K, \ldots, 1/K)$, WOWA becomes the OWA operator proposed by Yager in [12]. The OWA operator contains: the maximum, minimum, median, average and Hurwicz criteria as special cases.

We now apply the WOWA operator to the uncertain problem $\mathscr{P}$ and provide the interpretation of the vectors v and p. For a given solution $X \in \Phi$, let us define:

$$\text{WOWA}(X) = \text{wowa}_{(v,p)}(F(X, c_1), \ldots, F(X, c_K)).$$

Fig. 1 Three sample functions w^* for $K = 5$

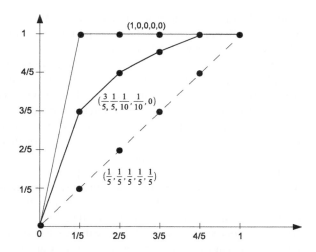

We thus obtain an aggregated value for X, by applying the WOWA criterion to the vector of the costs of X under scenarios in Γ. Given vectors v and p, we consider the following optimization problem:

$$\text{MIN-WOWA } \mathcal{P} : \quad \min_{X \in \Phi} \text{WOWA}(X).$$

The vector v models the level of risk aversion of decision maker. Namely, the more uniform is the weight distribution in v the less risk averse a decision maker is. In particular, $v^2 = (1/K, \ldots, 1/K)$ means that decision maker is risk indifferent while the vector $v^1 = (1, 0, \ldots, 0)$ means that decision maker is extremely risk-averse. On the other hand, p can be seen as a vector of subjective probabilities for scenarios. More precisely, p_j is a subjective probability of the occurrence of the state of the world which leads to scenario c_j (see, e.g. [10] for a description of the axiomatic definition of the subjective probability). Then, in particular, a risk indifferent decision maker aims to minimize the expected solution cost. In general, v and p define distorted probabilities and the WOWA criterion can be seen as a Choquet integral with respect to the distorted probabilities [3, 9]. Notice that MIN- WOWA $\mathcal{P}$ becomes the MIN- OWA $\mathcal{P}$ problem discussed in [6], when $p = (1/K, \ldots, 1/K)$.

2 Complexity of the Problem

Since MIN- MAX $\mathcal{P}$ is a special case of MIN- WOWA $\mathcal{P}$ with $v = (1, 0, \ldots, 0)$ and $p = (1/K, \ldots, 1/K)$, all the negative results known for MIN- MAX $\mathcal{P}$ remain valid for MIN- WOWA $\mathcal{P}$. Unfortunately, MIN- MAX $\mathcal{P}$ is typically NP-hard even when $K = 2$. This is the case for all basic network problems and for the selecting items problem, i.e. for the problem where $\Phi = \{X \subseteq E : |X| = p\}$ for some constant $p \in [n]$ [2, 7]. Furthermore, when K is part of the input, then for all these problems, MIN- MAX $\mathscr{P}$ is strongly NP-hard and also hard to approximate within any constant factor [4, 5]. The problem complexity becomes worse when the maximum criterion is replaced with the more general OWA one. It has been shown in [6], that all the basic network problems are then not at all approximable, when K is a part of the input. However, if the weights are nonincreasing, i.e. $v_1 \geq v_2 \geq \cdots \geq v_K$, then MIN- OWA $\mathscr{P}$ is approximable within $v_1 K$. Notice that this approximation ratio is in the interval $[1, K]$, as $v_1 \in [1/K, 1]$. The main goal of this paper is to generalize this result to the MIN- WOWA $\mathscr{P}$ problem.

It is easy to verify that the WOWA operator is monotone, i.e. $\text{wowa}_{(v,p)}(a)$ is nondecreasing with respect to each a_i in a. This fact immediately implies, that there exists an optimal solution X to MIN- WOWA $\mathscr{P}$, which is efficient (Pareto optimal), i.e. for which there is no solution Y such that $F(Y, c_j) \leq F(X, c_j)$ for each $j \in [K]$ with at least one strict inequality. Notice that each optimal solution to MIN- WOWA $\mathscr{P}$ must be efficient when all components of p and v are positive. For some problems, for example when $\mathscr{P}$ is the shortest path or minimum spanning tree, an optimal efficient solution can be found in pseudopolynomial time, provided that K is constant [1]. Hence, for constant K, MIN- WOWA $\mathscr{P}$ can be solved in pseudopolynomial time. However, the resulting algorithm is practically applicable only for small values of K. For larger values of K the approximation algorithm proposed in the next section can be more attractive.

3 Approximation Algorithm

In this section we construct an approximation algorithm for MIN- WOWA $\mathscr{P}$ under the assumptions that $v_1 \geq v_2 \geq \cdots \geq v_K$ and $\mathscr{P}$ is polynomially solvable. We will also assume that $p_j > 0$ for each $j \in [K]$. When $p_j = 0$ for some $j \in [K]$, then we can remove scenario c_j from Γ without changing the problem.

Lemma 1 *For any vector $a = (a_1, \ldots, a_K)$ and any sequence π of $[K]$ it holds* $\text{wowa}_{(v,p)}(a) \geq \sum_{j \in [K]} \omega_j a_{\pi(j)}$, *where* $\omega_j = w^*(\sum_{i \leq j} p_{\pi(i)}) - w^*(\sum_{i < j} p_{\pi(i)})$.

Proof Assume w.l.o.g. that $a_1 \geq a_2 \cdots \geq a_K$. Let $f_\pi(a) = \sum_{j \in [K]} \omega_j a_{\pi(j)}$. Consider any two neighbor elements $a_{\pi(i)}$ and $a_{\pi(i+1)}$ in π such that $a_{\pi(i)} \leq a_{\pi(i+1)}$. Let us interchange $a_{\pi(i)}$ and $a_{\pi(i+1)}$ in π and denote the resulting sequence by π'. We will show that $f_{\pi'}(a) \geq f_\pi(a)$ and the equality holds when $a_{\pi(i)} = a_{\pi(i+1)}$. This will complete the proof since we can transform π into $\sigma = (1, \ldots, K)$ by using a finite number of such element interchanges without decreasing the value of f_π and $f_\sigma(a) = \text{wowa}_{(v,p)}(a)$. It holds $f_{\pi'}(a) - f_\pi(a) = \omega'_i a_{\pi(i+1)} + \omega'_{i+1} a_{\pi(i)} - \omega_i a_{\pi(i)} - \omega_{i+1} a_{\pi(i+1)} = (\omega'_{i+1} - \omega_i) a_{\pi(i)} - (\omega_{i+1} - \omega'_i) a_{\pi(i+1)}$. It holds $\omega'_i + \omega'_{i+1} = \omega_i + \omega_{i+1}$ (see Fig. 2a), so $\omega'_{i+1} - \omega_i = \omega_{i+1} - \omega'_i = \alpha$. Hence $f_{\pi'}(a) - f_\pi(a) = \alpha(a_{\pi(i)} - a_{\pi(i+1)})$. Since w^* is concave, we have $\omega_{i+1}/p_{\pi(i+1)} \leq \omega'_i/p_{\pi(i+1)}$,

(a)   **(b)**

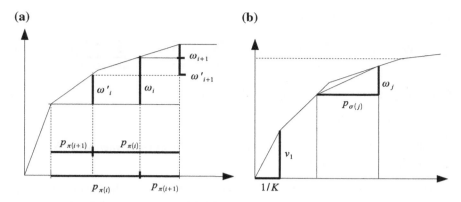

Fig. 2 Illustrations of the proofs of Lemmas 1 and 2

which implies $\alpha \leq 0$ since $p_{\pi(i+1)} > 0$. Hence $f_{\pi'}(\boldsymbol{a}) \geq f_\pi(\boldsymbol{a})$. Observe that $f_{\pi'}(\boldsymbol{a}) = f_\pi(\boldsymbol{a})$ if $a_{\pi(i)} = a_{\pi(i+1)}$, which means that the WOWA operator for non-increasing weights is symmetric, i.e. its value does not depend on the order of the elements in $\boldsymbol{a}$.   □

Lemma 2 *For any vector* $\boldsymbol{a} = (a_1, \ldots, a_K)$ *it holds* $\mathrm{wowa}_{(\boldsymbol{v},\boldsymbol{p})}(\boldsymbol{a}) \leq v_1 K \sum_{j \in [K]} p_j a_j$.

Proof Since w^* is concave and piecewise linear, it holds $\frac{\omega_j}{p_{\sigma(j)}} \leq \frac{v_1}{1/K} = v_1 K$ for each $j \in [K]$ (see Fig. 2b). In consequence, $\mathrm{wowa}_{(\boldsymbol{v},\boldsymbol{p})}(\boldsymbol{a}) = \sum_{j \in [K]} \omega_j a_{\sigma(j)} \leq \sum_{j \in [K]} v_1 K p_{\sigma(j)} a_{\sigma(j)} = v_1 K \sum_{j \in [K]} p_j a_j$.   □

Let $\hat{c}_i = \mathrm{wowa}_{(\boldsymbol{v},\boldsymbol{p})}(c_{1i}, \ldots, c_{Ki})$ be the aggregated cost of element $e_i \in E$ over all scenarios. Let $\hat{X}$ be an optimal solution for the costs $\hat{c}_i$, $i \in [n]$. The following theorem holds:

Theorem 1 *For any X, it holds* $\mathrm{WOWA}(\hat{X}) \leq K v_1 \cdot \mathrm{WOWA}(X)$.

Proof Let σ be a sequence of $[K]$ such that $F(\hat{X}, \boldsymbol{c}_{\sigma(1)}) \geq \cdots \geq F(\hat{X}, \boldsymbol{c}_{\sigma(K)})$ and $\omega_j = w^*(\sum_{i \leq j} p_{\sigma(i)}) - w^*(\sum_{i < j} p_{\sigma(i)})$. The definition of the WOWA operator and Lemma 1 imply the following inequality:

$$\mathrm{WOWA}(\hat{X}) = \sum_{j \in [K]} \omega_j \sum_{e_i \in \hat{X}} c_{\sigma(j)i} = \sum_{e_i \in \hat{X}} \sum_{j \in [K]} \omega_j c_{\sigma(j)i} \leq \sum_{e_i \in \hat{X}} \hat{c}_i. \qquad (1)$$

Using Lemma 2, we get $\hat{c}_i \leq v_1 K \sum_{j \in [K]} p_j c_{ji}$. Hence, from the definition of $\hat{X}$, we obtain

$$\sum_{e_i \in \hat{X}} \hat{c}_i \leq \sum_{e_i \in X} \hat{c}_i \leq K v_1 \sum_{e_i \in X} \sum_{j \in [K]} p_j c_{ji}. \qquad (2)$$

Since $v_1 \geq \cdots \geq v_K$ it follows that

$$\text{WOWA}(X) \geq \sum_{j \in [K]} p_j F(X, \mathbf{c}_j) = \sum_{j \in [K]} p_j \sum_{e_i \in X} c_{ji} = \sum_{e_i \in X} \sum_{j \in [K]} p_j c_{ji}. \quad (3)$$

Combining (1), (2) and (3) completes the proof.   □

Theorem 1 leads to the following corollary:

Corollary 1 *If $v_1 \geq \cdots \geq v_K$ and $\mathscr{P}$ is polynomially solvable, then* MIN- WOWA $\mathscr{P}$ *is approximable within $v_1 K$.*

The bound obtained in Corollary 1 is tight (see [6]). We get the largest ratio equal to K, when WOWA is the weighted maximum. On the other hand, when $v_1 = 1/K$ (WOWA is the expected value), then we get a polynomial algorithm for the problem.

4 Conclusions

In this paper we have proposed to use the WOWA criterion to choose a solution for a wide class of discrete optimization problems with uncertain costs. This criterion allows us to take both scenario probabilities and risk aversion of decision makers into account. Since the obtained problem is typically NP-hard even for two scenarios, we have proposed an approximation algorithm, which can be applied if the underlying deterministic problem $\mathscr{P}$ is polynomially solvable. We believe that better approximation algorithms can be designed for particular problems $\mathscr{P}$.

Acknowledgments The second author is partially supported by the National Center for Science (Narodowe Centrum Nauki), grant 2013/09/B/ST6/01525.

References

1. Aissi, H., Bazgan, C., Vanderpooten, D.: General approximation schemes for minmax (regret) versions of some (pseudo-)polynomial problems. Discret. Optim. **7**, 136–148 (2010)
2. Averbakh, I.: On the complexity of a class of combinatorial optimization problems with uncertainty. Math. Program. **90**, 263–272 (2001)
3. Grabish, M.: Owa operators and nonadditive integrals. In: Kacprzyk, J., Yager, R.R., Beliakov, G. (eds.) Recent Developments in the Ordered Weighted Averaging Operators: Theory and Practice. Studies in Fuzziness and Soft Computing, vol. 265, pp. 3–15. Springer, Berlin (2011)
4. Kasperski, A., Kurpisz, A., Zieliński, P.: Approximating the min-max (regret) selecting items problem. Inf. Process. Lett. **113**, 23–29 (2013)
5. Kasperski, A., Zieliński, P.: On the approximability of minmax (regret) network optimization problems. Inf. Process. Lett. **109**, 262–266 (2009)
6. Kasperski, A., Zieliński, P.: Combinatorial optimization problems with uncertain costs and the OWA criterion. Theor. Comput. Sci. **565**, 102–112 (2015)

7. Kouvelis, P., Yu, G.: Robust Discrete Optimization and its Applications. Kluwer Academic Publishers, Dordrecht (1997)
8. Luce, R.D., Raiffa, H.: Games and Decisions: Introduction and Critical Survey. Dover Publications Inc., New York (1957)
9. Ogryczak, W., Śliwiński, T.: On efficient WOWA optimization for decision support under risk. Int. J. Approx. Reason. **50**, 915–928 (2009)
10. Parmigiani, G., Inoue, L.Y.: Decision theory. Principles and Approaches, Wiley (2009)
11. Torra, V.: The weighted owa operator. Int. J. Intell. Syst. **12**, 153–166 (1997)
12. Yager, R.R.: On ordered weighted averaging aggregation operators in multi-criteria decision making. IEEE Trans. Syst. Man Cybern. **18**, 183–190 (1988)

Robust Single Machine Scheduling Problem with Weighted Number of Late Jobs Criterion

Adam Kasperski and Paweł Zieliński

Abstract This paper deals with a single machine scheduling problem with the weighted number of late jobs criterion, where some job parameters, such as: processing times, due dates, and weights, may be uncertain. This uncertainty is modeled by specifying a scenario set containing all vectors of the job parameters, called scenarios, which may occur. The min-max criterion is adopted to compute a solution under uncertainty. In this paper some of the recent negative complexity and approximability results for the problem are extended and strengthened. Moreover, some positive approximation results for the problem in which the maximum criterion is replaced with the OWA operator are presented.

1 Preliminaries

In a single machine scheduling problem with the weighted number of late jobs criterion, we are given a set of n independent, nonpreemptive, ready for processing at time 0 jobs, $J = \{J_1, \ldots, J_n\}$, to be processed on a single machine. For each job $j \in J$ a processing time p_j, a due date d_j and a weight w_j are specified. A schedule π is a permutation of the jobs representing an order in which the jobs are processed. We will use Π to denote the set of all schedules. Let $C_j(\pi)$ denote the completion time of job j in schedule π. Job $j \in J$ is *late* in π if $C_j(\pi) > d_j$; otherwise j is *on-time* in π. Set $U_j(\pi) = 1$ if job j is late in π and $U_j(\pi) = 0$ if j is on-time in π, $U_j(\pi)$ is called the *unit penalty* of job j in π. In the deterministic case, we wish to find a schedule $\pi \in \Pi$ which minimizes the value of the *cost function* $f(\pi) = \sum_{j \in J} w_j U_j(\pi)$. This problem is denoted by $1 || \sum w_j U_j$ in Graham's notation (see, e.g., [7]). The problem

A. Kasperski (✉)
Department of Operations Research, Wrocław University of Technology,
Wrocław, Poland
e-mail: adam.kasperski@pwr.edu.pl; adam.kasperski@pwr.wroc.pl

P. Zieliński (✉)
Department of Computer Science (W11/K2), Wrocław University of Technology,
Wrocław, Poland
e-mail: pawel.zielinski@pwr.edu.pl; pawel.zielinski@pwr.wroc.pl

© Springer International Publishing Switzerland 2016
M. Lübbecke et al. (eds.), *Operations Research Proceedings 2014*,
Operations Research Proceedings, DOI 10.1007/978-3-319-28697-6_39

$1 || \sum w_j U_j$ is weakly NP-hard [8]. However, its special cases $1 || \sum U_j$ (minimizing the number of late jobs) and $1 | p_j = 1 | \sum w_j U_j$ (minimizing the weighted number of late jobs with unit processing times) are polynomially solvable (see, e.g., [4]).

Suppose that all the job parameters may be ill-known. Every possible realization of the parameters, denoted by S, is called a *scenario*. We will use $p_j(S)$, $d_j(S)$ and $w_j(S)$ to denote the processing time, due date and weight of job j under scenario S, respectively. Without loss of generality, we can assume that all these parameters are nonnegative integers. Let *scenario set* $\Gamma = \{S_1, \ldots, S_K\}$ contain K explicitly listed scenarios. Now the job completion time, the unit penalty and the cost of schedule π depend on scenario $S \in \Gamma$, and we will denote them by $C_j(\pi, S)$, $U_j(\pi, S)$ and $f(\pi, S)$, respectively. In order to compute a solution we will use the min-max criterion, which is the most popular criterion in *robust optimization* (see, e.g. [11]). Namely, in the MIN-MAX $1 || \sum w_j U_j$ problem, we seek a schedule that minimizes the largest cost over all scenarios, that is

$$\min_{\pi \in \Pi} \max_{S \in \Gamma} f(\pi, S), \tag{1}$$

where $f(\pi, S) = \sum_{j \in J} w_j(S) U_j(\pi, S)$. We will also discuss its special cases, MIN-MAX $1 || \sum U_j$ and MIN-MAX $1 | p_j = 1 | \sum U_j$, in which $f(\pi, S) = \sum_{j \in J} U_j(\pi, S)$ is the number of late jobs in π under scenario S.

2 Single Machine Scheduling Problem with the Number of Late Jobs Criterion Under Uncertainty

In this section, we extend and strengthen the negative results which have been recently obtained in [1, 2] for some special cases of the MIN-MAX $1 || \sum w_j U_j$ problem, namely for MIN-MAX $1 || \sum U_j$ and MIN-MAX $1 | p_j = 1 | \sum U_j$. It has been proved in [2] that MIN-MAX $1 || \sum U_j$ with deterministic due dates and uncertain processing times is weakly NP-hard, if the number of processing time scenarios equals 2. We now show that if the number of processing time scenarios is a part of the input, then the problem is strongly NP-hard even if all jobs have a common deterministic due date.

Theorem 1 *When the number of processing time scenarios is a part of the input, then* MIN-MAX $1 || \sum U_j$ *is strongly NP-hard. This assertion remains true even when all the jobs have a common deterministic due date.*

Proof We show a polynomial time reduction from the following 3-SAT problem which is strongly NP-hard [6]. Given a set of boolean variables $x_1, \ldots, x_n$ and a set of clauses $C_1, \ldots, C_m$, where each clause contains exactly three distinct literals (variables or their negations). We ask if there is a truth assignment to the variables which satisfies all the clauses. Given an instance of 3-SAT, we create an instance of MIN-MAX $1 || \sum U_j$ in the following way. For each variable x_i we create two jobs J_{x_i}

Table 1 Processing time scenarios for the formula $(x_1 \vee \bar{x}_2 \vee \bar{x}_3) \wedge (\bar{x}_2 \vee \bar{x}_3 \vee x_4) \wedge (\bar{x}_1 \vee x_2 \vee \bar{x}_4) \wedge (x_1 \vee x_2 \vee x_3) \wedge (x_1 \vee x_3 \vee \bar{x}_4)$

	S_1	S_2	S_3	S_4	S_5	S_1'	S_2'	S_3'	S_4'	d_i
J_{x_1}	0	0	1	0	0	2	0	0	0	2
$J_{\bar{x}_1}$	1	0	0	1	1	2	0	0	0	2
J_{x_2}	1	1	0	0	0	0	2	0	0	2
$J_{\bar{x}_2}$	0	0	1	1	0	0	2	0	0	2
J_{x_3}	1	1	0	0	0	0	0	2	0	2
$J_{\bar{x}_3}$	0	0	0	1	1	0	0	2	0	2
J_{x_4}	0	0	1	0	1	0	0	0	2	2
$J_{\bar{x}_4}$	0	1	0	0	0	0	0	0	2	2

Schedule $\pi = (J_{x_1}, J_{\bar{x}_2}, J_{x_3}, J_{\bar{x}_4} | J_{\bar{x}_1}, J_{x_2}, J_{\bar{x}_3}, J_{x_4})$ corresponds to a satisfying truth assignment

and $J_{\bar{x}_i}$, so J contains $2n$ jobs. The due dates of all these jobs are the same under each scenario and equal 2. We form processing time scenario set Γ as follows. For each clause $C_j = (l_1, l_2, l_3)$ we construct scenario under which the jobs $J_{l_1}, J_{l_2}, J_{l_3}$ have processing time equal to 1 and all the remaining jobs have processing times equal to 0. Then, for each pair of jobs $J_{x_i}, J_{\bar{x}_i}$ we construct scenario S_i' under which the processing times of $J_{x_i}, J_{\bar{x}_i}$ are 2 and all the remaining jobs have processing times equal to 0. A sample reduction is shown in Table 1. We will show that the answer to 3-SAT is yes if and only if there is a schedule π such that $\max_{S \in \Gamma} f(\pi, S) \leq n$.

Assume that the answer to 3-SAT is yes. Then there exists a truth assignment to the variables which satisfies all the clauses. Let us form schedule π by processing first the jobs corresponding to true literals in any order and processing then the remaining jobs in any order. From the construction of the scenario set it follows that the completion time of the nth job in π under each scenario is not greater than 2. In consequence, at most n jobs in π is late under each scenario and $\max_{S \in \Gamma} f(\pi, S) \leq n$.

Assume now that there is a schedule π such that $f(\pi, S) \leq n$ for each $S \in \Gamma$ which means that at most n jobs in π are late under each scenario. Observe first that J_{x_i} and $J_{\bar{x}_i}$ cannot appear among the first n jobs in π for any $i \in [n]$; otherwise more than n jobs would be late in π under S_i'. Hence the first n jobs in π correspond to a truth assignment to the variables $x_1, \ldots, x_n$, i.e. when J_l is among the first n jobs, then the literal l is true. Since $f(\pi, S) \leq n$, the completion time of the n-th job in π is not greater than 2. We conclude that at most two jobs among the first n job have processing time equal to 1 under S, so there are at most two false literals for each clause and the answer to 3-SAT is yes. $\qquad\square$

We now discuss the MIN-MAX $1|p_j = 1|\sum U_j$ problem under due date uncertainty. It has been proved in [1], that when the number of due date scenarios is a part of the input and there are two distinct due date values, the problem is strongly NP-hard and it is not approximable within 2. We now extend this result, namely, we show that if the number of due date scenarios is a part of the input and there are two distinct due date values, the problem is not approximable within any constant factor.

Consider the following 0-1 SELECTING ITEMS problem. We are given a set of items $E = \{e_1, e_2, \ldots, e_n\}$ and an integer $p \in [n]$. For each item e_j, $j \in [n]$, there is a cost $c_j \in \{0, 1\}$. We seek a selection $X \subset E$ of exactly p items of the minimum total cost $f(X) = \sum_{e_j \in X} c_j$.

Proposition 1 *There is a cost preserving reduction from* 0-1 SELECTING ITEMS *problem to* $1 | p_j = 1 | \sum U_j$.

Proof Let $(E, p, (c_j)_{j \in [n]})$ be an instance of 0-1 SELECTING ITEMS. The corresponding scheduling problem is constructed as follows. We create a set of jobs $J = E$, $|E| = n$, with unit processing times. If $c_j = 1$ then $d_j = n - p$, and if $c_j = 0$, then $d_j = n$. Suppose that there is a selection X of p items out of E with the cost of C. Hence X contains exactly C items, $C \leq p$, with the cost equal to 1. In the corresponding schedule, we first process $n - p$ jobs from $J \setminus X$ and then the jobs in X in any order. It is easily seen that there are exactly C late jobs in π, hence the cost of schedule π is C. Let π be a schedule in which there are C late jobs. Clearly $C \leq p$ since the first $n - p$ jobs in π must be on-time. Let us form solution X by choosing the items corresponding to the last p jobs in π. Among these jobs exactly C are late, hence the cost of X is C. $\qquad\square$

We have arrived to the theorem that improves the lower bound for approximating MIN-MAX $1 | p_j = 1 | \sum U_j$ give in [1].

Theorem 2 *When the number of due date scenarios is a part of the input,* MIN-MAX $1 | p_j = 1 | \sum U_j$ *is not approximable within any constant factor unless* $P = NP$. *This assertion remains true even if there are two distinct values of the due dates in scenarios.*

Proof Proposition 1 shows that there is a cost preserving reduction from 0-1 SELECTING ITEMS to $1 | p_j = 1 | \sum U_j$. Therefore, there exists a cost preserving reduction from MIN-MAX 0-1 SELECTING ITEMS with K, 0–1 cost scenarios to MIN-MAX $1 | p_j = 1 | \sum U_j$ with K due date scenarios. Since the former problem is not approximable within any constant factor [9], the same results holds for the latter one. $\qquad\square$

3 Single Machine Scheduling Problem with the Weighted Number of Late Jobs Criterion Under Uncertainty

In this section we explore the MIN-MAX $1 | p_j = 1 | \sum w_j U_j$ problem under due date and weight uncertainty. We note first that if the jobs have a common deterministic due date, then $1 | p_j = 1 | \sum w_j U_j$ is equivalent to the SELECTING ITEMS problem discussed in [1, 3, 5], i.e. a generalization of 0-1 SELECTING ITEMS (see Sect. 2) in which the items have arbitrary nonnegative costs. It suffices to fix $E = J$, $c_j = w_j$, $j \in J$, and $p = n - d$, where d is a common due date. The same reduction allows us to transform any instance of SELECTING ITEMS to MIN-MAX $1|p_j = 1| \sum w_j U_j$ with a deterministic common due date. Hence, there exists a cost preserving reduction from MIN-MAX $1|p_j = 1| \sum w_j U_j$ with K weight scenarios to MIN-MAX SELECTING ITEMS with K cost scenarios and vice versa. Consequently, the results obtained in [1, 3] immediately imply the following theorem:

Theorem 3 *When the jobs have a common deterministic due date, then MIN-MAX $1|p_j = 1| \sum w_j U_j$ is NP-hard for two weight scenarios. Furthermore, it becomes strongly NP-hard and hard to approximate within any constant factor when the number of weight scenarios is part of the input.*

In [5], an LP-based $O(\log K / \log \log K)$ approximation algorithm for MIN-MAX SELECTING ITEMS has been proposed. Applying this algorithm to MIN-MAX $1|p_j = 1| \sum w_j U_j$ leads to the following result:

Theorem 4 *If the number of weight scenarios is a part of the input, then MIN-MAX $1|p_j = 1| \sum w_j U_j$ with a common deterministic due date is approximable within $O(\log K / \log \log K)$.*

We now consider the general case, in which both due dates and weights may be uncertain. We make use of the fact that $1|p_j = 1| \sum w_j U_j$ is a special case of the MINIMUM ASSIGNMENT problem. To see this, we can build an instance $(G = (V_1 \cup V_2, E), (c_{ij})_{(i,j) \in E})$ of MINIMUM ASSIGNMENT for given an instance of $1|p_j = 1| \sum w_j U_j$ in the following way. The nodes in V_1 correspond to job positions, $V_1 = [n]$, the nodes in V_2 correspond to jobs, $V_2 = J$, obviously $|V_1| = |V_2| = n$. Each node $j \in V_2$ is connected with every node $i \in V_1$. The arc costs c_{ij}, $(i, j) \in E$, are set as follows: $c_{ij} = w_j$ if $i > d_j$, and $c_{ij} = 0$ otherwise. There is one to one correspondence between the schedules and the assignments and the reduction is cost preserving. This fact still holds, when a scenario set Γ is introduced. In this case we fix $c_{ij}(S) = w_j(S)$ if $i > d_j(S)$ and $c_{ij}(S) = 0$ otherwise for each scenario $S \in \Gamma$. The reduction is then cost preserving under each scenario. In consequence, $1|p_j = 1| \sum w_j U_j$ with scenario set Γ belongs to the class of combinatorial optimization problems discussed in [10], which allows us to establish a positive result described in the next part of this section.

Let $v_1, \ldots, v_K$ be numbers such that $v_i \in [0, 1]$, $i \in [K]$, and $v_1 + \cdots + v_K = 1$. Given schedule π, let σ be a permutation of $[K]$ such that $f(\pi, S_{\sigma(1)}) \geq f(\pi, S_{\sigma(2)}) \geq \cdots \geq f(\pi, S_{\sigma(K)})$. The *Ordered Weighted Averaging* aggregation operator (OWA), introduced in [12], is defined as follows: $\text{OWA}(\pi) = \sum_{i \in [K]} v_i f(\pi, S_{\sigma(i)})$. We now consider the following MIN-OWA $1|p_j = 1| \sum w_j U_j$ problem:

$$\min_{\pi \in \Pi} \text{OWA}(\pi). \tag{2}$$

The choice of particular numbers v_i, $i \in [K]$, leads to well known criteria in decision making under uncertainty, among others: the maximum, the average and the Hurwicz pessimism—optimism criteria. Suppose that $v_1 \geq v_2 \geq \cdots \geq v_K$, such numbers are used if the idea of the robust optimization is adopted. Notice that this case contains both the maximum and the average criteria as special (boundary) cases. Indeed, if $v_1 = 1$ and $v_i = 0$ for $i \neq 1$, then we obtain the maximum criterion, the first extreme, and problem (2) becomes (1). If $v_i = 1/K$ for $i \in [K]$, then we get the average criterion—the second extreme. The following theorem holds:

Theorem 5 *If* $v_1 \geq v_2 \geq \cdots \geq v_K$ *then* MIN-OWA $1|p_j = 1| \sum w_j U_j$ *is approximable within* $v_1 K$.

Proof The result follows from the fact that the problem is a special case of MIN-OWA MINIMUM ASSIGNMENT, which can be approximated within $v_1 K$ [10]. □

When $v_1 = 1$, then MIN-OWA $1|p_j = 1| \sum w_j U_j$ becomes MIN-MAX $1|p_j = 1| \sum w_j U_j$. Thus, the latter problem, under due date and weight uncertainty, admits a K-approximation algorithm.

Acknowledgments This work was partially supported by the National Center for Science (Narodowe Centrum Nauki), grant 2013/09/B/ST6/01525.

References

1. Aissi, H., Aloulou, M.A., Kovalyov, M.Y.: Minimizing the number of late jobs on a single machine under due date uncertainty. J. Sched. **14**, 351–360 (2011)
2. Aloulou, M.A., Croce, F.D.: Complexity of single machine scheduling problems under scenario-based uncertainty. Oper. Res. Lett. **36**, 338–342 (2008)
3. Averbakh, I.: On the complexity of a class of combinatorial optimization problems with uncertainty. Math. Program. **90**, 263–272 (2001)
4. Brucker, P.: Scheduling Algorithms, 5th edn. Springer, Berlin (2007)
5. Doerr, B.: Improved approximation algorithms for the Min-Max selecting Items problem. Inf. Process. Lett. **113**, 747–749 (2013)
6. Garey, M.R., Johnson, D.S.: Computers and Intractability. A Guide to the Theory of NP-Completeness. W. H Freeman and Company (1979)
7. Graham, R.L., Lawler, E.L., Lenstra, J.K., Rinnooy Kan, A.H.G.: Optimization and approximation in deterministic sequencing and scheduling: a survey. Ann. Discret. Math. **5**, 169–231 (1979)
8. Karp, R.M.: Reducibility among combinatorial problems. Complexity of Computer Computations, pp. 85–103 (1972)
9. Kasperski, A., Kurpisz, A., Zieliński, P.: Approximating the min-max (regret) selecting items problem. Inf. Process. Lett. **113**, 23–29 (2013)
10. Kasperski, A., Zieliński, P.: Combinatorial optimization problems with uncertain costs and the OWA criterion. Theor. Comput. Sci. **565**, 102–112 (2015)
11. Kouvelis, P., Yu, G.: Robust Discrete Optimization and its applications. Kluwer Academic Publishers (1997)
12. Yager, R.R.: On ordered weighted averaging aggregation operators in multi-criteria decision making. IEEE Trans. Syst. Man Cybern. **18**, 183–190 (1988)

Price of Anarchy in the Link Destruction (Adversary) Model

Lasse Kliemann

Abstract In this model of network formation, players anticipate the destruction of one link, which is chosen according to a known probability distribution. Their cost is the cost for building links plus the expected number of other players to which connection will be lost as a result of the link destruction. We consider different equilibrium concepts (Nash equilibrium, pairwise Nash equilibrium, pairwise stability) and two different ways in which the probability distribution depends on the network.

We give proof sketches for bounds on the price of anarchy for the link destruction model (a.k.a. adversary model), a network formation game studied by the author since 2010. For details, we refer to [4, 5].

1 Equilibrium Concepts for Graphs

Let $G = (V(G), E(G))$ be an undirected, simple graph. We denote the *link* (a.k.a. *edge*) between v and w by $\{v, w\}$ for $v, w \in V(G)$. We write $G + \sum_{i \in [k]} \{v_i, w_i\} := (V(G), E(G) \cup \{\{v_1, w_1\}, \ldots, \{v_k, w_k\}\})$ and $G - \sum_{i \in [k]} \{v_i, w_i\} := (V(G), E(G) \setminus \{\{v_1, w_1\}, \ldots, \{v_k, w_k\}\})$ to add or remove one or multiple links. If no confusion can arise, we notationally do not distinguish between G and the set of links $E(G)$. Denote $e(G) := |E(G)|$ the number of links.

Let $n \in \mathbb{N}_{\geq 3}$ and $\mathscr{G}_n$ the class of all undirected, simple graphs on $[n] = \{1, \ldots, n\}$ as the vertex set. The numbers in $[n]$, i.e. the vertices of the graphs, will often be called *players*, since in the equilibrium concepts we are about to introduce, they can be considered as decision-making entities. Let $\alpha \in \mathbb{R}_{>0}$. For each player v, there is a *disutility function* $D_v : \mathscr{G}_n \longrightarrow \mathbb{R}$. A popular concrete disutility function is

L. Kliemann (✉)
Department of Computer Science, Kiel University,
Christian-Albrechts-Platz 4, 24118 Kiel, Germany
e-mail: lki@informatik.uni-kiel.de

© Springer International Publishing Switzerland 2016   285
M. Lübbecke et al. (eds.), *Operations Research Proceedings 2014*,
Operations Research Proceedings, DOI 10.1007/978-3-319-28697-6_40

$D_v(G) = \sum_{w \in [n]} \text{dist}_G(v, w)$ [1]. We will introduce our disutility function in Sect. 2. The *cost* for player v is $C_v(G) := \deg_G(v)\alpha + D_v(G)$, so in addition to disutility, each player pays α for each incident link. We call $G \in \mathscr{G}_n$ *pairwise stable* (PS) [2] if

$$C_v(G) \leq C_v(G - \{v, w\}) \qquad \forall\{v, w\} \in G$$
$$C_v(G) < C_v(G + \{v, w\}) \vee C_w(G) < C_w(G + \{v, w\}) \qquad \forall\{v, w\} \notin G$$

An interpretation is that each unordered pair $\{v, w\}$ of players forms a temporary coalition and together they decide whether the link $\{v, w\}$ should be built. If none of the two players object, then the link is built. If one objects, then the link is not built. Objection must be justified by the link being an *impairment*, i.e. strictly increasing cost. This way of establishing links is also called *bilateral link formation*. When, given any $G \in \mathscr{G}_n$ with $e = \{v, w\} \in G$, we move from G to $G - e$, we also speak of v *selling* the link e (then w is forced to sell as well).

We call $G \in \mathscr{G}_n$ a *pairwise Nash equilibrium* (PNE) [2] if

$$C_v(G) \leq C_v(G - \{v, w_1\} \ldots - \{v, w_k\}) \qquad \forall\{v, w_1\}, \ldots, \{v, w_k\} \in G$$
$$C_v(G) < C_v(G + \{v, w\}) \vee C_w(G) < C_w(G + \{v, w\}) \qquad \forall\{v, w\} \notin G$$

This is similar to PS, but players can also evaluate the effect of selling any number of their incident links. Obviously, if a graph is a PNE then it is also PS.

A variation is to assign *ownerships*: each link is *owned* by exactly one of its endpoints, and only the owner has to pay for it and only the owner can sell it. We use orientations to indicate ownership: if link $\{v, w\}$ is owned by v, we orient it in the way (v, w). Formally, an *orientated graph* is a directed graph $\overrightarrow{G}$ where for each unordered pair $\{v, w\}$ of vertices we have $(v, w) \in \overrightarrow{G} \implies (w, v) \notin \overrightarrow{G}$. We denote by $G := \{\{v, w\}; (v, w) \in \overrightarrow{G} \vee (w, v) \in \overrightarrow{G}\}$ the *underlying undirected graph*. Denote $\overrightarrow{\mathscr{G}}_n$ the class of all orientated graphs on $[n]$. The *cost* experienced by player v in $\overrightarrow{G} \in \overrightarrow{\mathscr{G}}_n$ is $C_v(\overrightarrow{G}) := \text{outdeg}_{\overrightarrow{G}}(v)\alpha + D_v(G)$, so in addition to disutility, each player pays α for each link that she owns. If v owns the link $e = \{v, w\}$, i.e. if $(v, w) \in \overrightarrow{G}$, and we change to $\overrightarrow{G} - (v, w)$, we say that v *sells* the link e. We call $\overrightarrow{G} \in \overrightarrow{\mathscr{G}}_n$ a *Nash equilibrium* (NE) if

$$C_v(\overrightarrow{G}) \leq C_v\left(\overrightarrow{G} - \sum_{i \in [k]} (v, w_i) + \sum_{s \in [t]} (v, u_s)\right)$$
$$\forall(v, w_1), \ldots, (v, w_k) \in \overrightarrow{G} \quad \forall\{v, u_1\}, \ldots, \{v, u_t\} \notin G$$

This concept gives the most freedom in terms of link building: any player can link (at the cost of α per link) to any other player even without the other player's consent, this is also called *unilateral link formation*. Also, each player can sell any of the links that she owns. We call $G \in \mathcal{G}_n$ a NE if there exists an orientation $\overrightarrow{G}$ of G that is a NE. This model has first gained attention through the work by Fabrikant et al. [1].

The *social cost* SC for a graph (undirected or orientated) is the sum over all players' costs. Even with ownerships (orientated graph), this sum only depends on the underlying undirected graph, so we define SC as a function on $\mathcal{G}_n$. We have:

$$\mathrm{SC}(G) = \begin{cases} 2e(G)\,\alpha + \sum_{v \in [n]} D_v(G) & \text{for PS and PNE} \\ e(G)\,\alpha + \sum_{v \in [n]} D_v(G) & \text{for NE} \end{cases}$$

The first term is called *total building cost* and the second term *total disutility*, denoted $\mathrm{TD}(G)$. So written shortly, $\mathrm{SC}(G) = \Theta(e(G)\,\alpha) + \mathrm{TD}(G)$. A graph $G \in \mathcal{G}_n$ is called *optimal* if it has minimum social cost among all graphs in $\mathcal{G}_n$; we denote its social cost by $\mathrm{OPT}(n, \alpha)$. For a fixed equilibrium concept (PS, PNE, or NE), the *price of anarchy* (PoA) [6] is

$$\mathrm{PoA}(n, \alpha) := \max_{G \in \mathcal{G}_n \text{ is equilibrium}} \frac{\mathrm{SC}(G)}{\mathrm{OPT}(n, \alpha)}.$$

Lower and upper bounds on the PoA in this model in terms of n and α have been the topic of extensive research since the work by Fabrikant et al. [1].

2 Link Destruction Model

A *destroyer*[1] is a map $\mathcal{D}$ on $\mathcal{G}_n$, assigning to each $G \in \mathcal{G}_n$ a probability measure $\mathcal{D}_G$ on the links of G, i.e. $\mathcal{D}_G(e) \in [0, 1]$ for each $e \in G$ and $\sum_{e \in G} \mathcal{D}_G(e) = 1$. Given a connected graph G, the *relevance* of a link $e \in G$ for a player v, denoted $\mathrm{rel}_G(e, v)$, is the number of vertices that can, starting at v, *only* be reached via e. The *separation* $\mathrm{sep}_G(e)$ of a link $e \in G$ is the number of ordered vertex pairs (v, w) for which there exist no v-w path in $G - e$. Disutility in the link destruction model for connected G is defined as $D_v(G) := \sum_{e \in G} \mathrm{rel}_G(e, v)\mathcal{D}_G(e)$, which is the expected number of players that v will no longer be able to reach when one link is destroyed in G, chosen randomly according to $\mathcal{D}_G$. The destroyer $\mathcal{D}$ is a parameter, like n and α. If G is disconnected, then we define disutility to ∞, resulting in optima, PS graphs, PNE, and NE all being connected. For connected G, total disutility can be written as $\mathrm{TD}(G) = \sum_{e \in G} \mathrm{sep}_G(e)\mathcal{D}_G(e)$. We focus on two destroyers, namely:

- The *uniform destroyer* $\mathcal{D}_G(e) = \frac{1}{e(G)}$ for all $e \in G$, i.e. the link to destroy is chosen uniformly at random.
- The *extreme destroyer* $\mathcal{D}_G(e) = \frac{1}{|G_{\max}|}$ if $e \in G_{\max}$ and $\mathcal{D}_G(e) = 0$ otherwise, where $G_{\max} := \{e \in G; \mathrm{sep}_G(e) = \mathrm{sep}_{\max}(G)\}$ and $\mathrm{sep}_{\max}(G) := \max_{e \in G}$

---

[1] In earlier work by the author, the term "adversary" was used instead of "destroyer". However, in the context of the PoA, "adversary" would be more suited to describe a system that would aim at maximizing the PoA. Hence the name was changed.

$\text{sep}_G(e)$. So the link to destroy is chosen uniformly at random from the set of links where each causes a maximum number of ordered vertex pairs to be separated. We have $\text{TD}(G) = \text{sep}_{\max}(G)$.

We summarize existence results, holding for both destroyers [4, 5]: if $\alpha > 2 - \frac{1}{n-1}$, then a star is PS, a PNE, and a NE; if $\alpha \leq \frac{1}{2}\lfloor \frac{n-1}{2} \rfloor$, then a cycle is PS, a PNE, and a NE. The ranges for α overlap if $n \geq 9$. All statements regarding PoA have to be understood as: "If equilibria exist (which is at least the case if $n \geq 9$, regardless of α), then the PoA satisfies…" Star and cycle are also optima, depending on α, with social cost $\Theta(n\alpha)$.

We can give a rough bound on the PoA for any fixed destroyer $\mathscr{D}$ and any equilibrium concept. Since optima are connected, they have at least $n - 1$ links and hence social cost $\Omega(n\alpha)$. Let G be a worst-case equilibrium. Using $\text{sep}_G(e) \leq n^2$ for all e, we get:

$$\text{PoA}(n, \alpha) \leq \frac{\mathscr{O}(e(G)\,\alpha) + \sum_{e \in G} \text{sep}_G(e)\mathscr{D}_G(e)}{\Omega(n\alpha)} = \mathscr{O}\left(\frac{e(G)}{n} + \frac{n}{\alpha}\right) \quad (1)$$

3 PoA for Nash Equilibrium

The PoA for NE and the two destroyers was shown to be $\mathscr{O}(1)$ by the author [3, 4]. Bounding total building cost in a NE works by recognizing that NE are pseudo-chord-free, where a *pseudo-chord* is a link $\{u, v\}$ with two internally vertex-disjoint u-v paths neither of which traverses $\{u, v\}$. Removing a pseudo-chord does not change any relevance value, since there still exist two alternative routes between u and v. This settles the case for the extreme destroyer. For the uniform one, cost increases slightly (at most by $\frac{1}{2}$) when a pseudo-chord is sold since the probability of destruction increases for the remaining links. But then it can be shown that there is always the option for one of the endpoints of improving cost by selling the pseudo-chord and building a different link instead, contradicting NE. So we do not have pseudo-chords in a NE. Next, a graph-theoretical result shows that a pseudo-chord-free graph on n vertices has only $\mathscr{O}(n)$ links. This yields an $\mathscr{O}(n\alpha)$ bound on total building cost in a NE. By (1), we obtain an $\mathscr{O}(1 + \frac{n}{\alpha})$ bound on the PoA.

An important tool for improving this bound to $\mathscr{O}(1)$ is the *bridge tree* $\tilde{G}$ of a connected graph G. It is obtained by replacing each maximal bridgeless connected subgraph of G, these are called *islands*, by a single vertex, and linking these new vertices according to their connections in G: for each two islands I_1 and I_2 we include the link $\{I_1, I_2\}$ in the bridge tree iff there exist $u \in I_1$ and $v \in I_2$ with $\{u, v\} \in G$. Alternatively, we can think of successively contracting each cycle in G to a single vertex, until the graph is cycle-free. Links inside of islands have relevance 0 for everyone. There is an obvious bijection between the links of $\tilde{G}$ and the bridges of G, and we will often speak of relevance or separation of a link in $\tilde{G}$ meaning in fact the corresponding bridge of G. When counting vertices in the bridge tree, we count $|I|$ for each vertex I of $\tilde{G}$, so we count vertices in the respective islands.

For the uniform destroyer, we can show easily that $\mathrm{TD}(G) \leq n \operatorname{diam}(\tilde{G})$. Then we show for a NE G that $\operatorname{diam}(\tilde{G}) = \mathcal{O}(\alpha)$ by the following argument. Let $P = (I_0, \ldots, I_\ell)$ be a path in $\tilde{G}$ and e a link on P half-way between I_0 and I_ℓ. Then w.l.o.g. at least $\frac{n}{2}$ players (we omit floor and ceiling when they are unimportant for our arguments) are located beyond e from the view of any player $v \in I_0$, giving high relevance for v to the $\frac{\ell}{2}$ links from I_0 up to e, namely the sum of those relevances is $\Omega(n\ell)$. Player v can put all those links on a cycle by building a new link, reducing all those relevances to 0. A computation shows that her savings in disutility are at least $\frac{1}{e(G)+1}\Omega(n\ell)$, which is $\Omega(\ell)$ by $e(G) = \mathcal{O}(n)$. In a NE, those saving cannot exceed α, hence $\ell = \mathcal{O}(\alpha)$.

For the extreme destroyer, let $\overrightarrow{G}$ be a NE. We call the links in $G_{\max}$ the *critical* links. The whole proof is long and consists of a detailed case analysis. We only sketch the case $|G_{\max}| = 1$ here, since it relies on unilateral link formation and does not work for PS or PNE. In the bridge tree, let $e = \{I_0, I_1\}$ be the critical link with I_1 in the component with a minimum number of vertices in $G - e$. Denote $T_1, \ldots, T_N$ all the subtrees below I_0 when rooting the tree at I_0, enumerated so that $I_1 \in V(T_1)$ and $n_2 \geq \cdots \geq n_N$, where $n_i := |V(T_i)|$ for all i. Since e alone has maximum separation, $n_1 > n_2$. Moreover, $\frac{1}{2}n \geq n_1$. We restrict here to the case that we can find $v_1 \in V(T_1)$ and $v_2 \in V(T_2)$ such that building $\{v_1, v_2\}$ does not induce any critical links inside of T_1 nor T_2. Then any of the new critical links can cut off only n_3 players from v_1. We consider the change in disutility for v_1 when building $\{v_1, v_2\}$, i.e. adding (v_1, v_2) to $\overrightarrow{G}$, which is upper-bounded by α by the NE property. We have $\alpha \geq D_{v_1}(G) - D_{v_1}(G + \{v_1, v_2\}) = (n - n_1) - n_3 \geq (n - n_1) - \frac{1}{3}n \geq \frac{1}{2}n - \frac{1}{3}n = \frac{1}{6}n$. It follows $\mathrm{TD}(G) = \mathrm{sep}_{\max} \leq n^2 = \mathcal{O}(n\alpha)$.

So for both destroyers, we have $\mathcal{O}(n\alpha)$ for the social cost of a NE, with a ratio of $\mathcal{O}(1)$ to the optimum.

4 PoA for Pairwise Stability and Pairwise Nash Equilibrium

The PoA for PS and PNE and the two destroyers was studied by the author in 2013 [5]. Absence of pseudo-chords can be shown almost in the same way as for NE. For the extreme destroyer, it works the same. For the uniform destroyer, the difficulty is the at most $\frac{1}{2}$ increase in disutility. In a NE, we can argue that a player would sell the pseudo-chord and build a more beneficial link instead, but for PS and PNE this argument does not work since we also have to show that the other endpoint has no objection against a new link. For simplicity, we restrict to $\alpha > \frac{1}{2}$ in the following, in which case players would always sell a pseudo-chord to save α in building cost. By (1), we obtain an $\mathcal{O}(1 + \frac{n}{\alpha})$ bound on the PoA. For the extreme destroyer, it is tight up to constants. The lower bound is given by a graph with exactly one critical link e and numbers of players adjusted so that the players in the smaller component of $G - e$ wish to put e on a cycle by adding a link, but players in the larger component of $G - e$ are less enthusiastic about it: they will only lose connection to the smaller component when e is destroyed. This example is PS and a PNE.

For the uniform destroyer, we show an $\mathcal{O}(1)$ bound for PS, which implies the same bound for PNE. By pseudo-chord-freeness, we are left with bounding total disutility. The obvious idea, namely using $\mathrm{TD}(G) \leq n \operatorname{diam}(\tilde{G})$, does not work since $\operatorname{diam}(\tilde{G})$ can be $\sqrt{n}$ in a PS graph, while $\alpha = 1$. For a path P let $\mathrm{rel}_G(P, v) := \sum_{e \in E(P)} \mathrm{rel}_G(e, v)$ the sum of relevances along P. The lower bound construction for the diameter uses a graph G consisting of a cycle with a path P of length $\sqrt{n}$ attached, and $n \geq 4$ and $\alpha = 1$. Let v be a player on the cycle and w the player at the end of the attached path. Then $D_v(G) = \frac{\mathrm{rel}_G(P,v)}{n} \leq \frac{\sqrt{n}\sqrt{n}}{n} = 1 \leq \alpha$ and $D_w(G) \geq \frac{\sqrt{n}\,(n-\sqrt{n})}{n} = \sqrt{n} - 1 \geq 1 = \alpha$. Adding the link $\{v, w\}$ will let disutility drop to 0 for everyone, but this is still an impairment for v, although (desperately, for large n) desired by w. So this graph is not a NE, but it can be easily seen to be PS. However, its social cost is within a constant of the optimum, providing no interesting lower bound on the PoA.

The key for proving the $\mathcal{O}(1)$ bound lies in recognizing that the above example is essentially the worst case. Let G be PS. Generally, in a tree T, there is a vertex u such that if T is rooted at u, in each sub-tree there are at most $\frac{n}{2}$ vertices. This also works for the bridge tree, so let R be an appropriate island and root the bridge tree at R. Let $P = (I_0, \ldots, I_\ell)$ be a path and $v \in I_0$ and $w \in I_\ell$. Then (i) $\mathrm{rel}_G(P, v) = \mathcal{O}(n\alpha)$ or (ii) $\mathrm{rel}_G(P, w) = \mathcal{O}(n\alpha)$ since otherwise neither v nor w could object to the addition of $\{v, w\}$. If $I_0 = R$, we have $\mathrm{rel}_G(e, w) \geq \frac{n}{2}$ for all $e \in E(P)$, so in case (ii) we conclude $\ell = \mathcal{O}(\alpha)$. Moreover, there can be at most one subtree S in the rooted $\tilde{G}$ with a path starting at R for which (ii) does not hold: otherwise, two paths without property (ii) could be concatenated creating a path with neither (i) nor (ii). So disregarding S, we have diameter $\mathcal{O}(\alpha)$ in the bridge tree. Let $P = (R, \ldots, I)$ be a longest path in S. By similar arguments as before, it can be shown that around P in S, only paths of length $\mathcal{O}(\alpha)$ grow. Hence once we put P on a cycle by an additional link, the bridge tree has diameter $\mathcal{O}(\alpha)$. We have $e(G) \cdot \mathrm{TD}(G) = \sum_{e \in G} \mathrm{sep}_G(e) = \sum_{\substack{e \in G \\ e \notin E(P)}} \mathrm{sep}_G(e) + \sum_{e \in E(P)} \mathrm{sep}_G(e)$. The second sum is $\mathcal{O}(n^2\alpha)$ by a simple calculation. The first sum is the same as $\sum_{e \in G'} \mathrm{sep}_{G'}(e)$ with $G' := G + \{v, w\}$ for $v \in R$ and $w \in I$. By $\operatorname{diam}(\tilde{G}') = \mathcal{O}(\alpha)$, we have $\mathrm{TD}(G') = \mathcal{O}(n\alpha)$ which means $\sum_{e \in G'} \mathrm{sep}_{G'}(e) = \mathcal{O}(e(G')\,n\alpha)$. In total, it follows $\mathrm{TD}(G) = \mathcal{O}(n\alpha)$, which implies the $\mathcal{O}(1)$ bound on the PoA.

References

1. Fabrikant, A., Luthra, A., Maneva, E., Papadimitriou, C.H., Shenker, S.: On a network creation game. In: Proceedings of the 22nd Annual ACM SIGACT-SIGOPS Symposium on Principles of Distributed Computing, Boston, Massachusetts, USA, July 2003 (PODC 2003), pp. 347–351

(2003). doi:10.1145/872035.872088

2. Jackson, M.O., Wolinsky, A.: A strategic model of social and economic networks. J. Econ. Theory **71**(1), 44–74 (1996). doi:10.1006/jeth.1996.0108

3. Kliemann, L.: Brief announcement: the price of anarchy for distributed network formation in an adversary model. In: Proceedings of the 29th Annual ACM SIGACT-SIGOPS Symposium on Principles of Distributed Computing, Zurich, Switzerland, July 2010 (PODC 2010), pp. 229–230 (2010). doi:10.1145/1835698.1835749

4. Kliemann, L.: The price of anarchy for network formation in an adversary model. Games **2**(3), 302–332 (2011). doi:10.3390/g2030302

5. Kliemann, L.: The price of anarchy in bilateral network formation in an adversary model. In: Vöcking, B. (ed.) Proceedings of the 6th Annual ACM-SIAM Symposium on Algorithmic Game Theory, Aachen, Germany, October 2013 (SAGT 2013), no. 8146 in Lecture Notes in Computer Science (2013). arXiv:1308.1832

6. Papadimitriou, C.H.: Algorithms, games, and the Internet. In: Proceedings of the 33rd Annual ACM Symposium on Theory of Computing, Crete, Greece, July 2001 (STOC 2001), pp. 749–753 (2001). doi:10.1145/380752.380883. Extended abstract at ICALP 2001

Decision Support System for Week Program Planning

Benjamin Korth and Christian Schwede

Abstract This contribution aims to give an overview of the research issue on weekly program filling in the automotive industry and presents an approach to improve this process by using a logistic assistance system. This system supports a dispatcher at choosing customer orders for a production week and takes into account the large amount of restrictions. Finally, results of the heuristic used are presented.

1 Introduction

In the automotive industry, the weekly program filling is one important part of the week program planning process. The week program planning deals with the selecting a production week for every vehicle order to form the production program, aiming at fully utilize factory capacities of every production week while considering a huge number of restrictions (cf. [2] S.35ff, [1] S.190ff, [4] S.190ff).

To handle the complexity of this time-critical task, Fraunhofer Institute for Material Flow and Logistics has developed a *Logistics Assistance System* (LAS), which is a decision support system that helps the planners in sales and logistics to perform the weekly program filling. This paper first describes the research issue on weekly program filling, the LAS and the heuristics used for order dispatching within the weekly program filling. Then, the developed heuristic for order dispatching is validated and compared with the present processes at the automotive manufacturer. Finally, a conclusion is drawn and opportunities for future research are given.

B. Korth (✉) · C. Schwede (✉)
Fraunhofer Institut für Materialfluss und Logistik, Joseph-von-Fraunhofer-Straße 2-4,
44227 Dortmund, Germany
e-mail: benjamin.korth@iml.fraunhofer.de

C. Schwede
e-mail: christian.schwede@iml.fraunhofer.de

© Springer International Publishing Switzerland 2016
M. Lübbecke et al. (eds.), *Operations Research Proceedings 2014*,
Operations Research Proceedings, DOI 10.1007/978-3-319-28697-6_41

2 Research Issue—Procedure of Filling

The weekly program filling takes place before the week enters the frozen zone (often four weeks before production), which means, that the production program is finalized and must not be changed any further. Before that, demand and capacity management is performed to forecast the demands for every single market and to determine the production and supplier capacities accordingly. In addition, the already incoming customer orders are dispatched to particular weeks in response to these determined capacities. Due to the fact that forecasts and real market demands seldom match, orders often have more than 150 critical capacities that impede the dispatching to a certain week. Thus, at the time of program freezing the production week is seldom filled so that the full capacity of the assembly line is used. Hence, the target of the weekly program filling is to assign as many orders to the production week that is intended to be frozen, that the assembly line can be run at full capacity. Particularly, orders for upcoming weeks are pre-drawn, if this is possible considering the delivery date. The set of orders that are considerate for pre-drawing is called order stock. When filling the week program, production and supplier capacities must be monitored so that they are not exceeded. To achieve this, the OEMs (Original Equipment Manufacturer) use so called "virtual sight glasses" to display and control restrictions (cf. [2] S.35). Traditionally, a sight glass is a transparent tube, through which the filling of a liquid tank can be observed. In context of the weekly program filling, a sight glass can be available for whole vehicles, features or combination of features, and corresponds to a lower or upper limit and a discrete filling level for orders. If an order associated with the sight glass is dispatched to a certain week, the filling level is increasing. If the upper limit is reached, no order associated with the sight glass can be dispatched to the certain week and therefore must be dispatched to one of the following.

This task can be described formally as an integer optimization problem. The set of orders I is the order stock. The variable $x_i \in \{0, 1\}$ determines, if the order i is dispatched in the current week or not. Dispatching aims at maximizing the degree of capacity utilization of the plants capacities for the current week and thus at maximizing the number of orders dispatched in this week:

$$\max \sum_{i \in I} x_i \tag{1}$$

The restrictions $j \in J$ have an upper limit $b_i \in N_0$. The associations between orders from I and sight glasses from J are represented by the matrix A with the elements $a_{ij} \in \{0, 1\}$. Therefore, the constraints can be described for every sight glass j by the following inequality:

$$\sum_{i \in I} a_{ij} x_i \leq b_j \tag{2}$$

Aside from sight glasses for capacities, there are sight glasses for planned quotas for market countries. Markets get their capacities (volume and features) due to their forecasts and thus get rewarded or penalized for good or poor planning. Additionally, by using market sight glasses, an overreaching of markets is prevented.

The original process of order dispatching performed by the OEMs provides a complete re-dispatching of orders before the weekly program filling. Hence, all orders are detached and tried to re-dispatch to the earliest production week, following a given prioritization. After the re-dispatching, additional orders often cannot be dispatched to the earliest week due to critical restrictions, even though the assembly line is not being used at full capacity. The task of the planers in sales department is then to manually select individual orders that can be dispatched with minimal modification to the restrictions. The modifications are confirmed by the logistics department, only if it is feasible due to safety stock or possible special measures (extra shifts or special transports) and economically reasonable. Because of the large number of restrictions, it is often very difficult to identify necessary modifications. Sometimes, certain restrictions are forgotten by the sales department because they only become critical after dispatching the first orders of the selected subset. Hence, it can even happen that requested and confirmed capacities cannot be used due to other inevitable restrictions. Beside the restrictions, further criteria must be considered, such as type of orders, delivery reliability, balanced degree of capacity utilization, high productivity and planning reliability.

3   The Logistic Assistance System

To support the process of filling the weekly program, a decision support system (LAS) (cf. [5, 6]) was developed. This section describes its basic functions. The central dispatching algorithm is explained in the next section.

Facing the problem of systematically determining the necessary modifications to restrictions, the LAS enables a preview of the filling. Hence, a week program is created first in which all restrictions are considered as "soft" and can be suitable adjusted. The resulting program is completely filled, if there are enough orders in stock. The planer can now set hard limitations where production or logistic has no flexibilities and a request to implement the modification would be hopeless. Considering these hard restrictions, a new week program is created. The results are modifications of restrictions, which enable the maximization of production capacity utilization. The LAS supports the subsequently necessary negotiation about these modifications with the logistics department. Role specific competences allow to request and confirm modifications. If the requested modifications are not feasible, the planner can create further scenarios on the basis of new hard restrictions.

4 Dispatching Algorithm

This section describes the algorithm for filling the week program and for determining the required modifications. It selects orders in consideration of the target criteria out of the $n = |I|$ orders in the order stock I. The plant has a capacity of k orders, resulting in a solution space with size $\binom{n}{k}$. Depending on the factory and the vehicle model, the order stock could contain e.g. 10,000 orders from which up to 2,000 orders must be selected. This would result in $1,655 \times 10^{2171}$ possible combinations.

Additionally, e.g. 800 sight glasses must be considered while every order is associated with about 130 of them. For solution finding, multiple week programs must be created within a short time-frame. Therefore, a heuristic is applied that has a short runtime and provides better results than the present re-dispatching algorithm. The workflow of the heuristic is described next. Before dispatching an order, the residual capacities of the sight glasses are predicted. This is done by determining the proportion m_j of orders associated with a sight glass j to the amount of orders:

$$m_j = \frac{\sum_{i \in I} a_{ij}}{\sum_{i \in I} 1}$$

Assuming that the sight glasses are filled according to this proportions, the remaining capacities $remainingCapacity_j$ (3) for every sight glass can be predicted for a completely filled week without considering other restrictions.

$$remainingCapacity_j = (c_j - m_j o) \tag{3}$$

$$Score_i = \sum_j a_{ij} remainingCapacity_j \tag{4}$$

If there remains a high capacity for a particular sight glass, it should not be critical to fill the week program. The lower the remaining capacity, the more it should be avoided to dispatch orders that are associated with it. If the remaining capacity is negative and the sight glass has a hard restriction, an associated order is not allowed to be dispatched in this week at that moment.

For the assessment and selection of an order, the extrapolated capacities of the associated sight glasses are summed (4). The order with the highest value is dispatched if no associated restriction with a negative remaining capacity is declared "hard". To speed up the assessment, orders that are equal regarding to associated sight glasses were combined to so called representatives. Thereby the amount of orders that must be considered regarding their sight glasses can be reduced by 30–60 %. To take priorities and delivery dates into account, the orders of a representative are sorted due to these criteria. To improve the quality of the planning, the LAS used an optional exchange algorithm called "m to n exchange" that tries to replace a subset of orders in the considered week through a subset of orders from the order stock.

The exchange is done, if the new subset enables to put more orders in the week or to reduce the amount of necessary sight glass modifications. In this optimization step it is not allowed to soften any restriction.

5 Results

The performance of the dispatching heuristics will be compared with the present used procedure for re-dispatching. For this purpose, the criteria filling degree, proposed sight glass modifications and degree of capacity utilization are considered. Additionally, the dispatched orders are divided into the following four groups to compare the treatment of priority orders with time-critical delivery date orders:

1. Priority order and delivery date $\leq$ planning week
2. Priority order and delivery date $>$ planning week
3. No priority order and delivery date $\leq$ planning week
4. No priority order and delivery date $>$ planning week

Both procedures were applied to two experiments that include different planning scenarios. The first experiments simulate the re-dispatching at the beginning of the week in which the order is allocated to an expected planning week without adjustment to the limitation of sight glasses. The results are shown in Table 1. The number of dispatched orders per order group is specified for two model groups. The filling degree of the whole week increases by 9.7 % and 14.9 %, respectively. On average, dispatching heuristics reached an increase of 8–12 % relative to the procedure of the OEM. In the second experiment, the adjusted limitation of the sight glass for a manually created week program with 100 % capacity utilization is used as base. For the dispatching algorithm, these limitations were declared as hard. The dispatch of

Table 1 Experiment 1: Comparison of dispatching processes

Vehicle group A	Group 1	Group 2	Group 3	Group 4	Level
Group system:	81	23	910	287	1301/2861 (45,5 %)
Heuristic:	54 (−33,3 %)	17 (−26,1 %)	946 (+3,9 %)	562 (+95,8 %)	1579/2861 (55,2 %)
Vehicle group B	Group 1	Group 2	Group 3	Group 4	Level
Group system:	0	0	1008	26	1034/1448 (71,4 %)
Heuristic:	0 (± 0 %)	0 (± 0 %)	1013 (+0,5 %)	236 (+907,7 %)	1249/1448 (86,3 %)

Although the modification of limitations are not allowed, the heuristic can increase the amount of orders

Table 2 Experiment 2 (1): A week program was created with the LAS used only hard restrictions that were manual determined

SG	Manual filling	Optimization	Difference	% Difference
SG 0001	90	88	−2	−2, 2 %
SG 0002	6	2	−4	−66, 7 %
SG 0003	210	188	−22	−10, 5 %
…	…	…	…	…
20 SGs:			$\sum$ 53 Ø2, 65	Ø19, 7 %

It was found that 20 % of adjustments could have been avoided

Table 3 Experiment 2 (2): The lower modifications to restrictions was at expense to delivery reliability and to order priorities

Vehicle group C	Group 1	Group 2	Group 3	Group 4	Level
Group system:	13	39	524	193	769/769 (100 %)
Heuristic:	11 (−15,4 %)	28 (−28,2 %)	488 (−6,9 %)	242 (+25,5 %)	769/769 (100 %)

orders by heuristics shows that in average less than 20 % modifications are required (cf. Table 2). Table 3 shows that on the contrary, the savings of necessary modifications are at the expense of order criteria such as priority and delivery reliability.

6   Conclusion and Outlook

The process of week program filling can be effectively supported by the use of the LAS. The experiments show that capacities of the factory can be used more efficiently or that the number of necessary modifications could be lowered. Under certain circumstances, this has an impact on delivery reliability and the consideration of order priorities. Furthermore, the work with the LAS reduces the effort of program filling because many small, partially incorrect modification requests can be avoided. Instead, only a few requests are necessary to get the best result regarding the use of factory capacities. Under consideration that two employees are engaged full time in this task per vehicle model, the LAS offers a high potential of reducing work load. In addition to the determination of modifications, the LAS could support reviewing and confirmation of modifications requests. This is also a complex process in which the modifications have to be translated from sight glass to part level in order to be compared with capacities of production and logistic as well as possible adaptation measures. The integration of translation to part level by means of bill of material explosion in the LAS to balance capacities on part level as well as the use of simulative income forecast (see Eco2Las [3]) offers another great potential.

References

1. Herlyn, W.J.: PPS im Automobilbau: Produktionsprogrammplanung und -steuerung von Fahrzeugen und Aggregaten. Fahrzeugtechnik. Hanser, München (2012). ISBN: 3446428488
2. Herold, L.: Kundenorientierte Prozessteuerung in der Automobilindustrie (2005)
3. Klingebiel, K., et al.: Ressourceneffiziente Logistik. In: Neugebauer, R. (ed.) Handbuch Ressourcenorientierte Produktion, pp. 719–748. Carl Hanser (2014). ISBN: 978-3446430082
4. Klug, F.: Logistikmanagement in der Automobilindustrie. Springer, Berlin (2010). ISBN: 978-3-642-05292-7. doi:10.1007/978-3-642-05293-4
5. Kuhn, A., Toth, M.: Assistenzsysteme für die effektive Planung logistischer Netzwerke. In: Scholz-Reiter, B. (ed.) Technologiegetriebene Veränderungen der Arbeitswelt. Schriftenreihe der Hochschulgruppe für Arbeits- und Betriebsorganisation e.V. (HAB), pp. 257–278. Gito-Verl, Berlin (2008). ISBN: 9783940019493
6. Schwede, C, Toth, M, Wagenitz, A.: Funktions-übergreifende Zusammenarbeit in Unternehmen durch Logistische Assistenzsysteme. In: Spath, D. (ed.) Wissensarbeit - zwischen strengen Prozessen und kreativem Spielraum. Schriftenreihe der Hochschulgruppe für Arbeits- und Betriebsorganisation e. V. (HAB), pp. 435–447. GITO, Berlin (2011). ISBN: 394218351X

On Multi-product Lot-Sizing and Scheduling with Multi-machine Technologies

Anton V. Eremeev and Julia V. Kovalenko

Abstract We consider a problem of multi-product lot-sizing and scheduling where each product can be produced by a family of alternative multi-machine technologies. Multi-machine technologies require one or more machine at the same time. A sequence dependent setup time is needed between different technologies. The criterion is to minimize the makespan. Preemptive and non-preemptive versions of the problem are studied. We formulate mixed integer linear programming models based on a continuous time representation for both versions of the problem. Using these models, the polynomially solvable cases of the problem are found. It is proved that the problem without setup times is strongly NP-hard if there is only one product, and each technology occupies at most three machines. Besides that, problem cannot be approximated within a practically relevant factor of the optimum in polynomial time, if P $\neq$ NP.

1 Introduction

In practice, many scheduling problems involve tasks, machines and materials such as raw materials, intermediate and final products. Each task may consist in storage, loading/unloading or transformation of one material into another and may be preceded by a sequence-dependent setup time. One of the standard optimization criteria is to minimize the makespan, i.e. the time when the last task is completed.

This paper considers a multi-product lot-sizing and scheduling problem with *multi-machine technologies*, where a multi-machine technology requires more than one machine at the same moment of time, also known as a multi-processor task [3] in parallel computing scheduling. The problem is motivated by the real-life scheduling

A.V. Eremeev (✉)
Omsk Branch of Sobolev Institute of Mathematics SB RAS, 13 Pevtsov Str.,
644099 Omsk, Russia
e-mail: eremeev@ofim.oscsbras.ru

J.V. Kovalenko (✉)
Omsk Law Academy, 12 Korolenko Str., 644010 Omsk, Russia
e-mail: julia.kovalenko.ya@ya.ru

© Springer International Publishing Switzerland 2016
M. Lübbecke et al. (eds.), *Operations Research Proceedings 2014*,
Operations Research Proceedings, DOI 10.1007/978-3-319-28697-6_42

applications in chemical industry and may be considered as a special case of the problem formulated in [4].

An analysis of computational complexity of lot-sizing and scheduling problems with multi-processor tasks and zero setup times is carried out in [3, 5, 8]. In the present paper we consider a more general case where the non-zero setup times may be required as well. A similar multi-product lot-sizing and scheduling problem with setup times on unrelated parallel machines was studied in [2]. However on one hand, in [2] each technology involved just a single machine, on the other hand the lower bounds on the lot sizes were given.

2 Problem Formulation

Consider a plant producing k different products. Let $V_i > 0$ be the demanded amount of product i, $i = 1, \ldots, k$ and let m be the number of machines available at the plant. For each product i, $i = 1, \ldots, k$, there is at least one technology to produce this product. Let U be the set of all technologies, $d = |U|$, and each technology is characterized by the set of machines it simultaneously occupies $M_u \subseteq \{1, \ldots, m\}$, $u \in U$, and the product i it produces. While the product i is produced by technology u, all machines of the subset M_u are engaged and at any moment each machine of the plant may be engaged in not more than one technology.

Let $U_i \subseteq U$ denote the set of technologies that output product i, $i = 1, \ldots, k$, and $a_u > 0$ is the production rate, i.e. the amount of product i produced by u per unit of time, $u \in U_i$. It is assumed that a feasible schedule may assign to the same product i one or more technologies from U_i, $i = 1, \ldots, k$, i.e. the *migration* is allowed according to the terminology from [8]. For each machine l the setup times from technology u to technology q are denoted by s_{luq}, $s_{luq} > 0$ for all u, $q \in K_l$, where $K_l = \{u : l \in M_u, u \in U\}$ is the set of technologies that use machine l, $l = 1, \ldots, m$.

The problem asks to find for each product i, $i = 1, \ldots, k$, the set of technologies from U_i that will be utilized for production of i, to determine the lot-sizes of production using each of the chosen technologies and to schedule this set of technologies so that the makespan $C_{\max}$ is minimized and the products are produced in demanded volumes $V_1, \ldots, V_k$. The problem is considered in two versions: when preemptions of technologies are allowed (denoted $P|set_i$, pmtn, $s_{luq}|C_{\max}$) and when the preemptions are not allowed (denoted $P|set_i$, $s_{luq}|C_{\max}$).

In practice one often may assume that the setup times satisfy the triangle inequality $s_{luq} + s_{lqp} \geq s_{lup}$, $l = 1, \ldots, m$, $u, q, p \in K_l$. In what follows we denote the special case of preemptive scheduling with the triangle inequality assumption by $P|set_i$, pmtn, $\Delta s_{luq}|C_{\max}$.

The problems formulated above are strongly NP-hard because in the special case of $m = 1$ the metric shortest Hamilton path reduces to them and this problem is known to be NP-hard in the strong sense [7].

3 Problem Complexity in Case of Zero Setup Times

It was shown in [5, 8] that in case of zero setup times the problems formulated in Sect. 2 are intractable. These results are obtained using the graph coloring and fractional graph coloring problems. The results from [5, 8, 12] imply that even in the special case when each product has exactly one technology producing it and each machine suits only two technologies, problems $P|\text{set}_i$, $s_{luq} = 0|C_{\max}$ and $P|\text{set}_i$, pmtn, $s_{luq} = 0|C_{\max}$ can not be approximated within a factor $k^{1-\varepsilon}$ for any $\varepsilon > 0$, if P $\neq$ NP.

Here we claim that in the case of single product, when multiple technologies are allowed, the problems formulated in Sect. 2 are intractable as well:

Proposition 1 *Problems $P|\text{set}_i$, $s_{luq} = 0|C_{\max}$ and $P|\text{set}_i$, pmtn, $s_{luq} = 0|C_{\max}$ are strongly NP-hard even in the special case when the number of products $k = 1$, and all technologies have equal production rates, however each technology occupies at most 3 machines.*

Besides that, in the case of $k = 1$, the problems $P|\text{set}_i$, $s_{luq} = 0|C_{\max}$ and $P|\text{set}_i$, pmtn, $s_{luq} = 0|C_{\max}$ are not approximable within a factor $d^{1-\varepsilon}$ for any $\varepsilon > 0$, assuming $P \neq NP$.

4 Mixed Integer Programming Model

Let us define the notion of *event points* analogously to [6]. By event point we will mean a subset of variables in mixed integer programming (MIP) model, which characterize a selection of a certain set of technologies and their starting and completion times. In one event point each machine may be utilized in at most one technology. The set of all event points will be denoted by $N = \{1, \ldots, n_{\max}\}$, where the parameter $n_{\max}$ is chosen sufficiently large on the basis of a-priory estimates or preliminary experiments.

The structure of the schedule is defined by the Boolean variables w_{un} such that $w_{un} = 1$ if technology u is executed in event point n, and $w_{un} = 0$ otherwise. In case technology u is executed in event point n, the staring time and the completion time of technology u in this event point are given by the real-valued variables T^s_{un} and T^f_{un} accordingly. The variable $C_{\max}$ is equal to the time when the last technology is finished (the makespan).

Define the following notation:

let I be the set of all products, $|I| = k$;

let M be the set of machines, $|M| = m$;

$H = \sum_{i \in I} \max_{u \in U_i} \left\{ \frac{v_i}{a_u} \right\} + (k - 1) \cdot \max_{l \in M, \, u,q \in K_l} \{s_{luq}\}$ is an upper bound on makespan. The amount of time H is sufficient to produce all the demanded products. Then the MIP model for $P|\text{set}_i$, pmtn, $s_{luq}|C_{\max}$ problem is as follows:

$$C_{\max} \to \min, \tag{1}$$

$$T_{un}^f \leq C_{\max}, \ u \in U, \ n \in N, \tag{2}$$

$$\sum_{u \in K_l} w_{un} \leq 1, \ l \in M, \ n \in N, \tag{3}$$

$$T_{un}^s \geq T_{q\tilde{n}}^f + s_{lqu} - H \cdot (2 - w_{un} - w_{q\tilde{n}} + \sum_{q' \in K_l} \sum_{\tilde{n} < n' < n} w_{q'n'}), \tag{4}$$

$$l \in M, \ u, \ q \in K_l, \ n, \ \tilde{n} \in N, \ n \neq 1, \ \tilde{n} < n,$$

$$T_{un}^f \geq T_{un}^s, \ u \in U, \ n \in N, \tag{5}$$

$$T_{un}^f - T_{un}^s \leq w_{un} \cdot \max_{q \in U_i} \left\{ \frac{V_i}{a_q} \right\}, \ i \in I, \ u \in U_i, \ n \in N, \tag{6}$$

$$\sum_{n \in N} \sum_{u \in U_i} a_u \cdot (T_{un}^f - T_{un}^s) \geq V_i, \ i \in I, \tag{7}$$

$$T_{un}^s \geq 0, \ u \in U, \ n \in N, \tag{8}$$

$$w_{un} \in \{0, 1\}, \ u \in U, \ n \in N. \tag{9}$$

The objective function (1) and inequality (2) define the makespan criterion. Constraint (3) implies that in any event point on machine l at most one technology may be executed. Constraint (4) indicates that the starting time of technology u on machine l should not be less than the completion time of a preceding technology on the same machine, plus the setup time. Constraint (5) guarantees that all technologies may be performed only for non-negative time. If a technology u is not executed in the event point n (i.e. $w_{un} = 0$) then its duration should be zero—this is ensured by inequality (6). Constraint (7) bounds the amount of production according to the demand. Constraints (8)–(9) give the area where the variables are defined.

A MIP model for problem $P|\text{set}_i, \ s_{luq}|C_{\max}$ may be obtained from (1)–(9) by adding the inequality

$$\sum_{n \in N} w_{un} \leq 1, \ u \in U, \tag{10}$$

which ensures each technology is executed without preemptions.

These two models and their modifications for the triangle inequality case are studied experimentally in [10].

5 Polynomially Solvable Cases

In order to find an optimal solution to $P|\text{set}_i, \; s_{luq}|C_{\max}$ using model (1)–(10), it is sufficient to set $n_{\max} = d$ because the preemptions are not allowed. Denote $\mathscr{P}_{LP}$ the linear programming problem obtained by fixing all Boolean variables (w_{un}) in model (1)–(10). Here and below by *fixing* of the variables we assume assignment of some fixed values to them (which turns these variables into parameters). Problem $\mathscr{P}_{LP}$ with $n_{\max} = d$ involves a polynomially bounded number of variables, which means it is polynomially solvable (see e.g. [9]).

Let τ_{LP} be an upper bound on the time complexity of solving problem $\mathscr{P}_{LP}$. The problem $P|\text{set}_i, \; s_{luq}|C_{\max}$, where the number of technologies is bounded by a constant from above, we will denote by $P|\text{set}_i, \; s_{luq}, d = const|C_{\max}$. This problem reduces to $(n_{\max} + 1)^d$ problems of $\mathscr{P}_{LP}$ type with $n_{\max} = d$. Therefore the following theorem holds.

Theorem 1 *Problem $P|\text{set}_i, \; s_{luq}, d = const|C_{\max}$ is polynomially solvable within $O(\tau_{LP} \cdot d^d)$ time.*

To find an optimal solution to $P|\text{set}_i, \; \text{pmtn}, \; \Delta s_{luq}|C_{\max}$ problem, it suffices to set $n_{\max} = d^m$ in model (1)–(9). Indeed, the number of different sets of technologies that may be executed simultaneously does not exceed $\prod_{l=1}^{m} f_l \leq d^m$, where $f_l = |K_l| + 1$ if $|K_l| < d$, otherwise $f_l = d$. Besides that, there exists an optimal solution to problem $P|\text{set}_i, \; \text{pmtn}, \; \Delta s_{luq}|C_{\max}$ where each of the above mentioned sets of technologies is executed simultaneously at most once. This fact follows by the lot shifting technique which is applicable here since the setup times obey the triangle inequality.

Let $\mathscr{P}'_{LP}$ denote the linear programming problem obtained by fixing all Boolean variables (w_{un}) in MIP model (1)–(9). A problem $\mathscr{P}'_{LP}$ with $n_{\max} = d^m$ and the number of machines bounded above by a constant is polynomially solvable. Let τ'_{LP} denote an upper bound of the time complexity of solving $\mathscr{P}'_{LP}$. The problem $P|\text{set}_i, \; \text{pmtn}, \; \Delta s_{luq}|C_{\max}$, where the numbers of machines and products are bounded by a constant will be denoted by $Pm|\text{set}_i, \; \text{pmtn}, \; \Delta s_{luq}, k = const|C_{\max}$ in what follows. This problem reduces to $2^{dn_{\max}}$ problems of $\mathscr{P}'_{LP}$ type, where $n_{\max} = d^m$. The total number of technologies d does not exceed $k(2^m - 1)$, so the following result holds.

Theorem 2 *Problem $Pm|\text{set}_i, \; \text{pmtn}, \; \Delta s_{luq}, k = const|C_{\max}$ is polynomially solvable within $O(\tau'_{LP} \cdot 2^{(k(2^m-1))^{m+1}})$ time.*

A number of other polynomially solvable cases of problems $P| \text{set}_i, \; s_{luq}| C_{\max}$ and $P| \text{set}_i, \; \text{pmtn}, \; s_{luq}| C_{\max}$ with zero setup times may be found in [1, 5, 8, 11].

6 Conclusion

The problem of multi-product lot-sizing and scheduling with multi-machine tech-nologies is studied in preemptive and non-preemptive versions. Non-approxi-mability of the problem is shown and new NP-hard special cases with zero setup times are identified. MIP models are formulated for both versions of the problem using the event-points approach and continuous time representation. New polynomially solv-able special cases of the problem are found using the MIP models, under assumption that the number of technologies is bounded by a constant.

Further research appears to be appropriate in extending the obtained results to the version of the problem where technologies may involve several tasks which should be executed sequentially and each task is performed on a number of machines simultaneously.

Acknowledgments This research have been supported by the RFBR Grants 12-01-00122 and 13-01-00862.

References

1. Bianco, L., Blazewicz, J., Dell'Ohno, P., Drozdowski, M.: Scheduling multiprocessor tasks on a dynamic configuration of dedicated processors. Ann. Oper. Res. **58**, 493–517 (1995)
2. Dolgui, A., Eremeev, A.V., Kovalyov, M.Y., Kuznetsov, P.M.: Multi-product lot sizing and scheduling on unrelated parallel machines. IIE Trans. **42**(7), 514–524 (2010)
3. Drozdowski, M.: Scheduling multiprocessor tasks—an overview. Eur. J. Oper. Res. **94**, 215–230 (1996)
4. Floudas, C.A., Kallrath, J., Pitz, H.J., Shaik, M.A.: Production scheduling of a large-scale industrial continuous plant: short-term and medium-term scheduling. Comp. Chem. Eng. **33**, 670–686 (2009)
5. Hoogeven, J.A., van de Velde, S.L., Veltman, B.: Complexity of scheduling multiprocessor tasks with prespecified processors allocations. Discr. Appl. Math. **55**, 259–272 (1994)
6. Ierapetritou, M.G., Floudas, C.A.: Effective continuous-time formulation for short-term scheduling: I. Multipurpose batch process. Ind. Eng. Chem. Res. **37**, 4341–4359 (1998)
7. Itai, A., Papadimitriou, C.H., Szwarcfiter, J.L.: Hamilton paths in grid graphs. SIAM J. Comput. **11**(4), 676–686 (1982)
8. Jansen, K., Porkolab, L.: Preemptive scheduling with dedicated processors: applications of fractional graph coloring. J. Sched. **7**(1), 35–48 (2004)
9. Khachiyan, L.G.: A polynomial algorithm in linear programming. Dokladi Akademii Nauk SSSR **244**, 1093–1096 (1979)
10. Kovalenko, Ju. V.: A continuous time model for scheduling with machine grouping by tech-nology. Math. Struct. Modell. **27**(1), 46–55 (2013)
11. Kubale, M.: Preemptive version nonpreemptive scheduling of biprocessor tasks on dedicated processors. Eur. J. Oper. Res. **94**, 242–251 (1996)
12. Zuckerman, D.: Linear degree extractors and the inapproximability of max clique and chromatic number. Theory Comput. **3**, 103–128 (2007)

Condition-Based Maintenance Policies for Modular Monotic Multi-state Systems

Michael Krause

Abstract In this work, we consider a modular monotonic multi-state system, i.e., a system with several components evolving in different states, where the maintenance of a component cannot impair the performance of the system. A structure function provides the system performance depending on the states of its components. Furthermore, we suppose that for each component, the deterioration over time can be described by a deterministic or stochastic process with known properties. The amount of money to be spent on the components' maintenance is limited by a given budget. A loss in the system performance results in opportunity costs. We try to find a component-specific maintenance policy which minimizes the opportunity cost over a finite planning horizon.

1 Introduction

In many industries, maintenance cost diminish the EBIT of a company significantly. However, maintenance still does not receive the attention it deserves. In many cases, only simple maintenance policies, such as age or block replacement [1, 9], have been applied to the companies' facilities. Moreover, in the literature the research focus is set on systems, where the components or even the whole system have only two possible states each ("up" and "down", see [3, 7]).

In this work, we are concerned with the maintenance of modular monotonic multi-state systems. Modular means that the system consists of several components, which may be maintained individually. Multi-state systems are the generalization of binary systems [8] where there may be more than two states for each component and the entire system. Roughly speaking, the monotonicity of a system postulates that a maintenance activity does not deteriorate the system's performance. The relationship between the states of the components and the performance of the system is specified by the structure function of the system. Furthermore, we suppose that a state of a

M. Krause (✉)
Operations Management Group, Clausthal University of Technology,
38678 Clausthal-Zellerfeld, Germany
e-mail: michael.krause@tu-clausthal.de

© Springer International Publishing Switzerland 2016
M. Lübbecke et al. (eds.), *Operations Research Proceedings 2014*,
Operations Research Proceedings, DOI 10.1007/978-3-319-28697-6_43

component represents its performance and that for each component, the deterioration over time can be described by a deterministic or stochastic process with known properties. The performance of a component can be improved by maintenance, the degree of improvement depending on the component's age and the amount of money spent on the maintenance activity. The total expense for maintenance is limited by a given budget. The loss in the system performance is penalized by an opportunity cost rate. The problem under study consists in finding a component-specific maintenance policy which minimizes the opportunity cost over a discrete planning horizon.

The remainder of this paper is organized as follows. At first, we develop a general deterministic model (Sect. 2). We motivate why the associated dynamic program can only be solved for two-period instances (Sect. 3). In Sect. 4, we apply approximate dynamic programming (ADP) to the stochastic version of the problem, where we substitute a deterministic deterioration function by a stochastic process. Our conclusions are given in Sect. 5.

2   A Deterministic Model

Before presenting a general model for the deterministic version of our problem, we first give some definitions in Sect. 2.1. Next, we develop the model step-by-step in Sect. 2.2.

2.1   Definitions

Let J be the set of all components j, where $n := |J|$. Then we define the following:

Definition 1 Let $\mathscr{S}_j$ be the set of all states of component j. We call $\mathbf{s} = (s_j)_{j \in J}$ the *system state*, i.e., the combination of all component states. Without loss of generality we assume that each state $s_j \in \mathscr{S}_j$ is expressed as a percentage, thus the set of all system states is $\mathscr{S} = \times_{j \in J} \mathscr{S}_j \subseteq \mathbb{R}^n_{\geq 0}$.

Definition 2 The *structure function* f provides the system's performance as a function of the system state $\mathbf{s}$:
$$f : \mathscr{S} \to \mathbb{R}. \tag{1}$$

Definition 3 A component j is *relevant*, if there exist two system states $\mathbf{s}, \mathbf{s}' \in \mathscr{S}$ with $s_k = s'_k$ for all components $k \neq j$ such that $f(\mathbf{s}) \neq f(\mathbf{s}')$. A system is called *monotonic* if

- all components are relevant and
- the structure function is componentwise increasing.

2.2 Development of the Model

We consider a monotonic system with components $j \in J$ and countable or uncountable state sets $\mathscr{S}_j$ evolving over a finite planning horizon comprising time periods $t = 1, \ldots, T$ of equal length. Let x_{jt} denote the decision variable specifying the payment for a maintenance activity of component j in period t. The objective consists in minimizing the total opportunity cost incurred over the planning horizon, i.e.,

$$\min_{x_{jt}} \sum_{t=1}^{T} c \cdot (f^{max} - f(\mathbf{s}_{t+1}))$$

where c is the opportunity cost rate for performance losses, f^{max} stands for the maximum system performance, and $\mathbf{s}_{t+1} = (s_{j(t+1)})_{j \in J}$ denotes the vector of all component performances at the end of period t. Without loss of generality we may assume that $f^{max} = f(\mathbf{s}_1)$. Furthermore, we may also state the objective in the equivalent form

$$\max_{x_{jt}} \sum_{t=1}^{T} f(\mathbf{s}_{t+1})$$

Starting with budget $B_1 = B$ at the beginning of the first period $t = 1$, the remaining budget B_{t+1} at the end of period t is given by $B_{t+1} = B_t - \sum_{j \in J} x_{jt}$.

We suppose that the evolution of the performance of a component j from period t to period $t + 1$ depends on the component's age a_{jt} and on amount x_{jt}. The age of j can be partially or completely reset to zero depending on its current age a_{jt} and amount x_{jt}. With appropriate transition functions α_j and γ_j, the evolution of component j can be described via the equations

$$a_{j(t+1)} = \alpha_j(a_{jt}, x_{jt}) \quad (t = 1, \ldots, T)$$
$$s_{j(t+1)} = \gamma_j(s_{jt}, a_{jt}, x_{jt}) \quad (t = 1, \ldots, T)$$

For the transition functions α_j and γ_j we establish the following conventions:

1. Functions α_j are strictly increasing in a_{jt} and strictly decreasing in x_{jt}.
2. $\alpha(a_{jt}, 0) = a_{jt} + 1$ and $\lim_{x_{jt} \to \infty} \alpha(a_{jt}, x_{jt}) = 0$ for all a_{jt}.
3. Functions γ_j are strictly decreasing in a_{jt} and strictly increasing in s_{jt} and x_{jt}.
4. $\lim_{x_{jt} \to \infty} \gamma_j(s_{jt}, a_{jt}, x_{jt}) = 1$ for all s_{jt}, a_{jt}.
5. Functions γ_j are concave in x_{jt}.

Convention 1 implies that the age of component j at the beginning of period $t + 1$ is positively correlated with its age at the beginning of the previous period t and that maintenance leads to a regeneration of j. Convention 2 means that if component j is not maintained in period t, the component is subject to natural aging. Furthermore, by spending a sufficiently large amount on maintenance, a component can be completely renewed. According to convention 3, an older component's performance decreases faster than a younger one. In addition, the performance at the beginning of period t and the amount spent on maintenance in period t have a positive impact on the performance of a component at the beginning of period $t + 1$. Convention 4 ensures that the maximum performance can be obtained by spending an arbitrarily large amount on maintenance. Finally, convention 5 expresses the diminishing marginal utility of the maintenance efforts.

By denoting the initial values of the performance and the age of the components $j \in J$ by s_j^0 and a_j^0, respectively, our maintenance model can be formulated as a dynamic program with stages $t = 1, \ldots, T$.

$$(DP) \begin{cases} \text{Max. } \sum_{t=1}^{T} f(\mathbf{s}_{t+1}) \\ \text{s.t. } s_{j1} = s_j^0 & (j \in J) \\ \quad a_{1j} = a_j^0 & (j \in J) \\ \quad B_1 = B \\ \quad a_{j(t+1)} = \alpha_j(a_{jt}, x_{jt}) & (j \in J; \ t = 1, \ldots, T) \\ \quad s_{j(t+1)} = \gamma_j(s_{jt}, a_{jt}, x_{jt}) & (j \in J; \ t = 1, \ldots, T) \\ \quad B_{t+1} = B_t - \sum_{j \in J} x_{jt} & (t = 1, \ldots, T) \\ \quad B_{t+1} \geq 0 & (t = 1, \ldots, T) \\ \quad x_{jt} \geq 0 & (j \in J; \ t = 1, \ldots, T) \end{cases}$$

The Bellman equation [4] decomposes the problem into a sequence of subproblems $P_t(S_t)$ on stages $t = 1, \ldots, T$:

$$P_t(S_t) \begin{cases} V_t(S_t) = \max_{x_{jt}} \{ \underbrace{C_t(S_t, \mathbf{x}_t) + V_{t+1}(S_{t+1}(S_t, \mathbf{x}_t))}_{=f(\mathbf{s}_{t+1})} \} \end{cases}$$

C_t denotes the *contribution* to the current stage t, $S_t = (\mathbf{a}_t, \mathbf{s}_t, B_t)$ is the state attained at stage t, and V_{t+1} is the *value function*, which sums up the contributions of all remaining periods $t = t + 1, \ldots, T$.

3   Solution Approach for a Serial System with Two Components

In this section we provide an exact solution for a system containing two components $j = 1, 2$ arranged in series. For such a system, we may use the following structure function, which implies that the performance of this monotonic system is always determined by the weakest of its components:

$$f(\mathbf{s}_{t+1}) = \min\{s_{1(t+1)}, s_{2(t+1)}\}.$$

On stage t we obtain subproblem

$$P_t(S_t) \begin{cases} \text{Max. } \min\{s_{1(t+1)}, s_{2(t+1)}\} + V_{t+1}(S_{t+1}(S_t, \mathbf{x}_t)) \\ \text{s.t. } s_{j1} = s_j^0, \ a_{1j} = a_j^0, \ B_1 = B & (j \in J) \\ \quad a_{j(t+1)} = \alpha_j(a_{jt}, x_{jt}) & (j \in J) \\ \quad s_{j(t+1)} = \gamma_j(s_{jt}, a_{jt}, x_{jt}) & (j \in J) \\ \quad B_{t+1} = B_t - \sum_{j \in J} x_{jt} \\ \quad B_{t+1}, x_{jt} \geq 0 & (j \in J) \end{cases}$$

We start the backward computation at stage $t = T > 1$ with $V_{T+1} = 0$. As the ages a_{jt} are not considered in the contribution C_t, the associated constraints can be neglected. Introducing auxiliary variables z_T, the problem can be stated as

$$P_T(S_T) \begin{cases} \text{Max. } z_T \\ \text{s.t. } z_T \leq \gamma_j(s_{jT}, a_{jT}, x_{jT}) & (j \in J) \\ \quad B_T - \sum_{j \in J} x_{jT} \geq 0 \\ \quad x_{jT}, z_T \geq 0 & (j \in J) \end{cases}$$

Applying the second-order sufficient optimality conditions (SSC) of Karush, Kuhn, and Tucker [2], we obtain the following result (given that $\gamma_j \in C^2 (j \in J)$):

1. if $\gamma_1(s_{1T}, a_{1T}, B_T) \leq \gamma_2(s_{2T}, a_{2T}, 0)$ then $x_{1T} = B_T$, $x_{2T} = 0$
2. if $\gamma_2(s_{2T}, a_{2T}, B_T) \leq \gamma_1(s_{T1}, a_{T1}, 0)$ then $x_{1T} = 0$, $x_{2T} = B_T$
3. otherwise, $z_T = \gamma_1(s_{1T}, a_{1T}, x_{1T}) = \gamma_2(s_{2T}, a_{2T}, x_{2T})$ and $x_{1T} + x_{2T} = B_T$, where x_{1T} and x_{2T} can be computed numerically.

Since in the third case, the amounts x_{jT} are not obtained analytically, every further step of the backward computation, i.e., solving problems $P(S_t)$ with $t = T-1, \ldots, 1$, cannot be achieved analytically.

4 Stochastic Model and Approximate Dynamic Programming

To tackle the problem when the problem size grows and randomness is added to the deterioration of the components' performances, we need to apply a heuristic method, such as approximate dynamic programming (ADP) [5, 6, 10]. We can formulate a stochastic version of our problem as follows:

$$(SDP) \begin{cases} \text{Max. } \sum_{t=1}^{T} \mathbb{E}\{f(\hat{s}_{t+1})\} \\ \text{s.t. } \hat{s}_{j(t+1)} = \hat{s}_{jt} - \hat{R}_{j(t+1)}(\hat{s}_{jt}, a_{jt}) + \varphi(x_{jt}) & (j \in J; \ t = 1, \ldots, T) \\ \quad a_{j(t+1)} = \alpha_j(a_{jt}, x_{jt}) & (j \in J; \ t = 1, \ldots, T) \\ \quad B_{t+1} = B_t - \sum_{j \in J} x_{jt} & (t = 1, \ldots, T) \\ \quad \hat{s}_{j1} = s_j^0, \ a_{j1} = a_j^0, \ B_1 = B & (j \in J) \\ \quad B_{t+1}, x_{jt} \geq 0 & (j \in J; \ t = 1, \ldots, T) \end{cases}$$

Components now are subject to a stochastic and observable deterioration of their performance with random variables $\hat{R}_{j(t+1)}(\hat{s}_{jt}, a_{jt})$, which depend on the current performance and age of component j.

The function $\varphi(x)$ is some concave, increasing function in order to represent the diminishing marginal value of the maintenance payments. Experiments with small instances of this model have shown that the stochastic version of the problem can be solved by some basic ADP-algorithm, like the method given in [10, p. 141].

5 Conclusions

We discussed the problem of optimizing condition-based maintenance policies for modular and monotonic multi-state systems. First, we developed a deterministic model, which was formulated as a dynamic program. Using a simple series system with two components we motivated the need of applying heuristic methods. Finally, we proposed a generalization of our a model including stochastic deterioration. Preliminary results show that basic variants of approximate dynamic programming using a lookup table approximation of the value function provide good solutions to small instances of this model. Our next step consists in substituting the lookup table approximation by neural networks, which avoid the drawbacks associated with the discretization of the state spaces.

References

1. Barlow, R.E., Proschan, F.: Mathematical Theory of Reliability. SIAM, New York (1996)
2. Bazaraa, M.S., Sherali, H.D., Shetty, C.M.: Nonlinear Programming: Theory and Algorithms, 3rd edn. Wiley (2006)
3. Beichelt, F., Tittmann, P.: Reliability and Maintenance: Networks and Systems. Taylor & Francis, Boca Raton (2012)
4. Bellman, R.: Dynamic Programming. Princeton University Press, Princeton (1957)
5. Bertsekas, D.P.: Dynamic Programming and Optimal Control – Vol. II: Approximate Dynamic Programming. Athena Scientific, Nashua (2012)
6. Bertsekas, D.P., Tsitsiklis, J.N.: Neuro-Dynamic Programming. Athena Scientific, Belmont, Mass. (1996)
7. Gertsbakh, I.: Reliability Theory–with Applications to Preventive Maintenance. Springer, Berlin (2005)
8. Lisnianski, A., Frenkel, I., Ding, Y.: Multi-state System Reliability Analysis and Optimization for Engineers and Industrial Managers. Springer, London (2010)
9. Nakagawa, T.: Maintenance Theory of Reliability. Springer, London (2005)
10. Powell, W.B.: Approximate Dynamic Programming–Solving the Curses of Dimensionality, 2nd edn. Wiley, Hoboken, USA (2011)

An Optimization Model and a Decision Support System to Optimize Car Sharing Stations with Electric Vehicles

Kathrin S. Kühne, Tim A. Rickenberg and Michael H. Breitner

Abstract An increasing environmental awareness, rising energy cost, progressing urbanization, and shortage of space cause to rethink individual mobility behavior and personal car ownership in cities. Car sharing is a sustainable mobility concept that allows individuals to satisfy their mobility needs without owning a car and addresses modern mobility. Car sharing is particularly suitable to cover medium-range distances and can be linked to the public transport of major cities (intermodal mobility). Within this context, the integration of electric vehicles represents an opportunity to further protect the environment and potentially save energy cost. In order to create an efficient car sharing transportation network, the location of stations, the number of vehicles and the availability of electric fast charging infrastructure are critical success factors. We provide a decision support system (DSS) to plan and optimize car sharing stations for electric vehicles. An optimization model and the DSS OptCarShare 1.1 enable to optimize stations and visualize results. Parameters, such as the annual lease payment for charging infrastructure, the expected travel time of consumers, the charging time of electric vehicles dependent on available charging infrastructure, affect the decision variables such as the number of car sharing stations, vehicles and fast chargers. On the basis of evaluations and benchmarks for the cities of Hannover and Zürich, we establish generalizations for the parameters of the model. The results show a high impact of fast chargers (half an hour to fill 80 % of the battery) on the current model and the optimal solution.

K.S. Kühne (✉) · T.A. Rickenberg · M.H. Breitner
University of Hannover, Königsworther Platz 1, 30167 Hannover, Germany
e-mail: kuehne@iwi.uni-hannover.de

T.A. Rickenberg
e-mail: rickenberg@iwi.uni-hannover.de

M.H. Breitner
e-mail: breitner@iwi.uni-hannover.de

© Springer International Publishing Switzerland 2016
M. Lübbecke et al. (eds.), *Operations Research Proceedings 2014*,
Operations Research Proceedings, DOI 10.1007/978-3-319-28697-6_44

313

1   Introduction

Automobile traffic is one major factor of air pollution and noise annoyance in cities. A good alternative to private cars is car sharing which allows to remain mobile without owning a car while saving cost and emissions. In this concept, individuals, especially young adults share vehicles which are property of an organization [3]. Car sharing is particularly suitable to cover medium-range distances and can be linked to the public transport of major cities such as e.g. Hannover or Zürich. It thus fills the gap between public transport and private automobile [7].

Car sharing in connection with electric vehicles has great potential with regard to sustainability. It can not only protect the environment (less CO2, noise and required parking area), it represents cost security for customers and their mobility needs [4]. Since 2005 with the increase in the sales figures, electric cars have become a serious alternative to conventionally propelled vehicles [1]. Electric vehicles differ in the range and the maximum speed, many of them have an average of around 150 Km range with a maximum speed of about 130 km/h [2]. An important component of a pure electric car is the battery (lithium-ion) which typically needs to be charged for eight hours on a conventional wall socket. These batteries may also be subjected to a fast charge, which takes about 0, 5 h to fill 80 % of the battery, but this is associated with high investments. With regard to the cost of a car sharing organization and the satisfaction of the customer demand, the limited range and long charging times respectively expensive fast charging infrastructure represent challenges.

The protection of the environment and scarce natural resources as well as limited parking space caused by urbanization are urgent topics and are basis for the idea to refine an optimization model for car sharing stations by Rickenberg et al. [6]. The question, how many fast chargers need to be positioned will be addressed with an enhanced optimization model. We pursue the following research questions:

RQ 1: What factors of electric vehicles need be considered to optimize the location and size of car sharing stations? and

RQ 2: What influence do exogenous parameters have on the decision variables?

2   Optimization Model and Decision Support System

The objective of this model is to determine optimal locations of candidate car sharing stations i ($i = 1, \ldots, m$) as well as to optimize the number of vehicles with fast charging ($f_i^{fc} \in \mathbb{N}$) and with regular charging infrastructure ($f_i^{rc} \in \mathbb{N}$). The minimization of total cost and the satisfaction of the customer have the highest priority. The maximum distance of any demand point j ($j = 1, \ldots, n$) to the next car sharing station must not exceed a definite limit (*maxd*). In our case, any period (t^{period}) is 24 h and is related to the normal-distributed demand n_j, which is also given for one day (24 h). A vehicle is available when the travel time (t_i^t) and the appropriate charging time (tc_i) is over. To calculate the charging time a coefficient (γ) and the expected travel time (normal-distributed) are needed. There are two varieties for γ, one for the fast charging and one for regular charging infrastructure. At any station are limited parking lots ($maxp_i$) available and limited fast chargers ($maxfcs$) possible. Since electric cars of one single type are used in this model, a homogenous fleet is assumed. The mathematical problem can be formulated as follows:

$$Min.\ Z = \sum_{i=1}^{m}[f_i^{rc}(kf + ka + kl^{rc}) + f_i^{fc}(kf + ka + kl^{fc}) + y_i * ks] \quad (1)$$

$$d_{ij} * z_{ij} \leq maxd \quad \forall i;\ j \quad (2)$$

$$\sum_{i=1}^{m} z_{ij} = 1 \quad \forall\ j \quad (3)$$

$$y_i = z_{ij} \quad \forall\ i;\ j \quad (4)$$

$$f_i^{rc}\frac{t^{period}}{t_i^t + tc_i^{regular}} + f_i^{fc}\frac{t^{period}}{t_i^t + tc_i^{fast}} \geq \sum_{j=1}^{n} n_j * z_{ij} \quad \forall\ i \quad (5)$$

$$f_i^{rc} + f_i^{fc} \leq maxp_i \quad \forall\ i \quad (6)$$

$$f_i^{fc} \leq maxfcs \quad \forall\ j \quad (7)$$

$$y_i * v_i \leq a \quad \forall\ i \quad (8)$$

$$w_i \geq minb * y_i \quad \forall\ i \quad (9)$$

$$z_{ij};\ y_i; \in \{0, 1\} \quad \forall\ i;\ j \quad (10)$$

$$f_i^{rc};\ f_i^{fc} \in \mathbb{N} \quad \forall\ i \quad (11)$$

The objective function (1) describes the incurred cost of a car sharing organization which are to be minimized. The cost is accumulated annual fees for renting vehicles (kf), parking lots (ka), charging infrastructure (kl^{rc} and kl^{fc}) as well as annual cost to maintain stations. Restriction (2) implies that the distance between a demand point and a station must not exceed a maximum value and constraint (3) assigns every demand point to a station but only if the station is actually built (4). The fulfillment of the demand is ensured by restriction (5). The variables $tc_i^{regular}$ and tc_i^{fast} are calculated as follows: $tc_i^{regular} = \gamma^{regular} * t_i^t$ and $tc_i^{fast} = \gamma^{fast} * t_i^t$. The coefficient γ describes the charging time per travel hour dependent on the maximum range of electric vehicles, the average speed and the charging time. It is calculated as follows: $\frac{max\ range}{average\ speed} = max\ travel\ time$ and accordingly $\gamma^{regular}$ and $\gamma^{fast} = \frac{max\ charging\ time}{max\ travel\ time}$.

The total number of vehicles (also the number of associated parking lots) must be smaller than the maximum number of parking lots for each station (6) [5]. Restriction (7) guarantees the electricity supply. Parameter v_i is defined as follows: v_i = free parking lots around station i/registered vehicles around station i * 100 %. The smaller parameter v_i, the higher is the shortage of parking. Due to (8), the actual shortage of parking cannot be bigger than the default shortage of parking (a). Parameter w_i is defined as follows: w_i = population at station i/area at station i [6]. Because of (9) a minimum level of population density within an area is reached. In equations (10) and (11) are the binary variables and decision variables defined.

Based on the optimization model, we implement the decision support system (DSS) OptCarShare 1.1 to enable the optimization of stations and visualization of results. The DSS, the underlying model, and sample data pools are available online at http://archiv.iwi.uni-hannover.de/CarSharing/.

3 Benchmarks in Hannover and Zürich

Influence of charging-infrastructure—charging time and infrastructure cost:
We run several benchmarks by using different values for the parameters and show thereby the applicability of the optimization model. We choose the German city Hannover and the Swiss city Zürich since both have an appropriate size, population density and well public transportation to allow efficient car sharing [5, 6].

The initial values for the benchmarks are $i^{Hannover} = 100$, $j^{Hannover} = 30$, $i^{Zürich} = 200$, $j^{Zürich} = 50$, $kf = 25{,}000€$, $ka = 7{,}000€$, $t^{period}=24\,h$, $GAP=3$ %, $maxp_i = 5$, $maxfcs = 2$, $minb = 1{,}200$, $maxd = 1\,km$. Parameter γ is not a fixed value and varies with the average speed, maximum range or required charging time. Our initial values are: $\gamma^{regular} = \frac{4}{3}$ and $\gamma^{fast} = \frac{1}{12}$ with a maximum range of 150 km, average speed of 25 km/h and a required charging time of 0.5 h or 8 h. We ignore the low cost of regular charging infrastructure and consider only the annual fees of fast charging infrastructure. There are different providers and types of charging infrastructure, which is still in development phase and therefore the cost of the fast chargers could decrease within the coming years. It is even possible that the regular charging infrastructure could improve and thus the gap between these two possibilities could diminish. Here, the smaller the charging coefficient, the more efficient the regular charging infrastructure works. We vary the charging coefficient $\gamma^{regular}$ and the cost kf^c to investigate their influence, which can be seen from the following table (Tables 1 and 2).

A reduction of $\gamma^{regular}$, which implies shorter charging time, results in less total cost, since less vehicles with fast charging infrastructure are needed to meet customer demand. For example in Zürich and annual fees of the fast chargers of 26,000€, the amount of car sharing stations always remains equal at 17, while the number of vehicles related to fast or regular charging infrastructure varies. For $\gamma^{regular} = 4/3$, the total cost is 1,181,000€, but the cost decrease to 959,000€ for $\gamma^{regular} = 2/3$. Because of less charging time for vehicles even with regular charging infrastructure,

Table 1 Influence of charging coefficient $\gamma^{regular}$ and cost of fast charging infrastructure kl^{fc}

Hannover	$kl^{fc} = 5{,}000$EUR				$kl^{fc} = 26{,}000$EUR				$kl^{fc} = 35{,}000$EUR			
	s	f_i^{fc}	f_i^{rc}	cost	s	f_i^{fc}	f_i^{rc}	cost	s	f_i^{fc}	f_i^{rc}	cost
$\gamma^{regular} = 4/3$	8	12	0	452,000	8	10	3	684,000	8	7	9	765,000
$\gamma^{regular} = 3/3$	8	12	0	452,000	9	5	11	651,000	8	2	17	686,000
$\gamma^{regular} = 2/3$	9	11	1	474,500	9	2	14	573,000	8	0	18	584,000
Zürich	$kl^{fc} = 5{,}000$EUR				$kl^{fc} = 26{,}000$EUR				$kl^{fc} = 35{,}000$EUR			
	s	f_i^{fc}	f_i^{rc}	cost	s	f_i^{fc}	f_i^{rc}	cost	s	f_i^{fc}	f_i^{rc}	cost
$\gamma^{regular} = 4/3$	14	18	3	776,000	17	18	3	1,181,000	16	10	18	1,262,000
$\gamma^{regular} = 3/3$	16	17	4	773,000	17	12	12	1,157,000	18	1	32	1,141,000
$\gamma^{regular} = 2/3$	16	16	5	768,000	17	3	24	959,000	19	0	30	979,000

Table 2 Max. distance to station *maxd* and expected travel time

Hannover	maxd = 0.5 km				maxd = 1 km				maxd = 1.25 km			
	s	f_i^{fc}	f_i^{rc}	cost	s	f_i^{fc}	f_i^{rc}	cost	s	f_i^{fc}	f_i^{rc}	cost
Low travel time	11	2	9	373,000	5	1	4	170,000	4	1	3	137,000
Med. travel time	14	10	3	479,000	7	7	1	298,000	4	7	0	263,000
High travel time	14	16	2	670,000	9	14	0	527,000	7	13	1	520,000
Zürich	maxd = 0.5 km				maxd = 1 km				maxd = 1.25 km			
	s	f_i^{fc}	f_i^{rc}	cost	s	f_i^{fc}	f_i^{rc}	cost	s	f_i^{fc}	f_i^{rc}	cost
Low travel time	26	4	21	846,000	11	1	10	368,000	9	1	7	270,000
Med. travel time	29	12	16	985,000	14	11	3	517,000	11	9	1	413,000
High travel time	31	24	11	1,470,000	15	19	3	814,000	14	19	1	934,000

more vehicles of this type will be deployed. The effect is higher, the higher the annual fees of the fast chargers are.

Influence of different driving time profiles and max. distance to station:
The distance to the next car sharing station is an important factor since consumers do not want to walk a long way, e.g. from public transport stations and also from home, to satisfy their mobility needs [8]. For that reason, we vary the parameter $maxd$ and compare it against three different expected travel time profiles.

The total number of vehicles increases, the higher the average travel time is. The vehicles are longer on the roads and therefore need longer charging time which results in a lower availability of the vehicles. The more vehicles are deployed, the higher the total cost for the car sharing organization. For low average travel time, significantly more vehicles are used with regular charging infrastructure, because the vehicles only need a brief time to be charged even with regular charging infrastructure to be available for the next customer. However, with higher travel time, more vehicles with fast charging infrastructure are required. To meet the demand, a certain number of vehicles is needed. For low $maxd$, e.g. 0.5 km for Zürich at the medium travel time, the total cost is 985,000€, but for higher $maxd$ (1.25 km), the cost decrease to 413,000€, although the demand and travel time are still the same.

4 Generalization, Limitations and Conclusion

Concerning RQ2 and based on the benchmarks of Hannover and Zürich, the influences of selected parameters can be generalized to establish a general relationship between the exogenous parameters and the resulting effect on decision variables.

Number of car sharing stations: With higher expected travel time as well as an increasing coefficient of regular charging infrastructure, the number of car sharing stations increases since more vehicles are needed. If the coefficient of the fast charging is lower, the number of car sharing station is reduced by having less vehicles.

Charging infrastructure: Analogous to the number of stations, the number of vehicles with fast charging infrastructure increases with higher travel time since the efficient fast chargers allow to reduce the charging time. However, the more expensive the fast charging infrastructure is, the less vehicles will be deployed with this infrastructure and in consequence the amount of regular charging infrastructure increases. If the coefficient $\gamma^{regular}$ decreases, the number of vehicles with regular charging infrastructure increases and the number of vehicles with fast charging infrastructure decreases due to their higher cost. With less fast chargers or more efficient regular charging infrastructure, the total cost decreases. The effect of γ^{fast} is opposed to $\gamma^{regular}$. The more vehicles are required, the higher is the total cost. While minimizing total cost, it is affected by the mentioned parameters. If $\gamma^{regular}$ increases, less vehicles are needed and in consequence the total cost decreases.

$$t_i^t(\uparrow) \Rightarrow \sum_i y_i(\uparrow) \qquad \gamma^{regular}(\downarrow) \Rightarrow \sum_i y_i(\uparrow) \qquad \gamma^{fast}(\uparrow) \Rightarrow \sum_i y_i(\downarrow) \qquad (12)$$

$$t_i^t(\uparrow) \Rightarrow \sum_i f_i^{fc}(\uparrow) \qquad \gamma^{regular}(\downarrow) \Rightarrow \sum_i f_i^{fc}(\downarrow) \qquad kl^{fc}(\uparrow) \Rightarrow \sum_i f_i^{fc}(\downarrow) \qquad (13)$$

$$t_i^t(\uparrow) \Rightarrow \sum_i f_i^{rc}(\downarrow) \qquad \gamma^{regular}(\downarrow) \Rightarrow \sum_i f_i^{rc}(\uparrow) \qquad kl^{fc}(\uparrow) \Rightarrow \sum_i f_i^{rc}(\uparrow) \qquad (14)$$

$$t_i^t(\uparrow) \Rightarrow Z(\uparrow) \qquad \gamma^{regular}(\downarrow) \Rightarrow Z(\downarrow) \qquad kl^{fc}(\uparrow) \Rightarrow Z(\uparrow) \qquad (15)$$

Concerning the limitations, this model can be used for strategic and tactical planning since a homogenous fleet is assumed and no operative factors (booking management, max range, etc.) are considered. Furthermore, we regard a normal distributed demand and do not consider peaks and off-peaks. Here, only one vehicle can be assigned to a fast charger, but in reality it could be possible to share a fast charger.

To conclude, the optimization model and the DSS are a first approach to support the planning of car sharing stations for electric vehicles. The results of the benchmarks of Hannover and Zürich show that fast chargers and the charging infrastructure in general heavily determine the amount of stations, cars, and total cost.

References

1. Dijk, M., Orsato, R.J., Kemp, R.: The emergency of an electric mobility trajectory. Energy Policy **52**, 135–145 (2012)
2. Dudenhöffer, F., Bussmann, L., Dudenhöffer, K.: Elektromobilität braucht intelligente Förderung. Wirtschaftsdienst **92**(4), 274–279 (2012)
3. Duncan, M.: The cost saving potential of carsharing in a US context. Transportation **38**, 363–382 (2011)
4. Martin, E., Shaheen, S.: The impact of carsharing on houshold vehicle ownership. ACCESS Mag. **38**, 22–27 (2011)
5. Olivotti D., Rickenberg T.A., Breitner M.H.: Multikonferenz Wirtschaftsinformatik 2014, Car Sharing in Zürich—Optimization and Evaluation of Station Location and Size (2014)
6. Rickenberg T.A., Gebhardt A., Breitner M.H.: A decision support system for the optimization of car sharing stations. In: Proceedings of the 21st European Conference on IS (2013)
7. Shaheen, S.: Carsharing in Europe and North America: past, present, and future. Trans. Q. **52**(3), 35–52 (1998)
8. Stillwater, T., Mokhtarian, P., Shaheen, S.: Carsharing and the Built Environment. Trans. Res. Rec. **21**(10), 27–34 (2009)

Optimising Energy Procurement for Small and Medium-Sized Enterprises

Nadine Kumbartzky and Brigitte Werners

1 Introduction

In recent years, energy prices have increased drastically. In Germany, purchase prices (excluding sales tax) for industrial consumers have risen by 28 % for natural gas and by 42 % for electricity from 2007 to 2013. SMEs find themselves under increasing pressure as they have to cope with higher energy costs. To remain competitive, SMEs need to utilise potential energy cost savings. In addition to changing consumption habits, a sustainable reduction of energy costs can be realised by decreasing purchase costs. These costs depend on purchase prices which in turn depend on the chosen procurement strategy. Thus, we develop a mixed integer program to determine an optimal selection of purchase contracts which minimises procurement costs. A similar problem has been formulated in [2, 3] concerning short-term electricity procurement for large-scale consumers. In [1, 5, 6], a quantitative model to assist in the procurement of a local gas supply company is proposed. Since procurement procedures for SMEs might differ from the ones used by energy supply companies or large energy consumers, a two-stage optimisation model that identifies appropriate procurement strategies for SMEs is presented.

The remainder of the paper is structured as follows. In Sect. 2, energy procurement of SMEs is introduced and different procurement strategies are illustrated. The developed quantitative optimisation model is outlined in Sect. 3. Subsequently, Sect. 4 discusses computational results of an exemplary case study. Final conclusions are drawn in Sect. 5.

N. Kumbartzky (✉) · B. Werners
Department of Management and Economics, Chair of Operations
Research and Accounting, Ruhr University Bochum, Bochum, Germany
e-mail: nadine.kumbartzky@rub.de

B. Werners
e-mail: or@rub.de

© Springer International Publishing Switzerland 2016   321
M. Lübbecke et al. (eds.), *Operations Research Proceedings 2014*,
Operations Research Proceedings, DOI 10.1007/978-3-319-28697-6_45

2 Energy Procurement of SMEs

To purchase the volume of energy needed in day-to-day operations, SMEs conclude contracts with energy supply companies or other energy providers (e.g. traders, electricity producers, or gas distribution companies). In general, there exist two different types of purchase contracts: baseload and open contracts. Primary features of a baseload contract are that the cumulative amount of energy provided is fixed before delivery and that it is consumed at a constant level throughout the contract duration. Purchase prices are also settled before the actual commencement of contract and need to be paid for the contracted cumulative amount of energy. The actual quantity consumed can differ from the contracted one only if negotiated with the supplier and might lead to an additional charge for the surplus quantity.

On the contrary, the volume of energy purchased from open contracts can vary over time. Due to this flexibility, purchase prices contain an additional margin of risk. Thus, they are typically higher than those of baseload contracts. Boundaries for minimum and maximum quantity of energy per period and/or for the cumulative amount are often fixed. For instance, this can be done by applying take-or-pay clauses that require the buyer to either take a minimum amount (e.g. 80 % of the cumulative volume) or pay for this amount even if it is not consumed (see [7]).

If the amount received by an open contract is large enough, energy suppliers frequently offer the possibility of a delivery in tranches. Here, the cumulative amount is split up into several tranches that can be procured consecutively within the contract period. The price of a single tranche is calculated using a predefined formula that usually incorporates current market prices. Therefore, prices can vary over time and favourable market developments can be utilised. Since energy procurement is still flexible, purchase prices also contain a margin of risk.

SMEs can purchase energy according to one of the following procurement strategies:

1. traditional full supply contract
2. full supply contract with delivery in tranches
3. baseload delivery in combination with one open contract
4. structured procurement concept.

The easiest strategy is to sign a traditional full supply contract enabling SMEs to consume energy by an open contract provided by a single supplier. However, to benefit from a transparent pricing on market terms while keeping the flexibility of an open contract, a full supply contract with delivery in tranches might be preferable.

If SMEs seek to partially avoid the margin of risk included in full supply contracts, a baseload delivery in combination with one open contract to fulfill the residual demand can be chosen. Finally, the most complex strategy is a structured procurement concept. It is characterised by an efficient combination of several baseload and open contracts (with or without delivery in tranches) that might differ regarding contract length, date of contract conclusion, contracted volume, and supplier.

If a structured procurement concept is chosen, SMEs need to assign the total amount of energy required during the contract period to several contracts and have to fix each contracted volume. Therefore, SMEs face volume risk which is characterised by a deviation of the contracted volume from the actual amount consumed. This kind of risk is also present in baseload contracts. Another form of risk occurring in procurement decisions is price risk, i.e. the risk of losses due to unforeseen or unfavourable market developments. By choosing a structured procurement concept, SMEs are confronted with price risk since they need to decide when to conclude a contract. This also applies to full supply contracts with delivery in tranches.

It results from the above that purchase prices depend on the specific type of contract. Thus, SMEs can reduce energy purchase costs by choosing an appropriate procurement strategy.

3 Two-Stage Optimisation Model

A two-stage MIP is used to determine an optimal selection of purchase contracts according to one of the introduced procurement strategies. In order to obtain solutions that are robust against uncertain parameter realisations, a minimax relative regret approach is applied. This means inherent uncertainty of energy demand and purchase prices of open contracts with delivery in tranches is handled by introducing a finite number of scenarios. Each scenario captures one possible realisation of all time-dependent uncertain parameters. Similar to the model in [5], first-stage decisions at $t = 1$ include contract conclusions and cumulative volume of baseload and open contracts since these decisions have to be made at the beginning of the contract period. Second-stage decisions are used to react on uncertain fluctuations in demand and market prices. Thus, the actual quantities purchased via both contract types as well as shortfall and excess quantities are drawn at the second stage for every point in time $t \in T$.

In the following, we exemplarily present some parts of the optimisation model that is used for determining the optimal solution for each scenario $s \in S$. Similar to [6] and [7], baseload contracts are modelled as follows: Let J be the set of baseload contracts. Each baseload contract $j \in J$ is characterised by the contract period $[\underline{t}_j, \overline{t}_j]$, take-or-pay level λ_j, and by the minimum and maximum capacity per period v_j^{min} and v_j^{max}, respectively. Decision variables are the cumulative volume v_j^{cum}, quantity $x_{j,t,s}$ consumed in period t and scenario s, excess capacity of take-or-pay volume $e_{j,s}$ in scenario s, and take-or-pay shortfall $f_{j,s}$ in scenario s. Besides, a binary variable z_j is introduced modelling the conclusion of a contract, i.e. $z_j = 1$ if contract j is concluded, and 0 otherwise. In each scenario $s \in S$, the objective is

$$\min \ Z_s^* = \sum_{j=1}^{J} c_j^{fix} z_j + \sum_{j=1}^{J} \sum_{t=1}^{T} c_j^{var} x_{j,t,s} + \sum_{j=1}^{J} \left(c_j^{var} f_{j,s} + c_j^{ex} e_{j,s} \right), \quad (1)$$

where c_j^{fix} denotes fixed costs, c_j^{var} are variable costs, and c_j^{ex} are costs for exceeding the contractually fixed cumulative volume. Thus, procurement costs are composed of four components: fixed costs for entering a contract, variable costs for the actual amount of energy received, variable costs to pay for a take-or-pay shortfall (if present), and an additional charge for the excess capacity if the actual volume consumed exceeds the contractually fixed quantity. For an optimal selection of baseload contracts, the following constraints have to be fulfilled:

$$v_j^{min} z_j \leq x_{j,t,s} \leq v_j^{max} z_j \ \forall \, j \in J, \ \forall \, \underline{t}_j \leq t \leq \bar{t}_j, \ \forall \, s \in S \qquad (2)$$

$$\sum_{t \in T} x_{j,t,s} - v_j^{cum} \leq e_{j,s} \ \forall \, j \in J, \ \forall \, s \in S \qquad (3)$$

$$\lambda_j v_j^{cum} - \sum_{t \in T} x_{j,t,s} \leq f_{j,s} \ \forall \, j \in J, \ \forall \, s \in S. \qquad (4)$$

Constraints (2) guarantee that the actual amount of energy received in period t (during delivery) and scenario s is bounded by the minimum and maximum capacity, if contract j is concluded. Constraints (3) and (4) determine excess quantities and take-or-pay shortfalls, respectively.

Open contracts $k \in K$ are modelled in an analogous manner taking into account that variable costs $c_{k,t,s}^{var}$ depend on uncertain market conditions. Furthermore, the constraint

$$\sum_{j=1}^{J} x_{j,t,s} + \sum_{k=1}^{K} y_{k,t,s} = d_{t,s} \ \forall \, t \in T, \ \forall \, s \in S \qquad (5)$$

assures that the uncertain demand $d_{t,s}$ is fulfilled at all times either by baseload or open contracts.

To identify the optimal procurement strategy, the above model is solved to determine all scenario optima Z_s^*. These values are used to model a minimax relative regret approach as in [8]. This ensures that the resulting optimal procurement strategy leads to promising solutions close to optimality in all scenarios considered.

4 Case Study

In this section, an exemplary case study is conducted for an industrial laundry with an annual energy consumption of 15 GWh. We assume that power accounts for 8 % and natural gas for 92 % of the annual energy consumption. Since the structure of power and gas procurement contracts is quite similar, we exemplarily optimise the procurement of natural gas considering a time horizon of one year.

To determine an optimal selection of purchase contracts, predictions of the uncertain gas demand and market prices are needed. A forecast of future demand is made using temperature-dependent load profiles. To account for prediction errors, positive and negative deviations from the forecasted demand of 3 and 5 %, respectively, are incorporated. Scenarios for uncertain price movements of open contracts with delivery in tranches are generated assuming an expected yearly price increase of 2.5 %. The case study is comprised of 2 open contracts without delivery in tranches, 2 open contracts with delivery in tranches (assigned to quarters and months, respectively) and monthly baseload contracts. The specific contract parameters are fixed using real-world data.

The minimax relative regret model is solved using the standard optimisation software Xpress Optimization Suite [4]. In order to evaluate the obtained solution, expected annual purchase costs are calculated. For the industrial laundry, the optimal procurement strategy is to choose a structured procurement concept with expected annual purchase costs of approximately 641,200 €. To compare different strategies, the model is also solved for each of the remaining procurement concepts.

The results show that a full supply contract as well as a baseload delivery in combination with one open contract lead to notably higher expected procurement costs. However, an open contract with delivery in tranches is a possible alternative if an increase of 0.7 % of expected procurement costs compared to a structured procurement concept is acceptable. The two results are close to each other due to the assumption that the considered laundry works in two shifts between 6 a.m. and 10 p.m. Since only a small proportion of gas is needed during night time, baseload contracts have a limited share of the cumulative amount of energy consumed.

The results also indicate that the maximum scenario costs of a structured procurement concept are only slightly higher than, for example, those of a full supply contract, whereas minimum scenario costs are significantly lower. Therefore, implementing a structured procurement concept offers great potential cost savings.

We further investigate the quality of the solution obtained by the optimisation model. For this purpose, 500 additional price and demand scenarios are generated. For each procurement concept, the obtained first-stage decisions, that means contract conclusions and cumulative volumes, are fixed. Then the optimisation model is solved

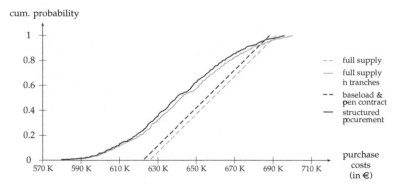

Fig. 1 Cumulative distribution function of annual purchase costs for each procurement strategy

for each of the 500 scenarios separately to obtain optimal second-stage decisions and resulting procurement costs.

Figure 1 shows the cumulative distribution function of purchase costs for each procurement strategy. It reveals that in 97 % of the scenarios, the structured procurement concept leads to the lowest annual purchase costs with a mean of 640,900 €.

5   Conclusion

In this paper, a two-stage MIP for determining an optimal selection of purchase contracts for SMEs is presented. It is shown that SMEs can reduce its purchase costs by choosing an appropriate procurement strategy. Uncertainty regarding future energy demand and prices of open contracts with delivery in tranches is included. The optimisation model is solved using a minimax relative regret approach. Computational results of the conducted case study indicate that a structured procurement concept has a high potential to significantly reduce energy costs, in particular when comparing it to traditional full supply contracts. However, SMEs have to decide whether they are willing to take over volume and price risk. Furthermore, the effort of time and personnel required to implement such a strategy as well as the company-specific knowledge of energy markets has to be included into the decision-making process.

References

1. Aouam, T., Rardin, R., Abrache, J.: Robust strategies for natural gas procurement. Eur. J. Oper. Res. **205**(1), 151–158 (2010)
2. Beraldi, P., Violi, A., Scordino, N., Sorrentino, N.: Short-term electricity procurement: a rolling horizon stochastic programming approach. Appl. Math. Model. **35**(8), 3980–3990 (2011)
3. Carrión, M., Philpott, A.B., Conejo, A.J., Arroyo, J.M.: A stochastic programming approach to electric energy procurement for large consumers. IEEE Trans. Power Syst. **22**(2), 744–754 (2007)
4. FICO: FICO Xpress Optimization Suite: Getting Started with Xpress (2009)
5. Koberstein, A., Lucas, C., Wolf, C., König, D.: Modelling and optimising risk in the strategic gas purchase planning problem of local distribution companies. J. Energy Markets **4**(3), 47–68 (2011)
6. Schmidt, S.: Optimale Vertragsauswahl in der strategischen Gasbeschaffung, Schriftenreihe QM—Quantitative Methoden in Forschung und Praxis, vol. 28. Kovač, Hamburg (2012)
7. Schultz, C.R.: Modeling take-or-pay contract decisions. Decis. Sci. **28**(1), 213–224 (1997)
8. Werners, B., Wülfing, T.: Robust optimization of internal transports at a parcel sorting center operated by deutsche post world net. Eur. J. Oper. Res. **201**(2), 419–426 (2010)

Approximation Schemes for Robust Makespan Scheduling Problems

Adam Kurpisz

Abstract Makespan Parallel Machine and Flow Shop Scheduling belong to the core of the polynomial optimization problems. Both problems are well studied and they are known to be NP-hard, thus no optimal polynomial time algorithm exists under certain theoretical assumptions. In this paper we present a *Polynomial Time Approximation Scheme* for the generalized MIN- MAX version of the problems and the *Competitive Ratio Approximation Scheme* for the online counterpart of considered problems. All the presented algorithms work in linear time in the input size.

1 Introduction

In this paper, we investigate two robust makespan scheduling problems. In a considered robust approach we are given a constant-size scenario set with possible realizations of each element cost. The objective is to minimize the cost under the worst possible scenario. The first considered problem is a Min-Max Makespan Parallel Machine Scheduling Problem (MIN- MAX $Pm||C_{max}$). The problem is equivalent to the know Vector Scheduling Problem [2]. The latter one is Min-Max Makespan Flow Shop Scheduling Problem (MIN- MAX $Fm||C_{max}$). In this paper we extend the idea presented in [3] and we prove that a simple merging rule can reduce the number of jobs to a constant depending only on ε, such that the optimum differs from the original one by at most a $1 + O(\varepsilon)$ factor. As a consequence we provide a *Polynomial Time Approximation Scheme* (PTAS) for both problems. The running time of the algorithm is linear which improves the previous known results [5, 6]. The second result of the paper concerns the online counterpart of the problems. Up to now there were several approximation results with a given *competitive ratio* [1, 8]. Our contribution is a *Competitive Ratio Approximation Scheme* (CRAS) [4, 7].

A. Kurpisz (✉)
Faculty of Fundamental Problems of Technology, Wrocław University
of Technology, Wybrzeże Wyspiańskiego 27, 50-370 Wrocław, Poland
e-mail: adam.kurpisz@pwr.wroc.pl

© Springer International Publishing Switzerland 2016
M. Lübbecke et al. (eds.), *Operations Research Proceedings 2014*,
Operations Research Proceedings, DOI 10.1007/978-3-319-28697-6_46

Our approach uses several classic transformations of the input instance which may increase the objective by a factor of *at most* $1 + O(\varepsilon)$. Throughout this paper, when we describe this type of transformation, we say it produces $1 + O(\varepsilon)$ loss.

2 Min Max Parallel Machine Scheduling Problem

In the MIN MAX $Pm||C_{max}$ we have m identical parallel machines and a set of n jobs $J = \{J_1, \ldots, J_n\}$ waiting for processing. Each machine can process at most one job at a time. Preemption is not allowed. Processing times are described with a set of scenarios $\Gamma = \{S_1, \ldots, S_K\}$. Each job J_j can be described with a vector $\{p_j^1, \ldots, p_j^K\}$ of possible processing times. K and m are constant. Let $C_{max}^S(\pi)$ be the maximum completion time for a schedule π under scenario S. We are looking for π^* such that:

$$OPT = C_{max}(\pi^*) = \min_{\pi} \max_{S \in \Gamma} C_{max}^S(\pi) \tag{1}$$

We start with setting the lower and the upper bound. We have:

$$LB = \frac{1}{m} \max_{S \in \Gamma} \sum_{j=1}^{n} p_j^S \leq OPT \leq \max_{S \in \Gamma} \sum_{j=1}^{n} p_j^S = UB.$$

Lemma 1 *With $1 + \varepsilon$ loss, for every job J_j under any scenario S_k we can round the $p_j^{S_k}$ down to the nearest value $\left(\max_{S \in \Gamma} p_j^S \right) \cdot \frac{\varepsilon}{mK} \cdot i, \ i = 0, 1, \ldots, \frac{mK}{\varepsilon} - 1$.*

Proof First, let us notice that decreasing the processing times cannot increase the optimum. Now, let P_{S_k} be the subset of jobs, for which the largest processing time is under scenario S_k i.e. $P_{S_k} := \{J_j : p_j^{S_k} = \max_{S \in \Gamma} p_j^S\}$. Let us consider any schedule π. Now we replace the rounded processing times with the original ones sequentially, starting from P_{S_1}. Since $\sum_{J_j \in P_{S_1}} p_j^{S_1} \leq m \cdot C_{max}^{S_1}(\pi)$, thus replacing the processing time of jobs from P_{S_1} may increase the makespan under any scenario by at most $\frac{\varepsilon}{mK} m \cdot C_{max}^{S_1}(\pi)$. Iterating this for any P_S may increase the makespan by at most

$$\frac{\varepsilon}{mK} m \cdot \left(\sum_{S \in \Gamma} C_{max}^S(\pi) \right) \leq \frac{\varepsilon}{mK} mK \max_{S \in \Gamma} C_{max}^S(\pi) = \varepsilon \max_{S \in \Gamma} C_{max}^S(\pi)$$

and the claim follows.   □

For any job J_j we define the *job profile* to be K-tuple $< \Pi_{1,j}, \ldots, \Pi_{K,j} >$ such that $p_j^{S_k}$ is equal to $\frac{\varepsilon}{mK} \cdot \Pi_{k,j} \max_{S \in \Gamma} p_j^S$. The value $d_j := \max_{S \in \Gamma} p_j^S$ is called the *scale factor* of the job J_j. Any job is fully described with its *job profile* and *scale factor*.

Lemma 2 *The number of distinct job profiles is at most $l := \left(\frac{mK}{\varepsilon} \right)^K$.*

Proof Every *job profile* has K entries with possible $\frac{mK}{\varepsilon}$ values. □

Now we perform the grouping step. Let $\delta := \frac{\varepsilon LB}{mK}$. We partition the set of jobs into two disjoint subsets of *large* $L := \{J_j : d_j > \delta\}$ and *tiny* jobs $T := \{J_j : d_j \leq \delta\}$. Next we partition the set of tiny jobs into l subsets $(T_1, \ldots, T_l)$ of jobs having the same profile. For every set T_ϕ we take any two jobs $J_a, J_b \in T_\phi$ with scale factors $d_a, d_b \leq \frac{\delta}{2}$ and merge them together. We repeat this step as long as only one job with scale factor smaller than $\frac{\delta}{2}$ is left. After grouping step, applied for each T_ϕ, there are at most l tiny jobs left with scale factor smaller than $\frac{\delta}{2}$.

In the following we prove, that an optimal solution for the instance with a merged jobs, differs from the optimum of the original instance by at most a $1 + \varepsilon$ factor. Thus we can get a PTAS by enumerating all possible solutions of the reduced instance.

Lemma 3 *With* $1 + O(\varepsilon)$ *loss, the number of jobs can be reduced in linear time to be at most* $\Delta = \min\{n, O\left(\left(\frac{mK}{\varepsilon}\right)^K\right)\}$.

Proof The number of *large* jobs is bounded by a constant $\frac{UB \cdot K}{\frac{\varepsilon LB}{mK}} = \frac{m^2 K^2}{\varepsilon}$, thus we can enumerate all possible assignments of *large* jobs to the machines. One can choose the assignment corresponding to the optimal schedule. For this assignment let t_i^S denotes the time when a machine i finishes processing the *large* jobs under scenario S. The LP for the problem, denoted by LP_1 takes the form:

$$\min \ C \ s.t. \quad t_i^S + \sum_{J_j \in T} x_{ij} p_j^S \leq C, \ i = [m], \ S \in \Gamma \tag{2}$$

$$\sum_{i=1}^m x_{ij} = 1, \qquad j \in T; \tag{3}$$

$$x_{ij} \in \{0, 1\}, \qquad i = [m], \ j \in T \tag{4}$$

where $x_{ij} = 1$ means that job J_j has been assigned to machine i.

LP_1 can be seen as assigning each job (maybe fractionally) to some machine. For each $\phi \in [l]$ let us consider a merged job with a job profile ϕ and a scale factor equal to $\sum_{J_j \in T_\phi} d_j$. Now, instead of deciding on which machine each job will be scheduled, for each job profile ϕ we can decide how big part of a merged job for this profile will be scheduled on each machine. Let $y_{\phi i}$ be the variable describing the fraction of merged job from T_ϕ, processed on machine i, this results in the following LP_2:

$$\min \ C \ s.t. \quad t_i^{S_k} + \sum_{\phi=1}^l y_{\phi i} \sum_{J_j \in T_\phi} d_j \frac{\varepsilon}{mK} \Pi_{k,j} \leq C, \ i \in [m], \ k \in [K] \tag{5}$$

$$\sum_{i=1}^m y_{\phi i} = 1, \qquad \phi \in [l]; \tag{6}$$

$$y_{\phi i} \geq 0, \qquad i \in [m], \ \phi \in [l]. \tag{7}$$

It is easy to see that any feasible set of values x_{ij} for LP_1 can be transformed to a feasible set of values $y_{\phi i}$ for LP_2. The last thing is to get rid of the fractional jobs from LP_2. Let $y_{\phi i}^*$ be the optimal solution to LP_2. The number of preempted jobs is at most $f_\phi - 1$ where $f_\phi = |\{y_{\phi i}^* : y_{\phi i}^* > 0, i = 1, \ldots, m\}|$. We remove all the preempted jobs and schedule them at the end of any machine. This increase the makespan by at most $\delta \sum_{\phi=1}^{l} (f_\phi - 1)$. A basic feasible solution of LP_2 has the property that the number of positive variables is at most the number of rows in the constraint matrix, $mK + l$, thus $\sum_{\phi=1}^{l} (f_\phi - 1) \le mK$. In order to bound the total increase by εLB we have to choose δ such that $\delta \le \frac{\varepsilon LB}{mK}$. The number of jobs after merging can be bounded by $\frac{UB \cdot mK}{\delta/2} + l \le \frac{2m^2 K^2}{\varepsilon} + \left(\frac{mK}{\varepsilon}\right)^K = O\left(\left(\frac{mK}{\varepsilon}\right)^K\right)$, thus the claim follows. $\square$

3   Min Max Flow Shop Scheduling Problem

In the MIN MAX $Fm||C_{max}$ we are given a set of jobs $J = \{J_1, \ldots, J_n\}$, which must be processed sequentially on each of m machines $M_1, \ldots, M_m$. Each job completes on machine M_i before it starts on M_{i+1}. Each machine can execute at most one job at a time. A *schedule* is a permutation of jobs. The processing times are uncertain and are described by a constant-size set of scenarios $\Gamma = \{S_1, \ldots, S_K\}$. Thus p_{ji}^S is the processing time of job J_j on machine M_i under scenario S. Similar as in previous section we are looking for a schedule π^* satisfying (1).

Once again we start with setting the lower and the upper bound:

$$LB = \frac{1}{m} \max_{S \in \Gamma} \sum_{j=1}^{n} \sum_{i=1}^{m} p_{ji}^S \le OPT \le \max_{S \in \Gamma} \sum_{j=1}^{n} \sum_{i=1}^{m} p_{ji}^S = UB.$$

Lemma 4 *With $1 + \varepsilon$ loss, for every job J_j under scenario S we can round the p_{ji}^S, down to the nearest value* $\left(\max_{S \in \Gamma} \max_i p_{ji}^S\right) \cdot \frac{\varepsilon}{m^2 K} \cdot k, \ k = 0, 1, \ldots, \frac{m^2 K}{\varepsilon} - 1.$

Proof The proof is analogous to the proof of Lemma 1. The additional m factor appears because replacing job processing time results in increasing its m parts. $\square$

For any J_j we define the *job profile* to be mK-tuple $< \Pi_{1,1,j}, \ldots, \Pi_{K,m,j}, >$ such that $p_{ji}^{S_k}$ is equal to $\frac{\varepsilon}{m^2 K} \cdot \Pi_{k,i,j} \max_{S \in \Gamma} \max_i p_{ji}^S$. The value $d_j := \max_{S \in \Gamma} \max_i p_{ji}^S$ is the *scale factor*.

Lemma 5 *The number of distinct job profiles is at most* $l := \left(\frac{m^2 K}{\varepsilon}\right)^{mK}$.

Proof Every *job profile* has mK entries with possible $\frac{m^2 K}{\varepsilon}$ values. $\square$

Now we perform the grouping step described in previous section with $\delta := m \left(\frac{\varepsilon}{6m^2 K^2}\right)^{\alpha+1} LB$, for α- constant, defined later.

Lemma 6 *With $1 + O(\varepsilon)$ loss, the number of jobs can be reduced in linear time to be at most* $\Delta = \min\{n, 2K \left(\frac{6m^2 K^2}{\varepsilon}\right)^{m/\varepsilon} + \left(\frac{m^2 K}{\varepsilon}\right)^{mK}\}$.

Proof Let $\sigma = \frac{\varepsilon}{6m^2K^2}$ and $\alpha = 0, 1, \ldots, m/\varepsilon - 1$. First we partition the set of *large* jobs into two subsets L_1, L_2 such that $L_1 = \{J_j : m\sigma^\alpha LB < d_j\}$, $L_2 = \{J_j : m\sigma^{\alpha+1}LB < d_j \le m\sigma^\alpha LB\}$. Now we can choose α such that $\sum_{J_j \in L_2} \sum_{i \in [m]} \sum_{S \in \Gamma} p_{ji}^S \le \varepsilon K \cdot LB$, see [5]. Thus with $1 + O(\varepsilon)$ loss the jobs from L_2 can be scheduled at the end.

Next we can bound the number of jobs from L_1 by $\frac{K \cdot UB}{m\sigma^\alpha LB} = \frac{K}{\sigma^\alpha}$ which is a constant. Using the techniques from e.g. [5] we can partition the time $[0, UB]$ into constant number of intervals and with $1 + O(\varepsilon)$ loss provide that each job from L_1 starts at the beginning of some interval. Thus we can enumerate all possible schedules of jobs from L_1. The last thing is to schedule jobs from T. To do so we divide the time into intervals such that t_{vi}^S ($v = 0, \ldots, |L_1|, i \in [m], S \in \Gamma$) defines the empty space between consecutive jobs $v, v + 1 \in L_1$ on machine i under scenario S. Moreover, $t_{i0}^S = 0$ and $t_{i|\mathscr{L}_1|+1}^S = C$ is a parameter optimized by the LP_3. Let l_{vi}^S be the length of interval t_{vi}^S. The task is to schedule jobs from T within the intervals.

We formulate LP_3 as follows. For each job $J_j \in T$ we use the set of decision variables $x_{j\tau} \in [0, 1]$ for tuples $\tau = (\tau_1, \ldots, \tau_m) \in A$, where $A = \{(\tau_1, \ldots, \tau_m)|0 \le \tau_1 \le \tau_2 \le \ldots \le \tau_m \le |L_1|\}$. Now $x_{j\tau}$ represents the fraction of job J_j processed according to $\tau = (\tau_1, \ldots, \tau_m)$, i.e. the job J_j on machine i is processed in interval τ_i.

Let $L_{v,i}^S = \sum_{J_j \in T} \sum_{\tau \in A|\tau_i = v} x_{j\tau} p_{ji}^S$ be the load of jobs from T in t_{vi}^S. By Lemma 4 with a small loss we have that $L_{v,i}^{S_k} = \sum_{J_j \in T} \sum_{\tau \in A|\tau_i = v} x_{j\tau} \frac{\varepsilon}{m^2K} \Pi_{k,i,j} d_j$. Let LP_3:

$$min \ C \ s.t. \quad L_{v,i}^S \le l_{v,i}^S \quad i \in [m], \ S \in \Gamma, \ v = 0, \ldots, |L_1| \qquad (8)$$

$$\sum_{\tau \in A} x_{j\tau} = 1, \ J_j \in T, \qquad (9)$$

$$x_{j\tau} \ge 0, \quad J_j \in T, \ \tau \in A, \qquad (10)$$

Similar to previous section, for each ϕ let us consider a merged job with a profile ϕ and a scale factor equal to $\sum_{J_j \in T_\phi} d_j$. Let $y_{\phi\tau}$ be the variable describing the fraction of merged job from T_ϕ, processed according to τ. Let

$$L_{v,i}^{*S_k} = \sum_{\tau \in A|\tau_i = v} \sum_{\phi=1}^{l} y_{\phi\tau} \sum_{J_j \in T_\phi} \frac{\varepsilon}{m^2K} \Pi_{k,i,j} d_j,$$

this results in the following linear programming formulation LP_2:

$$min \ C \ s.t. \quad L_{v,i}^{*S} \le l_{v,i}^S, \quad i \in [m], \ S \in \Gamma, \ v = 0, \ldots, |L_1|, \qquad (11)$$

$$\sum_{\tau \in A} y_{\phi\tau} = 1, \ \phi = 1, \ldots, l; \qquad (12)$$

$$y_{\phi\tau} \ge 0, \quad \phi \in [l], \ , \tau \in A, \qquad (13)$$

With a notation and argumentation from the previous section we get $\sum_{\phi=1}^{l} (f_\phi - 1) \le mK(|L_1| + 1) \le 2mK\frac{K}{\sigma^\alpha}$, thus the increase of the makespan can be bounded by $2mK\frac{K}{\sigma^\alpha} \cdot m\sigma^{\alpha+1}LB = 2m^2K^2\sigma LB$. Now using the Sevastianov algorithm for each interval t_{vi}^S we can find a schedule of the length at most $l_{vi}^S + (mK + 1)m\sigma^{\alpha+1}$ LB, this increase the total length at most by $(|L_1| + 1)(mK + 1)m\sigma^{\alpha+1}LB \leq$ $4m^2K^2\sigma LB$. The total increase can be bounded by $2m^2K^2\sigma LB + 4m^2K^2\sigma LB =$ $6m^2K^2\sigma LB$. In order to bound the total increase by εLB we choose σ such that $\sigma \leq \frac{\varepsilon}{6m^2K^2}$. The number of jobs after merging can be bounded by $\frac{UB \cdot K}{\delta/2} + l \leq$ $2K\left(\frac{6m^2K^2}{\varepsilon}\right)^{\frac{m}{\varepsilon}} + \left(\frac{m^2K}{\varepsilon}\right)^{mK}$. $\square$

4   Approximation Scheme for Online Problems

In this section we conclude that based on the constructed PTASs we can construct an approximation scheme for the online counterpart of these problems.

In the considered online setting there is a release date associated to every job at which the full information of the job arrives. The released information is ill-known i.e. the *task master* reveals the constant-size set of scenarios with possible processing times. The *scheduler* assigns jobs to machines without knowing the future jobs information. At any time, the task master may stop issuing any further job, the scheduler completes the solution. At this time jobs takes one of the possible, described with scenarios, processing values. The chosen scenario may be considered as a choice of a task master. Now, we compute the *competitive ratio* [4] of the length of the schedule for the worst case scenario to the optimal offline Min-Max schedule.

Our goal is to construct a *Competitive Ratio Approximation Scheme* (CRAS) introduced in [4, 7], which algorithmically constructs an online algorithm with a competitive ratio arbitrarily close to the best competitive ratio for a given problem.

We notice that the construction of a PTAS for both considered problems fulfil all required properties to construct a CRAS. For more details we refer the reader to [7].

Acknowledgments This work was partially supported by the National Center for Science, grant 2013/09/B/ST6/01525. The Ph.D. dissertation of the author was supported by the National Center for Science within Ph.D. scholarship based on the decision number DEC-2013/08/T/ST1/00630.

References

1. Bansal, N., Vredeveld, T., van der Zwaan, R.: Approximating vector scheduling: almost matching upper and lower bounds. In: Pardo, A., Viola, A. (eds.) LATIN, Lecture Notes in Computer Science, vol. 8392, pp. 47–59. Springer (2014)
2. Chekuri, C., Khanna, S.: On multidimensional packing problems. SIAM J. Comput. **33**(4), 837–851 (2004)
3. Fishkin, A.V., Jansen, K., Mastrolilli, M.: Grouping techniques for scheduling problems: simpler and faster. Algorithmica **51**(2), 183–199 (2008)
4. Günther, E., Maurer, O., Megow, N., Wiese, A.: A new approach to online scheduling: approximating the optimal competitive ratio. In: Khanna, S. (ed.) SODA, pp. 118–128. SIAM (2013)

5. Kasperski, A., Kurpisz, A., Zielinski, P.: Approximating a two-machine flow shop scheduling under discrete scenario uncertainty. Eur. J. Oper. Res. **217**(1), 36–43 (2012)
6. Kasperski, A., Kurpisz, A., Zielinski, P.: Parallel machine scheduling under uncertainty. In: Greco, S., Bouchon-Meunier, B., Coletti, G., Fedrizzi, M., Matarazzo, B., Yager, R.R. (eds.) IPMU (4), Communications in Computer and Information Science, vol. 300, pp. 74–83. Springer (2012)
7. Kurpisz, A., Mastrolilli, M., Stamoulis, G.: Competitive-ratio approximation schemes for makespan scheduling problems. In: Erlebach, T., Persiano, G. (eds.) WAOA, Lecture Notes in Computer Science, vol. 7846, pp. 159–172. Springer (2012)
8. Zhu, X., Li, Q., Mao, W., Chen, G.: Online vector scheduling and generalized load balancing. J. Parallel Distrib. Comput. **74**(4), 2304–2309 (2014)

The Price of Fairness for a Small Number of Indivisible Items

Sascha Kurz

Abstract We consider the price of fairness for the allocation of indivisible goods. For envy-freeness as fairness criterion it is known from the literature that the price of fairness can increase linearly in terms of the number of agents. For the constructive lower bound a quadratic number of items was used. In practice this might be inadequately large. So we introduce the price of fairness in terms of both the number of agents and items, i.e., key parameters which generally may be considered as common and available knowledge. It turns out that the price of fairness increases sublinearly if the number of items is not too much larger than the number of agents. For the special case of conformity of both counts, exact asymptotics are determined. Additionally, an efficient integer programming formulation is given.

1 Introduction

Fair division, i.e., the problem of dividing a set of goods between several agents, is studied since ancient times, see e.g. [2]. Also in the context of global optimization fairness aspects cannot be faded out completely. As argued by Bertsimas et al. [1], harming a certain fairness criterion may lead to the situation that a globally optimal solution is not implementable by selfish agents. So, from a practical point of view it is vital to take possible barriers for the execution of an optimized plan into account. Perceived fairness is indeed an important issue. However, considering additional fairness requirements comes at a certain cost. So, several authors have studied the price of fairness as a measurement of the costs of ensuring a certain kind of fairness among the agents.

Here we consider the allocation of m indivisible items among n agents with respect to the fairness criterion of envy-freeness, which roughly means that no agents wants to swap the assigned allocation with another agent. Additionally we assume additive utility functions, i.e., the utility of a bundle is just the sum of the utilities of the

S. Kurz (✉)
University of Bayreuth, 95440 Bayreuth, Germany
e-mail: sascha.kurz@uni-bayreuth.de

© Springer International Publishing Switzerland 2016
M. Lübbecke et al. (eds.), *Operations Research Proceedings 2014*,
Operations Research Proceedings, DOI 10.1007/978-3-319-28697-6_47

335

elements in the bundle, summing up to one for all agents. This setting as well as other fairness criteria and types of items have been studied, see e.g. [3].

While our theoretical setting is rather narrow, our contribution lies in highlighting that the number of items has a significant impact on the price of fairness. For a complete instance with all information, i.e., all utility functions are known, one can compute the unique value of the price of fairness. A possible justification for the study of worst case bounds for the price of fairness are situations where it should be more generally decided whether a fairness criterion should be incorporated into a certain procedure or not. Giving bounds for the maximum value of the price of fairness as a function of certain key parameters, which generally may be considered as common and available knowledge, may be beneficial for such decisions. The number of agents n is an obvious example of such a key parameter. Here we argue that the number of items m is also an interesting key parameter.

In our setting, the price of fairness can be as large as $\Theta(n)$ if we allow a large number of items. If the number of items is restricted to a small number, compared to the number of agents, then it turns out that the worst-case bound decreases to $\Theta(\sqrt{n})$. Even the smallest case possible, admitting envy-free allocations, $m = n$ is far from being innocent. Nevertheless we determine its exact value for all n up to a constant and give a fast-to-solve ILP formulation.

2 Basic Notation and Definitions

Let $\mathcal{J} = \{1, \ldots, n\}$ be a set of agents and $\mathcal{I} = \{1, \ldots, m\}$ be a set of indivisible items. Each agent $j \in \mathcal{J}$ has a non-negative and additive utility function u_j over the subsets of $\mathcal{I}$ with $u_j(\emptyset) = 0$ and $u_j(\mathcal{I}) = 1$. An allocation $\mathcal{A} = (A_1, \ldots, A_n)$ is a partition of $\mathcal{I}$ meaning that the elements of A_j are allocated to agent j. We call an allocation envy-free, if we have $u_j(A_j) \geq u_j(A_{j'})$ for all $j, j' \in \mathcal{J}$, i.e., no agent evaluates a share of one of the other agents higher than his own share. Depending on the utility functions there may be no envy-free allocation at all, consider e.g. $u_j(\{1\}) = 1$ and $u_j(\{i\}) = 0$ for all $i \neq 1$. As a global evaluation of an allocation we use the sum of the agents utilities, i.e., $u(\mathcal{A}) = \sum_{j=1}^{n} u_j(A_j)$. By $\mathcal{A}^\star$ we denote an allocation maximizing the global utility u and similarly by $\mathcal{A}_f^\star$ we denote an envy-free allocation, if exists, maximizing u. With this the price of envy-freeness for n agents $p_{envy}(n)$ is defined as the supremum of $u(\mathcal{A}^\star)/u(\mathcal{A}_f^\star)$. Obviously we have $p_{envy}(1) = 1$. For $n > 1$ the authors of [3] have shown $\frac{3n+7}{9} \leq p_{envy}(n) \leq n - \frac{1}{2}$. Besides $p_{envy}(2) = \frac{3}{2}$ no exact value is known.

The construction for the lower bound of $p_{envy}(n)$ uses $\Omega(n^2)$ items so that one can ask if the price of fairness decreases if the number of items is restricted to a sub-quadratic number of items, which seems to be more reasonable in practice. So we define $p_{envy}(n, m)$ as the supremum of $u(\mathcal{A}^\star)/u(\mathcal{A}_f^\star)$, where the number of items equals m. In any envy-free allocation we have $u_j(A_j) \geq \frac{1}{n}$ since otherwise $u_j(\mathcal{I}) < 1$. Thus $|A_j| \geq 1$ so that we can assume $m \geq n$. The first case $m = n$ is studied in the next section. Obviously we have $p_{envy}(n, m) \leq p_{envy}(n, m + 1) \leq p_{envy}(n)$ for all $m \geq n \geq 1$. The case of a *small* (cf. Theorem 2) number of items is considered in Sect. 4.

3 The Smallest Case: One Item per Agent

As an abbreviation we use $x_{ij} = u_j(\{i\})$ for all $1 \leq i, j \leq n$. The maximum utility $u(\mathscr{A}^\star)$ can be easily determined as $\sum_{i=1}^{n} \max_j x_{ij}$ in linear time. As argued before in any envy-free allocation of an instance with $n = m$ each agent is assigned exactly one item. W.l.o.g. we assume that item j as assigned to agent j for all $1 \leq i \leq n$, i.e., we have $x_{jj} \geq x_{ij}$ for all $1 \leq i, j \leq n$. Using a matching algorithm the existence of an envy-free allocation for $m = n$ can be checked in polynomial time. For this special case all envy-free allocations have the same utility.

The problem of determining worst case examples, i.e., $p_{envy}(n, n)$ can be formulated as an integer linear programming problem:

$$\max \sum_{i=1}^{n} \sum_{j=1}^{n} z_{ij} - \alpha \sum_{i=1}^{n} x_{ii} \tag{1}$$

$$x_{ij} \in \mathbb{R}_{\geq 0} \quad \forall 1 \leq i, j \leq n \qquad \sum_{i=1}^{n} x_{ij} = 1 \quad \forall 1 \leq j \leq n \tag{2}$$

$$x_{jj} \geq x_{ij} \quad \forall 1 \leq i, j \leq n \tag{3}$$

$$y_{ij} \in \{0, 1\} \quad \forall 1 \leq i, j \leq n \qquad \sum_{j=1}^{n} y_{ij} = 1 \quad \forall 1 \leq i \leq m \tag{4}$$

$$z_{ij} \in \mathbb{R}_{\geq 0} \quad \forall 1 \leq i, j \leq n \qquad z_{ij} \leq \min(y_{ij}, x_{ij}) \quad \forall 1 \leq i, j \leq n \tag{5}$$

Here inequalities (2) specify the non-negative utilities of agent j for item i, which sum up to one. The envy-freeness of the allocation given by $A_j = \{j\}$ is guaranteed by Inequality (3). In an optimal assignment item i is assigned to agent j iff $y_{ij} = 1$, see inequalities (4). The auxiliary variables z_{ij} measure the contribution to the global welfare, see inequalities (5). If the target function (1) admits a non-negative value, then we have $p_{envy}(n, n) \geq \alpha$ and $p_{envy}(n, n) < \alpha$ otherwise. We can already conclude that the supremum is attained in the definition for the price of fairness.

Using a bisection approach we were able to exactly determine $p_{envy}(n, n)$ for all $n \leq 9$, i.e., $p_{envy}(n, n) = 1, 1, \frac{8}{7}, \frac{4}{3}, \frac{60}{43}, \frac{3}{2}, \frac{63}{40}, \frac{72}{43}, \frac{9}{5}$. It turned out that the optimal solution for $n \geq 2$ have a rather special structure. The x_{ij} all were either equal to zero or to $\frac{1}{k_j}$, where $2 \leq k_j \leq n$ is an integer. Even more, at most three different k_j-values are attained for a fixed number n, where one case is always $k_j = n$. In the next subsection we theoretically prove this empirical observation.

3.1 Special Structure of the Optimal Solutions for **m = n**

For the ease of notation we use $\tau : \{1, \ldots, n\} \to \{1, \ldots, n\}$, mapping an item i to an agent j, representing an optimal assignment, i.e., $y_{i\tau(i)} = 1$ for all $1 \le i \le n$. By $u^\star(x) = \sum_{i=1}^n x_{i\tau(i)}$ we denote the welfare of an optimal assignment and by $u_f^\star(x) = \sum_{i=1}^n x_{ii}$ the welfare of an optimal envy-free assignment. In the following we always assume that x represents utilities from an example attaining $p_{envy}(n, n)$. We call an agent j *big* if $j \in \mathrm{im}(\tau)$ and *small* otherwise.

Lemma 1 *If agent j is small, then we have $x_{ij} = \frac{1}{n}$ for all $1 \le i \le n$.*

Proof If $x_{jj} \le \frac{1}{n}$ then we have $x_{ij} = \frac{1}{n}$ for all $1 \le i \le n$ so that we assume $x_{jj} > \frac{1}{n}$. Consider x' arising from x by setting $x'_{ij} = \frac{1}{n}$ for all $1 \le i \le n$. With this we have $u_f^\star(x') < u_f^\star(x)$ and $u^\star(x') \ge u^\star(x)$.

Lemma 2 *If agent j is big, then we have $x_{ij} = x_{jj}$ for all i with $\tau(i) = j$.*

Proof We set $w = \sum_{i:\tau(i)=j} x_{ij}$ and $k = |\{i \mid \tau(i) = j\}|$. W.l.o.g. assume $x_{jj} > \frac{w}{k}$. Consider x' arising from x by setting $x'_{ij} = \frac{w}{k}$ for all $1 \le i \le n$ with $\tau(i) = j$. With this we have $u_f^\star(x') < u_f^\star(x)$ and $u^\star(x') \ge u^\star(x)$.

Lemma 3 *If agent j is big, then we can assume $x_{ij} = 0$ or $x_{ij} = \frac{1}{n}$ for all i with $\tau(i) \ne j$ w.l.o.g.*

Proof skipped

Thus we can assume w.l.o.g. that $x_{\star j}$ consists of zeros and k_j times the entry $1/k_j$, where k_j is a positive integer. If $k_j < n$, then all k_j items with utility $1/k_j$ are assigned to agent j in an optimal solution. If $\tau(i) \ne j$, then $x_{ij} \in \{0, 1/n\}$. We can further assume that there is at most one big agent j with $k_j = n$.

3.2 An Improved ILP Formulation and an Almost Tight Bound for **p_envy(n, n)**

Given the structural result from the previous subsection we can reformulate the ILP to:

$$\max \sum_{i=1}^n \frac{r_i}{i} - \alpha \sum_{i=1}^n \frac{s_i}{i}$$

$$s_i \in \mathbb{Z}_{\ge 0} \quad \forall 1 \le i \le n \qquad \sum_{i=1}^n s_i = n$$

$$r_i \in \mathbb{Z}_{\ge 0} \quad \forall 1 \le i \le n \qquad \sum_{i=1}^n r_i = n \qquad r_i \le i \cdot s_i \quad \forall 1 \le i \le n$$

Here s_i counts how often we have $k_j = \frac{1}{i}$ and r_j counts how often we have $x_{i\tau(i)} = \frac{1}{j}$. Having this ILP formulation at hand the exact values of $p_{envy}(n, n)$ can be computed easily for all $n \leq 100$. We observe that in each case at most three values of the vector s are non-zero—going in line with our previous empirical findings.

Lemma 4 *If* $x_{jj} = \frac{1}{k}$ *and* $x_{j'j'} = \frac{1}{k+g}$, *where* $k, g \in \mathbb{N}$ *and* $k + g < n$, *then* $g \leq 1$.

Proof skipped

Theorem 1 $p_{envy}(n, n) \leq \frac{1}{2}\sqrt{n} + O(1)$.

Proof Choose k such that $k_j \in \{k - 1, k, n\}$ for all $j \in \mathscr{J}$ and set $a = |\{j \mid k_j = k\}|$, $b = |\{j \mid k_j = k - 1\}|$. With this we have $u^\star(x) = a + b + c/n$, where $c = n - ak - b(k - 1)$ and $u_f^\star(x) = a/k + b/(k - 1) + (n - a - b)/n$. Next we set $d = a + b$ and $\tilde{k} = (ak + b(k - 1))/d$. Since $c/n \leq 1$, $d/\tilde{k} \leq a/k + b/(k - 1)$, and $\tilde{k} \leq n/d$ we have

$$\frac{u^\star(x)}{u_f^\star(x)} \leq \frac{d + 1}{\frac{d}{k} + \frac{n-d}{n}} \leq \frac{n(d + 1)}{d^2 + n - d} \leq \frac{n(d + 1)}{d^2 + n} =: g(d).$$

For $d \in \{0, n\}$ we have $g(d) = 1$. The unique local maximum of $g(d)$ in $(0, n)$ is at attained at $d = -1 + \sqrt{1 + n}$. Thus $p_{envy}(n, n) = \frac{u^\star(x)}{u_f^\star(x)} \leq g(d) \leq \max\left(1, \frac{1}{2}\sqrt{n} + \frac{1}{n} + 1\right)$.

Lemma 5 $p_{envy}(n, n) \geq \frac{1}{2}\sqrt{n} - \frac{1}{2}$.

Proof Set $a = k = \lfloor \sqrt{n} \rfloor$ and x' with a rows of the form $(\frac{1}{k}, \ldots, \frac{1}{k}, 0, \ldots, 0)$ and $n - a$ rows of the form $(\frac{1}{n}, \ldots, \frac{1}{n})$. With this we have

$$p_{envy}(n, n) \geq \frac{u^\star(x')}{u_f^\star(x')} \geq \frac{a + \frac{n-ak}{n}}{\frac{a}{k} + \frac{n-a}{n}} \geq \frac{a}{2} \geq \frac{\sqrt{n}}{2} - \frac{1}{2}.$$

Thus we can state $p_{envy}(n, n) = \frac{1}{2}\sqrt{n} + \Theta(1)$.

4 Bounds for $p_{envy}(n, m)$ for a Small Number of Items

If the number of items is not too large, i.e., $m \leq n + c\sqrt{n}$ for a constant c, then we can utilize our results for the case $m = n$ in order to deduce an $\Theta(\sqrt{n})$-bound for the price of fairness.

Theorem 2 *If* $m \in n + \Theta(\sqrt{n})$ *with* $m \geq n$ *then* $p_{envy}(n, m) \in \Theta(\sqrt{n})$.

Proof Consider a utility matrix x with $p_{envy}(n, m) \leq u^{\star}(x)/u_f^{\star}(x) + \varepsilon$ for a small constant $\varepsilon \geq 0$. Choose a constant $c \in \mathbb{R}_{\geq 0}$ with $m = c + \sqrt{n}$. By $S \subseteq \mathscr{J}$ we denote the set of agents to which a single item is assigned in the optimal envy-free allocation and set $s = |S|$. All other agents get at least two items so that $n - s \leq 2c\sqrt{n}$ and $s \geq n - c\sqrt{n}$. Now consider another utility matrix x' arising from x as follows. For each agent in S copy the utility row from x. Replace the remaining agents from $\mathscr{J} \setminus S$ by $m - s \geq n - s$ new agents having utility $1/m$ for each item. With this we have $u_f^{\star}(x) \geq u_f^{\star}(x') - (m - s) \cdot \frac{1}{m} \geq u_f^{\star}(x') - \frac{3c}{\sqrt{n}}$ and $u^{\star}(x) \leq u^{\star}(x') + 2c\sqrt{n}$ since each agent $j \notin S$ could contribute at most 1 to $u^{\star}(x)$. Thus we have

$$\frac{u^{\star}(x)}{u_f^{\star}(x)} \leq \frac{u^{\star}(x') + 2c\sqrt{n}}{u_f^{\star}(x') - \frac{3c}{\sqrt{n}}} \leq \frac{1}{1 - \frac{3c}{\sqrt{n}}} \cdot \left(\frac{u^{\star}(x')}{u_f^{\star}(x')} + 2c\sqrt{n} \right)$$

due to $u_f^{\star}(x') \geq 1$. Since the number of agents coincides with the number of items in x', the right hand side of the last inequality is in $O(\sqrt{n})$. The lower bound follows from the case $m = n$.

5 Conclusion

We have introduced the price of fairness in terms of the number of agents and the number of items. As a special case we have considered the allocation of indivisible goods with respect to envy-freeness as a fairness criterion and *normalized* additive utility functions. It turned out that the price of fairness is significantly lower if only a small number of items has to be allocated compared to the case of a large number of items. Up to a constant we have determined the exact value of the price of fairness for the special case when the number of items coincides with the number of agents. In order to determine the exact value we have given an efficient ILP formulation.

We close with some open questions: Can further values of $p_{envy}(n, m)$, where $m > n$, be computed exactly? Can the ILP approach be extended to $m > n$? What is the price of fairness in our setting for $m \in \Theta(n)$ (or more generally, for $m \in \Theta(n^{\alpha})$) with $\alpha < 2$? What happens for other fairness criteria?

References

1. Bertsimas, D., Farias, V.F., Trichakis, N.: The price of fairness. Oper. Res. **59**(1), 17–31 (2011)
2. Brams, S.J., Taylor, A.D.: Fair Division: From Cake-Cutting to Dispute Resolution. Cambridge University Press (1996)
3. Caragiannis, I., Kaklamanis, C., Kanellopoulos, P., Kyropoulou, M.: The efficiency of fair division. Theory Comput. Syst. **50**(4), 589–610 (2012)

Robustness Concepts for Knapsack and Network Design Problems Under Data Uncertainty

Manuel Kutschka

Abstract This article provides an overview of the author's dissertation (Kutschka, Ph.D. thesis, RWTH Aachen University, 2013 [10]). In the thesis, we consider mathematical optimization under data uncertainty using MIP techniques and following the robust optimization approach. We investigate four robustness concepts, their parametrization, application, and evaluation. The concepts are Γ-robustness, its generalization multi-band robustness, the novel more general submodular robustness, and the two-stage recoverable robustness. We investigate the corresponding robust generalizations of the knapsack problem (KP) presenting IP formulations, detailed polyhedral studies including new classes of valid inequalities, and algorithms. In particular, for the submodular KP, we establish a connection to polymatroids and for the recoverable robust KP, we develop a nontrivial compact reformulation and carry out detailed computational experiments. Further, we consider the Γ-robust and multiband brobust generalizations of the network design problem (NDP) presenting MIP formulations, new detailed polyhedral insights with new classes of valid inequalities, and algorithms. For example, we derive alternative formulations for these robust NDPs by generalizing metric inequalities. Furthermore, we present representative computational results for the Γ-robust NDP using real-life measured uncertain data from telecommunication networks based on our work with the German ROBUKOM project.

1 Introduction

Mathematical optimization strives for providing theory, models, and methods to tackle complex real-world problems and obtain relevant solutions in practice. The knapsack (KP) and the network design problem (NDP) are two prominent problem structures occurring in many applications in practice. The understanding of these

M. Kutschka (✉)
Aviation Division, INFORM GmbH, Pascalstraße 35, 52076 Aachen, Germany
e-mail: mkutschka@googlemail.com

M. Kutschka
RWTH Aachen University, Lehrstuhl II für Mathematik, 52056 Aachen, Germany

© Springer International Publishing Switzerland 2016
M. Lübbecke et al. (eds.), *Operations Research Proceedings 2014*,
Operations Research Proceedings, DOI 10.1007/978-3-319-28697-6_48

(sub)problems allows a more accurate mathematical model of the original problem. However, aspects as data uncertainty are oftentimes simplified or ignored. Thus solutions become suboptimal or even infeasible for the original problem. Robust optimization is one approach to take uncertainty into account. The data uncertainty is modeled implicitly by an uncertainty set. The robust optimization problem asks to find an optimal solution that is feasible for any possible data realization in this uncertainty set. In particular, robust linear optimization offers several advantages over other approaches. The definition of an uncertainty set does not rely on the knowledge of probability distributions and is thus often better suited to applied problems where only a finite discrete set of historical data is available, if any. In addition, robust solutions are feasible for all realizations in the uncertainty set by definition. Further, the complexity of robust linear programs does not increase compared to the original non-robust linear program under mild conditions.

Contributions. The main contributions of the thesis are the following.

- The introduction and study of the concept of submodular robustness.
- A detailed investigation of the recoverable robust KP; in particular with a Γ-robust scenario set and the k-removal recovery rule.
- A detailed investigation of the submodular robust KP introducing the classes of submodular robust $(1, k)$-configuration and weight inequalities.
- A study of the structure of covers and their extendability for robust KPs.
- A detailed investigation of the Γ-robust NDP including new classes of strong inequalities (e.g., Γ-robust cutset, Γ-robust envelope, Γ-robust arc residual capacity, and Γ-robust metric inequalities) and algorithms solving the corresponding separation problems as well as the Γ-robust NDP problem itself.
- A first-time investigation of the multi-band brobust NDP including MIP formulations, polyhedral studies yielding new classes of valid inequalities (multi-band brobust cutset and multi-band brobust metric inequalities), and corresponding separation algorithms. In particular, we point out by examples how results of the Γ-robust NDP can be generalized to the multi-band brobust setting.
- Representative extensive computational studies for two recoverable robust KPs and one robust NDP with application to telecommunications.
- Practical decision support methods to determine the right parameters for a robust approach, evaluate robustness by different realized robustness measures, and visualize the quality of a robust solution by its robustness profile.

Outline. This article follows the three-parted structure of the author's thesis.

2   Part I: Concepts

In the first part of this thesis, we introduce the four considered robustness concepts.

Γ**-robustness**. Introduced in [4, 5], the popular Γ-robustness uses a budget of robustness to control conservatism by its robustness parameter Γ and offers a computational tractable Γ-robust counterpart of a LP by exploiting strong LP duality.

Multi-band robustness. A generalization of the concept of Γ-robustness can be obtained by using multiple deviation intervals, so-called bands. For each band and each uncertain data coefficient an associated deviation value is assumed. Thus, a "histogram-like" discrete distribution can be specified. Following this approach, the concept of Γ-robustness is the less detailed special case with only one nominal value (band) and one negative and one positive deviation band. The idea of multi-band robustness goes back to portfolio optimization [6] and has not been formulated as a general robustness concept until recently in [8, 9].

Submodular robustness. A new and more general robustness concept is submodular robustness. Here, the constraints of the robust counterparts are described by submodular functions yielding submodular knapsack constraints. Given a base set $N := \{1, \ldots, n\}$, a function $f : 2^N \to \mathbb{R}_{\geq 0}$ is called *submodular* if $f(X) + f(Y) \geq f(X \cup Y) + f(X \cap Y)$ holds for all $X, Y \subseteq N$. Submodular robustness generalizes Γ-robustness and multi-band robustness.

The submodular robust uncertainty set is the polymatroid of the corresponding submodular function. Although many studies exist, to our knowledge polymatroids have not been related to robust optimization except for [2] on mean-risk minimization where submodular functions and polymatroids are considered for a stochastic optimization problem. Note, a linear function can efficiently be optimized over a polymatroid using a greedy algorithm. Therefore the worst-case realization of a submodular robust uncertainty set can be determined efficiently explaining this well-known observation for the special case of Γ-robustness.

Recoverable robustness. Recoverable robustness, introduced in [11], is a recent two-stage robust approach that can be seen as a deterministic alternative to stochastic programming with limited recourse.

It can be sketched as follows: after the first-stage decision the realization of the uncertain data is observed. Then, the previous decision may be altered according to a given adjustment rule taking the realization into account. This second-stage adjustment is called recovery as the first-stage decision may become infeasible by the realization. Both, the first stage decision and its second stage adjustment inflict costs. An optimal recoverable robust solution minimizes the overall costs, i.e., the first stage costs and the worst-case second stage costs.

3 Part II: Robust Knapsack Problems

In the second part of this thesis, we consider the robust counterpart of the KP for each of the four robustness concepts. For each resulting robust KP, we present mathematical formulations, study the corresponding polyhedral solution sets identifying strong classes of valid inequalities, and develop algorithms solving the occurring separation problems as well as the robust knapsack problem itself.

The Γ-robust knapsack problem (Γ-RKP). We consider the Γ-RKP in this thesis as "standard" example of a robust KP and a benchmark for our studies of other robust variants of the KP. Later, it will turn out to be the special case of more general concepts. Besides we consider covers for this problem and the resulting class of cover inequalities which is valid for the knapsack polytope. It can be tightened by lifting to the so-called extended cover inequalities. In this thesis, we are able to obtain even stronger inequalities which we call strengthened extended cover inequalities by exploiting the structure of the worst-case realizations.

The multi-band robust knapsack problem (mb-RKP). Applying the multi-band robustness concept to the KP with uncertain item weights results in the mb-RKP; a generalization of the Γ-RKP with a detailed uncertainty model. So far, the mb-RKP has only been studied implicitly: in [8, 12] the multi-band brobust counterpart of a general linear constraint, and thus a multi-band brobust knapsack constraint, is described. But the mb-RKP itself is not introduced nor polyhedrally analyzed.

In this thesis, we give a corrected compact formulation, a first time study of the mb-RKP polytope including its dimension, trivial facets, as well as multi-band brobust (extended) cover inequalities. Again, we are able to tighten the latter to multi-band brobust strengthened extended cover inequalities. We provide new exact ILP-based separation algorithms for all considered classes of valid inequalities.

The submodular (robust) knapsack problem (SMKP). The most general robust KP we study is the SMKP. It has been introduced in [3] investigating in particular extended covers and specific submodular functions.

In this thesis, we introduce the classes of submodular robust $(1, k)$-configuration inequalities and submodular robust weight inequalities generalizing the corresponding classes of the non-robust KP. We show that SMKP polytope restricted to the item set defining a submodular robust $(1, k)$-configuration is completely described by submodular robust $(1, k)$-configuration inequalities and non-negativity and thus generalizing a result for the KP.

The recoverable robust knapsack problem (RRKP). In contrast to the other considered robust KPs, the RRKP implements a two-stage approach to handle the KP with uncertain item weights.

In this thesis two particular RRKPs are considered: the K, l-RRKP with discrete scenarios and the k-RRKP with Γ-scenarios (k/Γ-RRKP). We provide compact MIP formulations for these problems, a polyhedral study of the corresponding polytopes including their dimensions, trivial facets, and ((strengthened) extended) cover inequalities. We give an ILP-based exact separation algorithms. In addition, we study the problem to determine the worst-case realization of the k/Γ-RRKP and give an exact combinatorial algorithm solving this problem efficiently. Furthermore, we develop different algorithms to solve the k/Γ-RRKP exploiting our novel compact problem formulation, and robustness cuts.

Computations. We complement our studies of robust KPs by extensive computations on the RRKP evaluating the gain of recovery, the effectiveness of cover constraints, and our new compact formulation of the RRKP with Γ-robust scenario set.

4 Part III: Robust Network Design Problems

The Γ-robust and multi-band brobust NDPs are studied in this part of the thesis.

The Γ-robust network design problem (Γ-RNDP). The Γ-robustness concept has first been applied to the Γ-RNDP in [1].

In this thesis, we study the properties of Γ-RNDP polyhedra. For the link flow polyhedron, we present a detailed study following the general method to consider smaller (sub)networks or relaxations polyhedrally and transfer back the results to the original polyhedron. First, we consider cutset-based inequalities which we identify by studying a lower dimensional projection of the Γ-robust cutset polyhedron and a related auxiliary polyhedron. By the latter, we identify the new class of Γ-robust envelope inequalities and are able to completely describe this auxiliary polyhedron. By lifting, we derive strong inequalities for the original Γ-RNDP polyhedron which are facet-defining under mild conditions. Second, we consider the Γ-robust single arc design problem and its related polyhedron. Here, we identify two new classes of inequalities valid for the original Γ-RNDP polyhedron. Third, we consider Γ-robust metric inequalities. We present a capacity formulation of the Γ-RNDP and generalize the so-called "Japanese Theorem" to the Γ-robust setting. Moreover, we show metric inequalities (together with nonnegativity) completely describe the Γ-RNDP capacity polyhedron. We also investigate special subclasses of Γ-robust metric inequalities valid for the original Γ-RNDP polyhedron. Finally, we present separation algorithms including (i) exact algorithms and a heuristic for Γ-robust cutset and envelope inequalities, (ii) an exact combinatorial polynomial algorithm for Γ-robust arc residual capacity inequalities, and (iii) exact separation algorithms for Γ-robust length and metric inequalities, and selected subclasses of Γ-robust metric inequalities. Further, we present a polynomial time separation algorithm for violated Γ-robust metric inequalities.

The multi-band brobust network design problem (mb-RNDP). So far, only a simplified preliminary version of multi-band robustness has been considered in connection with the NDP in [7]. Recently a technical paper [12] was published on a NDP with multiple intervals following a similar but less general concept.

In this thesis, we present a compact MIP formulation of the mb-RNDP, and study the corresponding multi-band brobust network design link-flow polyhedron (analogously to the Γ-RNDP). We successfully further generalize the "Japanese Theorem" and obtain a capacity formulation. Further, we investigate the multi-band brobust cutset polyhedron obtaining the generalization of Γ-robust cutset inequalities and Γ-robust envelope inequalities and use these as examples how results for the Γ-RNDP can be generalized to the mb-RNDP. We also present exact separation algorithms for multi-band brobust cutset, length, and metric inequalities including a polynomial separation algorithm for violated multi-band brobust metric inequalities.

Computations. We conclude this part of the thesis with a detailed experiments on the Γ-RNDP using real traffic measurements of telecommunication networks. We address the parametrization of the Γ-robustness using historical data, propose a Pareto front analysi giving a practical example. Our studies include (i) the comparison of different formulations, (ii) the evaluation of nine different classes of valid inequalities, (iii) the comparison of different algorithms to solve the Γ-RNDP, (iv) the investigation of the scalability to large instances, and (v) the evaluation of the quality of the obtained robust network designs.

5   Conclusions

In this thesis we present a comprehensive study of four different robustness concepts with application to two important reoccurring optimization problems, the KP and the NDP. We provide mathematical formulations, high quality polyhedral studies with new classes of strong inequalities, and (separation) algorithms for each of them. Our investigations are accompanied by extensive computational experiments on representative data (real traffic measurements from telecommunication) and the development of analysis methods (realized robustness measures, robustness profile, and Pareto front analysis) to be used as decision support in practice.

References

1. Altin, A., Amaldi, E., Belotti, P., Pinar, M.C.: Provisioning virtual private networks under traffic uncertainty. Networks **49**(1), 100–155 (2007)
2. Atamtürk, A., Narayanan, V.: Polymatroids and mean-risk minimization in discrete optimization. Oper. Res. Lett. **36**(5), 618–622 (2008)
3. Atamtürk, A., Narayanan, V.: The submodular knapsack polytope. Discrete Optim. **6**(4), 333–344 (2009)
4. Bertsimas, D., Sim, M.: Robust discrete optimization and network flows. Math. Program. **98**(1), 49–71 (2003)
5. Bertsimas, D., Sim, M.: The Price of Robustness. Oper. Res. **52**(1), 35–53 (2004)
6. Bienstock, D.: Histogram models for robust portfolio optimization. J. Comput. Finance **11**, 1–64 (2007)
7. Bienstock, D., D'Andreagiovanni, F.: Robust wireless network planning. In: Proceedings AIRO 2009, the 40th Annual Conference of the Italian Operational Research Society, pp. 131–132 (2009)
8. Büsing, C., D'Andreagiovanni, F.: New results about multi-band uncertainty in robust optimization. In: Klasing, R. (ed.) Experimental Analysis—SEA 2012, LNCS, vol. 7276, pp. 63–74. Springer (2012)
9. Büsing, C., D'Andreagiovanni, F.: Robust optimization under multi-band uncertainty—Part I: Theory (2013). arXiv:1301.2734v1
10. Kutschka, M.: Robustness concepts for knapsack and network design problems under data uncertainty: gamma-, multi-band, submodular, and recoverable robustness. Ph.D. thesis, RWTH Aachen University (2013). ISBN: 978-3-95404-593-8 (Cuvillier Verlag, 2013)

11. Liebchen, C., Lübbecke, M.E., Möhring, R.H., Stiller, S.: The concept of recoverable robustness, linear programming recovery, and railway applications. In: R. Ahuja, R. Möhring, C. Zaroliagis (eds.) Robust and Online Large-Scale Optimization, LNCS, vol. 5868, pp. 1–27. Springer (2009)
12. Mattia, S.: Robust optimization with multiple intervals. Technical Report R. 7 2012, Istituto di Analisi dei Sistemi ed Informatica (IASI), Consiglio Nazionale delle Ricerche (CNR), viale Manzoni 30, 00185 Rome, Italy (2012)

An Insight to Aviation: Rostering Ground Personnel in Practice

Manuel Kutschka and Jörg Herbers

Abstract Numerous dynamic, interdependent processes exist at an airport. These processes are highly affected by uncertain events as changing flight schedules, delays, or weather conditions. Naturally a flexible workforce management is needed to support such operation. Airlines, airports, and ground handlers provide the necessary workforce to meet this demand. But legal requirements, union agreements and company policies define the flexibility of workforce planning and utilization in practice. Nevertheless a valid (monthly) roster matching the supply with demand under all these requirements has to be prepared usually several weeks before the day of operation. In this paper we discuss the optimization challenges to create monthly rosters for ground personnel at an airport. We give examples of typical constraints, point out characteristics of different work areas at an airport, and how this affects the rostering. Further we present how rostering is solved by our branch-and-price solution methodology in practice. Using this approach, we report on our real world experience with optimized rostering in airport ground handling.

1 Introduction

Ground handling at airports comprises a number of different services. Ramp handling includes services like baggage and cargo loading, unloading and transportation, pushback of aircraft, deicing, fueling and cabin cleaning. In addition, passenger services typically include check-in, boarding and ticketing.

All of these work areas are faced with an important characteristic of air traffic, namely the considerable level of variations in flight volume within the day, the week and over the year. As an example, daily traffic at many airports is governed by two to three more or less sharp peaks that translate into corresponding workforce demands during limited time periods. In such situations, providing sufficient staff for

M. Kutschka (✉) · J. Herbers
Aviation Division, INFORM GmbH, Pascalstraße 35, 52076 Aachen, Germany
e-mail: manuel.kutschka@inform-software.com; mkutschka@googlemail.com

J. Herbers
e-mail: joerg.herbers@inform-software.com

© Springer International Publishing Switzerland 2016   349
M. Lübbecke et al. (eds.), *Operations Research Proceedings 2014*,
Operations Research Proceedings, DOI 10.1007/978-3-319-28697-6_49

peak periods almost necessarily leads to idle times in off-peak times. Flexible shift models (including part-time employees on short shifts) are typical means to address this challenge, but ground operators will also try to run operations at sufficient service levels without fully covering peak demand.

Workforce management in the aviation and other service industries typically consists of different long- to short-term processes. At a strategic level, recruiting and training of staff needs to be planned. At a tactical level, annual leave has to be assigned (e.g. based on fairness criteria) and actual rosters with shift timings generated and published. Operational scheduling is governed by short-term changes both in flight-induced demand (e.g. due to extra flights) and by employee availability (e.g. short-term sickness) that need to be compensated e.g. by additional shifts and overtime.

All of these processes ultimately aim at providing the right amount of staff at the right times, with the right qualifications and at appropriate locations of the airport. The uncertainty both of flight schedule information (e.g. delays) as well as absences like sickness form an integral part of the challenges that ground operators face every day.

2   Related Work

Personnel scheduling is a very active research area in Operations Research. Correspondingly, there is a vast body of scientific articles on the topic. An excellent survey and classification of problems can be found in [5]. Van den Bergh et al. [10] give a literature review and classification of articles by the problems considered, types of constraints, objective functions, solutions approaches and application areas. Further bibliographies can be found in [6, 7].

An application at an aircraft refueling stations is described in [1], using a continuous tour scheduling formulation for a mixed workforce and employing a tabu search methodology. Chu [3] describes the solution of a shift scheduling and tour scheduling problem at Hongkong International Airport, using a goal programming approach solved by heuristics. Brunner and Stolletz [2] solve a discontinuous tour scheduling problem for check-in operations by a stabilized branch-and-price algorithm.

A complete IT system for ground staff rostering is described by Dowling et al. [4] who solve a tour scheduling problem by a simulated annealing approach. Jacobs et al. [9] describe a computer system developed for United Airlines that solves a tour scheduling problem using a combined column generation/simulated annealing algorithm.

3 Monthly Rostering of Ground Personnel

In this paper, we focus on tactical rostering and in particular the creation of monthly rosters. We start with a rather informal definition.

Definition 1 Given demands as set of shifts with corresponding qualification sets and required staffing levels, a set of employees with qualifications, and further rostering rules. The *Monthly Rostering Problem (MRP)* asks to find an assignment of shifts to the employees such that qualification restriction and the further rostering rules are obeyed and the staffing levels are met at the best.

Referring to the terminology of [5], the MRP therefore assumes that the problems of demand modeling and shift scheduling have been solved before. The MRP therefore corresponds to the shift assignment problem, see also [7, 10].

In practice, the underlying basic assignment problem becomes challenging by further rostering rules. These side constraints are imposed by different stakeholders, e.g. law, union agreements, and company policies. Moreover, some of them must be obeyed as hard rules while others may be treated as soft rules. We encounter several reoccuring types of rules in our customer projects:

- *Working time rules* introduce restriction on the minimal/maximal daily/weekly/ monthly working time. Additional rules exist, e.g. on overtime handling or the minimum rest time between shifts.
- *Day on/off rules* impose restrictions or "patterns" of allowed sequences of working days and days off. Often these restrictions are even more specific and define e.g., early or late shifts.
- *Rules on rest days* and leave, e.g. specify the minimum or maximum number of days off in a certain period, the handling of days off in lieu of overtime or single days off.
- *Employee participation* is becoming more important nowadays, e.g. staff member availabilities, rostering of teams, carpooling, social rules, work preferences, and fairness concepts play an increasing role.
- *Ergonomic* have to be taken into account, e.g. shifts on consecutive days should use same start times or rotate forward, the number of consecutively worked shifts should be limited, and long consecutive shifts (in particular night shifts) are known to have an impact on employee health and fatigue.

In addition to these rules, the workforce is often composed of heterogeneous contracts (e.g. part-time staff, temporary workforce etc.) as well as different qualification levels. Mastery of all of these challenges is crucial in achieving cost efficiency in an increasing competitive environment.

In the following, we formulate the MRP more formally. Let E be the set of employees, S_e the set of possible rosters or schedules for employee $e \in E$. Let D be the set of demands, T the discretized planning period (here: days of a month), and A the set of activities, i.e. combinations of shifts and qualification sets. Furthermore, we denote by $A_{d,t}$ the activities that can be accounted for demand $d \in D$ on day $t \in T$, and for $e \in E$ by $A_{s,t}$ the allowed activities on day $t \in T$ in the roster $s \in S_e$.

We introduce four types of decision variables: the binary variable $x_{e,s}$ which equals 1 if employee $e \in E$ is assigned roster $s \in S_e$ and 0 otherwise, the binary variable $z_{a,t}$ which equals 1 if the activity $a \in A$ is worked on day $t \in T$ and 0 otherwise. The non-negative variables u and o are slack variables to represent under- and overcoverages of the demands, respectively. Their usage is penalized by costs p_d^u and p_d^o, respectively. The target level of demand $d \in D$ is denoted by n_d.

Now, we can formulate the MRP as a mixed integer program:

$$\min \quad \sum_{e \in E, s \in S_e} c_{e,s} x_{e,s} + \sum_{d \in D} p_d^u u_d + \sum_{d \in D} p_d^o o_d \tag{1a}$$

$$\text{s.t.} \quad \sum_{t \in T, a \in A_{d,t}} z_{a,t} + u_d - o_d = n_d \qquad \forall d \in D \tag{1b}$$

$$\sum_{e \in E, s \in S_e \,:\, a \in A_{s,t}} x_{e,s} - z_{a,t} = 0 \qquad \forall a \in A, t \in T \tag{1c}$$

$$\sum_{s \in S_e} x_{e,s} = 1 \qquad \forall e \in E \tag{1d}$$

$$u, o \geq 0, \ x \in \{0, 1\}^{E \times S_e}, \ z \in \{0, 1\}^{A \times T} \tag{1e}$$

where $c_{e,s}$ is the "cost" of assigning the employee e to roster s. These cost coefficients incorporate all hard and soft rules by penalty costs. Note that formulation (1a)–(1e) consists of exponentially many variables.

Solving the MRP. We follow a branch-and-price approach with temporal decomposition to solve (1a)–(1e). The pricing problem we need to solve in the branch-and-price turns out to be a resource-constrained shortest path problem (RCSPP) on an

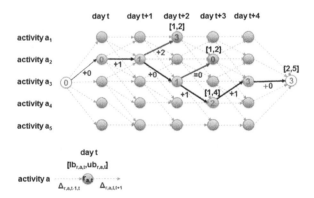

Fig. 1 The pricing problem as a RCSPP on an auxiliary network. This acyclic network is layered by the day. It contains a node for each activity, an arc for each allowed sequence of consecutive activities as well as node labels, node bounds and arc costs for multiple resources. A resource-constrained shortest path defines a sequence of daily activities and thus a roster variable with cost in the master problem

auxiliary network, see Fig. 1. The RCSPP is solved exactly using a dynamic program. To speed up the pricing process, a faster heuristic is applied first. Moreover, the auxiliary network is preprocessed to reduce its size and tighten the resource windows.

4 Practical Experience

Next we report on our real world experiences with monthly rostering of ground personnel based on past customer projects using INFORM's GroundStar RosterPlus.

INFORM [8] is a company founded as university spin-off in 1969 and located in Aachen, Germany, that specializes in intelligent planning and logistics decision-making software based on Operations Research methods. Today, aviation is its largest division with about 200 employees.

GroundStar RosterPlus is INFORM's optimization solution for staff rostering. It implements the mathematical model and solution approach sketched in the previous section. To represent the various customer-specific roster rules, it uses a flexible modeling language which is interpreted by the system to produce an appropriate optimization model for the branch-and-price algorithm.

We consider seven instances of the MRP for ground personnel. The data is based on several customer projects and spans three different work areas (cabin cleaning, ramp and passenger services).

Our results are summarized by the following reference values in Table 1: name (Instance), the considered employee group (Work Area), the number of employees (#Empl.), an assessment of the additional rostering rules imposed by law, union agreements and company policies (Rules), an indication on the heuristic decomposition strategy (Decomposition), and the subsequent time to solve the optimization problem by branch-and-price (Time), the average RSCPP network size (#Nodes and #Arcs), and the number of solved RSCPPs (#Solved).

In Table 1, we observe that work areas as cabin (cleaning) and ramp services have rather simple rostering rules while passenger services ask for more complex rules. The latter is mostly because of a higher number of part-timers with heterogeneous qualification sets. In addition, achieving good acceptance of rosters in this area is more difficult and requires more social rules to be modeled. The differences in computation times are partially due to different parameterization of the temporal problem decomposition. Clearly, the decomposition strategy affects the RCSPP network size and thus the overall solution time. For example, for eager or medium decomposition (leading to smaller core problems to be solved), we achieve optimization times of less than 90 s the lazy strategy yields solution times of about 1.5 h for these instances. This compares to manual roster creation that often takes planners 10 working days or more which is still wide-spread practice at airports, ground handlers and airports.

Table 1 Tactical rostering of ground personnel in practice using GroundStar RosterPlus

Instance	Work area	#Empl.	Rules	Decomposition	Time	RCSPP		#Solved
						#Nodes	#Arcs	
C1	CABIN	130	Simple	Eager	5.6	15	29	4540
R1	RAMP	271	Simple	Eager	30.1	37	153	9922
R2	RAMP	291	Simple	Medium	81.5	123	586	10120
R3	RAMP	313	Complex	Low	5379.0	70	293	24171
P1	PASSENGER	294	Complex	Eager	46.3	33	156	11330
P2	PASSENGER	295	Complex	Eager	59.3	37	203	11797
P3	PASSENGER	304	Complex	Lazy	5667.3	111	965	23898

5 Conclusions

In this paper, we have considered rostering processes at an airport. In particular, the MRP has been investigated. We have stated a mixed integer programming formulation and discussed the type of rostering rules occuring in practice. Further, we have described our approach to solve the MRP which is implemented in INFORM's GroundStar RosterPlus. Finally, we have shown real-world examples.

References

1. Alvarez-Valdez, R., Crespo, E., Tamarit, J.: Labour scheduling at an airport refuelling installation. J. Oper. Res. Soc. **50**, 211–218 (1999)
2. Bruner, J.O., Solletz, R.: Stabilized branch and price with dynamic parameter updating for discontinuous tour scheduling. Comput. Oper. Res. **44**, 137–145 (2014)
3. Chu, S.C.K.: Generating, scheduling and rostering of shift crew-duties: applications at the Hong Kong International Airport. Eur. J. Oper. Res. **177**, 1764–1778 (2007)
4. Dowling, D., Krishnamoorthy, M., Mackenzie, H., Sier, D.: Staff rostering at a large international airport. Ann. Oper. Res. **72**, 125–147 (1997)
5. Ernst, A.T., Jiang, H., Krishnamoorthy, M.: Staff scheduling and rostering: a review of applications, methods and models. Eur. J. Oper. Res. **153**, 3–27 (2004)
6. Ernst, A.T., Jiang, H., Krishnamoorthy, B., Owens, B., Sier, D.: An annotated bibliography of personnel scheduling and rostering. Ann. Oper. Res. **127**, 21–144 (2004)
7. Herbers, J.: Models and algorithms for ground staff scheduling on airports. Ph.D. thesis, RWTH Aachen, Aachen (2005)
8. INFORM GmbH, Pascalstr. 23, 52076 Aachen, Germany. http://www.inform-software.com
9. Jacobs, L., Brusco, M.J., Bongiorno, J., Lyons, D.V., Tang, B.: Improving personnel scheduling at airline stations. Oper. Res. **43**, 741–751 (1995)
10. Van den Bergh, J., Beliёna, J., De Bruecker, P., Demeulemeester, E., De Boeck, L.: Personnel scheduling: a literature review. Eur. J. Oper. Res. **226**, 367–385 (2013)

Consensus Information and Consensus Rating

A Note on Methodological Problems of Rating Aggregation

Christoph Lehmann and Daniel Tillich

Abstract Quantifying credit risk with default probabilities is a standard technique for financial institutes, investors or rating agencies. To get a higher precision of default probabilities, one idea is the aggregation of different available ratings (i.e. default probabilities) to a so called 'consensus rating'. But does the concept of 'consensus rating' really make sense? What is a 'real' consensus rating? This paper tries to clarify under which conditions a consensus rating exists. Therefore, the term of 'consensus information' is used. This leads to a concept that deals with the precision of aggregated rating characteristics. Within this framework the problem of misinformation resp. contradictory information is addressed. It is shown that consensus information is not necessarily more informative than individual information. Furthermore, the aggregation aspects are discussed from a statistical perspective.

1 Introduction

The problem of aggregating ratings seems to be a quite new one in the literature. One article in this field is [1, p. 76], who state:

> [...] but to the best of our knowledge there is no literature discussing how to combine different (heterogeneous) ratings of a company into a common rating.

In [1] a model is developed that combines different ratings into a so called 'consensus rating'. In contrast, in this paper it should be discussed what a consensus rating is and in which cases this concept makes sense.

Firstly, we introduce some notation. The credit default of debtor i is modeled by a default variable Y_i, which takes the value 1 in the case of default of debtor i and 0 otherwise, where $i = 1, \ldots, n$. Thus, $P(Y_i = 1)$ is the unconditional default probability of debtor i. Characterizing the creditworthiness of a debtor with an

C. Lehmann (✉) · D. Tillich (✉)
TU Dresden, Dresden, Germany
e-mail: Christoph.Lehmann@tu-dresden.de

D. Tillich
e-mail: Daniel.Tillich@tu-dresden.de

© Springer International Publishing Switzerland 2016
M. Lübbecke et al. (eds.), *Operations Research Proceedings 2014*,
Operations Research Proceedings, DOI 10.1007/978-3-319-28697-6_50

estimated default probability is the main target of a rating process. Typically, several rating characteristics are used for calculating an estimated default probability. These rating characteristics are contained in the subject specific real random vector $\mathbf{X}_i \overset{\text{def}}{=} (X_{i1}, X_{i2}, \ldots, X_{iK})$ with realization $\mathbf{x}_i \overset{\text{def}}{=} (x_{i1}, x_{i2}, \ldots, x_{iK}) \in \mathbb{R}^K$, $i = 1, \ldots, n$. Thus, the probability of interest is the conditional default probability $P(Y_i = 1 | \mathbf{X}_i = \mathbf{x}_i)$.

In the following we assume that all rating agencies are using the same vector $\mathbf{X}_i$. This assumption is needed in the sequel for reasonable set operations. It is not a crucial assumption because further variables can be added to the vector $\mathbf{X}_i$ without difficulty. At the first sight the assumption seems counterintuitive, but in our framework the differences do not lie in the rating characteristics $\mathbf{X}_i$ themselves but in the information about them. This is discussed more detailed in Sect. 2.

2 Information Set and Consensus Information

In reality, the complete vector of realizations $\mathbf{x}_i = (x_{i1}, x_{i2}, \ldots, x_{iK}) \in \mathbb{R}^K$ is unknown in the sense that some of its components are not exactly known or even completely unknown. In the following, an institution that assigns ratings is called 'rating agency'. This also could be a bank which evaluates the debtors.

Definition 1 The information set $I_{rik} \subseteq \mathbb{R}$ contains all information of rating agency r about the kth rating characteristic of debtor i.

Example 1 Three rating agencies $r = 1, 2, 3$ have information for debtor i about sex (rating characteristic k) and income class (rating characteristic k').

(a) Rating agencies 1 and 2 have the information that debtor i is female. Rating agency 3 does not know the sex of debtor i. Therefore the informaton sets are $I_{1ik} = \{1\}$, $I_{2ik} = \{1\}$, $I_{3ik} = \{0, 1\}$, where female is coded by 1 und male by 0.
(b) The rating agencies use different income classes. Rating agency 1 has the information that the income of debtor i lies between 1000 and 2000, i.e. the information set is $I_{1ik'} = [1000, 2000]$. Analogously, the information sets of rating agencies 2 and 3 are $I_{2ik'} = [1500, 2200]$ and $I_{3ik'} = [1100, 1800]$.

Definition 2 The information sets I_{rik} for all rating characteristics $k = 1, \ldots, K$ are combined in the pooled information set

$$I_{ri} \overset{\text{def}}{=} I_{ri1} \times I_{ri2} \times \cdots \times I_{riK}, \qquad i = 1, \ldots, n, \quad r = 1, \ldots, R.$$

Concerning the information of an arbitrary rating agency there are several situations.

1. **Zero information**: There is no information about all the rating characteristics, i.e. $I_{rik} = \mathbb{R}$ for all $k = 1, \ldots, K$, thus $I_{ri} = \mathbb{R}^K$.

2. **Complete information**: For all rating characteristics there is exactly one value x_{ik}. Thus every information set is a singleton, i.e. $I_{rik} = \{x_{ik}\}$ for all $k = 1, \ldots, K$ and $I_{ri} = \{\mathbf{x}_i\}$.
3. **Incomplete information**: At least one of the rating characteristics is not known exactly, i.e. for at least one $k = 1, \ldots, K$ the information set I_{rik} is not a singleton.

Because of different information sets, rating agencies estimate different conditional probabilities $P(Y_i = 1|\mathbf{X}_i \in I_{ri}) = P(Y_i = 1|X_1 \in I_{ri1}, X_2 \in I_{ri2}, \ldots, X_K \in I_{riK})$. In order to estimate the same probability, the information sets of the rating agencies should be merged. Therefore, assumptions about the information sets are needed. These assumptions are discussed in the following subsections.

2.1 No Contradictory Information

Assumption 1 There is no contradictory information, i.e. $\bigcap_{r=1}^{R} I_{ri} \neq \emptyset$.

Thereby it follows:

- The information sets are overlapping, in particular $I_{rik} \cap I_{r'ik} \neq \emptyset$ for all r, r', i, k.
- The information of the rating agencies differ referring to their precision, i.e. the size (cardinality) of the information sets.

With Assumption 1 the case of complete information is desirable. For getting close to the complete information as much as possible, within this framework different rating agencies should interchange and combine their information. This leads to a consensus information.

Definition 3 Let Assumption 1 hold. The consensus information set of the ith debtor for the kth component of the rating characteristics $I_{ik}^{\cap}$ and for all rating characteristics $I_i^{\cap}$ are defined as

$$I_{ik}^{\cap} \overset{\text{def}}{=} \bigcap_{r=1}^{R} I_{rik} \quad \text{resp.} \quad I_i^{\cap} \overset{\text{def}}{=} \bigcap_{r=1}^{R} I_{ri}. \tag{1}$$

Thus, $I_i^{\cap}$ is the Cartesian product of the $I_{ik}^{\cap}$, i.e. $I_i^{\cap} = I_{i1}^{\cap} \times I_{i2}^{\cap} \times \cdots \times I_{iK}^{\cap}$. The consensus information sets depend on the number of rating agencies involved. From Assumption 1, it follows $\emptyset \neq I_{ik}^{\cap} \subseteq I_{rik} \subseteq \mathbb{R}$ for all k and $\emptyset \neq I_i^{\cap} \subseteq I_{ri} \subseteq \mathbb{R}^K$. Hence, the consensus information set $I_i^{\cap}$ is at least as precise as every agency specific information set I_{ri}. If the condition $I_i^{\cap} \subsetneq I_{ri}$ is fulfilled for all $r \in \{1, 2, \ldots, R\}$, the intersection leads to a higher precision of the consensus information in comparison to every single information set.

Example 2 The information sets from Example 1 lead to the following consensus information sets: $I_{ik}^{\cap} = \{1\}$, $I_{ik'}^{\cap} = [1500, 1800]$, $I_i^{\cap} = \{1\} \times [1500, 1800]$.

The consensus information set $I_i^{\cap}$ is more precise than every agency specific information set I_{ri}.

The term 'consensus information' is already used in [1, p. 77]:

> [...] combining different rating information stemming from different rating sources, by deriving appropriate consensus information, i.e., consensus ratings, which incorporate the information of several rating sources.

By (1) the term 'consensus information' is defined and can be distinguished from the term 'consensus rating', that follows in Sect. 3. So we sharpen both terms, which cannot be distinguished clearly in [1]. In the case of rating aggregation the term of 'consensus' seems to be used the first time in [1]. Especially for 'consensus' they refer to [3, p. 591], who define

> consensus as the degree of agreement among point predictions aimed at the same target by different individuals [...]

Following [3] we use the term consensus only if the same target, i.e. the same conditional default probability, is addressed.

2.2 Contradictory Information

In Sect. 2.1 it is assumed that there is no contradictory information, i.e. that all the information sets are overlapping. In contrast, contradictory information means that the intersection $I_{ik}^{\cap} = \bigcap_{r=1}^{R} I_{rik}$ could be the empty set. This would imply that there is at least one misinformation and the approach from above is not reasonable anymore. How to get a consensus information in this case? One possibility is the union of the information sets $I_{ik}^{\cup} \overset{\text{def}}{=} \bigcup_{r=1}^{R} I_{rik}$ which in general leads to more imprecise information than before. A combination of both approaches is

$$I_{ik}^{*} \overset{\text{def}}{=} \begin{cases} I_{ik}^{\cap}, & \text{if } I_{ik}^{\cap} \neq \emptyset, \\ I_{ik}^{\cup}, & \text{if } I_{ik}^{\cap} = \emptyset. \end{cases}$$

The aggregation approaches above can be used for every scale. More specific types of aggregation are the different types of mean concepts, e.g. mode, median or (weighted) average. Another concept could be an interval from minimum to maximum. The use of these concepts depends on the respective scale. They are illustrated by the following examples.

Example 3 Five rating agencies have the following information wether an enterprise has its own division for risk controlling: no, no, no, no, yes.

- The intersection $I_{ik}^{\cap}$ is the empty set.
- Does it make sense to take the value 'no' because the majority of the observations is 'no'? This is the mode.
- Or is it better to choose 'unknown', because the first four agencies may have the same (erroneous) source for this information? This would be the case of the union $I_{ik}^{*} = I_{ik}^{\cup}$.

The following example demonstrates the case of a rating characteristic, whose realizations can take every value in an interval.

Example 4 Given the income classes from Example 1(b): What is the consensus information set?

- The intersection leads to the interval $I_{ik'}^{\cap} = [1500, 1800] = I_{ik'}^{*}$.
- The union leads to the interval $I_{ik'}^{\cup} = [1000, 2200]$.
- The choice of minimum and maximum of the available values leads to the information set $[1000, 2200]$.
- The average of the lower resp. upper endpoints of the intervals leads to

$$\left[\frac{1000+1500+1100}{3}, \frac{2000+2200+1800}{3} \right] = [1200, 2000].$$

It remains unclear which approach is to be preferred. But an important conclusion here is, that aggregation with contradictory information does not lead to an improved information in every case. Thus, the construction of a rating based on a consensus information in the case of contradictory information does not necessarily lead to a rating that is more precise than before. This is in contrast to [1, p. 77] who state

> that, from a general point of view, any suitably constructed consensus rating is more informative than the use of a single rating source.

3 Rating and Default Probability

Every information set I_{ri} leads to a conditional default probability $\pi(I_{ri}) = P(Y_i = 1 | X_i \in I_{ri}) \in [0, 1]$. A rating agency calculates an estimate $\hat{\pi}_r(I_{ri})$ for this unknown default probability $\pi(I_{ri})$. To get a consensus rating in the sense of estimating the same target, rating agencies have to use the same information set, e.g. the consensus information set. This information set can be reached with the aforementioned ideas. If the rating agencies do not use the same information set, they address different targets and therefore a consensus (rating) is not achievable.

But even if all the rating agencies use the same information set I_i, i.e. $I_{1i} = \cdots = I_{Ri} = I_i$, there are many aspects which lead to different estimates $\hat{\pi}_r(I_i) \neq \hat{\pi}_{r'}(I_i)$:

1. **Statistical model**: There are different models to calculate default probabilities (cf. [2]). A standard model is the logit/probit model. Within this model the variables, i.e. the elements of the information set, usually constitute a linear combination (linear predictor). Instead of a linear combination one can choose this predictor to be non-linear. Additionally, every rating agency uses (slightly) different variables in the models.
2. **Estimation**: There are different estimation methods, e.g. maximum likelihood, least squares, Bayesian methods etc. Another aspect is the estimation error of the parameters. This estimation error has an influence on the estimation error of the

resulting default probability. The use of confidence intervals can be a possibility to consider this problem.

3. **Data set**: The data sets that are used to estimate the model parameters will differ between the rating agencies.

4. **Time horizon**: Every rating refers to a time horizon, e.g. one month, one year etc. Additionally, there are two types of ratings that need to be distinguished, point-in-time and through-the-cycle. In order to compare and/or to aggregate ratings, these must have the same time horizon and must be of the same type.

4 Conclusion

The bottom line is that a real consensus rating is practically impossible, because the rating agencies not only need to have the same information (i.e. the consensus information) but also the same methods to calculate default probabilities. A slightly weaker concept of consensus rating could be the aggregation of different estimates, based on the consensus information. In this case it is ensured that the same default probability is estimated and such a concept is in line with [3]. As already denoted by [1, p. 77] an estimation based on the consensus information is to be preferred. Constructing a consensus information can be quite complicated as shown in Sect. 2. Especially in the case of contradictory information the resulting consensus information set is not necessarily more informative than the information sets of the single rating agencies. But the main problem is, that a rating agency strives to get more precise information than its competitors and especially interchange and recombination of information is not desired.

Still an open question is the aggregation of default probability estimates based on the same as well as on different information sets. One approach to aggregate ratings is suggested by [1], whose method can be applied in both cases. Especially in the case of different information sets the term 'compromise rating' seems to be more appropriate than 'consensus rating'.

References

1. Grün, B., Hofmarcher, P., Hornik, K., Leitner, C., Pichler, S.: Deriving consensus ratings of the big three rating agencies. J. Credit Risk **9**(1), 75–98 (2013)
2. McNeil, A.J., Frey, R., Embrechts, P.: Quantitative Risk Management—Concepts, Techniques, Tools. Princeton University Press, Princeton (2005)
3. Zarnowitz, V., Lambros, L.A.: Consensus and uncertainty in economic prediction. J. Polit. Econ. **95**(3), 591–621 (1987)

Integration of Prospect Theory into the Outranking Approach PROMETHEE

Nils Lerche and Jutta Geldermann

Abstract Outranking methods as a specific application of Multi-Criteria Decision Analysis (MCDA) are applied to structure complex decision problems as well as to elicitate the decision makers preferences. Therefore, a consideration of behavioral effects within outranking-methods seems to be meaningful. Several behavioral effects and biases have been identified in previous studies, however, only few approaches exist to consider such behavioral issues within the application of MCDA-methods explicitly. The prospect theory developed by Kahneman and Tversky (1979) represents one of the most prominent theories from behavioural decision theory. Their findings concerning the decision behaviour of humans, e.g. loss aversion or reference dependency, are broadly supported and confirmed through a variety of empirical research. Hence, the aim of the presented paper is to integrate these elements from prospect theory within the outranking approach PROMETHEE. For that purpose, an additional discrete reference alternative is incorporated. A case study concerning the sustainable usage of biomass for energy conversion illustrates the new developed method.

1 Introduction

An important theory concerning preferences and actual choice behaviour is prospect theory by Kahneman and Tversky [9]. Two elements from prospect theory are reference dependency and loss aversion. Since these elements deal about actual assessment behaviour concerning potential outcomes by humans, they seem to be of particular interest with respect to outranking-methods.

Hence, the aim of the presented paper [11] is to extend *Preference Ranking Organisation Method for Enrichment Evaluations* (PROMETHEE) about elements from

N. Lerche (✉) · J. Geldermann
Chair of Production and Logistics, Georg-August-University Göttingen,
Platz der Göttinger Sieben 3, 37073 Göttingen, Germany
e-mail: nils.lerche@wiwi.uni-goettingen.de; nlerche@gwdg.de

J. Geldermann
e-mail: geldermann@wiwi.uni-goettingen.de

© Springer International Publishing Switzerland 2016   363
M. Lübbecke et al. (eds.), *Operations Research Proceedings 2014*,
Operations Research Proceedings, DOI 10.1007/978-3-319-28697-6_51

prospect theory (further denoted as PT-PROMETHEE) and to validate this extended version within a case study. The paper is structured as follows: In Sect. 2, the theoretical background of prospect theory and the development of PT-PROMETHEE is explained. In Sect. 3, findings from an application within a case study are described. Finally, Sect. 4 summarizes the main results and concludes.

2 The Integration of Prospect into PROMETHEE

Prospect Theory was developed by Kahneman and Tversky [9] in order to explain human decision behaviour under risk. Two important findings from prospect theory are *reference dependency* and *loss aversion*. Reference dependency describes that humans assess a potential outcome relatively to personal reference, which they use as benchmark. Loss aversion follows from reference dependency and means that humans further divide potential outcomes into gains and losses, depending on the reference. Additionally, a potential loss has a larger effect on desirability than a gain of the same amount.

In prospect theory, this assessment behaviour of a decision maker is represented by a particular value function, which assigns a value $v(d)$ to every outcome. The reference is located in the origin and every outcome is assessed by its positive or negative difference d. Loss aversion is integrated through a steeper slope for potential losses, quantified by a loss aversion coefficient λ [12]. Originally, the value-function is S-shaped, but it can be also approximated by a piecewise linear value-function, which is further illustrated in Fig. 1 [9, 10].

PROMETHEE has been developed by Brans et al. [4] and important characteristics of PROMETHEE are that it is based on pairwise comparisons and considers weak preferences as well as incomparabilities. Moreover, several extensions have been developed in recent years, e.g. Fuzzy-PROMETHEE [7]. Two approaches have already been developed concerning the incorporation of prospect theory into PROMETHEE [3, 14]. However, in contrary to these approaches, the definition of a further discrete reference alternative a_r is added within PT-PROMETHEE to consider reference dependency and following steps are adjusted for the incorporation of loss aversion in a different way.

Through the integration of an additional discrete reference alternative a_r, an individual reference point can be defined which can serve as benchmark with regard to the pursued overall goal. Furthermore, it can be determined if the choice of one of the selectable alternatives is generally advantageous. In accordance to Bleichrodt et al. [2], the reference alternative is defined attribute-specific. This means, that a reference value must be determined for each criterion, which gives the decision maker the opportunity to define an individual reference alternative. Thus, the definition of the reference alternative a_r allows to check if a criterion and its measurement are adequate to address the overall goal.

Loss aversion is incorporated into PROMETHEE by adjusting the preference functions. For that purpose, the loss aversion coefficient λ is used. The transfer is

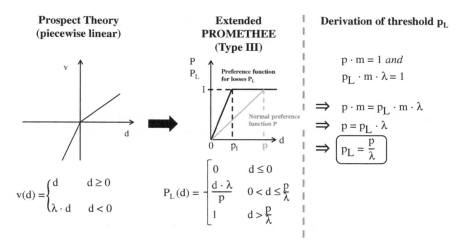

Fig. 1 Transfer of loss aversion into the preference functions of type III from PROMETHEE

based on the piecewise linear value function of prospect theory and the preference function of Type III (V-shape) from PROMETHEE. To be in accordance with the definition of preference functions, the logic of the piecewise linear value function is represented within the first quadrant. Consequently, a preference function for both gains and losses is obtained. Since a potential loss shall lead earlier to strict preference than a potential gain, the preference threshold for losses p_L needs to be smaller in case of loss aversion. With the slope defined by m, loss aversion can be transferred by dividing the normal preference threshold p by λ. The idea behind the transfer of λ into the preference function of type III is illustrated in Fig. 1. By applying the same procedure of dividing the thresholds p, q and σ by λ, also the other five generalised criteria defined by Brans et al. [4] can be adjusted.

To quantify λ a new approach of using linguistic expressions concerning loss aversion for a subsequent transfer on a quantitative scale was developed. For determination of the underlying quantitative scale findings from Heath et al. [8] as well as Abdellaoui et al. [1] are used. Based on their findings, a range from 1 to 4 was applied. The mean-value of 2.5 for loss aversion is chosen in orientation on the values of 2.25 identified by Tversky and Kahneman [13]. The case of no loss aversion is represented by $\lambda = 1$, since p_L is defined by p/λ and consequently, both normal preference functions and loss functions would be identically in this case. Additionally, also a value of 0.5 is included to offer also the expression of a contrary effect (risk seeking) to loss aversion.

Afterwards, the original procedure for calculation of outranking relations and flows, which has been developed by Brans et al. [4], is adjusted. Regarding any criterion i ($i = 1, \ldots, k$), a potential loss would occur, if the reference alternative a_r has a superior value compared to any other alternative. Thus, all pairwise comparisons of a_r with respect to any alternative a which result in a positive preference value represent potential losses and consequently, the preference function for losses $P_{Li}(d)$ must be applied for these comparisons:

$$P_{Li}(d) = P_{Li}(g_i(a_r) - g_i(a)) \tag{1}$$

Vice versa, a potential gain occurs if a selectable alternative a has a superior value compared to the reference alternative a_r. In this case, the normal preference function is applied, identically as it is for pairwise comparisons between selectable alternatives:

$$P_i(d) = P_i(g_i(a) - g_i(a_r)) \tag{2}$$

$$P_i(d) = P_i(g_i(a) - g_i(a')) \tag{3}$$

Based on this procedure, the calculation of outranking relations π is adjusted:

$$\pi(a_r, a) = \sum_{i=1}^{k} P_{Li}(g_i(a_r) - g_i(a)) \cdot w_i \tag{4}$$

$$\pi(a, a_r) = \sum_{i=1}^{k} P_i(g_i(a) - g_i(a_r)) \cdot w_i \tag{5}$$

$$\pi(a, a') = \sum_{i=1}^{k} P_i(g_i(a) - g_i(a')) \cdot w_i \tag{6}$$

Afterwards, the calculation of positive outranking flows Φ^+ and negative outranking flows Φ^- is adjusted:

$$\Phi^+(a_r) = \frac{1}{n-1} \sum_{a \in A} \pi(a_r, a) \tag{7}$$

$$\Phi^-(a_r) = \frac{1}{n-1} \sum_{a \in A} \pi(a, a_r) \tag{8}$$

$$\Phi^+(a) = \frac{1}{n-1} (\pi(a, a_r) \sum_{a' \in A} \pi(a, a')) \tag{9}$$

$$\Phi^-(a) = \frac{1}{n-1} (\pi(a_r, a) \sum_{a' \in A} \pi(a', a)) \tag{10}$$

The net flows Φ^{net} are determined as in original PROMETHEE:

$$\Phi^{net} = \Phi^+ - \Phi^- \tag{11}$$

Finally, both the partial and complete ranking can be determined [4]. However, in addition to common rankings as in original PROMETHEE, the reference alternative a_r is included yet, indicating if a selectable alternatives is generally advantageous.

3   Feedback from an Application of PT-PROMETHEE Within a Case Study

To gain feedback, PT-PROMETHEE was applied subsequently on an existing decision table [5, 6] concerning a sustainable concept for bioenergy. The alternatives are a small-scale biogas plant, a large-scale biogas plant as well as a bioenergy-village-concept and 39 criteria from environmental, economic, social and technical dimensions are considered. For determination of required information within PT-PROMETHEE, e.g. reference values or level of loss aversion, interviews with three experts, who also defined criteria and values within the original determination of the used decision table, were conducted.

Concerning reference dependency, the incorporation of an individual reference alternative was appreciated by the experts and they were very interested in defining reference values adequately. It was further mentioned that thinking about an adequate reference helps not only to focus on the underlying overall goal (sustainability), but also to check whether each criterion and its corresponding measurement unit are in accordance with the overall goal. However, determining reference values for qualitative criteria was very challenging.

With respect to the consideration of loss aversion, all experts were able for every criterion to declare whether loss aversion exists. In fact, a λ-value different to one occurs for most criteria. It is also remarkable, that the contrary effect to loss aversion has been applied. But, all experts mentioned that the expression of loss aversion is cognitively more challenging compared to defining the reference. Concerning the approach for quantification of λ, the experts were able to express their loss aversion on the linguistic scale. One expert also mentioned that the range of the scale is seen as adequate.

4   Conclusion

In conclusion, PT-PROMETHEE allows an incorporation of loss aversion from original prospect theory and gives the decision maker an opportunity to express it. Further it facilitates thinking about pursued goals and helps to gather additional information with respect to actual advantageousness of alternative through integration of an additional discrete reference alternative.

For further research, it would be meaningful to consider uncertainty, since prospect theory was originally developed concerning uncertain problems. For that purpose, an extension about the use of scenarios is currently developed and it is planned to test those approach within further workshops.

Acknowledgments This research is funded by a grant of the Ministry of Science and Culture of Lower Saxony and Volkswagenstiftung. It is part of the joint research project *Sustainable use of bioenergy: bridging climate protection, nature conservation and society* with a duration from 2009 till 2014.

References

1. Abdellaoui, M., Bleichrodt, H., Paraschiv, C.: Loss aversion under prospect theory: a parameter-free measurement. Manage. Sci. **53**, 1659–1674 (2007)
2. Bleichrodt, H., Schmidt, U., Zank, H.: Additive utility in prospect theory. Manage. Sci. **55**, 863–873 (2009)
3. Bozkurt, A.: Multi-criteria decision making with interdependent criteria using prospect theory. Masterthesis, Middle East Technical University, Ankara (2007). http://etd.lib.metu.edu.tr/upload/12608408/index.pdf
4. Brans, J.P., Vincke, Ph., Mareschal, B.: How to select and how to rank projects: the PROMETHEE method. Eur. J. Oper. Res. **24**, 228–238 (1986)
5. Eigner-Thiel, S., Schmehl, M., Ibendorf, J., Geldermann, J.: Assessment of different bioenergy concepts in terms of sustainable development. In: Ruppert, H., et al. (eds.) Sustainable Bioenergy Production—An Integrated Approach, pp. 339–384. Springer, Berlin (2013)
6. Geldermann, J., Eigner-Thiel, S., Schmehl, M., Ibendorf, J.: Multikriterielle Entscheidungsunterstützung bei der Auswahl von Biomassenutzungskonzepten. Presented at the conference Chancen und Risiken der Bioenergie im Kontext einer nachhaltigen Entwicklung. 24–25 January 2012, Göttingen, Germany
7. Geldermann, J., Rentz, O.: Multi-criteria analysis for the assessment of environmentally relevant installations. J. Ind. Ecol. **9**, 127–142 (2005)
8. Heath, C., Larrick, R.P., Wu, G.: Goals as reference points. Cogn. Psychol. **38**, 79–109 (1999)
9. Kahneman, D., Tversky, A.: Prospect theory: an analysis of decision under risk. Econometrica **47**, 263–292 (1979)
10. Korhonen, P., Moskowitz, H., Wallenius, J.: Choice behavior in interactive multiple-criteria decision making. Ann. Oper. Res. **23**, 161–179 (1990)
11. Lerche, N., Geldermann, J.: Integration of prospect theory into PROMETHEE—a case study concerning sustainable bioenergy concepts. Int. J. Multicriteria Decis. Making—Special Issue on New Developments in PROMETHEE Methods (under review)
12. Tversky, A., Kahneman, D.: Loss aversion in riskless choice: a reference-dependent model. Q. J. Econ. **106**, 1039–1061 (1991)
13. Tversky, A., Kahneman, D.: Advances in prospect theory: cumulative representation of uncertainty. J. Risk Uncertainty **5**, 297–323 (1992)
14. Wang, J.-Q., Sun, T.: Fuzzy multiple criteria decision making method based on prospect theory. Paper presented at the International Conference on Information Management, Innovation Management and Industrial Engineering, pp. 288–291. IEEE, Xian (2008)

Multistage Optimization with the Help of Quantified Linear Programming

T. Ederer, U. Lorenz, T. Opfer and J. Wolf

Abstract Quantified linear integer programs (QIPs) are linear integer programs (IPs) with variables being either existentially or universally quantified. They can be interpreted as two-person zero-sum games between an existential and a universal player on the one side, or multistage optimization problems under uncertainty on the other side. Solutions of feasible QIPs are so called winning strategies for the existential player that specify how to react on moves—certain fixations of universally quantified variables—of the universal player to certainly win the game. In order to solve the QIP optimization problem, where the task is to find an especially attractive winning strategy, we examine the problem's hybrid nature and combine linear programming techniques with solution techniques from game-tree search.

1 Introduction

Mixed-integer linear programming (MIP) [1] is the state-of-the art technique for computer aided optimization of real world problems. Nowadays, we are able to solve large MIPs of practical size, but companies observe an increasing danger of disruptions, which prevent them from acting as planned. One reason is that input data for a given problem is often assumed to be deterministic and exactly known when decisions have to be made, but in reality they are often afflicted with some kinds of uncertainties. Thus, there is a need for planning and deciding under uncertainty. Prominent solution paradigms for optimization under uncertainty are [2–6].

T. Ederer · U. Lorenz · T. Opfer
Fluid Systems, Technische Universität Darmstadt, Darmstadt, Germany
e-mail: thorsten.ederer@fst.tu-darmstadt.de

T. Ederer · U. Lorenz (✉) · T. Opfer · J. Wolf
Chair of Technology Management, University Siegen, Siegen, Germany
e-mail: ulf.lorenz@uni-siegen.de; lorenz@mathematik.tu-darmstadt.de

J. Wolf
e-mail: jan.wolf@fst.tu-darmstadt.de

© Springer International Publishing Switzerland 2016
M. Lübbecke et al. (eds.), *Operations Research Proceedings 2014*,
Operations Research Proceedings, DOI 10.1007/978-3-319-28697-6_52

Relatively unexplored are the abilities of linear programming extensions, as Quantified Linear (Integer) Programming (QIP) [7–9]. For ease of computation, we only consider 0/1 QIPs in the remainder of this paper.

2   Quantified Linear (Integer) Programming

We start with the major definitions for Quantified Linear (Integer) Programs (QLP/QIP), beginning with the continuous problem.

Definition 1 (*Quantified Linear Program*) Let there be a vector of n variables $x = (x_1, \ldots, x_n)^T \in \mathbb{Q}^n$, lower and upper bounds $l \in \mathbb{Z}^n$ and $u \in \mathbb{Z}^n$ with $l_i \leq x_i \leq u_i$, a coefficient matrix $A \in \mathbb{Q}^{m \times n}$, a right-hand side vector $b \in \mathbb{Q}^m$ and a vector of quantifiers $\mathcal{Q} = (\mathcal{Q}_1, \ldots, \mathcal{Q}_n)^T \in \{\forall, \exists\}^n$, let the term $\mathcal{Q} \circ x \in [l, u]$ with the component wise binding operator $\circ$ denote the *quantification vector* $(\mathcal{Q}_1 x_1 \in [l_1, u_1], \ldots, \mathcal{Q}_n x_n \in [l_n, u_n])^T$ such that every quantifier $\mathcal{Q}_i$ binds the variable x_i ranging over the interval $[l_i, u_i]$. We call $(\mathcal{Q}, l, u, A, b)$ with

$$\mathcal{Q} \circ x \in [l, u] : Ax \leq b \tag{QLP}$$

a *quantified linear program (QLP)*.

If the variables are forced to integer values, it is called an Quantified Interger Program (QIP). A QLP/QIP instance is interpreted as a two-person zero-sum game between an *existential player* setting the $\exists$-variables and a *universal player* setting the $\forall$-variables. Each fixed vector $x \in [l, u]$, that is, when the existential player has fixed the existential variables and the universal player has fixed the universal variables, is called *a game*. If x satisfies the linear program $Ax \leq b$, we say *the existential player wins*, otherwise *he loses* and *the universal player wins*. The variables are set in consecutive order according to the variable sequence. Consequently, we say that a player makes the move $x_k = z$, if he fixes the variable x_k to the value z. At each such move, the corresponding player knows the settings of $x_1, \ldots, x_{k-1}$ before taking his decision x_k. The most relevant term in order to describe solutions are so called strategies. Let $x \in \{0, 1\}^n$ for the rest of this paper.

Definition 2 (*Strategy*) A *strategy* $S = (V, E, c)$ is an edge-labeled finite arborescence with a set of nodes $V = V_\exists \,\dot{\cup}\, V_\forall$, a set of edges E and a vector of edge labels $c \in \mathbb{Q}^{|E|}$. Each level of the tree consists either of only nodes from $V_\exists$ or only of nodes from $V_\forall$, with the root node at level 0 being from $V_\exists$. The i-th variable of the QLP is represented by the inner nodes at depth $i - 1$. Each edge connects a node in some level i to a node in level $i + 1$. Outgoing edges represent moves of the player at the current node, the corresponding edge labels encode the variable allocations of the move. Each node $v_\exists \in V_\exists$ has exactly one child, and each node $v_\forall \in V_\forall$ has as two children, with the edge labels being the corresponding lower and upper bounds.

A path from the root to a leaf represents a game of the QLP and the sequence of edge labels encodes its moves. A strategy is called a *winning strategy* if all paths from the root node to a leaf represent a vector x such that $Ax \leq b$ [10]. If there is more than one winning strategy for the existential player, it can be reasonable to search for a certain (the 'best') one. We therefore modify the problem to include a linear objective function as shown in the following (where we note that transposes are suppressed when they are clear from the context to avoid excessive notation).

Definition 3 (*QLPs/QIPs with Objective Function*) Let $\mathcal{Q} \circ x \in [l, u] : Ax \leq b$ be given as in Definition 1 with the variable blocks being denoted by B_i. Let there also be a vector of objective coefficients $c \in \mathbb{Q}^n$. We call

$$z = \min_{B_1}(c^1 x^1 + \max_{B_2}(c^2 x^2 + \min_{B_3}(c^3 x^3 + \max_{B_4}(\ldots \min_{B_m} c^m x^m))))$$
$$\mathcal{Q} \circ x \in [l, u] : Ax \leq b$$

a QLP/QIP with *objective function* (for a minimizing existential player).

Note that the variable vectors $x^1, \ldots, x^i$ are fixed when a player minimizes or maximizes over variable block B_{i+1}. Consequently, we deal with a dynamic multi-stage decision process, similar as it is also known from multistage stochastic programming [2].

We investigate on how far linear programming techniques can be used in order to solve QIPs. As we will see, the simplex algorithm plays a key role similar as it does for conventional integer programs. Yasol, the ideas of which are presented in this paper, proceeds in two phases in order to find optimal solutions of 0/1-QIP instances.

- Phase 1: Determine the instance's feasibility, i.e. whether the instance has any solution at all. If it has any solution, present it. During this phase, Yasol acts like a QBF solver [11] with some extra abilities.
- Phase 2: Go through the solution space and find the provable optimal solution. The Alphabeta algorithm [12], which up to now has mainly been used in search trees of games like chess etc., performs this task.
  The alphabeta algorithm walks through the search space recursively and fixes variables to 0 or 1 when going into depth or relaxes the fixations again when returning from a subtree. Relaxing integrality as well as the universal-property results in an ordinary LP which often gives the opportunity to cut off parts of the search tree with the help of dual information, i.e. dual bounds or cutting planes for the original program. Algorithm 1 shows a basic alphabeta algorithm with the ability of non-chronologic back-jumping with the help of dual information, i.e. by solving an LP-relaxation, cf. line 2 and 5–8. The idea is quite old and goes back to Benders and has been described already in the seventies [13]. In [14] the technique has been combined with an implication graphs.

Algorithm 1: A basic alphabeta(int d, int a, int b) routine, sketched

---

1 compute LP-relaxation, solve it, extract branching variable or cut;
2 **if** *integer solution found* **then return** objective value ; // leaf reached
3 **if** x_i is an existential variable **then** score := $-\infty$; **else** score := $+\infty$;
4 **for** *val_ix from 0 to 1* **do** // search actually begins ...
5   **if** *level_finished(t)* **then** // leave marked recursion levels
6   **if** *x_i is an extistential variable* **then return** score ;
7   **else return** $-\infty$;
    end
8   assign(x_i, val[val_ix], ...);
9   v := alphabeta(d-1, fmax(a, score), b);
10   unassign(x_i);
11   **if** *x_i is an existential variable* **then**
12   **if** *v > score* **then** score := v; // existential player maximizes
13   **if** *score $\geq$ b* **then return** score ;
    else
14   ...anallogously ...;
    end
end

Algorithm 2: A local repetition loop extends the basic algorithm with conflict analysis and learning; replaces line 9 in Algorithm 1

---

1 **repeat**
2   **if** the current level is marked as finished **then** leave the recursion level;
3   **if** *propagate(confl, confl_var, confl_partner, false)* // unit prop. [11] **then**
4   **if** *x_i is an existential variable* **then**
5   v = alphabeta(t+1, lsd-1, fmax(a, score), b);
6   **if** v > score **then** score := v;
7   **if** *score $\geq$ b* **then** break;
        else
            ...analogously ...;
        end
    else
8   add reason, i.e. a constraint, to the database ; // [11]
9   returnUntil(out_target_dec_level) ; // set level_finished(...)
10   **if** x_i is an existential variable **then return** score; **else return** $-\infty$;
    end
until *there is no backward implied variable*;

The new cut possibly indicates that the search process can be directly continued several recursion levels above the current level. In this case, the superfluous levels are marked as finished—in Algorithm 1 that is indicated by level_finished(t)—and the alphabeta procedure breaks out of all loops and walks up the search tree with the help of the lines 5–7. For other design decisions like branching variable selection etc., we refer to [14]. Non-chronologic backtracking is realized with the help of Algorithm 2, which replaces line 9 of Algorithm 1. In fact, there is a loop around the alphabeta procedure call which is walked through as long as backward implied variables occur deeper in the search tree. The procedure propagate(.) performs the implication (bound propagation) of variables.

3 Computational Results

We compared the abilities of Yasol with the QBF-solver Depqbf and with Cplex and Scip utilizing DEPs (cf. [9]). The latter are then called DEPCplex and DEPScip. Depending on the type of test instances, i.e. QFB or 0/1-QIP with objective, we competed either with Cplex solving a DEP and with Depqbf (one of the leading QBF-solvers). Or, in the context of 0/1-QIPs, we passed on the QBF-solver and took into additional consideration CBC and Scip, which are the best open-source MIP solvers at the current time. The experiments on QBF instances were done on PCs with Intel i7 (3.6 GHz) processors and 64 GB RAM. Those on IP and QIP instances, where memory was no limiting factor, were executed on Intel i7 (1.9 GHz) with 8 GB RAM. All computations were limited to one thread, all other parameters were left at default.

Currently, our solver Yasol can solve QSAT instances reasonable well, however not as good as the best dedicated QBF-solvers. Cplex (V 12.5.0.1) is one of the leading MIP solvers and DEPCplex operates on the deterministic equivalents of the QBF instances which have been translated from boolean formulae format to MIP format. On a test collection of 797 instances, taken from the qbflib.org, Depqbf solves 674 instances, Yasol 584, but the DEP approach collapses. DEPCplex can solve only 478. Each program got a maximum of 10 min solution time for each instance. The solution time consists of the sum of all individual solution times for a specific program. Not solving the instance was punished with 600 s.

Table 1 shows the number of solved instances grouped by the number of universal variables. We see that the DEP approach becomes more and more unalluring, the more universal variables come into play. Experimental results draw a similar picture when we examine 0/1-QIPs with objective. In order to examine the effects, we generated

Table 1 Computational results

# Univ. var.			Depqbf	DEPCplex	Yasol
1–5 UV	Time (s)		39185	41133	73732
	# Solved		312/373	315/373	259/373
6–10 UV	Time (s)		1805	1351	2146
	# Solved		38/41	39/41	38/41
11–15 UV	Time (s)		7961	25297	11023
	# Solved		84/97	65/97	79/97
16–20 UV	Time (s)		2609	84685	5929
	# Solved		166/170	37/170	167/170
21+ UV	Time (s)		16856	69600[a]	39620
	# Solved		96/116	0/116	60/116
Σ	Time (s)		68416	194128	132350
	# Solved		696/797	456/797	603/797

[a]The small number is due to aborting, Cplex could not solve the instances with 64 GB memory

Table 2 0/1-IPs, taken from miplib.zib.de and from there artificially constructed 0/1-QIPs

No. UV	DEPCplex	Yasol	DEPScip	DEPCbc
0	59/59	24/59	48/59	34/59
(max 1 h)	19520 s	132129 s	55062 s	112958 s
1–4	138/138	103/138	137/138	138/138
(max 900 s)	42 s	39984 s	1072 s	354 s
10–14	46/46	46/46	34/46	23/46
(max 1200 s)	1854 s	972 s	23440 s	38867 s

artificial 0/1-QIP instances from some of the IP-instances. Compared with the MIP solvers Cplex, Scip, Yasol performs poor on pure IPs, as can be taken from Table 2, line 1. However, on the instance set with at least 10 universal variables (UV), the picture changes again, and Yasol takes the leadership (cf. line 3 of Table 2: '10 to 14'). Yasol seems to have by far less difficulties with the universal variables, however, has improvement potential for existential variables.

4 Conclusion

In this paper, we investigated for the domain of 0/1-QIPs with objective, in how far a direct search approach can compete with building a DEP and solving it with state-of-the-art MIP solvers. On some of instances, the prototypical implementation in the solver Yasol could solve 0/1-QIPs up to optimality, being even faster than the best commercial solvers on the corresponding DEPs. Next, modern cutting planes like implied-bounds, GMI, knapsack cover, . . . have to be integrated.

Acknowledgments Research partially supported by German Research Foundation (DFG), Lo 1396/2-1 and SFB 805.

References

1. Schrijver, A.: Theory of linear and integer programming. Wiley, New York (1986)
2. Birge, J.R., Louveaux, F.: Introduction to Stochastic Programming. Springer Series in Operations Research and Financial Engineering. Springer, New York (1997)
3. Ben-Tal, A., Ghaoui, L.E., Nemirovski, A.: Robust Optimization. Princeton University Press, Princeton (2009)
4. Liebchen, C., Lübbecke, M., Möhring, R., Stiller, S.: The concept of recoverable robustness, linear programming recovery, and railway applications. In: Robust and Online Large-Scale Optimization, pp. 1–27 (2009)
5. Bellmann, R.: Dynamic programming. Princeton University Press, Princeton (1957)
6. Kleywegt, A., Shapiro, A., Homem-De-Mello, T.: The sample average approximation method for stochastic discrete optimization. SIAM J. Opt. 479–502 (2001)

7. Subramani, K.: On a decision procedure for quantified linear programs. Ann. Math. Artif. Intell. **51**(1), 55–77 (2007)
8. Subramani, K.: Analyzing selected quantified integer programs. Springer LNAI **3097**, 342–356 (2004)
9. Ederer, T., Lorenz, U., Martin, A., Wolf, J.: Quantified linear programs: a computational study. In: Part I. ESA'11, pp. 203–214. Springer (2011)
10. Lorenz, U., Martin, A., Wolf, J.: Polyhedral and algorithmic properties of quantified linear programs. In: Annual European Symposium on Algorithms, pp. 512–523 (2010)
11. Zhang, L.: Searching for truth: techniques for satisfiability of boolean formulas. Ph.D. thesis, Princeton, NJ, USA (2003)
12. Feldmann, R.: Game tree search on massively parallel systems. Ph.D. thesis, University of Paderborn (1993)
13. Johnson, E., Suhl, U.: Experiments in integer programming. Discrete Appl. Math. **2**(1), 39–55 (1980)
14. Achterberg, T.: Constraint integer programming. Ph.D. thesis, Berlin (2007)

Analysis of Ambulance Location Models Using Discrete Event Simulation

Pascal Lutter, Dirk Degel, Lara Wiesche and Brigitte Werners

Abstract The quality of a rescue service system is typically evaluated ex post by the proportion of emergencies reached within a predefined response time threshold. Optimization models in literature consider different variants of demand area coverage or busy fractions and reliability levels as a proxy for Emergency Medical Service quality. But no comparisons of the mentioned models with respect to their real-world performance are found in literature. In this paper, the influence of these different model formulations on real-world outcome measures is analyzed by means of a detailed discrete event simulation study.

1 Introduction

Rescue and Emergency Medical Services (EMS) are an important part of public health care. The quality of a rescue service system is typically evaluated ex post by the proportion of emergencies reached within a predefined response time threshold. Coverage is one of the most accepted a priori quality criteria in EMS literature [1]. Since 1971 [9], different covering models and various extensions of these models are used to support ambulance location planning. The main challenge in ambulance location planning is to provide an adequate service level with respect to accessibility of an emergency within a predefined response time threshold and availability of ambulances [4]. Optimization models in literature consider different variants of demand area coverage, such as single coverage [2], double coverage [6] and empirically required coverage [5]. Other models use busy fractions [3] and reliability levels [7] as a proxy criterion for EMS quality. All those models support the decision maker on the strategic and tactical level of ambulance location planning, but differ regarding the specification of the objective functions as well as concerning input parameters and model assumptions. To the best of our knowledge, no systematic comparisons

P. Lutter (✉) · D. Degel · L. Wiesche · B. Werners
Faculty of Management and Economics, Chair of Operations
Research and Accounting,
Ruhr University Bochum, Bochum, Germany
e-mail: pascal.lutter@rub.de

© Springer International Publishing Switzerland 2016
M. Lübbecke et al. (eds.), *Operations Research Proceedings 2014*,
Operations Research Proceedings, DOI 10.1007/978-3-319-28697-6_53

of different ambulance location models exist in literature. The aim of this paper is to provide a comparison of the mentioned models concerning their suitability for decision support in strategic and tactical ambulance location planning. A discrete event simulation is used to systematically evaluate the resulting solutions of each covering concept. It is analyzed, which of those covering concepts provides the best proxy criterion for the real world performance measure. The remainder of the paper is structured as follows: First a brief overview of the selected ambulance location models is given. Technical details of the discrete event simulation are described and results of a real world case study are presented afterwards.

2 Ambulance Location Models

In this paper, daytime-dependent extensions of five well known models for ambulance location are considered: The (1) *Maximal Covering Location Problem* (MCLP) [2], the (2) *Double Standard Model* (DSM) [6], the (3) *Maximum Expected Covering Location Problem* (MEXCLP) [3], the (4) *Maximum Availability Location Problem* (MALP I/II) [7], and the (5) *Empirically Required Coverage Problem* (ERCP) [5]. To compare these models, a consistent constraint set is used and model assumptions are briefly summarized. For additional descriptions of these models see e.g. [1]. The aim of these models is to maximize the total demand served within a legal response time threshold of r minutes, given a limited number of p_t ambulances in period t. i indicates the planning squares or *demand nodes* ($i \in \mathscr{I}$), while d_{it} denotes the demand of node i in period $t \in \mathscr{T}$. To be able to serve an emergency at demand node i, at least one ambulance has to be available within the response time threshold r, e.g. positioned at node $j \in \mathscr{N}_{it} := \{j \in \mathscr{J} \mid \mathrm{dist}_{ijt} \leq r\}$, where dist_{ijt} describes the response time between node i and node j in period t. The integer decision variable $y_{jt} \in \mathbb{N}_0$ indicates the number of ambulances positioned at node j in period t, and the binary decision variable x_{it}^k is equal to 1 if demand node i is covered k times in period t. With the preceding notation, generic *covering constraints* are given by

$$\sum_{j \in \mathscr{N}_{it}} y_{jt} \geq \sum_{k=1}^{p_t} x_{it}^k \quad \forall i \in \mathscr{I}, \forall t \in \mathscr{T}. \tag{1}$$

Setting the right hand side of constraints (1) equal to 1 ensures that each demand node i can be reached within the response time threshold at least once if an ambulance is available. This single coverage may become inadequate when several emergencies occur at the same time and the assigned ambulances become busy. To hedge against parallel operations resulting in unavailability of ambulances, the mentioned models use different concepts and objective functions. All models ensure a sufficient number of ambulances located in $\mathscr{N}_{it}$ to serve each demand node i. Table 1 compares the covering constraints, the objective functions, and the assumptions of a priori information of the models. In addition to covering constraints, further constraints are used to ensure correct relocations and to restrict the number of

Table 1 Comparison of model constraints, objectives, and assumptions about required information

Model	Covering constraint	Objective	Required information
MCLP	$\sum_{j \in \mathcal{N}_{it}} y_{jt} \geq x_{it}^1$	$\max \sum_{i \in \mathcal{I}} \sum_{t \in \mathcal{T}} d_{it} x_{it}^1$	d_{it}
DSM	$\sum_{j \in \mathcal{N}_{it}} y_{jt} \geq x_{it}^1 + x_{it}^2$	$\max \sum_{i \in \mathcal{I}} \sum_{t \in \mathcal{T}} d_{it} x_{it}^2$	d_{it}
MEXCLP	$\sum_{j \in \mathcal{N}_{it}} y_{jt} \geq \sum_{k=1}^{p_t} x_{it}^k$	$\max \sum_{i \in \mathcal{I}} \sum_{t \in \mathcal{T}} \sum_{k=1}^{p_t} d_{it}(1 - q_t) q_t^{k-1} x_{it}^k$	d_{it}, q_t
MALP I	$\sum_{j \in \mathcal{N}_{it}} y_{jt} \geq \sum_{k=1}^{p_t} x_{it}^k$	$\max \sum_{i \in \mathcal{I}} \sum_{t \in \mathcal{T}} d_{it} x_{it}^{K_t}$	d_{it}, α, q_t, K_t
ERCP	$\sum_{j \in \mathcal{N}_{it}} y_{jt} \geq \sum_{k=1}^{p_t} x_{it}^k$	$\max \sum_{i \in \mathcal{I}} \sum_{t \in \mathcal{T}} d_{it} x_{it}^{K_{\ell_i t}}$	$d_{it}, K_{\ell_i t}$

$K_t := \lceil \ln(1 - \alpha)/\ln(q_t) \rceil$, $K_{\ell_i t}$ empirically required degree of coverage (see explanation below)

ambulances in use (see e.g. [5]). In the MCLP and the DSM a uniform single, respectively double coverage is maximized. Few information is needed, but the unavailability of ambulances due to parallel operations is ignored in the MCLP or simplified in the DSM by using a time and spatial fixed backup (double) coverage. In the MEXCLP it is assumed that each ambulance has the probability q, called *busy fraction*, of being unavailable and the expected covered demand is maximized. In the earliest version of the MEXCLP [3], it is assumed that each ambulance has the same probability of being busy. In order to compare the models on the basis of the same criterion, a time-dependent busy fraction q_t is used in the following analysis. In MALP I and MALP II [7], a chance constrained program is used to maximize the demand covered at least with a given probability α. The minimum number of ambulances required to serve demand node i with a reliability level of α in period t is determined by

$$1 - q_t^{\sum_{j \in \mathcal{N}_{it}} y_{jt}} \geq \alpha, \tag{2}$$

which can be linearized as $\sum_{j \in \mathcal{N}_{it}} y_{jt} \geq \lceil \ln(1 - \alpha)/\ln(q_t) \rceil =: K_t$ (with system unique busy fraction q_t in MALP I). In MALP II, the assumption of identical busy fractions is relaxed and the busy fractions q_{it} are calculated for each demand node i and period t. The problem of using demand node specific busy fractions q_{it} is that these values depend on the output of the model and are unknown a priori [1]. To overcome this difficulty, a more direct and data-driven way to determine the required coverage is used in the ERCP [5]. The empirical distribution function representing the number of parallel EMS operations per time unit and district is calculated. The 95 % quantile of the stochastic demand per district l and time period t is determined empirically in order to derive the required degree of coverage $K_{\ell_i t}$ and thus the necessary number of ambulances. This assures that there is a sufficient number of ambulances to cover all parallel operations in at least 95 % of all cases. To compare the mentioned models, all relevant model parameters, like busy fractions, reliability levels and the empirically required coverage levels are calculated using an identical data base which relies on the same spatial and time-dependent partitioning.

3   Discrete Event Simulation of EMS Systems

All previously described models consider different variants of demand coverage as a proxy criterion for EMS quality, defined as the proportion of calls served within the legal response time threshold. This real world outcome measure is mainly influenced by the positioning of ambulances implied by different objective functions and covering constraints. The solution quality, e.g. the quality of an EMS system can only be evaluated ex post. Discrete event simulation represents a common approach to analyze complex and dynamically changing environments like EMS systems. In this paper, a simulation approach is applied to compare the performance of the aforementioned models regarding the real world outcome measure. In the following, the main components of the simulation are described. The data generation process for the discrete event simulation consists of two main modules:

1. Generation of random events: A whole weekday is subdivided into 24 time intervals $t \in \{0, \ldots, 23\}$ with a length of $\Delta = 1$ h. For a given demand node i and a time interval $[t, t + \Delta)$, in the following indicated by t, the number of emergencies occurring within t can be approximated by a Poisson distribution P_λ with parameter λ. The average number of emergency calls per time interval t at a given weekday D is $P_{\lambda_{it}^D}$ with $\lambda_{it}^D := (\alpha_D/365) \cdot \sum_{\ell=1}^{365} d_{it}^\ell$. The parameter λ_{it}^D is used as an estimator for the parameter of the Poisson distribution, where d_{it}^ℓ denotes the historical number of emergencies occurring in period t in demand node i at day ℓ. The scaling factor α_D is determined empirically and serves as a correction term for introducing day-related seasonality. This is necessary since the total demand fluctuates within different weekdays. For each t and i, the quantity of emergencies d_{it} is sampled from previously specified Poisson distributions. Then, the emergencies are distributed according to the realization of a uniform random variable within the time interval t.
2. Travel time generation: The travel time is not constant for different time intervals t of the day, cf. [8]. Typically, higher traveling speeds are achieved in the evening, while lower speeds are observed around noon and during rush hours. To incorporate realistic driving speeds, a time-dependent random variable $v_t \sim N(\mu_t, \sigma_t)$ is used, where μ_t and σ_t are determined empirically. For each generated emergency, the travel time is sampled from $N(\mu_t, \sigma_t)$ and stored in the corresponding variable.

During the simulation, an emergency event is characterized by the time of occurrence, the associated demand node, the traveling speed of the associated ambulance and the emergency duration. The duration of each operation is sampled from the empirical distribution function. The simulation process works as follows: An ambulance is characterized by the assigned EMS station and an availability indicator. An ambulance is available, if it is currently not serving an emergency. For each emergency occurring, the selection of ambulances is performed by a predefined nearest-distance strategy: For a given emergency position i, all ambulance locations are sorted by increasing distances to i. Note, that the traveling distance depends on the location of ambulances which are an outcome of the tested optimization model. If there is an ambulance available at the nearest station, this vehicle is selected. Otherwise, the next station in the list is checked. The process repeats until an available ambulance has been found or all stations are checked. If no ambulance is available, a queue of unfulfilled requests is being built. Whenever an ambulance is assigned to serve an emergency, the travel time is generated and the vehicle is blocked for the duration time of the operation. The simulation process terminates after serving all emergencies. Finally, dividing the overall number of emergencies served on time by all emergencies occurring gives the desired real world quality measure.

4   Case Study and Results

A real world case study for evaluating model performances is conducted by specifying all required model parameters (demand, busy fractions and empirically required coverage) on the basis of a data set from a German city containing more than 20,000 operations per year. In all models the number of ambulances in time period t is given by the parameters p_t and all demand points are considered as potential ambulance locations. The average emergency demand over 1 year is visualized in the first picture of Fig. 1. The first objective is to maximize the model specific objective function. The second objective is to cover a maximal number of demand areas at least once. The third objective aims at minimizing the number of vehicle locations. To hedge against dual degeneracy in location models, a lexicographic approach is applied. The coverage induced by the solutions of the models are visualized in Fig. 1. Demand nodes are colored from light to dark gray and visualize the number of zero (light gray)

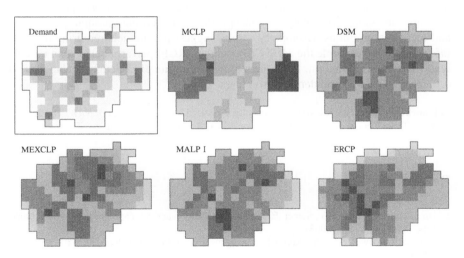

Fig. 1 Emergency demand and degree of coverage induced by the solutions of different models

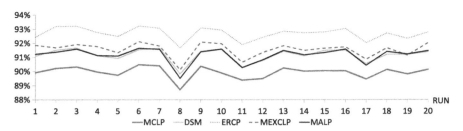

Fig. 2 Proportion of calls served within the time threshold of 8 min during 1 year (20 simulation runs)

to sixfold (dark gray) coverage. Results, i.e., the proportion of calls served within the legal response time threshold of 8 min for 20 simulation runs for each model solution are shown in Fig. 2. Data driven covering models like MEXCLP, MALP and ERCP outperform fixed covering models (MCLP, DSM) with respect to the real world EMS performance measure. Fixed covering models provide inadequate coverage of areas with high, resp. low, number of parallel operations due to disregarding the availability of ambulances. Instead, data driven approaches locate ambulances as needed by considering demand volume as well as criteria for ambulance unavailability.

5   Conclusion and Outlook

In this paper a discrete event simulation study is conducted to evaluate different ambulance location models. Based on the simulation study exemplary results of different coverage concepts concerning their influence on real world performance measures are shown. All analyzed concepts differ concerning the input parameters and model assumptions. Exemplary results suggest that models requiring detailed information (for example the MEXCLP and the ERCP) perform better than models ignoring these information. In the next step, studies are extended systematically to different typical city and demand structures.

Acknowledgments This research is financially supported by *Stiftung Zukunft NRW*. The authors thank staff members of *Feuerwehr und Rettungsdienst Bochum* for providing detailed insights.

References

1. Brotcorne, L., Laporte, G., Semet, F.: Ambulance location and relocation models. Eur. J. Oper. Res. **147**(3), 451–463 (2003)
2. Church, R., ReVelle, C.: The maximal covering location problem. Pap. Reg. Sci. **32**(1), 101–118 (1974)

3. Daskin, M.: A maximum expected covering location model: formulation, properties and heuristic solution. Transp. Sci. **17**(1), 48–70 (1983)
4. Daskin, M., Dean, L.: Location of health care facilities. In: Operations Research and Health Care, pp. 43–76. Springer (2004)
5. Degel, D., Wiesche, L., Rachuba, S., Werners, B.: Time-dependent ambulance allocation considering data-driven empirically required coverage. Health Care Manage. Sci. 1–15 (2014)
6. Gendreau, M., Laporte, G., Semet, F.: Solving an ambulance location model by tabu search. Location Sci. **5**(2), 75–88 (1997)
7. ReVelle, C., Hogan, K.: The maximum availability location problem. Transp. Sci. **23**(3), 192–200 (1989)
8. Schmid, V., Doerner, K.: Ambulance location and relocation problems with time-dependent travel times. Eur. J. Oper. Res. **207**(3), 1293–1303 (2010)
9. Toregas, C., Swain, R., ReVelle, C., Bergman, L.: The location of emergency service facilities. Oper. Res. **19**(6), 1363–1373 (1971)

Tight Upper Bounds on the Cardinality Constrained Mean-Variance Portfolio Optimization Problem Using Truncated Eigendecomposition

Fred Mayambala, Elina Rönnberg and Torbjörn Larsson

Abstract The mean-variance problem introduced by Markowitz in 1952 is a fundamental model in portfolio optimization up to date. When cardinality and bound constraints are included, the problem becomes NP-hard and the existing optimizing solution methods for this problem take a large amount of time. We introduce a core problem based method for obtaining upper bounds to the mean-variance portfolio optimization problem with cardinality and bound constraints. The method involves performing eigendecomposition on the covariance matrix and then using only few of the eigenvalues and eigenvectors to obtain an approximation of the original problem. A solution to this approximate problem has a relatively low cardinality and it is used to construct a core problem. When solved, the core problem provides an upper bound. We test the method on large-scale instances of up to 1000 assets. The obtained upper bounds are of high quality and the time required to obtain them is much less than what state-of-the-art mixed integer softwares use, which makes the approach practically useful.

1 Introduction

We study the Cardinality constrained Mean-Variance model (CMV) with n assets, for which $\boldsymbol{\mu} = (\mu_1, \ldots, \mu_n)^{\mathrm{T}}$ are expected returns, $\boldsymbol{\Sigma}$ is a positive semi-definite ($\succeq 0$) covariance matrix of the returns, μ_P is the minimum return of the portfolio,

F. Mayambala
Makerere University, 7062 Kampala, Uganda
e-mail: fmayambala@cns.mak.ac.ug

E. Rönnberg (✉) · T. Larsson (✉)
Linköping University, SE-58183 Linköping, Sweden
e-mail: elina.ronnberg@liu.se

T. Larsson
e-mail: torbjorn.larsson@liu.se

© Springer International Publishing Switzerland 2016
M. Lübbecke et al. (eds.), *Operations Research Proceedings 2014*,
Operations Research Proceedings, DOI 10.1007/978-3-319-28697-6_54

l_i and u_i are lower and upper bounds, respectively, on portions to be held in each asset, if held at all, and $\mathbf{x} = (x_1, x_2, \ldots, x_n)^{\mathrm{T}}$ are fractions of the capital invested in the assets. Letting K be the required cardinality of the portfolio and $\delta_i, i = 1, \ldots, n$, be variables indicating if an asset is held or not, the model can be stated as

$$\min \quad \mathbf{x}^{\mathrm{T}} \boldsymbol{\Sigma} \mathbf{x} \tag{CMV}$$

$$\text{s.t.} \quad \boldsymbol{\mu}^{\mathrm{T}} \mathbf{x} \geq \mu_P \tag{1}$$

$$\mathbf{e}^{\mathrm{T}} \mathbf{x} = 1 \tag{2}$$

$$l_i \delta_i \leq x_i \leq u_i \delta_i, \quad i = 1, \ldots, n \tag{3}$$

$$\sum_{i=1}^{n} \delta_i = K \tag{4}$$

$$\delta_i \in \{0, 1\}, \quad i = 1, \ldots, n. \tag{5}$$

It is assumed to be feasible. The continuous relaxation of (CMV) is denoted by (RCMV), and its feasible set is assumed to satisfy Slater's constraint qualification.

The problem (CMV) is non-convex and NP-hard [5], and since it is computationally challenging, a number of solution methods have been suggested in the literature. Examples of exact algorithms are the branch-and-bound methods in Borchers and Mitchell [4], Bonami and Lejeune [3], and Bertsimas and Shioda [1], as well as the branch-and-cut methods in Bienstock [2] and Frangioni and Gentile [7]. Relaxation algorithms are proposed in Shaw et al. [9] and in Murray and Shek [8]. Most commonly found in the literature are the heuristic algorithms, see for example Chang et al. [5], Soleimani et al. [10], and Fernandez and Gomez [6].

We use eigendecomposition of the covariance matrix and a core problem to provide high quality upper bounds for (CMV). These can be obtained within a reasonable amount of computing time compared to state-of-the-art softwares.

2 Theory and Solution Principle

To construct a relaxation of (CMV), an eigendecomposition of $\boldsymbol{\Sigma}$ is made, giving

$$\boldsymbol{\Sigma} = \sum_{i=1}^{n} \lambda_i P_i P_i^{\mathrm{T}},$$

where P_i is the ith eigenvector (of unit length) of $\boldsymbol{\Sigma}$ and λ_i is the corresponding eigenvalue. Assume without loss of generality that $\lambda_1 \geq \lambda_2 \geq \cdots \geq \lambda_n$. Let the set $I \subseteq \{1, 2, \ldots, n\}$ and define

$$\boldsymbol{\Sigma}_I = \sum_{i \in I} \lambda_i P_i P_i^{\mathrm{T}}.$$

Let $e : \mathbb{R}^n \to \mathbb{R}$ with $e(\mathbf{x}) = \mathbf{x}^{\mathrm{T}}(\boldsymbol{\Sigma} - \boldsymbol{\Sigma}_I)\mathbf{x}$. Then the objective of (CMV) can be written as $\mathbf{x}^{\mathrm{T}}\boldsymbol{\Sigma}\mathbf{x} = \mathbf{x}^{\mathrm{T}}\boldsymbol{\Sigma}_I\mathbf{x} + e(\mathbf{x})$, so that the function e represents the error introduced when replacing the total n pairs of eigenvectors and eigenvalues with $|I|$ out of them. This function is convex on $\mathbb{R}^n$ because $\boldsymbol{\Sigma} - \boldsymbol{\Sigma}_I = \sum_{i \notin I} \lambda_i P_i P_i^{\mathrm{T}} \succeq 0$. A first-order Taylor series expansion of $e(\mathbf{x})$ at $\bar{\mathbf{x}} \in \mathbb{R}^n$ gives the linear function $\bar{e} : \mathbb{R}^n \to \mathbb{R}$ with $\bar{e}(\mathbf{x}) = \bar{\mathbf{x}}^{\mathrm{T}}(\boldsymbol{\Sigma} - \boldsymbol{\Sigma}_I)\bar{\mathbf{x}} + 2\left[(\boldsymbol{\Sigma} - \boldsymbol{\Sigma}_I)\bar{\mathbf{x}}\right]^{\mathrm{T}}(\mathbf{x} - \bar{\mathbf{x}})$. Using the function $\bar{e}$, instead of e, yields the following approximation of (RCMV).

$$\begin{aligned} \min \quad & \mathbf{x}^{\mathrm{T}}\boldsymbol{\Sigma}_I\mathbf{x} + \bar{e}(\mathbf{x}) \\ \text{s.t.} \quad & (1),\ (2),\ (3),\ (4) \text{ and } \delta_i \in [0, 1],\ i = 1, \ldots, n \end{aligned} \qquad \text{(ACMV)}$$

Let $v(\cdot)$ denote the optimal objective value of an optimization problem. The following result shows that (ACMV) is a relaxation of (RCMV).

Proposition 1 *For any $I \subseteq \{1, 2, \ldots, n\}$ and $\bar{\mathbf{x}} \in \mathbb{R}^n$, $v(ACMV) \leq v(\text{RCMV})$ holds.*

Proof Because e is convex, $\mathbf{x}^{\mathrm{T}}\boldsymbol{\Sigma}\mathbf{x} = \mathbf{x}^{\mathrm{T}}\boldsymbol{\Sigma}_I\mathbf{x} + e(\mathbf{x}) \geq \mathbf{x}^{\mathrm{T}}\boldsymbol{\Sigma}_I\mathbf{x} + \bar{e}(\mathbf{x})$ holds for any $\mathbf{x} \in \mathbb{R}^n$, which implies that $v(\text{ACMV}) \leq v(\text{RCMV})$. $\qquad \square$

The problem (ACMV) is hence always a relaxation of (RCMV), and therefore also of (CMV). However, under a certain condition, (ACMV) becomes equivalent to (RCMV) with respect to the optimal objective values, as shown below.

Proposition 2 *If the linearization point $\bar{\mathbf{x}}$ is chosen as optimal in (RCMV), then $\bar{\mathbf{x}}$ is also optimal in (ACMV) and $v(\text{ACMV}) = v(\text{RCMV})$ holds.*

Proof Note that the two problems (RCMV) and (ACMV) have the same feasible set and that they are both convex. Further, from $\nabla (\bar{e}(\mathbf{x}))_{\mathbf{x}=\bar{\mathbf{x}}} = \nabla (e(\mathbf{x}))_{\mathbf{x}=\bar{\mathbf{x}}}$ it follows that $\nabla \left(\mathbf{x}^{\mathrm{T}}\boldsymbol{\Sigma}_I\mathbf{x} + \bar{e}(\mathbf{x})\right)_{\mathbf{x}=\bar{\mathbf{x}}} = \nabla \left(\mathbf{x}^{\mathrm{T}}\boldsymbol{\Sigma}\mathbf{x}\right)_{\mathbf{x}=\bar{\mathbf{x}}}$, which implies that any KKT point for (RCMV) is also a KKT point for (ACMV). This proves the first conclusion. Further, if $\bar{\mathbf{x}}$ solves (ACMV), then $v(\text{ACMV}) = \bar{\mathbf{x}}^{\mathrm{T}}\boldsymbol{\Sigma}_I\bar{\mathbf{x}} + \bar{e}(\bar{\mathbf{x}}) = \bar{\mathbf{x}}^{\mathrm{T}}\boldsymbol{\Sigma}_I\bar{\mathbf{x}} + \bar{\mathbf{x}}^{\mathrm{T}}(\boldsymbol{\Sigma} - \boldsymbol{\Sigma}_I)\bar{\mathbf{x}} = \bar{\mathbf{x}}^{\mathrm{T}}\boldsymbol{\Sigma}\bar{\mathbf{x}} = v(\text{RCMV})$. $\qquad \square$

Our next result further justifies the usefulness of the problem (ACMV) as an approximation of (RCMV).

Theorem 1 *Suppose that $\mathbf{x}_R^*$ is optimal in (RCMV) and that (ACMV) is constructed for the linearization point $\bar{\mathbf{x}} = \mathbf{x}_R^*$. Let $\mathbf{x}_I^*$ be optimal in (ACMV). Then $\mathbf{x}_I^*$ is feasible in (RCMV) and $\mathbf{x}_I^{*\mathrm{T}}\boldsymbol{\Sigma}\mathbf{x}_I^* - v(\text{RCMV}) \leq 2\max_{i \notin I} \lambda_i$.*

Proof Since all constraints of (RCMV) are included in (ACMV), the feasibility of $\mathbf{x}_I^*$ follows. We next note that Proposition 2 gives that $\mathbf{x}_R^*$ is optimal in (ACMV). Suppose that $\mathbf{x}_I^* \neq \mathbf{x}_R^*$. Consider the convex quadratic function $\varphi : \mathbb{R}^n \to \mathbb{R}$ with $\varphi(\mathbf{x}(t)) = \mathbf{x}(t)^{\mathrm{T}}\boldsymbol{\Sigma}_I\mathbf{x}(t) + \bar{e}(\mathbf{x}(t))$, where $\mathbf{x}(t) = \mathbf{x}_R^* + t(\mathbf{x}_I^* - \mathbf{x}_R^*)$, that is, the objective value in (ACMV) along the line passing through $\mathbf{x}_R^*$ and $\mathbf{x}_I^*$. The derivative of φ becomes $\varphi'(t) = 2(\mathbf{x}_I^* - \mathbf{x}_R^*)^{\mathrm{T}}\boldsymbol{\Sigma}_I(\mathbf{x}_I^* - \mathbf{x}_R^*)t + 2\mathbf{x}_R^{*\mathrm{T}}\boldsymbol{\Sigma}(\mathbf{x}_I^* - \mathbf{x}_R^*)$. Since φ is convex and both $\mathbf{x}_R^*$ and $\mathbf{x}_I^*$ are optimal in (ACMV), it follows that φ is constant on $\mathbb{R}$, so that $\varphi'(t) \equiv 0$.

Therefore, $(\mathbf{x}_I^* - \mathbf{x}_R^*)^{\mathrm{T}} \boldsymbol{\Sigma}_I (\mathbf{x}_I^* - \mathbf{x}_R^*) = 0$ and $\mathbf{x}_R^{*\mathrm{T}} \boldsymbol{\Sigma}(\mathbf{x}_I^* - \mathbf{x}_R^*) = 0$ must hold. In the case that $\mathbf{x}_I^* = \mathbf{x}_R^*$ holds, these two equalities are obviously also true. We then obtain that

$$(\mathbf{x}_I^* - \mathbf{x}_R^*)^{\mathrm{T}} (\boldsymbol{\Sigma} - \boldsymbol{\Sigma}_I)(\mathbf{x}_I^* - \mathbf{x}_R^*) = (\mathbf{x}_I^* - \mathbf{x}_R^*)^{\mathrm{T}} \boldsymbol{\Sigma}(\mathbf{x}_I^* - \mathbf{x}_R^*) = \mathbf{x}_I^{*\mathrm{T}} \boldsymbol{\Sigma} \mathbf{x}_I^* - \mathbf{x}_R^{*\mathrm{T}} \boldsymbol{\Sigma} \mathbf{x}_R^*$$
$$= \mathbf{x}_I^{*\mathrm{T}} \boldsymbol{\Sigma} \mathbf{x}_I^* - v(\mathrm{RCMV}).$$

Using that the Rayleigh quotient of a symmetric matrix is bounded from above by the largest eigenvalue, the fact that both $\mathbf{x}_I^*$ and $\mathbf{x}_R^*$ fulfil the restrictions $\mathbf{e}^{\mathrm{T}}\mathbf{x} = 1$ and $\mathbf{x} \geq 0$, and that $\|\mathbf{x}\|_2 \leq \|\mathbf{x}\|_1$, we further obtain that

$$(\mathbf{x}_I^* - \mathbf{x}_R^*)^{\mathrm{T}}(\boldsymbol{\Sigma} - \boldsymbol{\Sigma}_I)(\mathbf{x}_I^* - \mathbf{x}_R^*) \leq \left(\max_{i \notin I} \lambda_i\right) \|\mathbf{x}_I^* - \mathbf{x}_R^*\|_2^2 =$$
$$= \left(\max_{i \notin I} \lambda_i\right)\left(\|\mathbf{x}_I^*\|_2^2 - 2\mathbf{x}_I^{*\mathrm{T}}\mathbf{x}_R^* + \|\mathbf{x}_R^*\|_2^2\right) \leq \left(\max_{i \notin I} \lambda_i\right)\left(\|\mathbf{x}_I^*\|_2^2 + \|\mathbf{x}_R^*\|_2^2\right) \leq 2\max_{i \notin I} \lambda_i.$$

$\square$

The intuition we get from Theorem 1 is that an optimal solution to (ACMV) is near-optimal in (RCMV) and that its degree of near-optimality is governed by the largest eigenvalue among those that are not taken into account when constructing (ACMV). The corollary below follows directly from Theorem 1.

Corollary 1 *Consider a fixed cardinality of the set I, say k, and let $\mathbb{X}_I^*$ be the set of portfolios that are optimal in* (ACMV). *Then*

$$\min_{I:|I|=k} \max_{\mathbf{x}_I^* \in \mathbb{X}_I^*} \left(\mathbf{x}_I^{*\mathrm{T}} \boldsymbol{\Sigma} \mathbf{x}_I^* - v(\mathrm{RCMV})\right)$$

is achieved for the choice $I = \{1, 2, \ldots, k\}$.

Hence, in order to ensure the best possible degree of near-optimality of an optimal solution to (ACMV) in the problem (RCMV), the former problem shall be constructed using a number of largest eigenvalues and corresponding eigenvectors.

We further note that the convex quadratic objective of (RCMV) can have up to n positive eigenvalues, while the convex quadratic objective of (ACMV) can have at most $|I| \leq n$. This means that if $|I| \ll n$, then an optimal solution to (ACMV) can be expected to have a lower cardinality than an optimal solution of (RCMV), which justifies the use of (ACMV) for constructing a core problem for (CMV).

To form the core problem, we first solve the problem (RCMV). This provides a lower bound to the optimal value of (CMV) and a linearization point $\bar{\mathbf{x}} = \mathbf{x}_R^*$ to be used in (ACMV). Next we find an optimal solution to (ACMV) and identify a pre-specified number of largest values of the variables δ_i. The corresponding set of assets, denoted P, are the only ones considered in the core problem. We thus define the core problem as (CMV) with the additional restriction $\delta_i = 0$, $i \in \{1, \ldots, n\}\backslash P$.

3 Numerical Results and Conclusion

We here present numerical results for the upper bounds on the solutions to (CMV). Three data sets are used. The first is of size 225 assets and adopted from the OR-Library (http://people.brunel.ac.uk/~mastjjb/jeb/orlib/portinfo.html). The second and third data sets are of sizes 500 and 1000 assets, respectively, and are obtained from NYSE data for a period of 3100 trading days between 2005 and 2013. We get three problem instances from each of the data sets by varying the target return, μ_P. Table 1 shows the results of solving the standard Mean-Variance (MV) version of the nine instances. Here, $Card(\mathbf{x}^*)$ is the cardinality of the solution.

To construct instances of (CMV), we use $l_i = 0.01$, $u_i = 0.25$ and $K = 5$. The results for the solution of (CMV) using CPLEX 12.5, when allowed to run for a maximum of 3600 s, are shown in Table 2. The lower bound (LBD) and the upper bound (UBD) on the optimal value were recorded at termination. If no value of LBD is given, then UBD is proven optimal after the time given in the table.

Results from using our core problem approach are shown in Tables 3, 4 and 5. It can be concluded from these tables that already relatively small values of $|P|$ and k

Table 1 Problem instances

n	Problem instance	μ_P	v(MV)	$Card(\mathbf{x}^*)$
225	225a	0.0003	3.06030e-4	14
	225b	0.00165	3.61687e-4	12
	225c	0.003	5.15395e-4	8
500	500a	0.0003	2.60013e-5	38
	500b	0.00165	2.31520e-5	31
	500c	0.003	1.71429e-3	10
1000	1000a	0.0003	1.36468e-5	64
	1000b	0.00165	2.04368e-4	28
	1000c	0.003	1.30006e-3	13

Table 2 CPLEX results

Problem instance	LBD	UBD	CPU time
225a		3.19549e-4	1.3
225b		3.86179e-4	1.7
225c		5.35929e-4	1.4
500a	3.87805e-5	3.97803e-5	–
500b		3.31154e-4	504
500c		2.45698e-3	16.3
1000a	1.41139e-5	2.84444e-5	–
1000b	2.88558e-4	3.63203e-4	–
1000c		2.18773e-3	310

Table 3 $n = 225$

	$\lvert P \rvert$	k	UBD	CPU time
225a	$2K$	5	3.19549e-4	0.60
		15	3.19549e-4	0.62
		40	3.19549e-4	0.65
	$4K$	5	3.19549e-4	0.63
		15	3.19549e-4	0.65
		40	3.19549e-4	0.68
225b	$2K$	5	3.86809e-4	0.65
		15	3.86809e-4	0.65
		40	3.86179e-4	0.68
	$4K$	5	3.86179e-4	0.69
		15	3.86179e-4	0.70
		40	3.86179e-4	0.77
225c	$2K$	5	5.36947e-4	0.64
		15	5.35929e-4	0.66
		40	5.35929e-4	0.69
	$4K$	5	5.35929e-4	0.65
		15	5.35929e-4	0.66
		40	5.35929e-4	0.70

Table 4 $n = 500$

	$\lvert P \rvert$	k	UBD	CPU time
500a	$2K$	5	8.54622e-5	4.24
		15	6.92900e-5	4.27
		40	4.88882e-5	4.32
	$4K$	5	5.48636e-5	4.36
		15	4.08877e-5	4.63
		40	3.97803e-5	4.68
500b	$2K$	5	3.44318e-4	4.32
		15	3.44318e-4	4.35
		40	3.31154e-4	4.81
	$4K$	5	3.44318e-4	4.47
		15	3.31154e-4	4.58
		40	3.31154e-4	4.98
500c	$2K$	5	2.45698e-3	4.77
		15	2.45698e-3	4.80
		40	2.45698e-3	4.85
	$4K$	5	2.45698e-3	4.82
		15	2.45698e-3	4.83
		40	2.45698e-3	4.97

Table 5 $n = 1000$

	$|P|$	k	UBD	CPU time
1000a	$2K$	5	4.77390e-5	33.66
		15	3.83896e-5	33.70
		40	3.37782e-5	40.12
	$4K$	5	4.47532e-5	34.04
		40	2.86704e-5	34.51
	$6K$	10	2.71492e-5	35.78
1000b	$2K$	5	4.92784e-4	39.15
		15	4.60694e-4	39.29
		40	4.02322e-4	40.13
	$4K$	5	4.14883e-4	39.41
		15	3.87652e-4	39.45
		40	3.63203e-4	40.27
1000c	$2K$	5	2.57059e-3	45.47
		15	2.18773e-3	45.51
		40	2.18773e-3	46.14
	$4K$	5	2.18773e-3	45.56
		15	2.18773e-3	45.62
		40	2.18773e-3	45.93

will yield high-quality solutions to the cardinality constrained mean-variance model, and that those values depend on the number of assets and the target return. Note that for the instances for which CPLEX could not verify optimality within 3600 s, our core problem approach finds at least as good feasible solutions as CPLEX, but in less than a minute.

We conclude that the proposed method can be used to obtain high quality upper bounds to cardinality constrained mean-variance portfolio optimization problems in a reasonable amount of time. A topic for further research is the inclusion of a strong lower bounding criterion.

Acknowledgments We acknowledge the Eastern African Universities Mathematics Programme and the International Science Programme at Uppsala University, for financial support.

References

1. Bertsimas, D., Shioda, R.: Algorithm for cardinality-constrained quadratic optimization. Comput. Optim. Appl. **43**, 1–22 (2009)
2. Bienstock, D.: Computational study of a family of mixed-integer quadratic programming problems. Math. Program. **74**, 121–140 (1995)

3. Bonami, P., Lejeune, M.A.: An exact solution approach for portfolio optimization problems under stochastic and integer constraints. Oper. Res. **57**, 650–670 (2009)
4. Borchers, B., Mitchell, J.E.: An improved branch and bound algorithm for mixed integer nonlinear programs. Comput. Oper. Res. **21**, 359–367 (1994)
5. Chang, T.-J., Meade, N., Beasley, J.E., Sharaiha, Y.M.: Heuristics for cardinality constrained portfolio optimisation. Comput. Oper. Res. **27**, 1271–1302 (2000)
6. Fernandez, A., Gomez, S.: Portfolio selection using neural networks. Comput. Oper. Res. **34**, 1177–1191 (2007)
7. Frangioni, A., Gentile, C.: Perspective cuts for a class of convex 0–1 mixed integer programs. Math. Program. **106**, 225–236 (2006)
8. Murray, W., Shek, H.: A local relaxation method for the cardinality constrained portfolio optimization problem. Comput. Optim. Appl. **53**, 681–709 (2012)
9. Shaw, D.X., Liu, S., Kopman, L.: Lagrangian relaxation procedure for cardinality-constrained portfolio optimization. Optim. Methods Softw. **23**, 411–420 (2008)
10. Soleimani, H., Golmakani, H.R., Salimi, M.H.: Markowitz-based portfolio selection with minimum transaction lots, cardinality constraints and regarding sector capitalization using genetic algorithm. Expert Syst. Appl. **36**, 5058–5063 (2009)

Scheduling Identical Parallel Machines with a Fixed Number of Delivery Dates

Arne Mensendiek and Jatinder N.D. Gupta

Abstract We consider the scheduling problem of a manufacturer that has to process a set of jobs on identical parallel machines where jobs can only be delivered at a given number of delivery dates and the total tardiness is to be minimized. In order to avoid tardiness, jobs have to be both, processed and delivered before or at their due dates. Such settings are frequently found in industry, for example when a manufacturer relies on a logistics provider that picks up completed jobs twice a day. The scheduling problem with fixed delivery dates where the delivery dates are considered as an exogenously given parameter for the manufacturer' scheduling decisions can be solved by various optimal and heuristic solution procedures. Here, we consider a variant of this problem where only the number of deliveries is fixed and the delivery dates can be set arbitrarily. For example, a manufacturer may be entitled to assign the logistics provider two pick-up times per day and decide on the exact times of these pick-ups. Then, the machine schedule and the delivery dates can be determined simultaneously which may significantly improve adherence to due dates. Our findings can provide valuable input when it comes to evaluating and selecting distribution strategies that offer a different extent of flexibility regarding the delivery dates.

1 Introduction

Many traditional machine scheduling models focus on the processing of jobs and neglect aspects of distribution. This may be justified by assuming that jobs are delivered immediately and instantaneously, or that products are sold under "ex works"-

A. Mensendiek (✉)
Department of Business Administration and Economics, Bielefeld University,
Bielefeld, Germany
e-mail: arne.mensendiek@googlemail.com

J.N.D. Gupta
College of Business Administration, University of Alabama in Huntsville,
Huntsville, USA
e-mail: guptaj@uah.edu

© Springer International Publishing Switzerland 2016
M. Lübbecke et al. (eds.), *Operations Research Proceedings 2014*,
Operations Research Proceedings, DOI 10.1007/978-3-319-28697-6_55

conditions and hence customers are responsible to pick up completed jobs. Yet, many manufacturing companies are responsible for both, production and delivery of finished goods, and better overall solutions may be found if distribution aspects are considered when making scheduling decisions. In this context, [1–3], among others, have discussed scheduling problems with fixed delivery dates where jobs can only be delivered at a given number of exogenously determined delivery dates. Similar problems can often be found in practice, in particular when a manufacturer relies on the timetable of a logistics provider to ship completed jobs to customers. Practical examples of such settings involve parcel services with fixed pick-up times and air- or sea freight transportation (for examples see [4–6]).

During the last decade, a rich body of literature on the integration of production and outbound distribution has emerged (see [7, 8] for reviews), involving coordinated decisions on production schedules and, for instance, the number of deliveries and vehicle routing. Compared to such integrated problems, fixed delivery dates offer the manufacturer very little flexibility to make distribution decisions. However, as the integration of production and distribution decisions can yield significantly better solutions [9, 10], a manufacturer may profit from making more flexible arrangements with logistics providers. Thus, in this paper we consider a variant of the fixed delivery date problem where the number of deliveries is exogenously given, but the delivery dates can be set arbitrarily by the manufacturer. To illustrate, instead of contracting a logistics provider who picks up completed jobs at 3 p.m. and 8 p.m. each day, a manufacturer may benefit from a more flexible arrangement where it is entitled to assign the provider two pick-up times per day. Hence, the delivery dates become decision variables in the problem studied here.

In the next section, we formally state the scheduling problem and its complexity. We then propose a mathematical programming formulation and conduct numerical experiments to evaluate the efficiency of the proposed formulation and estimate the tardiness reduction potential compared to fixed delivery dates.

2 Problem Statement and Characteristics

The scheduling problem with a fixed number of delivery dates can be described as follows: A set of jobs, J, has to be scheduled non-preemptively on m identical parallel machines to minimize the total tardiness. Each job $j = 1, \ldots, n$ has a given integer processing time p_j and a due date d_j. The due dates denote the times when customers wish to receive their orders so that a job has to be both, processed and delivered before its due date in order to avoid tardiness. C_j denotes the time when a job's processing is completed and D_j the time when it is delivered. The tardiness of a job is defined as $T_j = \max\{0, D_j - d_j\}$.

Jobs can only be delivered at one of the delivery dates $k = 1, \ldots, s$. The number of delivery dates is exogenously given, but the times when these deliveries occur, $\Delta_1 < \cdots < \Delta_s$, can be set arbitrarily. Transportation times are not considered here and the delivery capacity at each delivery date is assumed to be infinite. Then, each job will be delivered at the first delivery date after its completion, that is to say, at $D_j = \min_{k \in K} \{ \Delta_k | \Delta_k \geq C_j \}$.

The scheduling problem with fixed delivery dates is commonly denoted as $Pm|s = \bar{s}| \sum_j T_j$ [2], and consequently we denote the problem variant studied here as $Pm|s = \bar{s}, \Delta = var| \sum_j T_j$.

Theorem 1 *The problem $Pm|s = \bar{s}, \Delta = var| \sum_j T_j$ is NP-hard.*

Proof Restrict the problem by allowing only instances with $m = 2$, $s = 1$, $d_j = \frac{\sum_j p_j}{2} \forall j$ and ask if there exists a schedule with $Z = \sum_j T_j \leq 0$. The restricted problem obviously corresponds to the parallel machine makespan problem, $P2||C_{max}$, a known NP-hard problem [11, p. 238]. $\square$

For the following arguments, let us say that a job is in block k on machine i if it is processed on machine i and delivered at delivery date Δ_k, that is to say, $\Delta_{k-1} < C_j \leq \Delta_k$. An important property was formulated by [2] for the problem $Pm|s = \bar{s}| \sum_j T_j$, but it also holds for the scheduling problem where the delivery dates are arbitrary.

Property 1 *In order to identify an optimal solution for the scheduling problem, it is sufficient to identify an optimal assignment of jobs to blocks.*

While the NP-hardness of the scheduling problem suggests that there may not be an algorithm to solve all problem instances efficiently, the last property may permit to develop solution approaches that perform well on instances of practically relevant size. In fact, our previous research [3] suggests that mathematical programming performs very well on instances of the problem $Pm|s = \bar{s}| \sum_j T_j$ with up to 20 jobs, and elaborate heuristic solution procedures are primarily required for larger instances. Consequently, a mathematical programming formulation for $Pm|s = \bar{s}, \Delta = var| \sum_j T_j$ is proposed next.

3 Mathematical Programming Formulation

The scheduling problem with a fixed number of delivery dates can be formulated as an assignment problem where x_{ijk} is a binary variable to denote if job j is processed in block k on machine i and q is a sufficiently large number.

$$Z = \sum_j T_j \rightarrow \min_{x_{ijk}} \tag{1}$$

subject to

$$\sum_i \sum_k x_{ijk} = 1 \quad \forall j \tag{2}$$

$$T_j \geq 0 \quad \forall j \tag{3}$$

$$T_j \geq \Delta_k - d_j - q\left(1 - \sum_i x_{ijk}\right) \quad \forall j, k \tag{4}$$

$$\sum_j \sum_{k':k'\leq k} x_{ijk'} p_j \leq \Delta_k \quad \forall i, k \tag{5}$$

$$\Delta_k \geq \Delta_{k-1} \quad \forall k \tag{6}$$

$$x_{ijk} \in \{0, 1\} \quad \forall i, j, k \tag{7}$$

The objective (1) is to minimize total tardiness. Constraint (2) makes sure that each job is processed in exactly one block on one machine. Constraints (3) and (4) define the tardiness of a job. Constraint (5) ensures that the capacity of each block on each machine is not violated. This formulation requires that deliveries are indexed in ascending order of the delivery dates $\Delta_1, \ldots, \Delta_s$ which is achieved by constraint (6) (assuming $\Delta_0 = 0$). Note that a similar formulation for the fixed delivery date scheduling problem can easily be derived by eliminating constraint (6) and possibly reformulating constraint (4) without the use of the large number [3].

4 Numerical Results

The proposed mathematical programming formulation is evaluated in a numerical study. The purpose of this is twofold: Firstly, we are interested in evaluating the efficiency of this approach. Secondly, we contrast the total tardiness and the computational time required with the fixed delivery date scenario in order to estimate the potential of more flexible delivery dates.

Instances are randomly generated for different numbers of jobs, machines and delivery dates, denoted by $n/m/s$. We consider instances with 15/2/4, 15/2/6, 15/3/4, 18/3/4, 18/3/6 and 18/4/4. Job processing times are integers drawn from a uniform distribution over $[1, 100]$, and job due dates are drawn from a uniform distribution over $[\max\{0, \left(\sum_j \frac{p_j}{m}\right) \cdot (1 - \tau \mp 0.5\rho)\}]$ where $\tau \in \{0.2, 0.5, 0.8\}$ defines the average tardiness and $\rho \in \{0.2, 0.6, 1.0\}$ the due date range of the jobs. 10 instances are randomly generated for each tuple (τ, ρ), yielding a total of 540 instances.

To compare the results with a fixed delivery date setting, fixed delivery dates are generated as follows: For each instance, the last delivery date Δ_s is obtained by calculating the makespan of a LPT list schedule, that is to say, by sequencing jobs in descending order of their processing times and assigning each job to the next machine that becomes available. The remaining delivery dates $1, \ldots, s - 1$ are integers drawn from a uniform distribution over $[\min_j\{p_j\}, \Delta_s]$ and are arranged in ascending order.

The mathematical program was implemented and solved with the CPLEX 12.2 solver in AIMMS on a computer with a 2.83 GHz four core processor and 3.21 GB RAM. As the computations often required a significant amount of time, in particular for the setting with arbitrary delivery dates, we imposed a runtime limit of 3,600 s

Table 1 Tardiness reduction potential and computation times for scheduling with a fixed number of delivery dates

Instance set (τ, ρ)	Av. red. in total tardiness in (%)	$Pm\|s = \bar{s}\| \sum_j T_j$ av. comp. time in (s)	$Pm\|s = \bar{s}, \Delta = var\| \sum_j T_j$ av. comp. time in (s)
(0.2, 0.2)	43.13	1.72	1,611.61
(0.2, 0.6)	85.41	0.33	582.77
(0.2, 1)	84.96	0.11	7.00
(0.5, 0.2)	35.99	274.25	2,569.51
(0.5, 0.6)	38.37	133.05	2,520.69
(0.5, 1)	41.51	69.78	2,148.44
(0.8, 0.2)	24.12	718.76	3,600.38
(0.8, 0.5)	26.82	172.54	3,466.69
(0.8, 1)	32.20	339.51	3,175.67
All (540)	45.83	190.01	2,186.97

on the solver and treat the solutions as heuristic results. Yet, we note that the solver usually quickly converges towards an optimal solution and then requires a lot of time to validate the optimality. The results are summarized in Table 1.

The results provide several important insights. Firstly, the problem with a fixed number of delivery dates apparently requires more computational time than the fixed delivery date problem where in general the computation time increases in the number of jobs, machines and delivery dates. Secondly, problems with a low average tardiness ($\tau = 0.2$) and a high due date range ($\rho = 1$) can be handled quite efficiently with mathematical programming. For these problems the manufacturer can apparently achieve the highest percentage reduction in tardiness which can be explained by the observation that a low tardiness in the fixed delivery date scenario can often be reduced to zero if the delivery dates can be set arbitrarily. In contrast, if average tardiness is high then the machine capacity is simply insufficient to process all jobs on time and thus, many jobs would be tardy even if they could be delivered individually and instantaneously. Notwithstanding, the average reduction in total tardiness amounts to an average 28 % for the instances with a high average tardiness ($\tau = 0.8$), and approximately 46 % for all instances. This indicates that the possibility to set the delivery dates arbitrarily offers the manufacturing company a major opportunity to improve due date adherence.

5 Conclusion

In this paper, we have proposed and evaluated a mathematical programming formulation for the parallel machine scheduling problem where jobs can only be delivered at a given number of arbitrary delivery dates. In particular, our purpose is to contrast the computation times and total tardiness that can be attained in this setting with the fixed delivery date scenario. Our first insight is that allowing the manufacturer to set the delivery dates flexibly seems to make the problem harder to solve, at least with an assignment-based mathematical programming approach. Hence further research will be devoted to the development of alternative optimal and heuristic solution procedures. Second, the ability to set delivery dates flexibly allows the manufacturer to improve due date adherence significantly. This potential should be considered when it comes to making decisions such as the choice of logistics providers. However, in practice more flexible delivery dates will usually be associated with higher costs for the manufacturer, and further research on this trade-off seems promising.

References

1. Matsuo, H.: The weighted total tardiness problem with fixed shipping times and overtime utilization. Oper. Res. **36**(2), 293–307 (1988)
2. Hall, N.G., Lesaoana, M., Potts, C.N.: Scheduling with fixed delivery dates. Oper. Res. **49**(1), 134–144 (2001)
3. Mensendiek, A., Gupta, J.N.D., Herrmann, J.: Scheduling identical parallel machines with fixed delivery dates to minimize total tardiness. Eur. J. Oper. Res. **243** (2), 514–522 (2015)
4. Li, K.P., Ganesan, V.K., Sivakumar, A.I.: Synchronized scheduling of assembly and multi-destination air-transportation in a consumer electronics supply chain. Int. J. Prod. Res. **43**(13), 2671–2685 (2005)
5. Stecke, K.E., Zhao, X.: Production and transportation integration for a make-to-order manufacturing company with a commit-to-delivery business mode. Manuf. Serv. Oper. Manage. **9**(2), 206–224 (2007)
6. Ma, H.L., Chan, F.T.S., Chung, S.H.: Minimising earliness and tardiness by integrating production scheduling with shipping information. Int. J. Prod. Res. **51**(8), 2253–2267 (2012)
7. Chen, Z.-L.: Integrated production and distribution operations: taxonomy, models, and review. In: Simchi-Levi, D., Wu, S.D., Shen, Z.-J. (eds.) Handbook of Quantitative Supply Chain Analysis: Modeling in the E-Business Era, pp. 711–740. Springer, New York (2004)
8. Chen, Z.-L.: Integrated production and outbound distribution scheduling: review and extensions. Oper. Res. **58**(1), 130–148 (2010)
9. Chandra, P., Fisher, M.L.: Coordination of production and distribution planning. Eur. J. Oper. Res. **72**(3), 503–517 (1994)
10. Chen, Z.-L., Vairaktarakis, G.L.: Integrated scheduling of production and distribution operations. Manage. Sci. **51**(4), 614–628 (2005)
11. Garey, M.R., Johnson, D.S.: Computers and Intractability—A Guide to the Theory of NP-Completeness. Freeman, New York (1979)

Kinetic Models for Assembly Lines in Automotive Industries

Lena Michailidis, Michael Herty and Marcus Ziegler

Abstract We discuss a new model for automotive production based on kinetic partial differential equations. Some theoretical analysis as well as numerical results are presented.

1 Introduction

From a mathematical point of view an automotive assembly line is a graph with products moving along the arcs of this graph. Flows on structured media have been studied widely in the literature in the past years and appear in an almost infinite variety. Here, we will use a similar description of the underlying process as in [1, 3]. In [1] a general production flow problem on a general graph structure has been studied and a kinetic partial differential equation for high-volume part flows is derived. A transport (macroscopic) equation could also be obtained and used as long-time approximation to the kinetic dynamics. Similarly, in [3] a system of hyperbolic equations is derived from a kinetic partial differential equation describing a simple production process on a single line. Statistical information on the production process entered in coefficients of the final hyperbolic equations. In this paper, we discuss results for an assembly line with statistical information obtained by car manufacturing plants. Compared with [3] the underlying particle dynamic is more complicated and hyperbolic closure relations are used to derive the macroscopic hyperbolic models. The detailed proofs as well as additional numerical results and further discussion of the modeling are given in [2]. In this paper we discuss the same model as in [2] but add

L. Michailidis (✉)
Daimler AG, Leibnizstr. 2, 71032 Böblingen, Germany
e-mail: lena.michailidis@daimler.com

M. Herty (✉)
RWTH Aachen University, Templergraben 55, 52056 Aachen, Germany
e-mail: herty@igpm.rwth-aachen.de

M. Ziegler (✉)
Daimler AG, Wilhelm-Runge-Str. 11, 89081 Ulm, Germany
e-mail: marcus.ziegler@daimler.com

© Springer International Publishing Switzerland 2016
M. Lübbecke et al. (eds.), *Operations Research Proceedings 2014*,
Operations Research Proceedings, DOI 10.1007/978-3-319-28697-6_56

new numerical results and investigate further the dependence on the coefficients of the model on the data. Further, new data has been acquired to provide the probability distributions required for the model.

2 Mathematical Modeling

We are interested in the prediction of the long term behavior of the overall workload within a supply chain, depending on the local statistics of installed parts at each station. All variants of a single type are produced on the same production line within this plant. The production is organized in several steps—a fully automated pre-assembly and manual final production. At the final production, each station represents one production step and each production step requires essentially the assembling of a certain number of parts to a moving car body. Due to the many variants the number of assembly parts within a fixed station is highly volatile. Historic data is available for a total of $N = 17$ (out of a total of 30) stations from one belt section.

2.1 Microscopic Model

We model an assembly line by N stations $n = 1, \ldots, N$. The time a generic car body spends within each station is (currently) fixed and given by $T = 120$. Based on historical information we derive a discrete probability distribution function $a \rightarrow \Phi(a, n)$ for each station n. $\Phi(a, n)$ is the probability to assemble a parts at station n to the arriving car body. To have $\Phi(a, n)$ defined for all a we interpolate the discrete probability function defined for the values a_j, $j = 1, \ldots, M$ by $\Phi(a, n) = H(a) \sum_{j=1}^{M} \delta(a - a_j)\Phi(a_j, n)$. We have $\Phi(a, n) = 0$ for $a < 0$. The description by $\Phi(a, n)$ allows to treat all car bodies as non-distinguishable. So far, a, n are T dimensional quantities which are normalized in the following. Therefore, from now on $a \in [0, 1]$. A particle $i \in \{1, \ldots, k \in \mathbb{N}\}$ (resembling a car body) is moving along the assembly line and has a state X_i. The different stations are called S_n, for $n = 1, \ldots N$ and they are in the following considered as nodes in a directed graph. S_1 is the first and S_N the last station in the line. Similar to [1], we assume particles as non-distinguishable. Dimensionless time is denoted by $t \geq 0$. Each particle i has a state $X_i = X_i(t) = (x(t), \tau(t), a(t), n(t))_i \in \mathbb{R} \times \mathbb{R} \times \mathbb{N} \times \mathbb{N}$ in state space $X = (X_i)_{i=1}^{N} \subseteq \mathbb{R}^N$ and $n_i(t) \in \{1, \ldots, N\}$ denotes the station index of particle i at time t. $\tau_i(t) \in \mathbb{R}_0^+$ is the time elapsed within the current station, and $x_i(t) \in [0, 1]$ is the stage of completion of particle i along the assembly line. Within a small time interval $\Delta t > 0$ the state of each particle may change according to the following dynamics. If $\tau_i(t) \leq T$ the particle is in between two stations $n_i(t)$ and $n_i(t) + 1$. So the number of parts a_i and the station index do not change whereas the elapsed time τ and the stage of completion increases. Theoretically, the stage of completion of the current particle is linear and independent of the number of assembled parts and therefore set the velocity $v \equiv 1$. However, this is not observed in practice where assembly is also conducted outside the designated stations, which means for example that work starts before scheduled time. No model and no data is available to quantify this effect. In order to at least qualitatively asses this problem, we derive a model for a general (sufficiently smooth) function $a \to v(a)$. Summarizing, we obtain the following dynamics for a particle i and an elapsed time $\tau_i < T$. If $\tau_i(t) \leq T$:

$$n_i(t + \Delta t) = n_i(t), \; a_i(t + \Delta t) = a_i(t), \tag{1}$$

$$x_i(t + \Delta t) = x_i(t) + \Delta t \, v(a_i(t)), \; \tau_i(t + \Delta t) = \tau_i(t) + \Delta t. \tag{2}$$

For $\tau_i(t) \geq T$ the particle has arrived at the next station. Here, it will change state for a new number of parts to be installed. Within any time interval Δt the change of state upon arrival happens with probability $\omega \Delta t$, where ω is the so-called collision frequency. In the considered assembly line the frequency is $\omega = \frac{1}{T}$. In order to discuss more general models, we keep the general variable $\omega > 0$. If the particle changes state the new number of parts α is obtained by random sampling from the probability distribution $\Phi(a, n_i(t) + 1)$, i.e., $dP(\alpha = s) = \Phi(s, n_i(t) + 1)ds$. We also increase the stage of completion and reset the elapsed time τ_i to zero. If $\tau_i(t) \geq T$:

$$n_i(t + \Delta t) = n_i(t)(1 - \omega \Delta t) + (n_i(t) + 1)\omega \Delta t, \tag{3}$$

$$a_i(t + \Delta t) = a_i(t)(1 - \omega \Delta t) + \alpha \omega \Delta t, \; P(\alpha = s) = \Phi(s, n_i(t) + 1), \tag{4}$$

$$x_i(t + \Delta t) = x_i(t) + \Delta t \, v(a_i(t)), \; \tau_i(t + \Delta t) = (\tau_i(t) + \Delta t)(1 - \omega \Delta t). \tag{5}$$

A kinetic equation for $f(t, X)$ with $X = (x, \tau, a, n)$ is derived. We denote by $f(t, X)dX$ the probability to find a particle in state X at time t.

$$\partial_t(f(t, x, \tau, a, n)) + \partial_x(v(a)f(t, x, \tau, a, n)) = C(f) \tag{6}$$

with $C(f) = -\partial_\tau f - \omega H(\tau - T)f + \omega \Phi(a, n)\delta(\tau) \int f(x, \tau, a, n - 1)H(\tau - T) d\tau da$ and if not stated otherwise the integration in the collision operator is on the full domain. It remains to discuss the case $n = 1$. In the following, we study a *periodic problem*, i.e., we assume that the last station $n = N$ is equivalent to station $n = 0$ and have $f(t, x, \tau, a, 0) := f(t, x, \tau, a, N)$. Then, the high-dimensional kinetic Eq. 6 on phase space X is well-posed for all $n = 1, \ldots, N$. A full discretization is therefore computationally expensive. As in gas dynamics or production models [1], we therefore derive approximate, low-dimensional (macroscopic) models, capturing some qualitative properties of the kinetic dynamics. We first analyze the kernel of $C(f)$ which is similar to [1], however, there are some differences due to the different particle dynamics. As in [1] the kernel of C is decomposed in an invertible and non-invertible part similar to the previously given dynamics. Choosing $\rho(t, x) := \frac{1}{N} \sum_{n=1}^{N} \int f(t, x, \tau, a, n)dad\tau$ we obtain the desired steady-states and leading to the definition of the mass density of the system.

2.2 Macroscopic Equations

We are now interested in an evolution equation for ρ independent of knowledge on f. Different approaches are known in the literature to derive a closed form, we aim in deriving partial differential equations allowing for finite speed of propagation. Since the total number of particles should remain constant we have a conservation property of Eq. 6 for the total mass $\int \rho(t, x)dx$.

2.2.1 Moment Approximations

In order to derive new equations we use a moment approximation in the assembled parts a. We consider mass density ρ and instead of a projection to the kernel manifold, we derive an equation for the first moment $\rho u := \frac{1}{N} \sum_{n=1}^{N} \int v f d\tau da$. The equations are given by

$$\partial_t \rho + \partial_x (\rho u) = 0, \tag{7}$$

$$\partial_t (\rho u) + \partial_x \left(p[f] + \rho u^2 \right) = Q[f], \tag{8}$$

$Q[f] = \omega \frac{1}{N} \sum_{n=1}^{N} \int (\mathbb{E}_{\Phi(\cdot,n)} v P f(t, x, a, n-1, \tau) - v P f(t, x, a, n, \tau)) H(\tau - T) \cdot (\lambda_0(u) + \lambda_1(u)v) d\tau da$ and $p[f] = \frac{1}{N} \sum_{n=1}^{N} \int (v - u)^2 f d\tau da$. Note, if $v(a) = cst$ then $u(t, x) = cst\rho(t, x)$. During the next section we will derive a closure relation to approximate $p[f]$ and $Q[f]$.

2.2.2 Extended Equilibrium Function

Interested in a closure relation which represents the deterministic problem from the beginning, we apply a Grad closure procedure. An extended equilibrium function is constructed for example using the Grad closure. We define for given ρ and u

$$f^{eq}(t, x, \tau, n) := \frac{1}{T + \frac{1}{\omega}} \Phi(a, n) \begin{pmatrix} 0 & \tau \leq 0 \\ 1 & 0 < \tau \leq T \\ \exp(-\omega(\tau - T)) & \tau \geq T \end{pmatrix} \rho(t, x) (\lambda_0 + \lambda_1 v(a)).$$

Herein, λ_0, λ_1 are functions depending on ρ and u. Using the moment relations we obtain the following set of equations determining (λ_0, λ_1): $\rho (\lambda_0 + \lambda_1 \mathbb{E}(v)) = \rho$, $\rho (\lambda_0 \mathbb{E}(v) + \lambda_1 \mathbb{E}(v^2)) = \rho u$. The previous system is solved for (λ_0, λ_1) provided that $\mathbf{V}(v) = \mathbf{E}(v^2) - \mathbf{E}(v)^2$ is non-zero for $\lambda_0(u) = \frac{\mathbf{E}(v^2) - u\mathbf{E}(v)}{\mathbf{V}(v)}$, $\lambda_1(u) = \frac{u - \mathbf{E}(v)}{\mathbf{V}(v)}$. Hence, we may close (7) by evaluating p and Q at the extended equilibrium function f^{eq} with $Q[f^{eq}] = \frac{\rho u - \rho \mathbf{E}(v)}{\mathbf{V}(v)(T + \frac{1}{\omega})} \left(\frac{1}{N} \sum_{n=1}^{N} \mathbb{E}_{\Phi(\cdot,n)}(v) \mathbb{E}_{\Phi(\cdot,n-1)}(v) - \mathbb{E}(v^2) \right)$ and $p[f^{eq}] = -\rho u^2 + \frac{\rho(\mathbf{E}^2(v^2) - \mathbf{E}(v^3)\mathbf{E}(v)) + \rho u(\mathbf{E}(v^3) - \mathbf{E}(v)\mathbf{E}(v^2))}{\mathbf{V}(v)}$. Then, f^{eq} has the moments ρ and ρu and we obtain the following system for the evolution of $(\rho, \rho u)$:

$$\partial_t \rho + \partial_x(\rho u) = 0, \tag{9}$$

$$\partial_t(\rho u) + \partial_x(\rho c_1 + (\rho u)c_2) = (\rho u)c_3 + \rho c_4. \tag{10}$$

The constants c_i depend on the statistical information of $\Phi(a, n)$ and are given by $c_1 \mathbf{V}(v) = \mathbb{E}^2(v^2) - \mathbb{E}(v^3)\mathbb{E}(v)$, $c_2 \mathbf{V}(v) = \mathbb{E}(v^3) - \mathbb{E}(v)\mathbb{E}(v^2)$, $c_3 \mathbf{V}(v)(T + \frac{1}{\omega}) = \frac{1}{N}\sum_{n=1}^{N}\mathbb{E}_{\Phi(\cdot,n)}(v(\cdot))\mathbb{E}_{\Phi(\cdot,n-1)}(v(\cdot)) - \mathbb{E}(v^2)$, $c_4 = -\mathbb{E}(v)c_3$. Equation 9 is a linear hyperbolic balance law provided that $4c_1 + c_2^2 > 0$ holds. In this case the real eigenvalues are $\lambda_{1,2} = \frac{c_2}{2}\Psi\frac{1}{2}\sqrt{4c_1 + c_2^2}$ and there exists a full set of eigenvectors. If we assume, that $\Phi(a, n) = \Psi(a)$, then the Grad closure Ansatz is well-defined provided the variance is non-zero. This also implies that $\mathbf{V}(v) > 0$. $\Phi(a, n) = 0$ for $a < 0$, all moments are non-negative $E_n^i \geq 0$, the eigenvalues are real, $\mathbf{V}(v) > 0$ and since the source terms are linear an initial value problem for Eq. 9 is well-posed.

3 Experimental Results

We present in the following distribution probabilities $a \to \Phi(a, n)$ of assembled parts within different stations n. $\Phi(a, n)$ at 5 selected stations n (out of 17) is depicted in Fig. 1 (left) and we present results on the dependence of Φ on the transport velocity v as the assembly line runs independent of a. In practice the production within some stations n is not completed within T leading to possibly highly utilized stations, see Fig. 1 (right). This effect will be included in the presented model by using a non constant velocity function $v(a)$, dependent on the number of assembled parts. We consider the extended equilibrium function and compute the transport coefficients in Eq. 9. We set $T = 120\,[s]$ and $\omega = 1/T$ and the maximal number of assembled parts $a_{max} = 200$. We investigate a Greenshields like model with slope $0 < \kappa < 1$, compute the hyperbolicity property and the coefficients c_i for $i = 1, \ldots, 4$.

$$v(a) = 1 - \kappa\frac{a}{a_{max}} \tag{11}$$

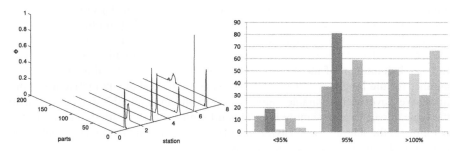

Fig. 1 Part distribution $\Phi(a, n)$ (*left*). Degree of capacity utilization at selected stations (*right*)

For the given data $\Phi(a, n)$, we observe that for all κ the eigenvalues are real, $\mathbf{V}(v) > 0$. and we numerically discretize (9) by standard first-order finite volume and operator splitting to treat the source term. All numerical results are obtained on an equidistant grid $x \in [0, 1]$ with $N_x = 400$ discretization points and Δt such that the CFL condition is fulfilled. In order to simulate the time evolution of the density ρ, we need to prescribe an initial station-averaged car distribution $\rho(t = 0, x)$. As initial data for $(\rho u)(t = 0, x)$, we consider Eq. 7 and density $\rho(0, x)$. With this choice and if we prescribe a constant car distribution $\rho(t, x = 0) = \rho_o$, then $(\rho u)(t, 0) = (\rho u)_o$ is constant. The eigenvalues $\lambda_{1,2}$ of (9) for different values of κ are positive and between $\approx \frac{1}{2}$ and 1 in all cases. Therefore, any perturbation of the state ρ_o will be transported towards $x = 1$. Assuming similar assembly times for each part, we study a production line at $\rho_0 = 95\%$ load and a perturbation of 0.1%. We prescribe a small perturbation at $x = 1\%$ as $\rho(0, x) = \rho_o - 0.1\% \exp(-(8x)^2)$. This leads to a perturbation in density and flow. The coefficients c_3 and c_4 exceed the order of the coefficients of c_1 and c_2 by at least one order. Therefore, the dynamics are mainly driven by the exponential growth of (ρu) over time. The model itself has no mechanism to prevent densities larger than one or less than zero. Due to the introduction of a velocity variable, we simulate Eq. 9 until time t^* where at one stage of completion x^* either the car density exceeds one. Within we consider two examples with initial load $\rho_o = 70\%$ and $\rho_o = 95\%$. For finding the ideal load of the assembly line the unknown parameter κ has to be estimated. In Fig. 1, we observe, at 4 out of 5 stations overload. The station averaged capacity utilization is between 20 and 40% for those stations. Identifying the capacity utilization with degree of completion (x) and assuming the initial perturbation is at position 1%, we find a suitable value for $\kappa \in (0.3, 0.6)$.

4 Conclusion

In this paper, we have represented an ansatz to model automotive assembly lines by using kinetic theory. To exhibit the same unidirectional flow of information as in the underlying particle model and reproduce the parabolic equations, we have derived a macroscopic model via hyperbolic conservation laws. To finally close the system we have used the Grad closure approach to derive an extended equilibrium function and to obtain the fluid dynamic equations (9), which are based on the statistical information of underlying data $\Phi(a, n)$. Within the numerically discretization of Eq. 9 and simulation of the density $\rho(t, x)$ we have filtered out of a characteristic number for the longtime behavior of the overall workload at automotive assembly lines. We have found suitable values for κ in the interval (0.3, 0.6) to estimate the typical load of an assembly line. This information may be used to quantify the workload at the stations as well as the rate of completion at stations and is a qualitative indicator of over- or underload at stations.

References

1. Herty, M., Ringhofer, C.: Averaged kinetic models for flows on unstructured networks. Kinet. Relat. Models **4**, 1081–1096 (2011)
2. Herty, M., Ziegler, M., Michailidis, L.: Kinetic part feeding models for assembly lines in automotive industries. Preprint (2014)
3. Unver, A., Ringhofer, C., Armbruster, D.: A hyperbolic relaxation model for product flow in complex production networks. J. Discrete Continuous Dyn. Syst. 791–800 (2009)

Optimization of Vehicle Routes with Delivery and Pickup for a Rental Business: A Case Study

Susumu Morito, Tatsuki Inoue, Ryo Nakahara and Takuya Hirota

Abstract Optimization of vehicle routes with delivery and pickup for a rental industry is considered. The company delivers to or pickups from customers rented products. Several types of products exist, and customers rent the specified number of products of the specific type. Time windows exist for delivery and pickup. There exist two sizes of vehicles, and their trips start from and end at depot and vehicles can make several trips during a day. Delivery must precede pickup on any trip of a vehicle. Capacity of vehicles depends on product type and also on how products are loaded on vehicles. Depending on demand quantity, split deliveries/pickups may be necessary. The company wants to minimize the total transportation cost. Based on the fact that the total number of distinct trips is rather small due to limited capacity of the vehicles, our solution strategy first enumerates all possible trips. Routes (i.e., collection of trips) are obtained by assigning trips to vehicles so that the total cost is minimized subject to constraints on demand, an upper limit on the number of trips per vehicle, and time compatibility of trips assigned to each vehicle. Since there exist many time compatibility constraints, the problem is first solved without them, we then check the compatibility and if necessary add compatibility constraints, and the problem is solved again until all routes become time compatible. Computational performance of the proposed solution approach is evaluated.

1 Introduction

Cost effective transportation of products for a rental business is considered, and optimization via integer programs is applied to generate vehicle schedules. Customers are classified into two groups; those to whom rental products should be delivered, and those from whom products should be picked up at the completion of the rental period.

S. Morito (✉) · T. Inoue · R. Nakahara · T. Hirota
Waseda University, 3-4-1 Ohkubo, Shinjuku 169-8555, Tokyo, Japan
e-mail: morito@waseda.jp

© Springer International Publishing Switzerland 2016
M. Lübbecke et al. (eds.), *Operations Research Proceedings 2014*,
Operations Research Proceedings, DOI 10.1007/978-3-319-28697-6_57

407

As in many other real vehicle routing (See, e.g., Toth and Vigo [1]), the problem of interest is a complicated one as such features as customer time windows, different types of vehicles as well as products, split deliveries due to limited capacity of vehicles, precedence relation between delivery and pickup (See, e.g., Toth and Vigo [2]), and multiple use of vehicles per day (See, e.g., Azi et al. [3]; sometimes called rotations) among others.

Despite complex appearance of the problem, it turned out to be manageable because of its moderate size in terms of the numbers of customers and vehicles, and also of the very limited capacity of the vehicles.

2 Vehicle Routing for a Rental Business

We consider optimization of daily vehicle routes for a business renting prefabricated unit houses. The goal is to generate a schedule minimizing the total cost of transportation the company pays to truck companies. Basic assumptions are listed below:

1. There exists a single depot from which products are delivered and to which products are returned.
2. Products are delivered to new rental customers, and are picked up from customers upon completion of rental.
3. There exist several product types and the number and the type of rented products are given for each delivery and pickup.
4. Generally there exist customer requested time windows for delivery and pickup.
5. A vehicle trip is a simple circuit originating from and ending at the depot, and a vehicle could make several trips within its working hours of a day.
6. There exist two types of vehicles, large and small.
7. Products can be loaded on vehicles either in a "fold" mode or in an "unfold" or "assembled" mode.
8. The number of products each type of vehicles can carry is limited, and is dependent on the product type as well as the loading mode.
9. The numbers of deliveries and pickups on a particular trip of a vehicle are limited to two and one, respectively.
10. Pickup should follow delivery in each trip.
11. Different types of products cannot be loaded on the same vehicle.
12. Travel time and time needed for "operations" at customer's site are all given constants.
13. The problem calls for a vehicle schedule that minimizes the cost of transportation as calculated based on the agreement between the company and truck companies.

We now describe how the cost of transportation is calculated at the company. Note that the cost is not a simple linear function of actual distance of the trips. The calculation depends on the sequence of customer visits and also on the number of trips for each vehicle. The cost is generally calculated for each customer based on the distance from depot to the customer and also on the type of vehicle used. The cost of a visit can be read from a given table of tariff. For the first (round)trip from depot, the first visit of the trip will be charged 100 % of the designated cost, whereas the second and possibly third visits of the trip are charged only 60 % of the cost. For the second and possibly third (round)trips, the trip's first visit will be charged 80 % (rather than 100 %) of the cost, whereas the second and possibly the third visits of the trip are charged 60 % of the designated cost. This means that the cost is less if more customers are visited in a single (round)trip, and also if more (round)trips are made for a vehicle.

3 Integer Programming Formulation and Solution Strategy

Typical sizes of the problem are approximately 60–70 customers (maximum of 100) including delivery and pickup, 60–70 vehicles consisting of 4t small trucks and 10t large trucks, vehicle capacity of 1–2 products for small trucks and at most 5 products for large trucks, approximately 100 products delivered per day in total. Because of limited vehicle capacity, we only consider the following trip patterns: delivery only, pickup only, delivery–pickup, delivery–delivery, delivery–delivery–pickup.

Because of the very limited capacity of vehicles and of limited patterns of trips, the number of distinct trips is moderate and manageable, and thus we opt to enumerate all feasible trips and then try to optimally assign trips to vehicles using integer programs. Some trips may not be assigned to the same vehicle due to time window for each customer. We check time compatibility of two or three trips as we only consider routes with at most three trips for each vehicle.

The following notations are used. N stands for the set of customers (including delivery and pickup), K the set of vehicles, R the set of all possible trips. $P(Q)$ denotes the set of 2-trip (3-trip) time compatibility constraints. Given a set of customers' transportation requirements (say, 10 products of type X to be delivered to customer Y at location Z), the required number of vehicles of each type is calculated first. Variable x_r^k takes value 1 when vehicle k selects trip r, and 0 otherwise. Variable y^k is 1 when vehicle k is used, and 0 otherwise. c_r^k is the cost of trip r for vehicle k. Finally, coefficient $g_{rp}(h_{rq})$ is 1 when trip r is involved in 2-trip (3-trip) time compatibility constraint $p(q)$, and 0 otherwise.

Since there exist no interaction between different types of vehicles, and thus the problem is separable for each type of vehicles. Therefore the following formulation assumes that all vehicles are of the same type. A vehicle routing and scheduling problem is now formulated:

$$(\text{IP}) \quad \min \sum_{k \in K} \sum_{r \in R} c_r^k x_r^k + \alpha \sum_{k \in K} y^k \tag{1}$$

$$\text{s.t.} \quad \sum_{r \in R} a_{ir} x_r^k \geq b_i, \ i \in N, \tag{2}$$

$$\sum_{r \in R} x_r^k \leq m y^k, \ k \in K, \tag{3}$$

$$\sum_{r \in R} g_{rp} x_r^k \leq 1, \ k \in K, p \in P \tag{4}$$

$$\sum_{r \in R} h_{rq} x_r^k \leq 2, \ k \in K, q \in Q \tag{5}$$

$$x_r^k \in \{0, 1\}, \ k \in K, r \in R, \tag{6}$$

$$y^k \in \{0, 1\}, \ k \in K, \tag{7}$$

Objective function (1) is minimization of the weighted sum of route cost and the number of vehicles used. Constraint (2) specifies the number of trips needed to satisfy customers' requirements, whereas constraint (3) gives the upper bound (currently $m = 3$) of the number of trips for each vehicle. Constraint (4) gives time compatibility constraints which prohibit selecting two time incompatible trips for each vehicle. Similarly, constraint (5) prohibits selecting three time incompatible trips for each vehicle.

Under the current model, we do not know before solving the model how many trips are made for each vehicle. Yet the real cost of transportation depends on the number of trips for a vehicle and their sequence as described earlier. In other words, the first term in our objective function is just an approximated cost of transportation. In order to introduce a mechanism to reduce the number of vehicles used and thus to increase the number of trips for a vehicle, we added the second term in our objective function.

Our solution strategy is to solve first the problem (IP) without constraints (4) and (5). We then check time compatibility of the resultant "optimal" solution. If the selected trips of the "optimal" solution are all time compatible on each vehicle, then the solution is in fact optimal for the original (IP). Otherwise, the solution is not time compatible for at least one vehicle, and we thus add time compatibility constraints so that the same solution is no longer feasible. We repeat this process of solving (IP) with only a subset of time compatibility constraints, that is, solving related problem until the solution becomes time compatible and thus optimal.

4 Computational Experiments

Table 1 shows the size of the particular instance to which computational results are presented below. Note that the total number of trips generated is 828.

Table 2 shows the computational results for the instance. The weighting factor α is changed in the range of 0.00–1.00. Travel cost means the exact cost of the routes to be

Table 1 Problem size

Customers		Vehicles		Trips	
Delivery	Pickup	Large	Small	Large	Small
53	8	23	44	351	477

Table 2 Computational results

	$\alpha = 0.00$	$\alpha = 0.20$	$\alpha = 0.40$	$\alpha = 0.60$	$\alpha = 0.80$	$\alpha = 1.00$
Transp. cost	1,933,500	1,903,500	1,909,500	1,910,000	1,912,000	1,919,000
♯ Vehicles	56	49	49	49	49	49
♯ Trips	65	64	64	64	64	64
CPU time (s)	690	1574	3415	4743	6999	9560
♯ Iterations	160	135	213	211	218	214
♯ Added constraints	1237	1427	1611	1870	1954	1967

paid to truck companies. The number of iterations, which is the number of (relaxed) integer programs solved, ranges around 150–200, during which on the average 1500 time compatibility constraints are identified and added. CPU time is ten minutes or more, and depends on the value of α. The total number of trips used remains roughly same without regard to the value of α. We note that generation of all feasible trips takes little time as compared with time to solve integer programs.

Just by slightly increasing α from 0 to positive, the number of vehicles used goes down, and the number of vehicles used will remain at the same level after that. Even though we could not verify, the number of vehicles used appears to be minimum. The transportation cost is a bit higher when $\alpha = 0$, but is more or less same for different values of positive α. We also note that CPU time to solve the problems with larger α tends to increase, and in fact we could not solve the problem with arbitrarily large α, i.e., the problem with the objective to minimize the number of vehicles used.

In our solution strategy, the size of (relaxed) integer programs grows as iterations proceed. This is because more time compatibility constraints are identified and are added to integer programs. Figure 1 shows that the number of added constraints (vertical axis) increases as iterations proceed. Reflecting the increased problem size, CPU time to solve an integer program increases as shown in Fig. 2.

Table 3 compares the transportation cost of the schedule generated by optimization and the schedule actually made by a planning personnel. The optimization yielded a schedule which is roughly 10 % less costly than the manual schedule.

Optimized routes for positive α tend to make more trips per day per vehicle. Optimized trips also tend to make more customer visits within a trip of a vehicle as compared with the manual schedule.

Fig. 1 Added # of
constraints

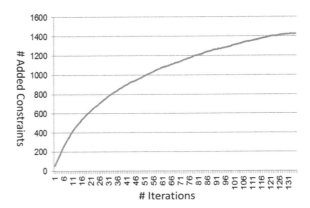

Fig. 2 CPU time per
iteration

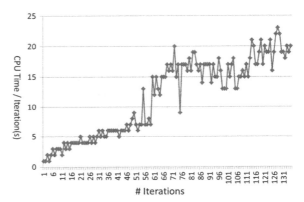

Table 3 Comparison of optimization and manual schedule

$\alpha = 0.20$	Transp. cost	♯ Vehicles	CPU time (s)	♯ Iters.	♯ Added consts.	Cost reduction (%)
Optimization	1,903,500	49	1574	135	2030	–
Manual	2,153,500	68	–	–	–	11.61

5 Conclusions and Future Work

We developed an optimization-based vehicle scheduling approach for a rental business, minimizing total transportation cost. We exploited the fact that the number of
distinct trips for vehicles is rather small (say, less than 1000 trips), and enumerated
all possible trips in advance, and generated an optimal schedule for each vehicle by
combining trips. We succeeded to produce, in 30–60 CPU minutes schedules whose
cost is 10 % less than manual schedule made by a planning personnel.

Considering the fact that there often exist last minutes changes of the transportation requirements, it is hoped to reduce CPU time of the optimization model, say down to a few minutes. One future direction is the set partitioning formulation with column generation as in Azi et al. [3].

References

1. Toth, P., Vigo, D. (eds.): The vehicle routing problem. SIAM (2002)
2. Toth, P., Vigo, D.: VRP with backhauls. In: Toth, P., Vigo, D. (eds.) The Vehicle Routing Problem. SIAM, pp. 195–224 (2002)
3. Azi, N., Gendreau, M., Potvin, J.Y.: An exact algorithm for a vehicle routing problem with time window and multiple use of vehicles. Eur. J. Oper. Res. **202**, 756–763 (2010)

An Ant Colony System Adaptation to Deal with Accessibility Issues After a Disaster

Héctor Muñoz, Antonio Jiménez-Martín and Alfonso Mateos

Abstract One of the main problems relief teams face after a natural or man-made disaster is how to plan rural road repair work to take maximum advantage of the limited available financial and human resources. In this paper we account for the accessibility issue, that is, to maximize the number of survivors that reach the nearest regional center in a minimum time by planning which rural roads should be repaired given the available financial and human resources. This is a combinatorial problem and we propose a first approach to solve it using an ant colony system adaptation. The proposed algorithm is illustrated by means of an example, and its performance is compared with the combination of two metaheuristics, GRASP and VNS.

1 Introduction

Natural disasters have a huge impact on human life, as well as on the economy and the environment. In spite of the advances in forecasting and monitoring the natural hazards that cause disasters, their consequences are often devastating.

The response activities during the emergency relief phase aim to provide assistance during or immediately after a disaster to ensure the preservation of life and of the basic subsistence needs of the victims [8]. Activities during the rehabilitation and reconstruction phases include decisions and actions taken after a disaster in order to restore or improve the living conditions of the affected community, but also activities related to mitigation and preparedness.

H. Muñoz · A. Jiménez-Martín · A. Mateos (✉)
ETSI Informáticos, Universidad Politécnica de Madrid,
Campus de Montegancedo S/N, 28660 Boadilla del Monte, Spain
e-mail: alfonso.mateos@upm.es; amateos@fi.upm.es

H. Muñoz
e-mail: hmunoz@cedint.upm.es

A. Jiménez-Martín
e-mail: antonio.jimenez@upm.es

© Springer International Publishing Switzerland 2016
M. Lübbecke et al. (eds.), *Operations Research Proceedings 2014*,
Operations Research Proceedings, DOI 10.1007/978-3-319-28697-6_58

One of the main problems relief teams face after a natural or man-made disaster is how to plan rural road repair work to take maximum advantage of the limited available financial and human resources. In this paper we account for the accessibility issue, which is defined by Donnges [3] as the degree of difficulty people or communities have in accessing locations for satisfying their basic social and economic needs. Specifically, we maximize the number of survivors that reach the nearest regional center (center of economic and social activity in the region) in a minimum time by planning which rural roads should be repaired given the available financial and human resources.

This is a combinatorial problem since the number of connections between cities and regional centers grows exponentially with the problem size. In order to solve the problem we propose using an *ant colony system* (ACS) adaptation, which is based on ants foraging behavior. Ants stochastically build minimal paths to regional centers and decide if damaged roads are repaired on the basis of pheromone levels, accessibility heuristic information and the available budget.

In Sect. 2, we introduce the mathematical model of the accessibility problem. In Sect. 3, we describe an adaptation of the ant colony system to deal with this combinatorial problem. An example illustrates the algorithm and is used for a performance comparison with GRASP and VNS in Sect. 4. Finally, some conclusions are provided in Sect. 5.

2 Problem Modeling

Let $G = (N, \varepsilon)$ be an undirected graph where $N = \{N_1 \cup N_2 \cup N_3\}$ is a node set and ε is an edge set. N is partitioned into three subsets: N_1, regional centers; N_2, rural towns; and N_3, road intersection points. Edges in ε represent roads with an associated binary level l_e (1 if the road is operational and 0 otherwise).

The subset of roads $\varepsilon_r \in \varepsilon$ is composed of roads that are not operational and can be repaired. The initial level for these roads is 0. There is a financial budget B and a manpower-time budget H allocated to road repair, whereas a financial cost c_e and a manpower requirement m_e are associated with each road $e \in \varepsilon_r$.

For each node $i \in N_2$ a measure of the accessibility is defined: the shortest traveling time from i to the closest regional center in N_1. Of course, this depends on which roads are singled out for repair. The time to traverse an edge is t_e when the road is operational and $t_e + M_e$ when it is not. M_e represents a penalty factor for using another means to traverse e (e.g., using animal-powered transport).

A weight w_i is defined for each node $i \in N_2$ to represent the importance of the node. The value of w_i is usually a function of the number of inhabitants of the rural town associated with node i. The objective consists of minimizing the weighted sum of the time to travel from each node $i \in N_2$ to its closest regional center in N_1.

Three types of decision variables have to be considered. First, the binary variables x_e indicate whether road $e \in \varepsilon_r$ is repaired ($x_e = 1$) or not ($x_e = 0$). Variable y_e^{ij} is assigned the value 1 when the road e is used on the path from $i \in N_2$ to $j \in N_1$ and

0 otherwise. Similarly, variable b_k^{ij} is given the value 1 when node k is visited on the path from $i \in N_2$ to $j \in N_1$.

A **mathematical integer program** for this accessibility problem, based on research by Campbell et al. [1] and Maya and Sorensen [7], is described below.

The *objective function* minimizes the weighted sum of the shortest paths for all $i \in N_2$ to the nearest regional center $j \in N_1$ as follows:

$$f(x) = min \sum_{i \in N_2} (w_i \times min_{j \in N_1} \{ \sum_{e \in \varepsilon} d_e y_e^{ij} \}) \tag{1}$$

where

$$d_e = \begin{cases} t_e + (1 - x_e)M_e, & \forall e \in \varepsilon_r \\ t_e, & \forall e \in \varepsilon \setminus \varepsilon_r \end{cases}. \tag{2}$$

The constraints to be considered in the optimization problem are as follows. First, the following constraints ensure, respectively, that there is exactly one road leaving i on the path from i to j, and that there is exactly one road entering j on the path from i to j, $\sum_{e \in \varepsilon(i)} y_e^{ij} = 1$, $\sum_{e \in \varepsilon(j)} y_e^{ij} = 1$ $\forall i \in N_2$, $\forall j \in N_1$, where $\varepsilon(i)$ is the set of roads adjacent to node i.

We must also ensure that the path from i to j is connected: $\sum_{e \in \varepsilon(k)} y_e^{ij} = 2b_k^{ij}$ $\forall k \in N \setminus \{i, j\}$, $\forall i \in N_2$, $\forall j \in N_1$.

Additionally, budget limitations regarding road repair have also to be taken into account, $\sum_{e \in \varepsilon_r} c_e x_e \leq B$, $\sum_{e \in \varepsilon_r} m_e x_e \leq H$, where B and H are the above financial and the person-hour budgets, respectively.

3   Ant Colony System Adaptation

The *ant colony system* (ACS) was first applied by Dorigo and Gambardela [4, 5] to the traveling salesman problem aimed at achieving performance improvements regarding the classical *ant system* (AS).

The **basic idea** of the algorithm that we propose is as follows. If N_2 consists of n_2 nodes, then n_2 paths have to be built taking into account the possibility of repairing roads given a financial and a manpower budget. The n_2 paths will be built simultaneously since they all have a share in the budget, and a repaired road is available for all paths. For this purpose, a common pheromone matrix will also be considered, and a set of 10 ants will be used for each node in N_2 when applying ACS.

In each iteration, 10 paths will be built on the basis of ACS from each node in N_2 to its closest regional center. Once we have identified the shortest paths for the nodes in N_2 we rank the non-operational roads in the shortest paths taking into account the number of ants that traverse them, the corresponding financial cost c_e and manpower requirement m_e, and the time saving if they are repaired (M_e). Then, we spend the road repair budget on the basis of the above ranking until there are no longer enough financial or human resources available to repair roads.

Now, the global pheromone trail is updated, as explained later, and a new iteration is carried out, in which 10 paths are again built from each node in N_2, the repair decision is made and the pheromones updated.

The **basic elements of the ACS** and its adaptation to the accessibility problem considered in this paper are described below. We denote by τ_{ij} and η_{ij} the associated pheromone trail and the heuristic information for the road (i, j), respectively. The adaptation of the ACS can be divided into two phases, an initialization and a construction phase.

In the **initialization phase**, pheromone levels are initialized as follows: $\tau_0 = \frac{1}{n \times IA}$, where n is the number of nodes in the accessibility problem, and IA refers to the insertion algorithm.

The *insertion algorithm* [7] allows us to build an feasible initial solution. It starts by computing matrix P, including the minimum distances between each pair of nodes in the considered accessibility problem, using the *Floyd-Warshall* algorithm [2]. Then, it iteratively repairs some roads taking into account the corresponding saving achieved in order to improve the total accessibility until no more improvements can be made within the remaining budgets.

Regarding the **path construction phase**, the *pseudorandom proportional* rule is used to decide which node to visit next when building a path from a node in N_2 to its closest regional center. An ant currently at node i chooses the road

$$
e = \begin{cases} argmax_{e \in \varepsilon(i)}\{[\tau_e], [\eta_e]^\beta\}, & \text{if } q \le q_0 \\ \\ J, & \text{otherwise} \end{cases} \tag{3}
$$

where $\varepsilon(i)$ is the set of roads adjacent to city i and J is randomly generated according to the following probabilities:

$$
\frac{[\tau_e] \times [\eta_e]^\beta}{\sum\limits_{l \in \varepsilon(i)} [\tau_l] \times [\eta_l]^\beta}. \tag{4}
$$

q is randomly generated from a uniform distribution in [0, 1] and q_0 models the degree of exploration and the possibility of concentrating the search around the best-so-far solution or exploring other paths. η_{ij} refers to the saving achieved by repairing the road e, that connects vertices i and j, computed as follows:

$$
saving(e) = \sum\limits_{l \in N_2} w_l(min_{k \in N_1}\{T[k, l]\} - min_{k \in N_1}\{T[k, i] + f_t(e, l) + T[j, l]\}),
$$
$$\tag{5}$$

where $T[i, j]$ represents the shortest traveling time from i to j, and the function $f_t(e, l)$ that gives the time to traverse road e when it is at level l.

Regarding the **pheromone update**, only the best-so-far ant adds pheromone in the *global pheromone trail update* after each iteration, $\tau_e = (1 - \rho) \times \tau_e + \rho \times \Delta\tau_e^{bs}$, $\forall e \in T^{bs}$, where $\Delta\tau_e^{bs} = 1/C^{bs}$, C^{bs} is the length of the path built by the best-so-far ant (T^{bs}) and ρ is a parameter representing the pheromone evaporation.

Note that the above global pheromone trail is updated taking into account the best-so-far ant in the paths for each node in N_2.

The *local pheromone trail update* is applied by ants immediately after having crossed a road e during the path construction, $\tau_e = (1 - \xi) \times \tau_e + \xi \times \tau_0$, where τ_0 is the initial value of pheromones and parameter ξ is experimentally fixed. This local pheromone trail update allows ants to explore not previously visited arcs, and prevents the algorithm from stagnating.

4   An Illustrative Example

We have used the accessibility problem at https://raw.githubusercontent.com/ekth0r/ndereba/master/resources/OR_SpectrumData/Data/2-40-2-55-25.txt to analyze the performance of the proposed adaptation of ACS against a combination of GRASP (*greedy randomized adaptive search procedure*) with VNS (*variable neighborhood search*) [6].

In the considered accessibility problem instance we have two regional centers ($N_1 = \{0, 1\}$), 40 rural towns, and two road cross points ($N_3 = \{42, 43\}$) and 55 roads, see Fig. 1. Traverse times, the financial cost and the manpower requirement associated with the repair of each road $e \in \varepsilon_r$, and node weights (w_i) accounting for their number of inhabitants are available at the cited link. The available budgets are 74 monetary units and 83 team hours. The penalty factor to traverse a non-operational road is 10.

Fig. 1 Accessibility problem instance

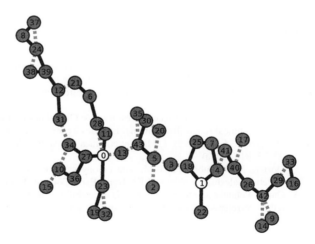

The values fixed for parameters in the ACS adaptation are as follows: $\beta = 2$, $q_0 = 0.9$, and $\xi = \rho = 0.1$. The initial pheromone level is $\tau_0 = 1.367382 \times 10^{-06}$. The number of iterations is 30 and the experiment was repeated 10 times.

In the solution the repaired roads are $\{(0, 28), (12, 21), (3, 5), (24, 38), (7, 41), (29, 33), (31, 34), (0, 13), (3, 18), (9, 42), (14, 42), (24, 37), (4, 41), (17, 40), (35, 43), (16, 29), (43, 13)\}$ and the traverse time associated with the paths from the 40 rural towns to their closest regional center is 715554.7 and the remaining budgets are 5.1 team hours and 9.5 monetary units.

If we compare this solution with the one derived in [6] using a combination of GRASP and VNS we realize that our solution is slightly outperformed the one proposed in [6] by 1471.2 time units, which constitutes only a 0.2 % of difference. Note that the roads repaired in the solution ($f^* = 712873.0$) with GRASP and VNS are $\{(4, 40), (12, 21), (13, 43), (29, 33), (0, 13), (9, 42), (24, 37), (3, 18), (35, 43), (2, 5), (17, 40), (14, 42), (0, 28), (31, 34), (3, 5), (4, 41), (5, 20)\}$.

5 Conclusions

We have proposed a first approach of an adaptation of the ant colony system to maximize the number of survivors that reach the nearest regional center in a minimum time after a natural disaster by planning which rural roads should be repaired given the available financial and human resources. The performance of the proposed algorithm has been compared with another metaheuristic, a combination of GRASP and VNS, and the solutions differ only in a 0.2 %.

Acknowledgments The paper was supported by the Spanish Ministry of Science and Innovation project MTM2011-28983-C03-03.

References

1. Campbell, A., Lowe, T., Zhang, J.: Upgrading arcs to minimize the maximum travel time in a network. Networks, **47**, 72–80 (2006)
2. Cormen, T.H., Leiserson, C.E., Rivest, R.L.: Introduction to Algorithms. MIT Press, Cambridge (1990)
3. Donnges, C.: Improving access in rural areas: Guidelines for integrated rural accessibility planning. Technical report, International Labour Organization (2003)
4. Dorigo, M., Gambardela, L.M.: Ant colonies for the traveling salesman problem. Biosystems, **43**, 73–81 (1997a)
5. Dorigo, M., Gambardela, L.M.: Ant colony system: a cooperative learning approach to the traveling salesman problem. IEEE Trans. Evol. Comput. **1**, 53–66 (1997b)
6. García-Peñuela, J.: Optimización de la accesibilidad mediante el uso de metaheurísticas en una red de carreteras dañadas. Proyecto Fin de Carrera, Universidad Politécnica de Madrid (2014)

7. Maya, P., Sorensen, K.: A GRASP metaheuristic to improve accessibility after a disaster. OR Spectrum, **33**, 525–542 (2011)
8. Moe, T., Pathranarakul, P.: An integrated approach to natural disaster management: public project management and its critical success factors. Disaster Prev. Man. **15**, 396–413 (2006)

Modelling and Solving a Train Path Assignment Model

Karl Nachtigall and Jens Opitz

Abstract We introduce a binary linear model for solving the train path assignment problem. For each train request a train path has to be constructed from a set of predefined path parts within a time-space network. For each possible path we use a binary decision variable to indicate, whether the path is used by the train request. Track and halting capacity constraints are taken into account. We discuss different objective functions, like maximizing revenue or maximizing total train path quality. The problem is solved by using column generation within a branch and price approach. This paper gives some modeling and implementation details and presents computational results from real world instances.

1 Introduction and Motivation

Freight train planning often suffers from the fact that passenger trains are scheduled first and the freight trains may only use the remaining capacity. As a result, those schedules are often of bad quality. The German Railway Company (DB Netz) changes the planning process as follows:

1. Passenger Trains and freight train slots between construction nodes are scheduled simultaneously.
2. Train assignment for freight train demand by connecting the slots to a full train path. Ad hoc requests will be solved by a greedy approach. Long term requests shall be handled by optimization.

For more details see [2, 5]. In this paper we look to the optimization problem in more detail.

K. Nachtigall (✉)
Faculty of Transportation and Traffic Sciences, Chair of Traffic Flow Science,
Dresden University of Technology, 01062 Dresden, Germany
e-mail: karl.nachtigall@tu-dresden.de; karl.nachtigall@t-online.de

J. Opitz (✉)
Faculty of Transportation and Traffic Sciences, Traffic Flow Science,
Dresden University of Technology, 01062 Dresden, Germany
e-mail: jens.opitz@tu-dresden.de

© Springer International Publishing Switzerland 2016   423
M. Lübbecke et al. (eds.), *Operations Research Proceedings 2014*,
Operations Research Proceedings, DOI 10.1007/978-3-319-28697-6_59

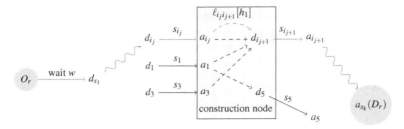

Fig. 1 Slot network and a train path for one request r

2 Basic Models

A train request r is defined by an origin node O_r with preferred departure time d_r and a destination node D_r.[1] In order to fulfill this demand a system of fright train slots is used. Each slot is some kind of placeholder, which might be used to run a fright train. The slots are constructed between pre-defined construction nodes and planned simultaneously with the passenger trains.

A train path for a request r can be modelled by a path

$$p = (w, s_{i_1}, \ell_{i_1 i_2}, s_{i_2}, \ldots, \ell_{i_{k-1} i_k}, s_{i_k})$$

within a time-space network, the so-called slot network (see Fig. 1). The slots $s_{i_1}, \ldots, s_{i_k}$ are linked by the halting positions $\ell_{i_1 i_2}, \ldots, \ell_{i_{k-1} i_k}$. Since more than one halting position may be accessible, there exist multiple arcs connecting consecutive slots. With each of those potential train paths p we associate a binary decision variable $x_{p,r}$, to indicate whether train request r uses path p:

$$x_{p,r} = \begin{cases} 1, & \text{if } r \text{ takes path } p \\ 0, & \text{otherwise} \end{cases}$$

For the stop of the train during $[a_{s_1}, d_{s_2}]$ there must be enough halting place capacity, which can be modelled by halting constraints (see Fig. 2).

Some of the potential train paths cannot be carried out simultaneously, because both paths are using one common slot or a pair of conflicting slots. For each slot $s \in \mathcal{S}$ we define by C_s the set of all train paths, which either contain s or a conflicting slot s', which cannot be used simultaneously with s. Then a feasible solution fulfills the slot conflict constraint $\sum_{p \in C_s} x_{p,r} \le 1$.

In general, not all requests can be fulfilled simultaneously. Minimizing the number of rejected requests performs bad, because a lot of the generated solutions have too

---

[1]The requirements for a request may be easily extended; e.g. intermediate stops and a desired time of arrival can be included into the model.

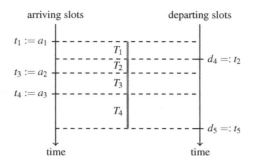

Fig. 2 Sort the sequence of all arrival and departure times of all slots with access to one halting position h by $t_1 < t_2 < t_3 < \cdots < t_k$. Then, the time interval system $T_j := [t_j, t_{j+1}[$ has the property, that the configuration of halting trains cannot be changed during each of those intervals. Hence, for each time interval T_j we define a halting constraint $H = (h, T_j, \kappa)$ with $\sum\limits_{p \in P(H)} x_{p,r} \leq \kappa_H$

much running and waiting time. The most promising approach will be to maximize total quality. This quality of a train path p for the request r is measured by the travel time $\tau_r(p) := a_p - d_r$ with respect to the arrival time a_p of the train path and the preferred departure time d_r. We use a detour factor ρ_{max} to define the quality of the train path by the objective coefficient $\omega_{p,r} := \rho_{max} \cdot \tau_r(p_r^*) - \tau_r(p)$.

Hence, total quality is maximized by the model

$$\sum_{p,r} \omega_{p,r} \cdot x_{p,r} \rightarrow \max \tag{1}$$

$$\forall C_s \in \mathscr{C} : \sum_{p \in C_s} x_{p,r} \leq 1 \tag{2}$$

$$\forall H \in \mathscr{H} \sum_{p \in P(H)} x_{p,r} \leq \kappa_H \tag{3}$$

$$x_{p,r} \in \{0, 1\} \tag{4}$$

3 A Branch Cut and Price Approach

Since the early 1990s, huge integer linear problems were challenging to solve in practice. The most significant advance in general methodologies occurred in 1991 when Padberg and Rinaldi [4] merged the enumeration approach of branch and bound algorithms with the polyhedral approach of cutting planes to create the technique branch cut and price or simply BCP. Figure 3 explains the principle working of BCP. The process of dynamically generating variables is called column generation (see [1]) and done by computing the reduced cost of the non-active column, which will

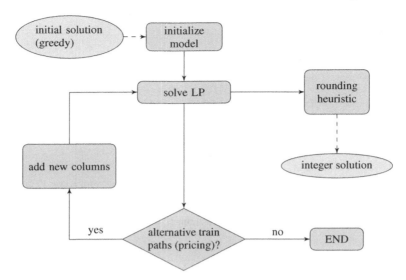

Fig. 3 Process flow of the branch-and-price algorithm

be added to the model if it has negative reduced cost. Lower bounds may either be calculated directly from a feasible integral solution, or, for the case that only a fractional solution of the LP-Relaxation is available, by applying a problem specific rounding heuristic. We use a rounding method by applying a greedy approach for the set of train requests with non-zero, fractional decision variables. Adding the most promising candidates to the model, will improve the actual best solution. Finding those variables is called the pricing problem and done by identifying those variables with minimum negative reduced cost. Reduced cost are calculated by using the dual prices, which are associated with each linear constraint of the underlying linear model. In our case, we have to deal with two types of constraints or equivalently, dual prices. This are

- Constraints of type (2) are assigned with dual prices α_s, which may be interpreted as that amount of cost, for which the solution could be potentially improved by relaxing all conflicts associated with slot s,
- Constraints of type (3) impose dual prices β_H, which measure the saturation of the associated halting constraint $H = (h, T, \kappa_H)$. Large values for β_H indicate high traffic congestion on the halting position h during the time period T.

The reduced cost for a non-active variable $x_{p,r}$ is given by

$$\tilde{\omega}_{p,r} := \sum_{s:p \in C_s} \alpha_s + \sum_{H:p \in P(H)} \beta_H - \omega_{p,r}$$

$$= \sum_{s:p \in C_s} \alpha_s + \sum_{H:p \in P(H)} \beta_H - \rho_{max} \cdot \tau_r(p_r^*) + \tau_r(p)$$

Searching for new columns with minimum negative cost can be formulated as a modification of a shortest path problem in the underlying slot network, the so-called regular language constrained shortest path problem (see [3]). To do this, the slots are labeled by the travel time plus dual conflict prices $\tilde{c}(s_i) = a_{s_i} - d_{s_i} + \alpha_s$ and the halting links are assigned with the $\tilde{c}(\ell_{i,i-1}) = d_{s_{i+1}} - a_{s_i} + \beta_{i,i-1}$, where $\beta_{i,i-1} = \sum\{\beta_H \mid H = (h, T, c_H)$ with $T \cap [a_i, d_{i+1}[\neq \emptyset\}$ summarizes the dual prices of all associated halting constraints.

The initial solution can be calculated by a greedy approach: All requests are sorted with respect to departure time. According to this sequence for each request the best possible train path will be assigned.We obtain an upper bound by using the well known concept of Lagrange relaxation.

4 Computational Results

The implemented method had been validated by a real world instance from DB Netz (See Fig. 4). Each instance has 18541 slots and a demand of 3425 freight trains requests. The first instance I (Table 1) has no restrictions for the halting places at the connection nodes. The second instance II (Table 2) considers the real track infrastructure for all halting postions and defines halting constraints for each possible connection of slots. Each train path p is qualified by the detour coefficient $\rho(p) = \frac{\tau_r(p)}{\tau_r(p_r^*)}$. The tables report the average and maximum value of the detour coefficient statics of each solution. During the optimization iteration, the number of rejected requests can be considerably reduced (see column 'rejected').

Table 1 Computational results for instance I without halting constraints

Iteration	Objective	Upper bound	CPU(s)	$\bar{\rho}$	max(ρ)	Rejected
1	2810.2065	∞	216	1.2987	2.4184	803
⋮	⋮	⋮	⋮	⋮	⋮	
11	3339.8515	3362.8113	6344	1.2427	2.4235	492

$\rho_{max} = 2.5$ maximum connection time $W_{max} = 1800\,s$

Table 2 Computational results for instance II with halting constraints

Iteration	Objective	Upper bound	CPU(s)	$\bar{\rho}$	max(ρ)	Rejected
1	2388.72	∞	638	1.27	2.42	1052
2	2391.39	6868.22	3366	1.27	2.42	1052
⋮	⋮	⋮	⋮	⋮	⋮	⋮
9	2910.90	3280.69	33907	1.21	2.38	776

$\rho_{max} = 2.5$ maximum connection time $W_{max} = 3600\,s$

Fig. 4 The figure represents
the space network for all
18541 slots between the
construction nodes

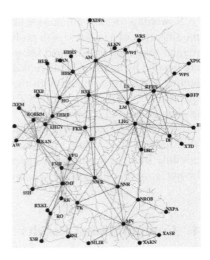

Performance and results of the method are principially satisfying. Instance I can be solved up to optimality, whereas instance II has a rather large duality gap of 12.5 %.

References

1. Barnhart, C., Johnson, E.L., Nemhauser, G.L., Savelbergh, M.W.P., Vance, P.H.: Branch-and-price: column generation for solving huge integer programs. Oper. Res. 316–329 (1998)
2. Feil, M., Poehle, D.: Why does a railway infrastructure company need an optimized train path assignment for industrialized timetabling?. In: Proceedings of International Conference on Operations Research, Aachen (2014) to appear
3. Jacob, R., Barrett, C., Marathe, M.: Formal language constrained path problems. SIAM J. Comput. **30**(3), 809–837 (2001)
4. Padberg, M., Rinaldi, G.: A branch-and-cut algorithm for the resolution of large- scale traveling salesman problems. **33**, 60–71 (1991)
5. Poehle, D., Feil, M.: What are new insights using optimized train paths assignment for the development of railway infrastructure?. In: Proceedings of International Conference on Operations Research, Aachen (2014) to appear

A New Approach to Freight Consolidation for a Real-World Pickup-and-Delivery Problem

Curt Nowak, Felix Hahne and Klaus Ambrosi

Abstract During courier and express providers' operational scheduling, vehicles are assigned to customer orders. This task is complex, combinatorially comprehensive, and contains aspects that defy modeling within reasonable effort, e.g. due to a lack of structured data. Hence, a fully automated solution cannot be achieved. In practice, human dispatchers often use dialog-oriented decision support systems (DSS). These systems generate recommendations from which the human dispatchers select the most profitable one, while additionally taking into account domain-specific knowledge. Solutions that consolidate the freight of multiple customer orders onto a single vehicle are usually particularly favorable. Generally, consolidating leads to a higher degree of vehicle capacity utilization, which in turn increases cost effectiveness and lowers the resulting environmental damage. We present a new recursive heuristic for this scenario based on the well-known savings algorithm. A central parameter of the algorithm limits the number of interdependent single tours. Through the appropriate setting of this parameter, one can control the results' complexity and ensure their transparency and acceptance by human dispatchers. Using real-world data benchmarks, we prove the effectiveness of our algorithm empirically.

1 Introduction

For large real-world pickup-and-delivery service providers, it is not possible to compute completely new dispatching plans when customer orders arrive dynamically. Instead, new orders are integrated as cost-efficiently as possible into the current planning. In this process, scheduling is preferred that achieves freight consolidations, i.e.

C. Nowak (✉) · F. Hahne · K. Ambrosi
Institut für Betriebswirtschaft und Wirtschaftsinformatik, Universität Hildesheim,
Marienburger Platz 22, 31141 Hildesheim, Germany
e-mail: nowak@bwl.uni-hildesheim.de

F. Hahne
e-mail: hahne@bwl.uni-hildesheim.de

K. Ambrosi
e-mail: ambrosi@bwl.uni-hildesheim.de

© Springer International Publishing Switzerland 2016
M. Lübbecke et al. (eds.), *Operations Research Proceedings 2014*,
Operations Research Proceedings, DOI 10.1007/978-3-319-28697-6_60

one vehicle carries the freight of more than one customer order at the same time. This offers several advantages, e.g. a more efficient utilization of the fleet's vehicles and in turn both direct cost savings, the possibility to handle more orders, and less environmental pollution.

Fully automated planning is not possible yet, due to the complexity of the given problem, the combinatorial multiplicity of its solution space, and—sometimes—due to the lack of (sufficiently structured) available data, e.g. special customer requests, precise freight dimensions, or simply traffic jams, etc. Therefore, planning is usually done by human dispatchers with the aid of a decision support system (DSS).

We analyzed past operational data from IN tIME Express Logistik GmbH, a pickup-and-delivery service provider handling transports throughout Europe, in order to improve their planning. The data comprises all information necessary to reconstruct past planning situations and in particular fleet data as well as order information such as freight details and the ordered vehicle category. If customer time windows allowed, reloads at the company's subsidiaries were an option for planning.

We present a new recursive heuristic that can be incorporated into a DSS for dispatch planning. Employed in multiple real-world benchmarks based upon the past operational data, it reveals considerable cost savings compared to former scheduling decisions.

2 The Recursive Savings Algorithm

For a better understanding of our contribution we will start with an example that illustrates the idea behind our algorithm before we present the pseudo-code.

The main principle is illustrated in the example in Fig. 1. Three customer orders (a, b, c) are depicted with individual pickup ($+$) and delivery ($-$) locations. For brevity, the given example offers only one subsidiary (s) for reloads. Let us assume that for a, b, and c all routes that are potentially feasible in terms of time, have been pre-calculated including those with reloads at s. The efficient storage of these routes is described in [3]. We will refer to trips that include a reload as *split routes* with two *parts*: the first part covers the trip between pickup and reload location and the second part covers the trip between reload and delivery location. Routes without reloads will be referred to as *direct trips*.

Starting with order a, the identified best match for consolidation is the second part of a split route for order b, so that a first trip is planned (marked as trip 1 in Fig. 1b). This yields a "loose end": the first part of b's split route. This is now the starting point for recursive calculations and is matched with the first part of a split route for order c. Hence, a second trip is planned (marked as trip 2 in Fig. 1c). The "loose end"—the second part of c's split route—is once more starting point for recursion, and since no other order is left for consolidation, a direct trip is planned (marked as trip 3 in Fig. 1d). The result is what we will refer to as a *route collective* consisting of three interdependent trips. A complete dispatching plan may include multiple route collectives. Note that a vehicle planned for trip 2 may also serve trip 1 or 3 subsequently. Therefore, a reload action for a's or c's freight might not even be necessary.

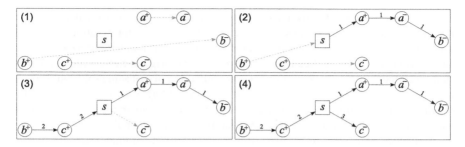

Fig. 1 Example of dispatch planning by the recursive savings algorithms. Depicted are three customer orders a, b, and c and the subsidiary s as potential reload location. *Dashed/solid arrows* denote not scheduled/scheduled trips. **d** shows the final route collective.

The *saving* (see [1]) of this route collective is the difference between the route collective's costs and the sum of the costs of the three direct trips for a, b, and c. Depending on the first customer order to incorporate (in our example: a), the result of the algorithm will be the route collective that provides the largest saving. The maximum number of recursions and thus the maximum number of customer orders within a route collective can be determined by a parameter called RMAX, i.e. after RMAX recursions, the plan for a remaining "loose end" will simply be a direct trip.

Algorithm 1 shows the detailed pseudo-code for a set of given undispatched customer orders. It is supplemented by Algorithm 2 which describes the subroutine for the consolidation in the stricter sense. While our example above only covers Algorithm 2, Algorithm 1 is used to create a complete dispatching plan for more than RMAX customer orders.

Algorithm 1 Recursive Savings Algorithm

Require: A set A of undispatched customer orders, maximum recursion depth RMAX

1: *routeCollectives* := ∅   ▷ a list for the partial results
2: **for all** $b \in A$ **do**
3:     Calculate partial plan *rc* for b using Algorithm 2 with parameters *rec* := 0, empty tabu list, RMAX
4:     *rc.saving* := *rc.saving* + directTrip(b).*costs*   ▷ equals zero, if b could not be consolidated
5:     **if** *rc.saving* $\geq$ 0 **then**
6:         Add *rc* to *routeCollectives*
7:     **end if**
8: **end for**

9: *dispatchingPlan* := ∅   ▷ container for the result
10: Sort *routeCollectives* descending by savings
11: **for all** *rc* $\in$ *routeCollectives* **do**
12:     Add *rc* to *dispatchingPlan*
13:     Remove all elements from *routeCollectives*, that contain (split) trips also contained in *rc*
14: **end for**
15: **return** *dispatchingPlan*

Ensure: A suboptimal dispatching plan is returned, that schedules all customer orders in A to plausible route collectives and direct trips respectively.

The pseudo-code contains a sub-routine called directTrip(b) for calculating the direct trip for (a part of) a customer order b. It also uses dot notation for object attributes such as directTrip(b).*costs* for the costs of b's direct trip, *rc.saving* for the saving value of a route collective *rc*, *a.parent* for the original route whose split partially resulted in route a, and *a.sibling* for the other part of that same split route. Thus, $a.parent = b.parent \iff a.sibling = b \iff b.sibling = a \iff$ both a and b resulted from reloading *a.parent* at a subsidiary. *rc.consequence* contains the necessary follow-up route collective with the plan for *rc*'s "loose end".

Algorithm 2 takes a (part of a) customer order b for a parameter and consists of three parts: **Part 1** calculates a direct trip for b and stores it as the currently best solution rc_{opt}. **Part 2** then tries to achieve a greater saving by consolidating b with another customer order and rc_{opt} is overwritten where applicable. **Part 3** eventually tries to consolidate b with split routes of customer orders to gain an even larger

---

Algorithm 2 Consolidation With a Maximum of RMAX Consequences

---

Require: A set of undispatched customer orders A, a (part of a) customer order b to schedule now, a tabu list of orders *tabu*, a current recursion depth *rec*, the maximum recursion depth RMAX

Part 1
1: $rc_{opt} :=$ directTrip(b)   ▷ i.e. no consolidation
2: $rc_{opt}.saving := -rc_{opt}.costs$ ▷ a direct trip results in costs, not in savings
3: **if** $rec \geq$ RMAX **then**
4:     **return** rc_{opt} ▷ end of recursion
5: **end if**

Part 2
6: **for all** $a \in A \setminus \{tabu \cup b\}$ **do**
7:     Create partial plan *rc* consolidating a and b
8:     $rc.saving :=$ directTrip(a).*costs* $- rc.costs$
9:     **if** *rc* is valid **and** $rc.saving > rc_{opt}.saving$ **then**
10:         $rc_{opt} := rc$
11:     **end if**

Part 3
12:         $children_a :=$ all plausible splits of a at all reload locations
13:         **for all** $a' \in children_a \setminus tabu$ **do**
14:             Create partial plan rc' consolidating a' and b ▷ $rc'.saving$ currently lacks "loose end"
15:             $rc'.saving :=$ directTrip($a'.parent$).*costs* $- rc'.costs$
16:             $tabuRec := tabu \cup$ all (split) orders in rc' as well as their respective parents
17:             $rc'.consequence :=$ result of recursive call of this procedure with parameters $a'.sibling$, $tabuRec$, $rec + 1$ and RMAX
18:             $rc'.saving := rc'.saving + rc'.consequence.saving$ ▷ $rc'.saving$ now covers "loose end"
19:             **if** rc' is valid **and** $rc'.saving > rc_{opt}.saving$ **then**
20:                 $rc_{opt} := rc'$
21:             **end if**
22:         **end for**
23: **end for**
24: **return** rc_{opt}

Ensure: A plausible suboptimal (partial) route collective is returned, that consolidates b with (a part of) another customer order and contains up to *RMAX* consequences, or else a direct trip of b if no valid consolidation could be found.

---

saving. This is the only part in which "loose ends" may occur. These need recursive planning and are incorporated within the resulting route collective either as another consolidation or as a direct trip. For m plausible parts of split routes Algorithm 2's complexity is $\mathcal{O}(m^{\text{RMAX}})$. Thus, for n customer orders (that result in m plausible parts of split routes) Algorithm 1's complexity is $\mathcal{O}(n \times m^{\text{RMAX}})$.

3 Benchmarks and Results

Using the provided past operational data, we constructed 2,066 snap-shots of the planning situations between 2008-01-01 04:00 and 2008-12-10 08:00 each four hours apart, ignoring daylight savings time such that only clock times 00:00, 04:00, 08:00, 12:00, and 16:00 are included. Within this time frame 208,286 customer orders were completely served according to the past data.

For each of these snap-shots we only considered customer orders whose dispatch had not yet started at the respective time of the snap-shot and whose real-world dispatch had *not* included consolidation. These lead to internal costs of €18.8 Mio. We then created a dispatching plan with route collectives by employing Algorithm 1. All route calculations—including those for trips that really *had* been driven—where performed on a digital road map of Europe provided by OpenStreetMap[1] from December 2012. We used further tabu lists to ensure that every customer order was scheduled exactly once, even if it occurred in multiple snap-shots. RMAX was set to 2 as previous tests suggested a best trade-off between speed and result quality.

The benchmark required time estimates both for journey times as well as for loading, reloading, and unloading. Each road category was assigned an average driving speed for every possible vehicle category. Hence, the journey time for a given route and vehicle category could be calculated by summing up the driving times for each contained road sector.

Two sets of times for freight handling (loading, reloading, and unloading) were extracted from the past data: B75 contains times such that 75 % of all respective freight handling actions took less or equal time; B90 is a more conservative estimation, covering 90 % of all recorded respective freight handling actions. This way our benchmark has no knowledge advantage over a human dispatcher, as would have been the case e.g. if we had calculated distinct handling times for every customer. This is important as we compare our results to the past real-world plannings.

All other assumptions were deliberately chosen so that the benchmark results act as a lower bound for the actual cost improvement. See [2] for details on further modeling.

---

[1] www.openstreetmap.org.

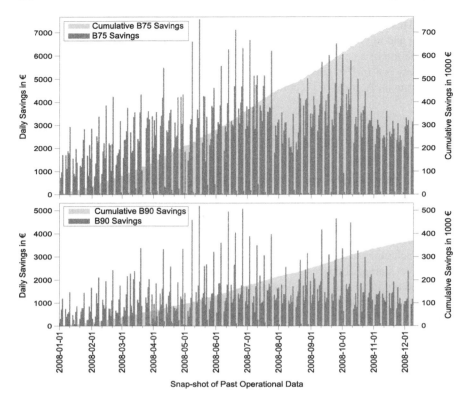

Fig. 2 Benchmark results for B75 and B90

Figure 2 shows the results for both B75 (top) and B90 (bottom). For B75, our approach saves a total of €764,919 (4.1 %) within the benchmark period; B90 with more conservative time estimations results in savings of €369,739 (2.0 %). All peaks in Fig. 2 belong to Fridays and are followed by two days with low savings due to weekend orders usually being known—and thus planned for—on Fridays already.

4 Conclusion

The algorithm shown provides an elegant and effective means for calculating dispatching plans with freight consolidations. Its resulting plans are ideal for usage in a DSS context since the size of the contained route collectives can be controlled by parameter RMAX. The empirical benchmarks based on real-world data reveal considerable cost improvements over planning by highly trained human dispatchers.

References

1. Clarke, G., Wright, J.W.: Scheduling of vehicles from a central depot to a number of delivery points. Oper. Res. **12**(4), 568–581 (1964)
2. Nowak, C.: Schnelle Wegsucheverfahren auf digitalen Straßenkarten—Entwicklung, Implementierung und Anwendungsbeispiel bei einem Logistikdienstleister. Dissertation, Universität Hildesheim (2014)
3. Nowak, C., Ambrosi, K., Hahne, F.: An application for simulations at large pickup and delivery service providers. In: Fischer, S. (ed.) Im Focus das Leben. Gesellschaft für Informatik, Bonn (2009)

High Multiplicity Scheduling with Switching Costs for Few Products

Michaël Gabay, Alexander Grigoriev, Vincent J.C. Kreuzen
and Tim Oosterwijk

Abstract We study several variants of the single machine capacitated lot sizing problem with sequence-dependent setup costs and product-dependent inventory costs. Here we are given one machine and $n \geq 1$ types of products that need to be scheduled. Each product is associated with a constant demand rate d_i, production rate p_i and inventory costs per unit h_i. When the machine switches from producing product i to product j, setup costs $s_{i,j}$ are incurred. The goal is to minimize the total costs subject to the condition that all demands are satisfied and no backlogs are allowed. In this work, we show that by considering the high multiplicity setting and switching costs, even trivial cases of the corresponding "normal" counterparts become non-trivial in terms of size and complexity. We present solutions for one and two products.

1 Introduction

The area of High Multiplicity Scheduling is still largely unexplored. Many problems that are easy in the normal scheduling setting become hard when lifted to their high multiplicity counterparts. In this work, we study a single machine scheduling problem with sequence dependent setup costs called *switching costs*, and under high multiplicity encoding of the input. In this problem, we have a single machine which can produce different types of products. Each day, only one type of product can be

M. Gabay
Laboratoire G-SCOP, Grenoble, France
e-mail: michael.gabay@g-scop.grenoble-inp.fr

A. Grigoriev · V.J.C. Kreuzen (✉) · T. Oosterwijk (✉)
School of Business and Economics, Maastricht University, Maastricht, Netherlands
e-mail: v.kreuzen@maastrichtuniversity.nl

T. Oosterwijk
e-mail: t.oosterwijk@maastrichtuniversity.nl

A. Grigoriev
e-mail: a.grigoriev@maastrichtuniversity.nl

© Springer International Publishing Switzerland 2016   437
M. Lübbecke et al. (eds.), *Operations Research Proceedings 2014*,
Operations Research Proceedings, DOI 10.1007/978-3-319-28697-6_61

produced. Overnight, the machine can be adjusted to produce another type of product the following day.

Related Work. High multiplicity scheduling problems have been investigated by several researchers. Brauner et al. [1] provided a detailed framework for the complexity analysis of high multiplicity scheduling problems. We refer the reader to this paper for an excellent survey of related work in this field. Madigan [3] and Goyal [2] both study a variant of our problem where setup costs only depend on the product which will be produced and holding costs are product-independent. The former proposes a heuristic for the problem, whereas the latter solves the problem to optimality for a fixed horizon.

2   The Model and Basic Properties

We model the general problem for multiple products as follows: we have a single machine that can produce a single type of product at any given time and we are given a set of products $J = \{1, \ldots, n\}$, and for each product $i \in J$, we are given a maximum production rate p_i, demand rate d_i and holding costs h_i per unit. Furthermore, we are given switching costs $s_{i,j}$ for switching from producing product i to producing product j. The problem is to find an optimal cyclic schedule S^* that minimizes the average costs per unit of time $\bar{c}(S^*)$. Note that for each product i, the rates d_i and p_i and costs h_i are assumed to be constant over time and positive. Observe that the input is very compact. Let m be the largest number in the input, then the input size is $\mathcal{O}(n \log m)$.

We distinguish three variants: The **Continuous** case, where the machine can switch production at any time; the **Discrete** case where the machine can switch production only at the end of a fixed unit of time e.g. a day; and the **Fixed** case, where the machine can switch production only at the end of a fixed unit of time, and each period in which the machine produces product i, a full amount of p_i has to be produced (in the other cases, we can lower production rates). We assume holding costs are paid at the end of each time unit.

We denote by LSP(A, n) with $A \in \{C, D, F\}$, $n \in \mathbb{N}$ the Lot-Sizing Problem of scheduling n products in the Continuous, Discrete or Fixed setting. Let $\pi_i^{[a,b]}$ denote the amount of product i produced during time interval $[a, b]$. Let $\pi_i^t = \pi_i^{[t-1,t]}$. Let x_i^t be a binary variable denoting whether product i is produced during time interval $[t-1, t]$. Let q_i^t denote the stock level for product i at time t. We explicitly refer to the stock for a schedule S as $q_i^t(S)$.

We now state some basic properties for the three variants.

Lemma 1 *All three variants of the Lot Sizing Problem are strongly NP-hard.*

Proof The lemma follows directly from a reduction from the Traveling Salesman Problem.   □

Lemma 2 *For all three variants of the problem, there exists a feasible schedule if and only if* $\sum_{i \in J} d_i / p_i \leq 1$.

Proof It is easy to see that d_i / p_i is the fraction of time product i needs to be scheduled on the machine and thus $\sum_{i \in J} d_i / p_i$ is at most 1. $\qquad\square$

Lemma 3 *Let S^* be an optimal schedule for LSP(C, n) or LSP(D, n), with $n \in \mathbb{N}$. S^* has no idle time.*

Proof If there is some idle time, we can simply decrease production rates to decrease holding costs. $\qquad\square$

3 Single Product Case

In most scheduling problems, scheduling a single product on a single machine is trivial. However, considering a high multiplicity encoding takes away some of the triviality of this seemingly simple problem.

Continuous Case. If a feasible schedule exists, we know that $p_1 \geq d_1$. In an optimal schedule, we produce to exactly meet demand, i.e. $\pi_1^{[a,b]} = d_1(b - a)$.

Discrete Case. If a feasible schedule exists, we know that $p_1 \geq d_1$. In an optimal schedule, we produce d_1 for every unit of time to exactly meet demand.

Fixed Case. The Fixed case for a single product is already non-trivial. We will prove the following theorem.

Theorem 1 *In an optimal schedule S^* for LSP(F, 1), $\pi_1^t > 0$ if and only if $q_1^{t-1} < d_1$.*

We first characterize the minimum cycle length for LSP(F, 1), followed by the costs of an optimal schedule. The proof shows that for an optimal schedule S^*, the inventory levels for the time units in the schedule are the multiples of $\gcd(p_1, d_1)$ smaller than p_1.

Lemma 4 *The minimum cycle length for LSP(F, 1) is*

$$l^* = \frac{p_1}{\gcd(p_1, d_1)} . \tag{1}$$

Proof Denote $\mathscr{G} = \gcd(p_1, d_1)$. Assume without loss of generality that $q_1^0 < p_1$. Since the cycle must be feasible, we have that $d_1 \leq p_1$.

Producing p_1 provides stock for $\lfloor p_1/d_1 \rfloor$ time units, with a leftover stock of p_1 mod d_1. Let stock at time t be $q_1^t = q_1^{t-1} + \pi_1^t - d_1$. The schedule is cyclic when $q_1^t = q_1^0$ for $t > 0$. For a minimum cycle length, we want to minimize over t such that $q_1^t = q_1^0 + \sum_{u=1}^t \pi_1^u - d_1 t = q_1^0$. Rewriting gives

$$t = \frac{\sum_{u=1}^t \pi_1^u}{d_1} = \sum_{u=1}^t x_1^u \frac{p_1}{d_1}.$$

Clearly, t is minimized when $\sum_{u=1}^{t} x_1^u = \frac{d_1}{\mathscr{G}}$, and thus $t = \frac{p_1}{\mathscr{G}} = l^*$.   □

Using this lemma we compute the costs of an optimal schedule.

Lemma 5 *The shortest optimal cyclic schedule S^* for* LSP(F, 1) *has unit costs of*

$$\bar{c}(S^*) = \frac{h_1}{2} (p_1 - \gcd(p_1, d_1)) . \tag{2}$$

Proof Denote $\mathscr{G} = \gcd(p_1, d_1)$. Assume without loss of generality that the initial stock $q_1^0 = 0$ (see Remark 1 in the appendix, available online at arxiv.org/abs/1504.00201). Let S^* be the optimal cyclic schedule with length l^*. Since S^* is cyclic, q_1^t has unique values for $t = 0, \ldots, l^* - 1$. Suppose $l^* > p_1/\mathscr{G}$. Then each q_1^t is a multiple of $\mathscr{G}$. Since $l^* > p_1/\mathscr{G}$ and each q_1^t has a unique value, there exists at least one t such that $q_1^t \geq p_1$, and thus the schedule is not optimal. Thus the length of the shortest optimal schedule is $l^* = p_1/\mathscr{G}$.

Since the total demand during the cycle is $d_1 l^*$ and each time unit of production produces p_1, we know that we produce during $d_1 l^*/p_1 = d_1/\mathscr{G}$ time units. Since q_1^t has a unique value for each $t < l^*$ and $q_1^0 = 0$, the stock values are all multiples of $\mathscr{G}$. Hence, the values of q_1^t are the multiples of $\mathscr{G}$ smaller than p_1. Since $p_1 = l^*\mathscr{G}$, the total stock for the cycle equals $\sum_{j=0}^{l^*-1} j\mathscr{G}$.

Thus the total costs of S^* are:

$$h_1 \sum_{j=0}^{l^*-1} j\mathscr{G} = h_1 \frac{1}{2}\mathscr{G}l^*(l^* - 1) = \frac{h_1 p_1}{2} \left(\frac{p_1}{\mathscr{G}} - 1\right) .   □$$

The optimal schedule S^* has length l^* as in Eq. (1), and total costs $l^*\bar{c}$ as in Eq. (2). The length of the cycle is linear in $p_1/\gcd(p_1, d_1)$, and Theorem 1 yields a polynomial delay list-generating algorithm.

4   Continuous Case with Two Products

Intuitively, the Continuous variant of the problem is less difficult than the Discrete one, which in turn is less difficult than the Fixed variant. In this section we show that for two products, even the Continuous case is already non-trivial. We represent a cyclic schedule of length C as a sequence:

$$[t_0, t_1]_{j_0}^{r_0}, [t_1, t_2]_{j_1}^{r_1}, \ldots, [t_s, C]_{j_s}^{r_s},$$

where $[t_i, t_{i+1}]_{j_i}^{r_i}$ denotes a *phase* of the schedule, such that no two consecutive phases share the same r_i and j_i, and in time interval $[t_i, t_{i+1}]$, product $j_i \in J$ is produced at rate $r_i \leq p_{j_i}$. A maximal sequence of consecutive phases of the same product j_i is called a *production period*, denoted by $[t_i, t_{i+1}]_{j_i}$.

We prove some structural results on the optimal schedule. The next lemma shows the machine produces every product i only at rates d_i and p_i to minimize holding costs.

Lemma 6 *Consider* LSP(C, n) *for any* $n \geq 2$. *There is an optimal cycle* S^* *such that for every product* $i \in J$, *every production period of* i *in* S^* *consists of at most two phases. For every production period, in the first phase the machine produces* i *at a rate of* d_i. *During the second phase* i *is produced at a rate of* p_i.

We call a schedule a *simple cycle* if there is exactly one production period for each product. The next lemma shows that in order to minimize holding costs, the optimal schedule for LSP(C, 2) is a simple cycle.

Lemma 7 *There exists an optimal schedule for* LSP(C, 2) *that is a simple cycle.*

Proof Let S^* be a minimal counterexample, i.e. $S^* = [0, t_1]_1, [t_1, t_2]_2, [t_2, t_3]_1, [t_3, C]_2$, where $t_1 \neq (t_3 - t_2)$. Now denote $A_1 = (t_1 + t_3 - t_2)/2$ and consider the following schedule,

$$S = [0, A_1]_1, [A_1, C/2]_2, [C/2, C/2 + A_1]_1, [C/2 + A_1, C]_2 ,$$

which is obtained from S^* by replacing the two production periods of each product by two production periods with averaged length. Since S^* is feasible, we have that $\pi_1^{[0,t_1]} + \pi_1^{[t_2,t_3]} \geq Cd_1$ and $\pi_2^{[t_1,t_2]} + \pi_2^{[t_3,C]} \geq Cd_2$. Let $\pi_1^{[0,A_1]} = d_1 C/2$ in S to cover the demand for product 1 during the first two production periods. Let the production during the other production periods be similar. Clearly, S is feasible. Note that $(t_2 - t_1) + (C - t_3) = (C/2 - A_1) + (C - C/2 - A_1)$, i.e. the sum of the lengths of the production periods for product i in S, is equal to that in S^*.

Now suppose there is in S^* a production period $[a, b]$ for product 1 with $q_1^a(S^*) > 0$. Then during the production period $[x, a]_2$, holding costs increase by $q_1^a(S^*)h_1(x - a)$ compared to S and thus $\bar{c}(S) < \bar{c}(S^*)$.

Next, suppose $q_i^a(S^*) = 0$ for every production period $[a, b]_i$. It is easy to see that holding costs for product 1 are only paid during production periods for 2 and during the non-empty phase where product 1 is produced at rate p_1. The same result holds for product 2. Note that the sum of the lengths of the production periods for product i in S, is equal to that in S^* and holding costs are linear. Hence, the area under the curve of the function of the holding costs over time, is the same in S as in S^*, thus $\bar{c}(S) \leq \bar{c}(S^*)$.

Observe that S consists of two simple cycles S' and S'' with $S' = S''$. Thus S' is a feasible simple cycle with the same unit costs as S. □

For the rest of this section we assume without loss of generality that $h_1 < h_2$, and we only consider simple cycles. Next we show that an optimal schedule for LSP(C, 2) consists of at most three phases.

Lemma 8 *There exists an optimal schedule for any* LSP(C, 2) *instance of the following form:*

$$S^* = [0, t_1]_1^{p_1}, [t_1, t_2]_2^{d_2}, [t_2, C]_2^{p_2}, \tag{3}$$

where the second phase is empty if and only if $d_1/p_1 + d_2/p_2 = 1$.

Proof Let S be an optimal cycle with four non-empty phases, i.e.

$$S = [0, t_1]_1^{p_1}, [t_1, t_2]_2^{d_2}, [t_2, C]_2^{p_2}, [C, t_3]_1^{d_1}.$$

Consider the schedule consisting of only the first three phases, i.e. we remove $[C, t_3]_1^{d_1}$. Note that $\pi_2^{[t_2, C]} = d_2(t_1 + (t_3 - C)) > d_2 t_1$. Hence the total amount of production for product 2 can be lowered by $(t_3 - C)d_2$, by decreasing the length of phase $[t_2, C]_2^{p_2}$. Let $\alpha = (t_3 - C)d_2/p_2$ and let

$$S^* = [0, t_1]_1^{p_1}, [t_1, t_2 + \alpha]_2^{d_2}, [t_2 + \alpha, C]_2^{p_2}.$$

Clearly S^* is feasible and $\bar{c}(S^*) < \bar{c}(S)$.

If $d_1/p_1 + d_2/p_2 = 1$ the schedule is tight and demand can only be met by producing at maximum rate, which implies $[t_1, t_2 + \alpha]_2^{d_2}$ is empty.

If $d_1/p_1 + d_2/p_2 < 1$, there has to be a phase in which the machine does not produce at maximum rate, to avoid overproduction. By Lemma 6 there are at most two phases of production at rate d_1 and d_2 respectively. Since $h_1 < h_2$, by the above reasoning we introduce only one phase where we produce d_2 in order to minimize costs. □

Using this result we calculate the optimal cycle length and corresponding costs. Let S^* be as in Eq. (3). The costs of the schedule as a function of the parameter t_1, are given as

$$\bar{c}(t_1) = \left(\frac{h_1(p_1 - d_1)}{2} + \frac{h_2 d_1 d_2}{2 p_1} \left(1 + \frac{d_2}{p_2 - d_2} \right) \right) t_1 + \left((s_{1,2} + s_{2,1}) \frac{d_1}{p_1} \right) \frac{1}{t_1},$$

which is minimized for

$$t^* = \sqrt{\frac{2(s_{1,2} + s_{2,1}) d_1}{h_1 p_1 (p_1 - d_1) + h_2 d_1 d_2 \left(1 + \frac{d_2}{p_2 - d_2} \right)}}.$$

The outcomes are summarized in the following theorem.

Theorem 2 *For* LSP(C, 2) *there exists an optimal schedule of length* $t^* p_1/d_1$ *with average costs* $\bar{c}(t^*)$.

References

1. Brauner, N., Crama, Y., Grigoriev, A., Van De Klundert, J.: A framework for the complexity of high-multiplicity scheduling problems. J. Comb. Optim. **9**(3), 313–323 (2005)
2. Goyal, S.K.: Scheduling a multi-product single machine system. J. Oper. Res. Soc. **24**(2), 261–269 (1973)
3. Madigan, J.G.: Scheduling a multi-product single machine system for an infinite planning period. Manag. Sci. **14**(11), 713–719 (1968)

Sampling-Based Objective Function Evaluation Techniques for the Orienteering Problem with Stochastic Travel and Service Times

Vassilis Papapanagiotou, Roberto Montemanni
and Luca Maria Gambardella

Abstract Stochastic Combinatorial Optimization Problems are of great interest because they can model some quantities more accurately than their deterministic counterparts. However, the element of stochasticity introduces intricacies that make the objective function either difficult to evaluate or very time-consuming. In this paper, we propose and compare different sampling-based techniques for approximating the objective function for the Orienteering Problem with Stochastic Travel and Service Times.

1 Introduction

Recently, there has been a growing interest and study of Stochastic Combinatorial Optimization Problems (SCOP). One of their important advantages is that they can give a more realistic solution in problems that encompass uncertain quantities such as travel times. In this paper, we compare different algorithms for the objective function of the Orienteering Problem with Stochastic Travel and Service Times (OPSTS).

Stochasticity makes the problems more complex to solve than their deterministic versions. Currently, there exist some exact methods for solving them but they work only for very small problem instances and so there is demand for designing efficient metaheuristics. However, even this task becomes difficult as computing the objective function many times is a hard problem or in our problem (OPSTS) very time consuming.

Metaheuristics based on Monte Carlo Sampling have become state-of-the-art approaches for many SCOPs such as the Probabilistic Traveling Salesman Problem (PTSP) [1], the Probabilistic Traveling Salesman Problem with Deadlines (PTSPD)

V. Papapanagiotou · R. Montemanni (✉) · L.M. Gambardella
IDSIA-SUPSI-USI, Via Galleria 2, CH-6928 Manno, Switzerland
e-mail: roberto@idsia.ch

V. Papapanagiotou
e-mail: vassilis@idsia.ch

L.M. Gambardella
e-mail: luca@idsia.ch

© Springer International Publishing Switzerland 2016
M. Lübbecke et al. (eds.), *Operations Research Proceedings 2014*,
Operations Research Proceedings, DOI 10.1007/978-3-319-28697-6_62

[2, 3] and recently OPSTS [4, 5]. However, we have observed that when we compute the objective function by using Monte Carlo Sampling in problems with deadlines such as the OPSTS, larger errors occur in the nodes where the deadline is likely to occur. This happens because Monte Carlo evaluation assigns the whole reward or penalty to a node and an error on a node will propagate to subsequent nodes. These errors sometimes are large enough to make the metaheuristic make worse choices than it otherwise would during the phases of exploration or exploitation.

Although orienteering problem variants have been well-studied in the past, the Orienteering Problem with Travel and Service Times (OPSTS), as far as we know was first proposed in [6]. In that paper, Campbell et al. propose two exact methods for solving a simplified version of the problem and the Variable Neighborhood Search (VNS) metaheuristic [7] for providing solutions. The objective function used in that paper is an analytical approximation of the probabilistic cost of travel times. In our paper we refer to that function as 'ANALYTICAL'.

In [4, 5] some alternative Monte Carlo sampling techniques for computing the objective function of OPSTS are presented and they are compared to the 'ANALYTICAL' in terms of time gain and error.

Using Monte Carlo Sampling to approximate the objective function is an idea that has been explored before for different problems and methods. The Probabilistic Traveling Salesman Problem and the Probabilistic Traveling Salesman Problem with Deadlines are two related problems where Monte Carlo Sampling has been applied before. In [8], the authors propose a Local Search algorithm for PTSP which uses Monte Carlo sampling in the objective function to approximate the cost of the solution.

In this paper, we study how to use Monte Carlo sampling and a mix of analytical methods with Monte Carlo sampling techniques in order to approximate the objective function of OPSTS faster and we examine its benefits when we use our techniques in the context of a metaheuristic. We extend our methods and present new results and experiments which are partially related to our previous works [4, 5].

2 Problem Definition

In this section, we give the formal definition of OPSTS as it was introduced in [6].

We denote $N = \{1, \ldots, n\}$ a set of n customers with the depot being node 0. In OPSTS a subset $M \subseteq N$ the set of customers selected to be served. There is a global deadline D for servicing customers without a penalty. We assume that the graph is full and therefore there is an arc (i, j) for all $i, j \in M$. Servicing a customer $i \in M$ before the deadline results in a reward r_i otherwise a penalty e_i is incurred. Let $X_{i,j}$ be a non-negative random variable representing the travel time from node i to node j and S_i a random variable representing the service time of customer i. It is assumed that we know the probability distribution of $X_{i,j} \forall i, j$ and that the service time S_i for the ith customer follows the same distribution as $X_{i-1,i}$ and can be added to the travel time $X_{i-1,i}$ and therefore need not be considered separately. In our case, the probability distribution of the random variables is the Γ distribution. Let the random variable A_i be the arrival time at customer i and $\bar{A}_i$ a realization of A_i. Now let $R(\bar{A}_i)$ be a function representing the reward earned at customer i when arriving to i at time $\bar{A}_i$. According to our definitions $R(\bar{A}_i) = r_i$ for $\bar{A}_i \leq D$, otherwise $R(\bar{A}_i) = -e_i$ (for $\bar{A}_i$).

A tour of the customers τ is defined as a sequence of customers $\in M$. The objective function of the problem is defined as the expected profit of the tour:

$$u(\tau) = \sum_{i \in \tau}[P(A_i \leq D)r_i - (1 - P(A_i \leq D))e_i] \qquad (1)$$

3 Objective Function Evaluators

Analytical Evaluation of the Objective Function
The analytical evaluation of the objective function is the one used in [6] and is used in this paper as the reference evaluator in order to measure errors and time gains. It is used to evaluate the 'ANALYTICAL' area (explained below).

Considering that the Cumulative Distribution Function (CDF) of A_i computes the probability $P(A_i \leq x)$ and that A_i is Γ distributed, we use the CDF F_{k_i} where k is the k parameter for the ith customer. Now, the Eq. 1 becomes:

$$u(\tau) = \sum_{i \in \tau}[F_{k_i}(D)r_i - (1 - F_{k_i}(D))e_i] \qquad (2)$$

Monte Carlo Evaluation (MC) of the Objective Function
To compute the objective function (1) we need to compute the probabilities involved. In Monte Carlo sampling we generate many samples which are graphs with different realizations of the arrival times (A_i) for every node. For each sample we compute its objective value (1) and the final objective value is the average of all objective values. To speed up the procedure, we precompute samples of travel times from every node to every other node. The precomputation is done once and the samples generated are reused throughout the computations. For more information the reader is referred to [4, 5].

Hybrid techniques for the evaluation of the Objective Function
To generate solutions for OPSTS, in this paper, the VNS metaheuristic is used [6, 7]. The metaheuristic generates some solutions to be evaluated by our objective function. Using Monte Carlo sampling (MC) speeds up the evaluation, however, because it has relatively high error when evaluating nodes in the deadline area (see [5]) it might affect the choices made by the metaheuristic and lead to worse solutions. This problem intensifies as the feasible solutions have more nodes. For example, in Fig. 1 we observe that using MC for enough time to find large feasible solutions makes the metaheuristic diverge from higher quality solutions. Therefore, there is the need

Fig. 1 Objective function
evaluation algorithms and
the final value reached by the
VNS metaheuristic [6, 7]
versus runtime, dataset with
66 customers, deadline set to
115, α for
MC-ANALYTICAL-MC
is 0.13

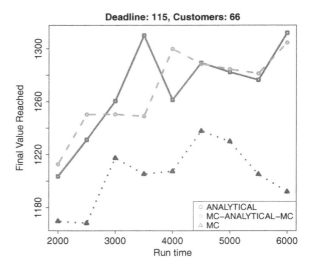

to reduce the evaluation error and have small impact on our speedup. We achieve
that by partitioning the solution. In this paper in order to partition the solution given,
we select an $0 < \alpha < 1$ and we consider that all nodes with arrival times between
$[(1 - \alpha) \cdot D, (1 + \alpha) \cdot D]$ belong to the 'deadline area' and are evaluated using the
'ANALYTICAL' method. The rest are evaluated using Monte Carlo sampling which
has higher evaluation speed. We call this evaluator MC-ANALYTICAL-MC. As it
was shown, for a given deadline, the 'deadline area' is defined by the parameter α
which we will call 'deadline area ratio' or 'area ratio' for the rest of the paper. In
Fig. 1 the α used was 0.13 which meant that the evaluation error was less than 2 %.
The reason will be explained in Sect. 4.

4   Experiments

In this section, a small set of experiments is presented to show how the hybrid method
behaves and how to tune it. The datasets and the dataset generation procedures are
the same as in [6] so that immediate comparison is possible. In this paper, we only
show results for a dataset with 66 customers (dataset 566 as seen in [6]), for more
results the reader is referred to [9]. Firstly, we show how to select the deadline area
ratio (mentioned as α above) for the MC-ANALYTICAL-MC, so that there is an
upper-bound on the error made in the individual computations.

   To select the deadline area we first decide on an upper-bound error threshold and
then we select the minimum deadline area such that for every deadline the error is less
than the threshold. In Table 1, we can see the results for the MC-ANALYTICAL-MC
method on a representative instance for different error thresholds. For example, if

Table 1 Time Gains and deadline area ratios for different error thresholds for MC-ANALYTIC-MC

Error thresholds	1 (%)	2 (%)	3 (%)	4 (%)
Time gains	27.29	30.94	32.18	33.05
Deadline area ratio (α)	21	13	9	5

Errors are measured with reference to the 'ANALYTICAL' method for the dataset of 66 customers used in the paper

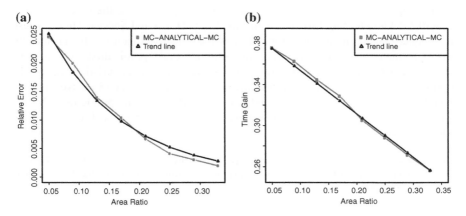

Fig. 2 Influence of area ratio α to relative error and time gain for deadline 40, for method MC-ANALYTICAL-MC, dataset with 66 customers. Errors and time gains are measured with reference to the 'ANALYTICAL' method. The trend lines (in *black*) are the fitted curves and model the trend. The fitted exponential curve is $y = e^{-3.56 \cdot x - 6.77}$ and the fitted line is $y = -0.57 \cdot x + 0.45$. **a** Residual standard error: 0.0025. **b** Multiple R-squared: 0.99

we want an error below 2 % we would select a deadline area ratio of 13 % and it would be 30.94 % faster in the evaluation. Time gains and errors are measured with reference to the 'ANALYTICAL' method. Time gains and errors are averages over all the evaluations done in 30 runs of the VNS metaheuristic.

The effect of the deadline area to relative error and time gains

By using the concept of the 'deadline area ratio' we can influence the relative error per evaluation and the time gains of our algorithm. In Fig. 2, we can see two graphs depicting deadline area ratio vs relative error and deadline area ratio vs time gains, respectively (dataset with 66 customers). We can observe that the error is decreasing with a higher rate than time gain as we increase the deadline area ratio. This fact adds to the usefulness of the pursuit for further improvement of hybrid methods of evaluation. To confirm our observations, we fitted exponential curves ($e^{k \cdot x + \beta}$) to model the error curves and lines to model the time gain curves. The average residual standard error for the error curves was 0.0025 and multiple R^2 for the time gain curves was 0.99. This means in the first case that there exist exponential curves that approximate the actual error curves with negligible error and in the second case, lines can replace the time gain curves and they would explain 99 % of the variance of time gains.

5 Conclusions

In this paper, we explored different techniques to approximate the objective function of the Orienteering Problem with Stochastic Travel and Service Times. These techniques can accelerate the computation of the objective function which is usually the performance bottleneck in Stochastic Combinatorial Optimization problems. We showed that Monte Carlo sampling error may make the metaheuristic behave suboptimally and we suggested a hybrid method for evaluating the objective function which is faster than the analytical and does not suffer from the problems of Monte Carlo sampling and we showed its behavior and how to tune it with respect to the 'deadline area ratio'.

Acknowledgments Vassilis Papapanagiotou was supported by the Swiss National Science Foundation through project 200020-134675/1: "New sampling-based metaheuristics for Stochastic Vehicle Routing Problems II".

References

1. Weyland, D., Bianchi, L., Gambardella, L.M.: New heuristics for the probabilistic traveling salesman problem. In: Proceedings of the VIII Metaheuristic International Conference (MIC 2009) (2009)
2. Brucker, P., Drexl, A., Möhring, R., Neumann, K., Pesch, E.: Resource-constrained project scheduling: notation, classification, models, and methods, Eur. J. Oper. Res. 112(1), 3–41, (1999). http://www.sciencedirect.com/science/article/pii/S0377221798002045
3. Weyland, D., Montemanni, R., Gambardella, L.M.: Hardness results for the probabilistic traveling salesman problem with deadlines. Combinatorial Optimization. Lecture Notes in Computer Science, vol. 7422, pp. 392–403. Springer, Berlin (2012)
4. Papapanagiotou, V., Weyland, D., Montemanni, R., Gambardella, L.M.: A sampling-based approximation of the objective function of the orienteering problem with stochastic travel and service times. In: 5th International Conference on Applied Operational Research, Proceedings, Lecture Notes in Management Science, pp. 143–152 (2013)
5. Papapanagiotou, V., Montemanni, R., Gambardella, L.M.: Objective function evaluation methods for the orienteering problem with stochastic travel and service times. J. Appl. Oper. Res. 6(1), 16–29 (2014)
6. Campbell, A.M., Gendreau, M., Thomas, B.W.: The orienteering problem with stochastic travel and service times. Ann. Oper. Res. 186, 61–81 (2011)
7. Sevkli, Z., Sevilgen, F.E.: Variable neighborhood search for the orienteering problem. Computer and Information Sciences-ISCIS 2006, pp. 134–143. Springer, Berlin (2006)
8. Birattari, M., Balaprakash, P., Stützle, T., Dorigo, M.: Estimation-based local search for stochastic combinatorial optimization. In: IRIDIA, Université Libre de Bruxelles (2007)
9. http://www.idsia.ch/papapanagio/. Accessed 9-Jul-2014

Optimized Pattern Design for Photovoltaic Power Stations

Martin Bischoff, Alena Klug, Karl-Heinz Küfer, Kai Plociennik
and Ingmar Schüle

Abstract The task of planning photovoltaic (PV) power plants is very challenging. The decision makers have to consider the local weather conditions, the land area's topography, the physical behavior of the technical components and many more complex aspects. We present an approach for optimizing one variant of the way routing problem for PV plants. We formulate the problem as an Integer Program (IP) which can be solved by standard solvers. In addition, we reformulate the IP as a maximum independent set problem on interval graphs. This graph-theoretic problem can be solved in polynomial time. Using the latter approach, we are able to generate a variety of solutions with different angles for the ways with little effort of time.

1 Optimization of Photovoltaic Power Plants

In times of increasing energy costs and decreasing incentives for green energy sources, the task of maximizing the economic efficiency of renewable energy power plants becomes more and more relevant. Especially the photovoltaic industry suffers from current political decisions and it is more important than ever to reach grid parity, so that the price for the produced energy can compete with fossil power plants.

There are several degrees of freedom for the optimization of PV plants. The planner can choose between different module and inverter technologies, the tables on which the modules are mounted can vary in size and tilt angle, the distances between the tables influence the shading of the modules, and many more (cf. [1, 2]). The problem is far too complex for being systematically optimized without computational support. Often, decisions have to be made based on rules of thumb and experience. In most cases, this results in a waste of potential for optimization.

A. Klug · K.-H. Küfer · K. Plociennik (✉) · I. Schüle
Fraunhofer ITWM, Fraunhofer Platz 1, 67663 Kaiserslautern, Germany
e-mail: kai.plociennik@itwm.fhg.de

M. Bischoff
Siemens AG, Otto-Hahn-Ring 6, 81739 Muenchen, Germany
e-mail: martin.bischoff@siemens.com

© Springer International Publishing Switzerland 2016   451
M. Lübbecke et al. (eds.), *Operations Research Proceedings 2014*,
Operations Research Proceedings, DOI 10.1007/978-3-319-28697-6_63

1.1 Analyzed Problem

The way routing on the available land area is one of several subproblems to be solved while planning photovoltaic layouts. One solution approach for this problem is addressed in this paper. These ways are needed for site access during construction and maintenance and must be placed under certain restrictions. In the underlying design concept, the ways are parallel stripes in the area which must be placed within certain distance thresholds, so that preferably many PV tables can be placed between them.

The degrees of freedom for this problem are the angle and the positions of the ways, which both depend on the form of the area and the size of the PV tables. There are also some additional influencing factors, like restricted parts of the area where no tables can be located. The objective is to place the ways so that the area where no tables can be placed is minimized.

Figure 1 shows an example of such a way placement. Here, the ways (illustrated in dark gray) are placed with a certain angle, deviating from north-south orientation, to better fit to the shape of the polygon. The available space that can be used in a later step for placing tables is shown in light gray. These "table columns" are parallelograms whose width corresponds to the width of the tables. In the discussed approach, only south- or north-facing PV tables are considered without azimuth angle variations. Hence, the upper and lower edges of the parallelograms must be parallel to the equator. The medium gray polygons mark obstacles, where the available area may not be used for placing PV tables.

Given such a placement of ways and table columns, a later step can be to fill the table columns with actual tables by applying a certain distance rule from south to north (or from north to south on the southern hemisphere).

Fig. 1 Example of a placement for the ways (*dark gray*) and table columns (*light gray*) for a given area (outer polygon) with obstacles (*medium gray*). Between the second and third table column from the *left*, a small gap is left open for a better area coverage

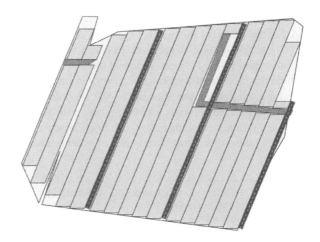

2 Mathematical Formulation

To systematically solve the problem described, we model the problem as an IP (Integer Program). Therefore, we discretize the area in which the tables will be placed in west-east direction. Since the actual placement of the tables on site generally involves tolerances and the distance between two discretization points can be chosen small, every practical solution can be found with this discretization.

In a first approach, we model the problem as an IP which maximizes the area covered by table columns. This program individually places the columns as well as the ways. The constraints ensure that no two items (i.e., ways or table columns) overlap and that the maximum distance between two ways is kept. The angle for the planning must be defined in advance. This IP can directly be solved by commercial solvers, although the computation time might become long. Hence, this approach is not suitable for an exploration of different way angles.

To avoid the high computational effort, we create fixed combinations of paths and columns, so called "columngroups" and formulate the placement of these items as a new IP. Such a columngroup is a collection of non-overlapping table columns (that may contain gaps in between) with an adjacent way. In Fig. 1, three columngroups each with five table columns are illustrated. The first columngroup contains a gap after the second table column from the left.

We first construct different types of columngroups with different numbers of table columns, such that they leave any reasonable gap between the columns or a column and a way. All of these constructions must satisfy the restriction that the distance between two ways is within the predefined limits. We denote the set of all feasible columngroups by J^*.

In the following IP, the factor L_i^j is the area which is coverd by the table columns of the columngroup of the form $j \in J^*$ placed at discretization point i. The set of all discretization points is denoted by I. The integer variable x_i^j specifies, whether or not this columngroup is placed

$$(\text{IP}) \quad \max \quad \sum_{j \in J^*} \sum_{i \in I} L_i^j \cdot x_i^j \tag{1}$$

$$\text{s.t.} \quad \sum_{j \in J^*} \sum_{k \in \widetilde{F}^j(i)} x_k^j \leq 1 \quad \forall i \in I \tag{2}$$

$$x_i^j \in \{0, 1\} \quad \forall i \in I, \forall j \in J^* \tag{3}$$

The set $\widetilde{F}^j(i)$ in the constraint (2) contains for a given discretization point i and a columngroup of the form j all discretization points k for which the chosen columngroup placed at point k covers the point i. Thus, the constraint ensures that the selected columngroups do not overlap. Note that the columns of a single columngroup are non-overlapping by construction. The objective function maximizes the area covered by the columns of the respective columngroups.

Fig. 2 Visualization of different intervals (*top*) that represent the west-east area coverage of columngroups and the corresponding interval graph (*bottom*). Two nodes are connected in the graph if the corresponding intervals overlap

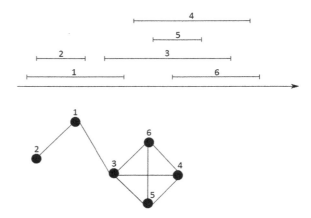

This IP can be reformulated as a graph-theoretical problem, the maximum weight independent set problem. For special graphs, such as interval graphs (cf. Fig. 2), this problem can be solved in polynomial time. An algorithm to solve the maximum weight independent set problem is described in [3]. The idea of our transformation is to model the overlapping of two columngroups placed at certain discretization points as an edge between two corresponding nodes in a weighted graph. Thus, the nodes correspond to the variables x_i^j in our above IP formulation. The weights of the nodes are given by the respective coefficients L_i^j in the objective function. A complete proof of this equivalence can be found in [4].

To summarize the results, we found a model of the discussed way placement problem, which can be solved in polynomial time. This enables us to apply the algorithm to different angles for a given area, allowing a more detailed analysis of the effect the angle has on the area covered by columns. Furthermore, a (commercial) solver for general IPs is not required for this type of problem.

3 Computational Results

We tested our independent set approach on several land areas of photovoltaic power plants in the real world. To this end, we implemented an algorithm for solving the independent set problem on interval graphs in C# and used a standard PC for execution. For different land areas, we executed our algorithm to compute an optimal way placement for different ways angles between reasonable minimum and maximum values, discretized in small steps. This approach is practicable since the computation times are reasonably small. This way, we analyzed the effects of different angles on the way placement problem. A selection of the results is presented in Table 1. More detailed results can be found in [4].

Table 1 Results of the maximum independent set approach for some land areas of PV plants in the real world

Area	Size (km²)	Ways angle (°)	Area coverage (%)	Running time (s)
Area 1	0.12	90	83.24	0.34
Area 1	0.12	143	81.61	0.31
Area 2	0.41	90	86.31	0.67
Area 2	0.41	75	86.66	0.72
Area 3	0.08	90	87.76	0.13
Area 3	0.08	76.07	83.71	0.25

From the table, it can be seen that a variation of the ways angle has a significant impact on the optimal objective value, i.e., the area coverage/useable land area for PV tables. For example, for area 3, the area coverage differs by around four percentage points for the two used ways angles. Note that not in every case vertical (north-south) ways are the best choice (cf. area 2). In these cases, choosing an angle which is parallel to one of the area's edges often gives better results.

In Fig. 3, the effect of different ways angles for a certain area is illustrated. It can be seen that even small changes to the angle can have significant effects on the area usage. The objective value "covered land area" differs by around three percent between maximum and minimum.

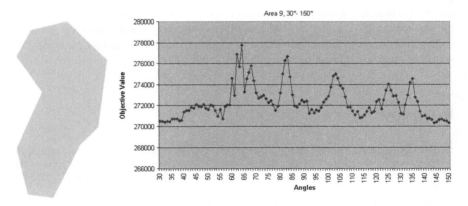

Fig. 3 *Left* Area for which we optimize the angle of the ways. *Right* Effect of different angles on the objective function for the illustrated area. It can be seen that the objective value is quite sensitive even to small changes of the angle. 90° refers to north-south ways

4 Summary

We considered a variant of the way routing problem for PV plants, which has practical significance for automatically creating layouts of large-scale ground-mounted photovoltaic power plants due to the high impact of the choice for the way routing on the plant's overall performance. We developed two solution approaches for the problem. The first one is based on an IP formulation of the problem and solving it using a commercial IP solver. The second approach is based on creating a certain interval graph representing the land area and the potential positions of the table columngroups, and solving the maximum weight independent set problem using a known polynomial-time algorithm. The two approaches are equivalent in terms of the objective function value of the optimal solution (maximize area coverage between the ways, i.e., maximize the space available for placing tables). Both approaches yield optimal solutions for the problem in acceptable time for real-world areas.

The overall approach of modeling and solving the way routing problem for PV plants in a mathematical way is superior to standard manual planning, since the latter wastes potential for optimization. Due to the large number of possibilities and complex interdependencies, good way routings are easily overlooked by the layout planner without such a systematic optimization approach.

We note that both discussed approaches are generic in a sense that in principle, they are integrateable in any PV plant layout planning concept, provided the concrete way routing guidelines can be formulated as an integer program or as a practically solvable graph-theoretic problem on a certain graph modeling the input. An interesting open question is hence, if and how other way routing rules used in practice can be handled by this approach. Also the inclusion of other design tasks, e.g., choosing good tilt angles and distances for the tables, can help in creating high-quality plant layouts and hence provide interesting research topics in the future.

References

1. Bischoff, M., Ewe, H., Plociennik, K., Schuele, I.: Multi-Objective Planning of Large-Scale Photovoltaic Power Plants. In: Operations Research Proceedings 2012, Selected Papers of the International Annual Conference of the German Operations Research Society (GOR), pp. 333–338 (2012)
2. Schuele, I., Plociennik, K., Ewe, H., Bischoff., M.: An economic evaluation of two layout planning concepts for photovoltaic power plants. In: Proceedings 28th European Photovoltaic Solar Energy Conference and Exhibition EU PVSEC 2013, pp. 4143–4147 (2013)
3. Frank, A.: Some polynomial algorithms for certain graphs and hypergraphs. In: Nash-Williams, C., Sheehan, J. (eds.) Proceedings of the Fifth British Combinatorial Conference 1975, pp. 211–226 (1975)
4. Klug, A.: Path planning and table pattern design in photovoltaic power stations. Diploma thesis at TU Kaiserslautern (2012)

What Are New Insights Using Optimized Train Path Assignment for the Development of Railway Infrastructure?

Daniel Pöhle and Matthias Feil

Abstract The train path assignment optimization algorithm generates an optimal solution for freight train service applications by connecting available slots between several construction nodes without conflicts. This method is not only used for a real timetable e.g. for the following year but also for timetable-based development of railway infrastructure in long-term scenarios. However, for infrastructure development the actual slot connections are not the main concern in this planning step. The railway infrastructure company rather wants to detect bottlenecks in the infrastructure and needs to get evidence for necessary developments of its railway network. By presenting results of a real world German railway network's test case, this paper shows which bottlenecks can be derived from an optimized slot assignment and which measures (in timetable and infrastructure) could eliminate the detected bottlenecks. Necessary key figures for discovering bottlenecks will be introduced, too. It is shown that shadow prices of the developed column generation method are a good indicator for the identification of bottlenecks. For the first time with the comparison of different scenarios one can deliver a clear monetary benefit for the removal of a single bottleneck, e.g. the revenue advantage of an additional track for dwelling of freight trains. Hence, using the developed optimization algorithm for train path assignment leads to new useful insights for a railway infrastructure company to develop its railway network.

D. Pöhle (✉) · M. Feil (✉)
Langfristfahrplan und Fahrwegkapazität, DB Netz AG, Theodor-Heuss-Allee 7,
60486 Frankfurt, Germany
e-mail: daniel.poehle@deutschebahn.com; d.poehle@gmx.de

M. Feil
e-mail: matthias.feil@deutschebahn.com

© Springer International Publishing Switzerland 2016
M. Lübbecke et al. (eds.), *Operations Research Proceedings 2014*,
Operations Research Proceedings, DOI 10.1007/978-3-319-28697-6_64

457

1 Introduction

Developing railway infrastructure is a long process. In terms of long-term planning, timetables for passenger trains are quite stable, and in some cases even governed by contracts (see [5]). Based on this, standardized train paths (called slots) are planned to provide capacity for freight trains. Using a rail freight traffic forecast (see [2]) as a prediction for train service applications, train path assignments are optimized. In so doing, O-D slot itineraries for all train service applications are optimized by connecting slots without conflicts as explained in [3] using a column generation approach. After optimization, there is a feasible solution for all predicted freight train service applications that have not been rejected due to a lack of capacity. However, for infrastructure development the actual slot assignments are not the main concern in this planning step. The railway infrastructure manager rather wants to detect bottlenecks in the infrastructure and must obtain evidence for necessary developments of its railway network. Hence, the interesting questions are:

- How is the line capacity utilized?
- How are freight trains routed?
- What transport time and quality arises for the different O-D routes?
- How many dwelling positions are required in a node for the connection of slots?
- Where and when do bottlenecks occur in the railway network?
- Is there spare capacity? Where and when does it appear?
- What measures can eliminate the identified bottlenecks?

With the existing macroscopic models (see e.g. [4]), used in long-term scenarios today, these questions cannot be answered satisfactorily, because no timetable information is considered. Having the optimized assignments, timetable-based conclusions are possible for the first time. This paper shows how to use the assignment solution to draw better (because timetable-based) conclusions for railway infrastructure development.

2 Determining the Potential for Railway Infrastructure Development

To determine whether there is a need to develop railway infrastructure due to bottlenecks, a top-down analysis is performed. First, the assignment for the total network is analyzed to determine where there is room for improvement. Subsequently, these lines and nodes are viewed in detail. To answer the questions from Sect. 1, aggregated key numbers as well as microscopic ones for single freight trains of the assignment are considered.

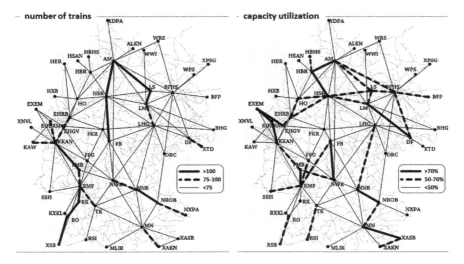

Fig. 1 Routing and line capacity utilization

2.1 Routing of Freight Traffic and Capacity Utilization

For an overview of the assignment it is adequate to compare the number of assigned slots on a line with the capacity utilization of the same line (see Fig. 1)[1]:

If the number of assigned slots of a line is considerably lower than the assigned slots of adjacent lines and the capacity utilization is high, this line is a potential bottleneck. In Fig. 1, this is the case for the line from Bebra (FB) to Wurzburg (NWR), for example. Generally, questioning all lines with particularly high or low capacity utilization is a good way to check whether the assignment appears plausible.

2.2 Quality and Transport Time

For a railway infrastructure manager it is fundamental to offer marketable freight train paths because there is always competition with other modes of transport, for example road transport or shipping. Therefore, a certain train path quality has to be achieved, which involves two aspects: travel time and detours. In Fig. 2, the assignments of the route from Cologne to Basel for different node levels[2] are compared with the used lines and distribution of transportation times (in hours). The bigger white rectangles

---

[1]This map shows a solution for node level 1, meaning only the line capacity restrictions are considered. It is also possible to include stricter node restrictions into the model, see [1]. In node level 2a, which is presented later, each node has a restricted number of dwelling positions.

[2]Level 1 has no node capacity constraints, in contrast, level 2a sets a global number of dwelling positions for each node.

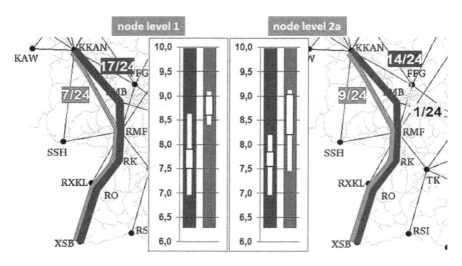

Fig. 2 Comparison of transportation times and used lines from Cologne to Basel

in the diagram include 50 % of the trains on this track. For node level 1 on the darker plotted track, 50 % of all trains from Cologne to Basel need between 7.5 and 7.9 h, for example. In today's timetable, the 50 % average travel times range between 7.7 and 9.6 h for this route. The application's O-D travel times are shorter and have a considerably smaller variance than today.

The additional node capacity constraints at level 2a result in a shift of traffic streams to other lines. One train even has to take a detour from Cologne (KKAN) via Friedberg (FFG) to Mannheim (RMF). This is caused by one of the passed nodes of the route from Cologne to Basel which lack in dwelling positions. Using the longer way via FFG, the train arrives later in the bottleneck node and can use a released dwelling position. Hence, the nodes' capacity utilization needs to be analyzed for the detection of bottlenecks, too.

2.3 Bottlenecks and Spare Capacity

Bottlenecks can result either from the lines' capacity or from the nodes' capacity or from the interdependencies between nodes' and adjacent lines' capacities. Even if the number of available slots exceeds the number of train service applications, bottlenecks can arise for a limited period of time due to an uneven distribution of the demand.

In practical scenarios in the size of the German railway network, the number of potential O-D-slot itineraries grows exponentially. Consequently, enumerating all of them would cause exponentially many binary decision variables in the optimization model. To solve large scenarios to optimality, Nachtigall and Opitz [3] use a column generation approach starting with a small set of slot itineraries. During the pricing further slot itineraries which could improve solution are generated. The column generation's pricing identifies variables with violated dual constraints using the shadow prices α_s and β_H:

$$\sum_{s:p\in C_s} \alpha_s + \sum_{H:p\in P(H)} \beta_H < \rho \cdot \tau(p_r^*) - \tau(p)$$

where ρ is a detour factor and $\tau(p)$ is the travel time for itinerary p. Looking at the shadow prices of a solution, it is possible to detect time periods where slots or dwelling positions run short. When the shadow price α_s for a slot s has a high value, the existence of one more slot at this time would improve the target value. If there are many slots with a shadow price greater than zero within a time window, it can be concluded that there is a bottleneck on this line at this time. Additionally, free assignment can be taken into consideration. In free assignment each train service application gets its best itinerary as if there were no other influencing applications. Figure 3 shows the distribution of slots and free assigned applications for the line from Mainz (FMB) to Mannheim (RMF) whose capacity utilization is greater than 70 % for one day.

There are several peaks observable where six or even eight applications want to choose the same slot. The affected slots' shadow prices are greater than zero, so having more slots at this time would improve the solution. Shadow prices β_H belong to the nodes' capacity and will have a value greater than zero, if an additional dwelling position would improve the target value. Again, the shadow prices are related to a point in time or a time window. Since it is not possible to provide a dwelling position only temporarily, an additional one is only worthwile if it is used continuously over the day. Therefore, the sum of each node's shadow prices over a day is an adequate key figure for the necessity of an additional dwelling position. For each node there is

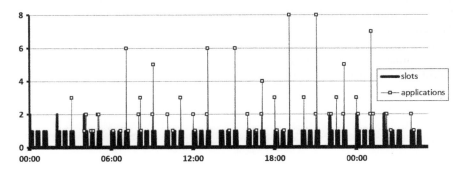

Fig. 3 Distribution of slots and free assigned applications for the line FMB-RMF

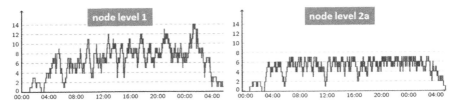

Fig. 4 Utilization of dwelling positions in the Mannheim node (RMF) for node level 1 and 2a

a graph showing the dwelling positions' utilization over time. In Fig. 4 it is contrasted for node level 1 and 2a of the Mannheim node[3]:

In this scenario it is obvious that Mannheim needs at least one additional dwelling position.[4] Hence, the Mannheim node is detected as a bottleneck in this scenario and it results in a detour for trains from Cologne to Basel, for example.

2.4 From Bottlenecks to Measures for the Development of Railway Infrastructure

After detecting the bottlenecks as described in the previous sections, corrective measures need to be found to eliminate the bottlenecks. If the number of slots on a line at a certain time is insufficient, the slots have to be modified. For that purpose measures affecting the timetable (e.g. shifting the passenger trains' departure time a little) or infrastructure (e.g. building an additional track or line) must be taken into account. For areas where slots overlap each other it is also possible to change their ratio if other routes have got spare capacities. The nodes' capacity can be improved by building more dwelling positions or increasing their total length so that they can be used by longer trains. With the help of additional assignment scenarios, where a solution for the bottleneck is included, the benefit of each measure can be determined. Furthermore, it is possible to identify the synergy effect of two coherent measures. Therefore the scenario's benefit with both measures is compared to the scenarios' benefits where each measure acts alone. Consequently, for the first time it is possible to determine a direct monetary gain for an additional infrastructure measure based on substantial timetable analyses.

---

[3] In node level 2a, the Mannheim node (RMF) has seven dwelling positions.

[4] The shadow price sum indicates the same: 2.79 in RMF compared with 0.00 in most other nodes.

3   Conclusion and Outlook

The effort of planning a long-term timetable and assigning a prediction of train service applications pays off and leads to a considerably better quality of long-term infrastructure planning. Firstly, all conclusion are based on a real timetable instead of analytical models with only traffic volumes. Consequently, the assignment provides a higher accuracy for all questions mentioned in Sect. 1 for developing the railway infrastructure. Secondly, it is possible to determine a direct monetary gain for an additional infrastructure based on the actual assignments.

Due to the intermeshed German railway network, there are a lot of slot over-lappings. Based on the assignment solution and detected bottlenecks, iteratively changing the slots' departure or interchanging overlapping slots will improve the assignment result considerably. The provided slots are adapted step by step to the needs of the train service applications. In terms of short-term planning, this is a very effective way to improve the assignments' solution and provide better train paths to the transportation companies. However, in long-term plannings the applications are only predicted and therefore the provided slots have to be robust to variations of demand. Hence, after a limited number of slot adaptions to improve the solution, the assignment has to be repeated with several scenarios containing a varied demand.

References

1. Feil, M., Pöhle, D.: Why does a railway infrastructure company need an optimized train path assignment for industrialized timetabling? In: International Conference on Operations Research, Aachen (2014)
2. Kübert, K.: Kapazitätsengpässe transparent machen. In: Deine Bahn, No. 5, pp. 38–41 (2012)
3. Nachtigall, K., Opitz, J.: Modelling and solving a train path assignment model. In: International Conference on Operations Research, Aachen (2014)
4. Oetting, A., Nießen, N.: Eisenbahnbetriebswissenschaftliches Werkzeug für die mittel- und langfristige Infrastrukturplanung. In: Proc.: 19. Verkehrswissenschaftliche Tage, Dresden (2003) Session 4e (98.1–98.8)
5. Weigand, W.: Von der Angebotsplanung über den Langfristfahrplan zur Weiterentwicklung der Infrastruktur. In: ETR, No. 7/8, pp. 18–25 (2012)

The Cycle Embedding Problem

Ralf Borndörfer, Marika Karbstein, Julika Mehrgardt,
Markus Reuther and Thomas Schlechte

Abstract Given two hypergraphs, representing a fine and a coarse "layer", and a cycle cover of the nodes of the coarse layer, the cycle embedding problem (CEP) asks for an embedding of the coarse cycles into the fine layer. The CEP is NP-hard for general hypergraphs, but it can be solved in polynomial time for graphs. We propose an integer programming formulation for the CEP that provides a complete description of the CEP polytope for the graphical case. The CEP comes up in railway vehicle rotation scheduling. We present computational results for problem instances of DB Fernverkehr AG that justify a sequential coarse-first-fine-second planning approach.

1 The Cycle Embedding Problem (CEP)

Let $G = (V, A, H)$ be a directed hypergraph with node set $V \subseteq E \times S$, i.e., a node $v = (e, s) \in V$ is a pair of an *event* $e \in E$ and *state* $s \in S$, arc set $A \subseteq V \times V$, and hyperarc set $H \subseteq 2^A$, i.e., a hyperarc consists of a set of arcs (this is different from most of the hypergraph literature). Each hyperarc $h \in H$ has cost $c_h \in \mathbb{Q}$. The following projections discard the state information:

R. Borndörfer · M. Karbstein · J. Mehrgardt · M. Reuther (✉) · T. Schlechte (✉)
Zuse Institute Berlin, Takustr. 7, 14195 Berlin, Germany
e-mail: reuther@zib.de

T. Schlechte
e-mail: schlechte@zib.de

R. Borndörfer
e-mail: borndoerfer@zib.de

M. Karbstein
e-mail: karbstein@zib.de

J. Mehrgardt
e-mail: mehrgardt@zib.de

© Springer International Publishing Switzerland 2016
M. Lübbecke et al. (eds.), *Operations Research Proceedings 2014*,
Operations Research Proceedings, DOI 10.1007/978-3-319-28697-6_65

$$[v] := e \qquad\qquad \text{for } v = (e, s) \in V \quad ([v]^{-1} = \{w \in V \mid [w] = [v]\}),$$
$$[a] := ([u], [v]) \qquad\quad \text{for } a = (u, v) \in A,$$
$$[h] := \{[a_i] \mid i = 1, \ldots, k\} \text{ for } h = \{a_i \mid i = 1, \ldots, k\} \in H.$$

We call $G = (V, A, H)$ the *fine (composition) layer* and $[G] := ([V], [A], [H])$ with $[V] := \{[v] \mid v \in V\}$, $[A] := \{[a] \mid a \in A\}$, and $[H] := \{[h] \mid h \in H\}$ the *coarse (configuration) layer*. W.l.o.g., we assume $[V] = E$. If $A = H$ then we equate $G = (V, A, H)$ with the standard graph $G = (V, A)$.

A set $K \subseteq A$ is a *cycle packing (partition)* in G if

1. $|\delta^-(v) \cap K| = |\delta^+(v) \cap K| \leq (=)1$, i.e., each node has at most (exactly) one incoming and at most (exactly) one outgoing arc and
2. there exists $H' \subseteq H$ such that $K = \bigcup_{h \in H'} h$ and $\forall a \in K \; \exists! h \in H' : a \in h$, i.e., the arc set K can be partitioned into hyperarcs; we say that $H(K) = H'$ is supported by K (there may be several supports).

K decomposes into a set of cycles $C^1, \ldots, C^k$. Let $C \in K \subseteq A$ be a cycle in G. We denote by $l(C) = |C|$ the *length* of cycle C. These definitions carry over to cycles and sets of cycles in $[G]$. It is easy to see that a cycle packing (partition) can only support hyperarcs h with $|h \cap \delta^-(v)| \leq 1$ and $|h \cap \delta^+(v)| \leq 1$ for all $v \in V$ and we henceforth assume that every $h \in H$ satisfies this property. We say that $[h] \in [H]$ *is embedded into* $h \in H$ and $h \in H$ *embeds* $[h]$. Our aim is to embed a coarse cycle partition into the fine layer.

Definition 1 Let $M \subseteq [A]$ be a cycle partition in $[G]$. The *CEP* is to find a cost minimal cycle packing $K \subseteq A$ in G such that

1. $\left| [v]^{-1} \cap V(K) \right| = 1$ for $[v] \in [V]$, i.e., the cycle packing M visits every event in exactly one state and
2. there exist fine and coarse supports $H(K)$ and $H(M)$ such that $[H(K)] = H(M)$, i.e., every hyperarc of $H(M)$ is embedded into a hyperarc of $H(K)$.

We call 1 the *uniqueness-condition* and 2 the *embedding-condition* and refer to the data of the cycle embedding problem as $(G, H, c, [G], M)$. Note that the embedding-condition 2 implies $[K] = M$. It further follows that the decomposition of K into cycles $C^1, \ldots, C^k$ gives rise to a decomposition of cycles $[C^1], \ldots, [C^k]$ for M.

An example for the CEP is illustrated in Fig. 1. We refer the reader to the master thesis of Mehrgardt [1] for further details and for proofs of the following results. For CEPs on standard graphs (the case $A = H$) the problem can be decomposed by considering each cycle of the coarse cycle partition M individually. For each such cycle one can define a "start node". Solving a shortest path problem for each state of the start node in the fine layer yields a polynomial time algorithm. In general, however, the problem is hard.

Theorem 1 *The cycle embedding problem can be solved in polynomial time for standard graphs; for hypergraphs, it is NP-hard.*

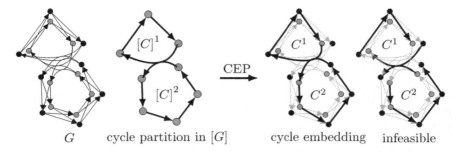

G cycle partition in $[G]$ cycle embedding   infeasible

Fig. 1 Example of the cycle embedding problem. *Left* A fine graph G and a hypercycle in $[G]$. *Right* A feasible cycle embedding and an infeasible set of hyperarcs which does not satisfy the uniqueness-condition 1

The CEP defined by the data $(G, H, c, [G], M)$ can be formulated as the following integer program. (Note that $h \in \delta^{+/-}(v) \Leftrightarrow \exists! a \in h : a \in \delta^{+/-}(v)$.)

$$\min \quad c^T x \tag{CEP}$$

$$\text{s.t.} \quad \sum_{h \in \delta^-(v)} x_h - \sum_{h \in \delta^+(v)} x_h = 0, \qquad \forall\, v \in V \tag{flow}$$

$$\sum_{h \in H : [h] = b} x_h = 1, \qquad \forall\, b \in [H](M) \tag{embedding}$$

$$x_h \in \{0, 1\} \qquad \forall\, h \in H.$$

There is a binary variable x_h for each hyperarc $h \in H$ indicating whether all arcs of h are contained in the cycle packing K. The (embedding)-constraints together with the integrality constraints define an assignment of every hyperarc in $[H](M)$ to exactly one hyperarc in H w.r.t. the given projection. The (flow)-constraints ensure that the solution is a cycle packing. Both conditions together ensure a unique assignment of events to states, i.e., the cycle packing K visits every event in exactly one state.

The uniqueness-condition 1 can be reformulated as follows.

Lemma 1 *Let K be a cycle packing in G with cycles $\{C^1, \ldots, C^k\}$ and let $[K]$ be a cycle partition in $[G]$. Then*

$$\left| [v]^{-1} \cap V(K) \right| = 1 \,\forall\, [v] \in [V] \quad \Leftrightarrow \quad \ell([C^i]) = \ell(C^i) \,\forall\, i \in \{1, \ldots, k\}. \tag{1}$$

Using this observation, we can come up with inequalities that prohibit cycles in G with different lengths than the corresponding cycles in $[G]$. Consider for some cycle $C \in M$ the set $\mathscr{U}(C) := \{\tilde{C} \text{ cycle in } G \mid [\tilde{C}] = C, \ell(\tilde{C}) \neq \ell(C)\}$ of cycles that project to the cycle C in $[G]$ but have a different length than C. Then, the cycles in

$\mathscr{U}(C)$ can be eliminated as follows:

$$\sum_{h:h\cap\tilde{C}\neq\emptyset} x_h \leqslant \ell(C) - 1 \qquad\qquad \forall\, C \in M,\ \tilde{C} \in \mathscr{U}(C). \qquad (2)$$

We call (2) *infeasible cycle constraints*. For the *basic CEP* these inequalities are all that it needed. More precisely, the CEP is a *basic CEP* if $H = A$ and $|S| = 2$, this is the simplest non-trivial problem variant.

Theorem 2 *The LP relaxation of (CEP) plus all infeasible-cycle constraints (2) provide a complete description for the basic CEP.*

The feasibility of a basic CEP can also be characterized combinatorially in terms of *switches*. A coarse arc $[(u, v)] \in [A]$ is a *switch w.r.t. state s* if $\delta^-(v) = \{(u, v)\}$, $u = (e, \tilde{s})$, $v = (f, s)$, $\tilde{s} \neq s$, i.e., each fine cycle containing node $v = (f, s)$ has to use arc b that switches from state $\tilde{s}$ to state s. The following theorem gives a complete characterization of the feasibility of the basic CEP that is easy to check:

Theorem 3 *A basic CEP has a feasible cycle embedding if and only if every coarse cycle has a state with an even number of switches.*

2 Application to Rolling Stock Rotation Planning

We aim at embedding a set of cycles, representing railway vehicle rotations computed in a coarse graph layer, into a finer graph layer with a higher level of detail. Our exposition resorts to a hypergraph based model of the *rolling stock rotation problem* (RSRP) proposed in our previous paper [2]. For ease of exposition, we discuss a simplified setting without maintenance and capacity constraints. In the following we define the (RSRP), introduce aspects of vehicle composition, and show how the results of Sect. 1 can be utilized in a two-step approach for the RSRP. Let V a set of *nodes*, $A \subseteq V \times V$ a set of directed *standard arcs*, and $H \subseteq 2^A$ a set of *hyperarcs*, forming an RSRP *hypergraph* that we denote by $G = (V, A, H)$. The nodes represent departures and arrivals of vehicles operating a set of timetabled passenger *trips* T, the arcs represent different ways to operate a timetabled or a deadhead trip by a single vehicle, the hyperarcs represent vehicle compositions to form trains. The hyperarc $h \in H$ *covers* trip $t \in T$ if every standard arc $a \in h$ represents t. We denote the set of all hyperarcs that cover $t \in T$ by $H(t) \subseteq H$. There are costs associated with the hyperarcs. The RSRP is to find a cost minimal set of hyperarcs $H_0 \subseteq H$ such that each timetabled trip $t \in T$ is covered by exactly one hyperarc $h \in H_0$ and $\bigcup_{h \in H_0} h \subseteq A$ is a set of *rotations*, i.e., a packing of cycles. The RSRP is NP-hard [2]. A key concept is the *orientation*, that describes the two options ($O = \{Tick, Tack\}$) how vehicles can be placed on a railway track. Deutsche Bahn Fernverkehr AG distinguishes the position of the first class carriage of the vehicle w.r.t. the driving direction. Tick (Tack) means that the first class carriage is located at the head (tail) of the vehicle. Every node of $v \in V$ has the form $v = (e, o)$, where e refers to an arrival or departure event of a vehicle operating a trip with an orientation $o \in O$. A hyperarc $h \in H$ models the connection of the involved tail and head nodes by a set of vehicles considering technical rules w.r.t. changes of the orientation. The hyperarcs of H are distinguished into those which implement a change of orientation caused by the network topology and those which implement an additional *turn around trip* that is necessary to establish dedicated orientations. The idea of our two step approach for the RSRP is to discard the orientations of the nodes in a first coarse planning step and to subsequently solve a CEP that arises from the solution of the first step in order to arrive at a solution in the fine layer including orientations. Evaluating this two-step approach investigates the question whether the topology of a railway network offers enough degrees of freedom to plan turn around trips subordinately. Let $[G] = ([V], [A], [H])$ be the hypergraph that arises if we discard the orientation by the projection procedure as it was defined for the CEP in Sect. 1. If we prevent ourselves from producing infeasible cycles in the first (coarse) step we increase the chance to end up with a CEP with a feasible solution in the second (fine) step. Forbidding such cycles is the main idea of our two-step approach. Let $[\mathbb{C}]$ be the set of all cycles in $([V], [A])$ that cannot be embedded if we consider the cycles as input for the basic CEP. Using a binary decision variable for every hyperarc $[h] \in [H]$, the model that we solve in the first step is the following coarse integer program:

$$\min \sum_{[h] \in [H]} \mathbf{c}_{[h]} x_{[h]}, \tag{MP}$$

$$\sum_{[h] \in [H](t)} x_{[h]} = 1 \qquad \forall t \in T, \tag{3}$$

$$\sum_{[h] \in [H]([v])^{in}} x_{[h]} - \sum_{[h] \in [H]([v])^{out}} x_{[h]} = 0 \qquad \forall [v] \in [V], \tag{4}$$

$$\sum_{[h] \in [H]([c])} x_{[h]} \leq |[C]| - 1 \qquad \forall [C] \in [\mathbb{C}], \tag{5}$$

$$x_{[h]} \in \{0, 1\} \qquad \forall [h] \in [H]. \tag{6}$$

The objective function of model (MP) minimizes the total cost. For each trip $t \in T$ the covering constraints (3) assign one hyperarc of $[H](t)$ to t. The Eqs. (4) are flow conservation constraints for each node $[v] \in [V]$ that define a set of cycles of arcs of $[A]$. Inequalities (5) forbid all cycles of $[\mathbb{C}]$. Finally, (6) states the integrality constraints for our decision variables. We solve model (MP) with the commercial solver for mixed integer programs `Cplex 12.4`. Since the number of inequalities (5) is exponential we handle these constraints dynamically by a separation routine that is based on an evaluation of integer solutions using Theorem 3 from the previous section, i.e., we check the switches in every cycle to see if it can be embedded or not. We evaluate this two-step approach by comparing two algorithmic variants to verify whether inequalities (5) are necessary and/or useful. The first variant "with ICS" is

Table 1 Comparison between embedding results of CEP models with or without separation of cycles that can not be embedded

Instance	T	With ICS				Without ICS		
		$|[H]|$	Cuts	Fine/coarse	Slacks	$|[H]|$	Fine/coarse	Slacks
1	267	1434	16	2/2	0	1434	2/4	199
2	617	3296	5	4/4	0	3292	4/5	208
3	617	3296	2	3/3	0	3292	4/5	14
4	617	3302	13	6/6	0	3292	2/5	285
5	617	3296	2	4/4	0	3294	3/4	29
6a	884	4779	2	5/5	0	4775	5/6	15
7	1443	78809	6	38/38	1	78807	39/44	63
8a	1443	29321	0	23/23	0	29321	23/23	0
9	1443	25779	2	33/33	1	25777	31/33	30
10	1443	14427	1	20/20	0	14421	21/22	12
11	1443	11738	2	17/17	0	11728	15/16	29
12	1713	14084	99	8/11	0	14074	1/11	1392
14	2319	15807	2	16/17	0	15787	17/17	43
15	2421	15829	4	18/19	0	15789	16/18	61
16	3101	40707	4240	40/42	74	40705	48/59	259
17	15	80	1	2/2	0	76	–/1	15

to solve the model MP as described and to solve the arising CEP in the second step. The only difference to the second variant "without ICS" is that we solve the model without constraints (5). Table 1 reports computational results for 16 instances with different numbers of trips (second column). Column "cuts" denotes the number of constraints (5) that were separated in the first step, while column "$|H|$" denotes the number of hyperarcs that appeared in the CEP of the second step. The columns "fine /coarse" and "slacks" report about the number of cycles that could be embedded vs. the number of cycles that were given to the CEP of the second step as well as the number of trips that could not be covered by the cycle packing in G. The instances were created to provoke that the arising CEPs are infeasible. Namely, we increased the time necessary to perform a turn around trip by approximately ten times the times in the real world data. This worst-case scenario substantially constrains the number of possible turn around trips. Our computational results show that by utilizing constraints (5) the RSRP can still be tackled by a two-step approach to produce feasible solutions. This gives evidence that railway vehicle rotation planning can indeed be done in two steps, embedding a coarse initial solution into a fine model layer.

References

1. Mehrgardt, J.: Kreiseinbettungen in Hypergraphen. Master's thesis, TU Berlin, Feb 2013
2. Markus Reuther, Ralf Borndörfer, and Thomas Schlechte.: A coarse-to-fine approach to the railway rolling stock rotation problem. In *14th Workshop on Algorithmic Approaches for Transportation Modelling, Optimization, and Systems, ATMOS 2014, September 11, 2014, Wroclaw, Poland*, pages 79–91, 2014

Dispatch of a Wind Farm with a Battery Storage

Sabrina Ried, Melanie Reuter-Oppermann, Patrick Jochem
and Wolf Fichtner

Abstract The combination of a wind farm with a battery storage allows to schedule the system in a more balanced way, alleviating natural wind power fluctuations. We present a mathematical model that optimizes the contribution margin (CM) of a system that consists of a wind farm and a lithium-ion battery storage from an operator's perspective. We consider the system to take part in the electricity stock exchange. We discuss adaptions of the model when additional participation at the minute reserve market is possible. We construct a test instance for the model for Germany and compare the optimal solutions to two reference cases. We evaluate if the gain of an integrated wind battery system compensates the investment and operating costs for the storage and we derive target prices for the battery system.

1 Introduction

Since 2012, the German renewable energy act (EEG) incentivizes direct marketing (DM) of electricity generated by renewable energies. Until July 2014, wind farm operators could choose between fix feed-in-tariffs (FIT) and DM, where a market premium was paid for the spread between average trading revenues and the FIT. By the end of 2013, more than 80 % of the electricity generated by wind power was traded through the direct marketing mechanism. An integrated wind battery system can be scheduled in a more balanced way and avoid throttling of the wind turbines due

S. Ried (✉)
IIP, Competence E, Karlsruhe Institute of Technology (KIT), Karlsruhe, Germany
e-mail: sabrina.ried@kit.edu

M. Reuter-Oppermann (✉)
KSRI, IOR, KIT, Karlsruhe, Germany
e-mail: melanie.reuter@kit.edu

P. Jochem · W. Fichtner
IIP, KSRI, KIT, Karlsruhe, Germany
e-mail: patrick.jochem@kit.edu

W. Fichtner
e-mail: wolf.fichtner@kit.edu

© Springer International Publishing Switzerland 2016
M. Lübbecke et al. (eds.), *Operations Research Proceedings 2014*,
Operations Research Proceedings, DOI 10.1007/978-3-319-28697-6_66

to grid bottlenecks. Furthermore, generation levelling facilitates additional revenue generation. This paper analyzes a 2013 installed wind battery system that takes part in different DM options and compares the results with two reference cases:

- Reference case 1: Average fix EEG FIT for wind energy
- Reference case 2: Revenues for wind energy from DM
- Wind farm with battery storage: Revenues are generated through the DM mechanism, where first the sole participation in the day-ahead market of the European energy exchange is considered (i), and second additional participation in the tertiary control market with minute reserve is possible (ii).

We present a mixed-integer linear program (MILP) that optimizes the CM for the direct marketing options (i) and (ii). We construct test instances and compare the optimal solutions to the reference cases. We evaluate whether the additional revenues in (i) and (ii) justify investing in the storage by a net present value (NPV) analysis.

2 Problem Formulation and Solution Approach

There are mainly two different approaches for an economic assessment of wind storage systems. MILP [1, 2] and stochastic dynamic programming models [3, 4]. Our MILP does not consider battery operating cost that we define to be fix, which allows a subsequent profitability analysis for different battery prices. Moreover, we assume perfect foresight on prices and wind power generation. The neglect of stochastics tends to result in an overestimation of the profitability. On the other hand, other model simplifications, such as excluding e.g. the intraday market and arbitrage through purchasing electricity, could influence the results in the opposite direction. Below, we describe the model (i) in detail and only explain the objective function and the most important changes in the constraints for the advanced model. The following notations for decision variables and parameters are used in model (i).

Decision variables:

X_t^{Spot}:   Energy that is sold on the spot market in period t [MWh]
P_t^c, P_t^d: Battery charging and discharging power in period t [MW]
C_t: Available battery capacity in period t [MWh]
W_t^u: Wind power used for trading and battery charging in period t [MW]
$B_t^c, B_t^d \in \{0, 1\}$: indicates if the battery is charged or discharged in period t

Parameters:

C_{min}, C_{max}:   Minimum and maximum battery capacity [MWh]
$P_{min}^{c/d}, P_{max}^{c/d}$: Minimum and max. battery charging and discharging power [MW]
D: Duration of one time period t [h], here 0.25 h
η^c, η^d: Battery charging and discharging efficiency [%]
p_t^{Spot}, p_t^{MP}: Spot market price and market premium in period t [€/MWh]

W_t^a: Available power of the wind farm in period t [MW]

$c_t^{W,var}$: Specific operating costs of wind farm in period t [€/MWh]

The formulation looks as follows:

$$\max CM = \sum_{t \in T} (p_t^{Spot} + p_t^{MP}) \cdot X_t^{Spot} - c_t^{W,var} \cdot W_t^u \cdot D \tag{1}$$

$$\text{s.t.} \qquad \frac{X_t^{Spot}}{D} + P_t^c = P_t^d + W_t^u \qquad \forall t \tag{2}$$

$$0 \le W_t^u \le W_t^a \qquad \forall t \tag{3}$$

$$X_{t+1}^{Spot} = X_t^{Spot} \qquad \forall t, z \le t \le z+4D, \tag{4}$$
$$z \in \{t \bmod 4 \ne 1\}$$

$$C_t = C_{t-1} + D \cdot (P_t^c \cdot \eta^c - P_t^d \cdot \frac{1}{\eta^d}) \qquad \forall t \tag{5}$$

$$C_{min} \le C_t \le C_{max} \qquad \forall t \tag{6}$$

$$P_{min}^{c,d} \le P_t^{c,d} \le P_{max}^{c,d} \cdot B_t^{c,d} \qquad \forall t \tag{7}$$

$$B_t^c + B_t^d \le 1 \qquad \forall t \tag{8}$$

$$B_t^{c,d} \in \{0, 1\} \qquad \forall t \tag{9}$$

$$X_t^{Spot} \ge 0 \qquad \forall t \tag{10}$$

The target function (1) maximizes the CM. While energy is balanced at all times (2), the generated wind power can remain unused (3). Equation (4) ensures that energy offered on the spot market remains constant within each 1 hour block. The battery charging state is modelled in (5). Moreover, there are boundaries for the battery size (6) and power rating (7). The constraints in (7) also ensure that the battery is only charged or discharged if the binary variable (9) is selected accordingly. The battery cannot be charged and discharged at once (8).

In a next step, we extended the model in order to enable additional participation in the minute reserve market. The battery can now be charged, when the system delivers negative minute reserve, and discharged, when the system delivers positive minute reserve. In order to allow for reservation of battery capacity for minute reserve, additional variables must be introduced. The target function of model (ii) is

$$\max CM = \sum_{t \in T} (p_t^{Spot} + p_t^{MP}) \cdot X_t^{Spot} + \sum_{t \in T} X_t^{pos} \cdot (\frac{1}{16} \cdot p_t^{C,pos} + d_t^{pos} \cdot p_t^{E,pos} \cdot D)$$
$$+ \sum_{t \in T} X_t^{neg} \cdot \left(\frac{1}{16} \cdot p_t^{C,neg} + d_t^{neg} \cdot p_t^{E,neg} \cdot D \right) - c_t^{W,var} \cdot W_t^u \cdot D \tag{11}$$

Here, $X_t^{pos/neg}$ is the reserved power for positive and negative minute reserve [in MW], $p_t^{C,pos/neg}$ is the capacity price [in €/MW] for reserved balancing power, and $p_t^{E,pos/neg}$ is the price for delivered energy [in €/MWh]. Parameter $d_t^{pos/neg}$ indicates the actually delivered minute reserve [in %].

3 Computational Results

In the following section, we present the input data and briefly describe and compare the results computed by the MILP with the reference cases.

3.1 Data

The test instance is created with 2013 data. The wind generation data from the transmission system operator 50 Hz is scaled to a wind park of 50 MW and yearly output of 2,700 kWh/kW. The usable battery size is set to 100 MWh; the battery can be charged and discharged at 50 MW [1, 2]. Due to the current progress in development and price decline, two lithium-ion batteries are chosen with a charging and discharging efficiency of 92.5 %, a depth of discharge of 80 % and cost of 600 and 1,000 €/kWh respectively. In the presented models, self-discharge as well as battery degradation are neglected. A lifetime of 20 years is assumed for both the battery and the wind farm. Yearly warranty cost of the battery is set to 2 % of the investment. The NPV is calculated with an interest rate of 6 % [3]. The wind farm is assumed to have investment cost of 1,000 €/kW and operating costs of 1.8 €-ct/kWh (maintenance and repair) [5]. Transaction costs for DM, taxes, EEG-levies, and grid fees are neglected. Spot and minute reserve market prices are available on [6] and [7].

3.2 Results for Wind-Battery System

When solely participating in the day ahead spot market, yearly revenues of 9.2 mn € can be realized. With variable operating cost of the wind farm, the CM is 6.8 mn €. However, taking into account investment and operating cost of the lithium-ion battery, the NPV is strongly negative between −142.4 and −80.2 mn € (Table 1).

Figure 1 shows generated wind power, electricity sold over the spot market, and the battery charging state, as well as spot market prices and revenues over the course of a day. Wind power generated during periods of low spot market prices or before periods of high spot market prices is used for charging the battery, whereas the battery is discharged at high spot market prices or before periods of low spot market prices. Through additional participation in the minute reserve market, yearly revenues increase to 11.1 mn €, the CM is 8.9 mn €, whereas the NPV is between −117.4 and −55.2 mn €. The maximum target battery price for a positive NPV is 80 €/kWh if

Table 1 Comparison of revenues, CM, and NPV

	Wind farm only		Combined wind and battery system			
	Ref. case 1	Ref. case 2	Spot market only (i)		With minute reserve (ii)	
Battery price in €/kWh	–	–	600	1,000	600	1,000
Yearly revenues in mn €	7.8	8.2	9.2		11.1	
Yearly CM in mn €	5.4	5.9	6.8		8.9	
NPV in mn €	4.3	9.6	−80.2	−142.4	−55.2	−117.4

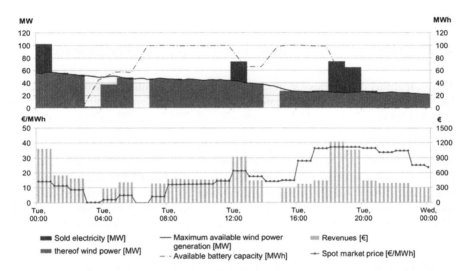

Fig. 1 Dispatch of the wind battery system and the dependency on spot market prices

electricity is traded on the spot market only, and increases to up to 240 €/kWh in case of additional participation in the minute reserve market.

3.3 Results for Reference Scenarios

The first reference case is a fix FIT for a 2013 installed 50 MW wind farm. The wind farm is assumed to apply for the energy system services bonus. The average FIT over 20 years is 5.83 €-ct/kWh. Compensation is calculated for 100 % of the generated electricity. The average yearly revenues would reach 7.8 mn €, the NPV is 4.3 mn €. Within the second reference scenario, the wind energy is traded over the day-ahead spot market. Assuming perfect foresight, as much energy as possible is sold, given that prices and market premium exceed the operating costs. Yearly revenues reach 8.2 mn €, the NPV is 9.6 mn €. With a market premium of zero, yearly revenues would reach 4.2 mn €, the NPV would be negative at −47 mn €.

3.4   Comparison

Through adding a lithium-ion battery system to a wind farm, the CM can be increased by 15–50 %. Taking battery investment into account, the NPV is strongly negative for lithium-ion battery prices between 600–1,000 €/kWh. This shows that trading electricity of a wind battery system was not economically viable in Germany in the year 2013. At hypothetical lithium-ion battery prices of 80–240 €/kWh, the market integration of wind battery systems might be more close to profitability.

4   Conclusions and Recommendations for Further Research

The profitability of batteries e.g. in combination with residential photovoltaic systems has been shown by recent publications [8, 9]. However, an economic viability of a wind battery system could not be shown with 2013 data. The results generated by the two MILP are mainly limited by perfect foresight. Yet, a further battery price decrease as well as the expected increasing market price fluctuations caused by a rising share of volatile renewable energy generation are indicating a future profitability of wind battery systems. Moreover, the latest EEG amendments will make alternative subsidy schemes become more attractive. In a next step, we will take into account uncertainties in wind forecasts and future price development and add other marketing options in order to deeper assess the profitability of the battery storage.

Acknowledgments This work has been funded by the European Commission (FP7 project MAT4BAT, grant no. 608931). We thank Lucas Baier for his support in the modeling part.

References

1. Kanngießer, A., Wolf, D., Schinz, S., Frey, H.: Optimierte Netz- und Marktintegration von Windenergie und Photovoltaik durch Einsatz von Energiespeichern. Konferenzbeitrag Internationale Energiewirtschaftstagung, Wien (2011)
2. Völler, S.: Optimierte Betriebsführung von Windenergieanlagen durch Energiespeicher. Ph.D. thesis, Wuppertal (2009)
3. Keles, D.: Uncertainties in Energy Markets and Their Consideration in Energy Storage Evaluation. KIT Scientific Publishing (2013)
4. Ding, H., Hu, Z., Song, Y.: Stochastic optimization of the daily operation of wind farm and pumped-hydro-storage plant. Renewable Energy **48**, 571–578 (2012)
5. Schlegl, T., Thomsen, J., Nold, S., Mayer, J.: Studie Stromgestehungskosten Erneuerbare Energien (2013)
6. Website of the European Power Exchange: Market data day ahead auction. www.epexspot.com (2014). Accessed 15 Feb 2014
7. Website for the auctioning of balancing services of the German transmission system operators. Balancing services data (2014). www.regelleistung.net. Accessed 01 Mar 2014

8. Kaschub, T., Jochem, P., Fichtner, W.: Interdependencies of home energy storage between electric vehicle and stationary battery. In: Proceedings of EVS27, Barcelona (2013)
9. Weniger, J., Tjaden, T., Quaschning, V.: Sizing of residential PV battery systems. In: Energy Procedia, 46, 7887 (2014)

Exact Algorithms for the Vehicle Routing Problem with Soft Time Windows

Matteo Salani, Maria Battarra and Luca Maria Gambardella

Abstract This paper studies a variant of the Vehicle Routing Problem with Soft Time Windows (VRPSTW) inspired by real world distribution problems. Soft time windows constraints are very common in the distribution industry, but quantifying the trade-off between routing cost and customer inconvenience is a hard task for practitioners. In our model, practitioners impose a minimum routing cost saving (to be achieved with respect to the hard time windows solutions) and ask for the minimization of the customer inconvenience only. We propose two exact algorithms. The first algorithm is based on standard branch-and-cut-and-price. The second algorithm uses concepts of bi-objective optimization and is based on the bisection method.

1 Introduction

The Vehicle Routing Problem (VRP) was proposed more than 50 years ago [6] and it is a challenging combinatorial optimization problem. One of the most studied real-world features are the so called *time windows constraints* [15]. Customers receiving goods often demand delivery within a time interval or *time window*. Time windows are classified as *Hard* (VRPHTW), if customers must be visited within the specified time interval, and *Soft* (VRPSTW), if time windows can be violated at the expense of customer inconvenience.

M. Salani (✉) · L.M. Gambardella
Istituto Dalle Molle di studi sull'Intelligenza Artificiale (IDSIA),
Scuola Universitaria Professionale della Svizzera italiana (SUPSI),
Università della Svizzera italiana (USI), Lugano, Switzerland
e-mail: matteo.salani@idsia.ch

L.M. Gambardella
e-mail: luca@idsia.ch

M. Battarra
University of Southampton, Southampton, UK
e-mail: m.battarra@soton.ac.uk

© Springer International Publishing Switzerland 2016   481
M. Lübbecke et al. (eds.), *Operations Research Proceedings 2014*,
Operations Research Proceedings, DOI 10.1007/978-3-319-28697-6_67

There is not a unique interpretation of time windows violations in the research community. Customer inconvenience for being visited too late and (possibly) too early with respect to the desired time window is typically quantified and the VRPSTW's objective function is modelled as a weighted combination of routing costs and a measure of the customer inconvenience. However, how to measure the customer inconvenience and how to quantify the relative weight of routing and customer inconvenience are still open research questions.

VRPSTWs have been presented in the pioneering articles by Sexton and Bodin [16, 17] and Sexton and Choi [18]. In Chang and Russell [5] a Tabu Search is developed for the same problem studied in Balakrishnan [1]. Taillard et al. [19] also proposed a Tabu Search for a VRPSTW, in which linear penalizations are applied only to late visits. Calvete et al. [3] applies goal programming to the VRPSTW with linear penalizations for early and tardy visits. Ibaraki et al. [10] extend the concept of time window and measure customer's inconvenience as a nonconvex, piecewise linear and time dependent function.

Fu et al. [9] acknowledge the need of a unified approach that model penalties associated to time windows. Figliozzi [8] proposed an iterative route construction and improvement algorithm for the same variant and compared its results with Balakrishnan [1], Fu et al. [9] and Chang and Russell [5].

Among the exact algorithms for VRPSTWs, we can mention Qureshi et al. [12], in which the authors solved by column generation a problem with semi soft time (tardy arrivals only). Bhusiri et al. [2] extend this algorithm to the variant in which both early and late arrival at a customer are penalizated, but the arrival time is bounded in an outer time window. Liberatore et al. [11] solves the VRPSTW with unbounded penalization of early and late arrival at the customers using a branch-and-cut-and-price technique.

Objective functions weighting routing costs and customer inconvenience suffer of typical drawbacks of weighted-sum multi objective optimization problems. As reported in Caramia and Dell' Olmo [4], the planner is often not "aware of which weights are the most appropriate to retrieve a satisfactorily solution, he/she does not know in general how to change weights to consistently change the solution".

In this paper, we model soft time windows constraints with in mind the needs of practitioners and their difficulties in comparing routing costs and customers inconvenience. Human planners accept time windows violations when the saving on routing costs is sufficiently big. We therefore compute a nominal solution in which hard time windows are imposed as a base of comparison for the planner. Then, the planner quantifies a desired saving on routing costs with respect to the nominal solution. The exact algorithm minimizes the time window violations to achieve this goal. We believe this variant will allow practitioners to make better use of VRPSTW software, because the parameter they are asked to define is just a measure of mileage saving and not a weight of routing cost and customer inconvenience. In the reminder, we propose two exact algorithms both based on branch-and-cut-and-price.

2   A Mathematical Model for Minimal Time Windows Violation

We recall the definition of the VRPHTW. A graph $G(V, A)$ is given, where the set of vertices $V = N \cup \{0\}$ is composed of a set of N vertices representing the customers and a special vertex 0 representing the depot. Non-negative weights t_{ij} and c_{ij} are associated with each arc $(i, j) \in A$; representing the traveling time and the transportation cost, respectively. Traveling times satisfy the triangle inequality. A positive integer demand d_i is associated with each vertex $i \in N$ and a capacity Q is associated with each vehicle of a set K. A non-negative integer service time s_i and a time window $[a_i, b_i]$, defined by two non-negative integers, are also associated with each vertex $i \in N$. The problem asks to find a set of routes with cardinality at most $|K|$, visiting all customers exactly once and respecting time windows and vehicles' capacity constraints. The objective is to minimize the overall routing cost.

In the VRPHTW, the vehicle has to wait until the opening of the time window a_i, in case of early arrival at customer's i location. In the VRPSTW, constant or proportional penalties are incurred for early or late service. The service may start any time between the arrival and the opening of the time window. However, both in the VRPHTW and in the VRPSTW, vehicles are allowed to wait at no cost before servicing the customer.

In our variant of VRPSTW, we propose an alternative model. The model assumes that the optimal value of the underlying VRPHTW, z^*, is known and is strictly positive. A cost saving is imposed by the planner as a maximum percentage $\beta < 1$ of z^*. The objective is to minimize the overall time windows violation. The model reads as follows:

$$g_\Theta = minimize \sum_{r \in \Theta} v^r x^r \tag{1}$$

$$s.t. \sum_{r \in \Theta} f_i^r x^r \geq 1 \quad \forall i \in N \tag{2}$$

$$\sum_{r \in \Theta} c^r x^r \leq \beta \cdot z^* \tag{3}$$

$$\sum_{r \in \Theta} x^r \leq |K| \tag{4}$$

$$x^r \in \{0, 1\} \quad \forall r \in \Theta \tag{5}$$

where Θ is the set of feasible routes in which the vehicle's capacity is not exceeded and v^r is the overall time window violation of route r. Constraint (3) states that the routing cost must be not greater than a fraction of the cost of the optimal VRPHTW solution. Constraint (4) imposes that no more than $|K|$ vehicles are used.

The routing cost improvement β is the only parameter required. Planners are likely to be comfortable defining parameter β as it directly relates to monetary savings.

3   Branch-and-Cut-and-Price Algorithm for the VRPSTW

Model (1)–(5) may contain a number of variables which grows exponentially with the size of the instance and cannot be dealt with explicitly. Therefore, to compute valid lower bounds, we solve the linear relaxation of the model recurring to a column generation procedure. To obtain feasible integer solutions we embed the column generation bounding procedure into an enumeration tree [7].

At each column generation iteration, the linear relaxation of the Restricted Master Problem (RMP, i.e., the model (1)–(5) where a subset of variables is considered) is solved. We search for new columns with negative reduced cost: $\overline{v}^r = v^r - \sum_{i \in N} f_i^r \pi_i - c^r \rho - \gamma$, where π_i is the nonegative dual variable associated to the ith constraint of the set (2), ρ is the nonpositive dual variable associated with the threshold constraint (3) and γ is the nonpositive dual variable associated with constraint (4). The pricing problem can be modeled as a resource constrained elementary shortest path problem (RCESPP). In our implementation, we extend the algorithms presented in Righini and Salani [13, 14] and Liberatore et al. [11].

4   Alternative Formulation and Bisection Search

The algorithm reported in Sect. 3 is effective to identify infeasible instances and a feasible solution, when available. On the other hand, it shows slow convergence in proving the optimality of the master problem because the computation time required by the exact pricing problem is cumbersome.

In this section we propose an alternative formulation and an alternative exact solution algorithm based on bisection search. The model minimizes the overall routing costs subject to a maximal permitted time windows violation. The model is solved with branch-and-cut-and-price and reads as follows:

$$h_\Theta = minimize \sum_{r \in \Theta} c^r x^r \tag{6}$$

$$s.t. \sum_{r \in \Theta} f_i^r x^r \geq 1 \quad \forall i \in N \tag{7}$$

$$\sum_{r \in \Theta} v^r x^r \leq g_{max} \tag{8}$$

$$\sum_{r \in \Theta} x^r \leq |K| \tag{9}$$

$$x^r \in \{0, 1\} \quad \forall r \in \Theta \tag{10}$$

The pricing problem searches for columns minimizing the following reduced cost: $\overline{c}^r = c^r - \sum_{i \in N} f_i^r \pi_i - v^r \psi - \gamma$, where π_i is the nonegative dual variable associated to the ith constraint of the set (7), ψ is the nonpositive dual variable associated with the time windows violation (8) and γ is the nonpositive dual variable associated with constraint (9). The pricing problem associated to this formulation is equivalent to that studied by Liberatore et al. [11] where the linear penalty for earliness and tardiness is adjusted with the dual variable of constraint (8).

Our algorithm is based on a bisection search on the value of the permitted violation g_{max}. At each iteration of the algorithm g_{max} represents either an upper or a lower bound to the optimal value of model (1)–(5), g_Θ^*.

Let $h_\Theta^*(g_{max})$ be the optimal solution of (6)–(10) for a given value of g_{max}. The algorithm exploits the following two properties:

1. If $h_\Theta^*(g_{max}) > \beta \cdot z_\Omega^*$, then g_{max} is a valid lower bound to g_Θ^*.
2. If $h_\Theta^*(g_{max}) \leq \beta \cdot z_\Omega^*$, then $\sum_{r \in \Theta} v^r x^r$ is a valid upper bound to g_Θ^*.

The Algorithm requires the existence of a feasible solution and the value of an upper bound g_{UB} to g_Θ^*. At each iteration, the range of possible values for the violation of time windows $(g_{LB,\Theta}^{it}, g_{UB,\Theta}^{it})$ is halved, as reported in algorithm 1; the value ε is strictly positive and is determined using the instance data.

We performed some preliminary computational experiments on the well known Solomon's data set. From the original set we derived 54 instances from the R and RC data set by adding a performance constraint of 1, 5 and 10 % with respect to the optimal solution without time windows violation. The bisection algorithm converges to an optimal solution for all instances while the branch and price algorithm failed to converge on 12 instances within an our of computation.

We observe that the solution of the pricing problem benefits from the alternative formulation adopted in the bisection algorithm. We believe that in other contexts, where branch and price algorithms applied to combinatorial optimization exhibit slow convergence, it may be beneficial to devise alternative formulations.

Algorithm 1 Bisection search

Require: β, z_Ω^*, g_{UB}
   $it := 0; g_{UB,\Theta}^{it} := g_{UB}; g_{LB,\Theta}^{it} := 0;$
   while $\left(g_{UB,\Theta}^{it} - g_{LB,\Theta}^{it} > \varepsilon \right)$ **do**
      $g_{max} := \left(g_{UB,\Theta}^{it} + g_{LB,\Theta}^{it} \right) /2;$
      $h_\Theta^{*,it} :=$ Solve (6)–(10);
      if $h_\Theta^{*,it} > \beta \cdot z_\Omega^*$ **then**
         $g_{LB,\Theta}^{it+1} := g_{max}; g_{UB,\Theta}^{it+1} := g_{UB,\Theta}^{it};$
      else
         $g_{UB,\Theta}^{it+1} := \sum_{r \in \Theta} v^r x^r; g_{lB,\Theta}^{it+1} := g_{LB,\Theta}^{it};$
      end if
      $it := it + 1;$
   end while

References

1. Balakrishnan, N.: Simple heuristics for the vehicle routeing problem with soft time windows. J. Oper. Res. Soc. **44**, 279–287 (1993)
2. Bhusiri, N., Qureshi, A., Taniguchi, E.: The trade-off between fixed vehicle costs and time-dependent arrival penalties in a routing problem. Transp. Res. Part E: Logistics Transp. Rev. **62**, 1–22 (2014)
3. Calvete, H., Gal, C., Oliveros, M., Sánchez-Valverde, B.: Vehicle routing problems with soft time windows: an optimization based approach. Monograf'ias del Seminario Matemático García de Galdeano **31**, 295–304 (2004)
4. Caramia, M., Dell' Olmo, P. (eds.): Multi-objective Management in Freight Logistics. Springer, London (2008)
5. Chang, W.C., Russell, R.: A metaheuristic for the vehicle-routeing problem with soft time windows. J. Oper. Res. Soc. **55**, 1298–1310 (2004)
6. Dantzig, G.B., Ramser, J.H.: The truck dispatching problem. Manage. Sci. **6**, 80–91 (1959)
7. Desaulniers, G., Desrosiers, J., Solomon, M. (eds.): Column generation. GERAD 25th Anniversary Series. Springer (2005)
8. Figliozzi, M.: An iterative route construction and improvement algorithm for the vehicle routing problem with soft time windows. Transp. Res. Part C: Emerg. Technol. **18**, 668–679 (2010)
9. Fu, Z., Eglese, R., Li, L.: A unified tabu search algorithm for vehicle routing problems with soft time windows. J. Oper. Res. Soc. **59**, 663–673 (2008)
10. Ibaraki, T., Imahori, S., Kubo, M., Masuda, T., Uno, T., Yagiura, M.: Effective local search algorithms for routing and scheduling problems with general time-window constraints. Transp. Sci. **39**, 206–232 (2005)
11. Liberatore, F., Righini, G., Salani, M.: A column generation algorithm for the vehicle routing problem with soft time windows. 4OR: A Quarterly J. Oper. Res. **9**, 49–82 (2011)
12. Qureshi, A., Taniguchi, E., Yamada, T.: An exact solution approach for vehicle routing and scheduling problems with soft time windows. Transp. Res. Part E: Logistics Transp. Rev. **45**, 960–977 (2009)
13. Righini, G., Salani, M.: Symmetry helps: bounded bi-directional dynamic programming for the elementary shortest path problem with resource constraints. Discret. Optim. **3**(3), 255–273 (2006)
14. Righini, G., Salani, M.: New dynamic programming algorithms for the resource constrained shortest path problem. Networks **51**, 155–170 (2008)
15. Schrage, L.: Formulation and structure of more complex/realistic routing and scheduling problems. Networks **11**, 229–232 (1981)
16. Sexton, T., Bodin, L.: Optimizing single vehicle many-to-many operations with desired delivery times: I. scheduling. Transp. Sci. **19**, 378–410 (1985a)
17. Sexton, T., Bodin, L.: Optimizing single vehicle many-to-many operations with desired delivery times: Ii. routing. Transp. Sci. **19**, 411–435 (1985b)
18. Sexton, T., Choi, Y.M.: Pickup and delivery of partial loads with "soft" time windows. Am. J. Math. Manage. Sci. **6**, 369–398 (1986)
19. Taillard, E., Badeau, P., Gendreau, M., Guertin, F., Potvin, J.Y.: A tabu search heuristic for the vehicle routing problem with soft time windows. Transp. Sci. **31**, 170–186 (1997)

Impact of Heat Storage Capacity on CHP Unit Commitment Under Power Price Uncertainties

Matthias Schacht and Brigitte Werners

Abstract Combined heat and power (CHP) plants generate heat and power simultaneously leading to a higher efficiency than an isolated production. CHP unit commitment requires a complex operation planning, since power is consumed in the moment of generation. The integration of a heat storage allows a partially power price oriented plant operation, where power is generated especially in times of high market prices. Consequently, an efficient plant operation depends on the accuracy of the anticipated power prices and the flexibility due to storage capacity. This contribution analyzes the effects of short-term uncertainties in power prices on the CHP unit commitment for different heat storage capacities. A simulation study is run to evaluate the financial impact of an inaccurate power price anticipation. Results show that the storage capacity affects the sensitivity of the solution due to stochastic influences. The isolated consideration of long-term uncertainties might result in a suboptimal choice of heat storage capacity. It is recommended, to explicitly consider short-term uncertainties when supporting strategic planning of heat storage capacities.

1 Introduction

The German electricity market for residual power is characterized by high uncertainties with respect to prices and residual load. The result of the priority feed-in of renewable energies into the grid is significantly determined by stochastic weather conditions. This stochastic residual power load can be fulfilled—among others—by very efficient combined heat and power (CHP) plants, which generate power and heat in one process. Thereby fuel efficiency can be increased by up to 40 % [3] which is why this technology is promoted by the German cogeneration protection law in order to achieve the German climate objectives [6].

M. Schacht (✉) · B. Werners
Faculty of Management and Economics, Chair for Operations Research
and Accounting, Ruhr University Bochum, 44780 Bochum, Germany
e-mail: matthias.schacht@rub.de

B. Werners
e-mail: or@rub.de

© Springer International Publishing Switzerland 2016
M. Lübbecke et al. (eds.), *Operations Research Proceedings 2014*,
Operations Research Proceedings, DOI 10.1007/978-3-319-28697-6_68

Unit commitment (UC) deals with the question which generation unit should run at which output level on every hour. UC for CHP plants is especially challenging and complex since demand and supply of power and heat have to be matched while these products are generated in one process [8]. Additionally, demand patterns for heat and power are often asynchronous. Complexity increases because uncertainties have to be considered. For the UC the crucial stochastic influence can be seen in the market for power sales [1].

Whereas power cannot be stored efficiently in a large scale system, heat can be stored in a storage. The integration of a storage device allows a partially power price oriented plant operation, where power is generated especially in times of high market prices [2]. In times of volatile power prices, this results in a significant financial benefit (added value) [2, 6, 8]. Up to now, the impact of short-term uncertainties in power prices on the UC for CHP plants with storage devices has not been focused in research, while short-term uncertainties for only power plants have been studied [1]. Long-term power price uncertainties have been incorporated in order to determine the optimal storage capacity [2]. Without consideration of short-term risks, the value of a heat storage might be determined incorrectly.

In the following the UC problem for a CHP unit with heat storage is presented.

2 CHP Unit Commitment Under Power-Price Uncertainties

The objective of UC for CHP plants is the coordination of heat and power generation, satisfying heat demand at all times, while maximizing trading revenues of power sales subtracted by its generation costs. Without heat storage, plant operation follows the demand pattern for heat in the moment of production since heat cannot be traded like power due to its physical limitations. The corresponding power output is produced without any consideration of power prices on the market.

The integration of a heat storage allows the anticipation of future power prices in order to run the plant in the most profitable hours with respect to revenues from the spot-market. The UC for CHP plants with a heat storage can be modeled as a mathematical optimization problem. The operation region of a CHP plant is described by its extreme points—characterized by power output P_i, heat output Q_i and corresponding costs C_i in each extreme point i. For a convex operation region, each feasible operation point is modeled as a convex combination of these extreme points (2–5), shown in [5]. The objective function maximizes the operational profit (1) which is the difference between the market revenues for power (generated power p_t to corresponding power price pr_t in periode t) and the occuring costs for the plant operation $\sum_{i \in I} C_i \cdot x_{it}$. Heat demand q_t^d has to be satisfied by heat production q_t or by the amount of discharged heat from the storage c_t^-.

$$\max \quad \sum_{t \in T} \left(p_t \cdot pr_t - \sum_{i \in I} C_i \cdot x_{it} \right) \tag{1}$$

$$\sum_{i \in I} C_i \cdot x_{it} = c_t \qquad \forall t \in T \tag{2}$$

$$\sum_{i \in I} Q_i \cdot x_{it} = q_t \qquad \forall t \in T \tag{3}$$

$$\sum_{i \in I} P_i \cdot x_{it} = p_t \qquad \forall t \in T \tag{4}$$

$$\sum_{i \in I} x_{it} = 1 \qquad \forall t \in T \tag{5}$$

$$q_t - c_t^+ + c_t^- = q_t^d \qquad \forall t \in T \tag{6}$$

$$\ell_{t-1} + c_t^+ - c_t^- = \ell_t \qquad \forall t \in T \setminus \{1\} \tag{7}$$

$$x_{it}, c_t^+, c_t^-, \ell_t \geq 0 \qquad \forall i \in I, t \in T \tag{8}$$

The storage level ℓ_t in period t is equal to the period before plus the amount of charged heat c_t^+ subtracted by the discharged heat c_t^- in period t (7). The model can be extended by limiting to maximum heat storage, charging and discharging as well as start-ups with corresponding costs [8].

The integration of a heat storage enables the UC to decouple the moment of production and moment of demand. This allows a plant operation in times of favorable power prices. Information about future power prices are used in order to decide wether to run the plant at the current period or not. Therefore, the anticipation of future power prices influences the storage policy for the current period. If power prices are inaccurately anticipated, revenues of the plant operation are negatively affected. The anticipation of power prices and the impact of uncertain power prices is explained in Fig. 1. The solid line shows the market price for power with the corresponding optimal usage of the heat storage (gray area). In times of (relatively) high market prices the plant is run and the storage is charged. This allows a shut-down in times of low market prices, while the demand for heat can be satisfied by the storage. The plant operation might differ significantly if price forecasts are inadequate. That is why forecasts are being updated in order to gain the best information about future price deployments and the UC is being rescheduled on the basis of the latest information.

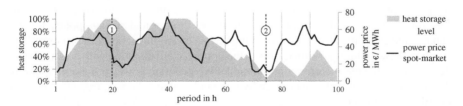

Fig. 1 Usage of a heat storage with respect to power price and effects of short-term uncertainties

There are two extreme situations where an incorrect anticipated power price might have a high impact on the trading revenues. In point ① the heat storage is filled, since high revenues were gained due to high power prices in the hours before. A sharp decrease of power price was expected and a shut down for the plant was planned. Assuming that at this point the forecast was updated and that the real power price will be significantly higher in the following hours, UC would adjust the plant operation and run the unit to gain high revenues. But this is impossible because the storage is fully charged so that the plant cannot be operated. In contrast to a situation where future power prices are underestimated, ② shows a situation, where an overestimation of the power price strongly influences revenues. Under the assumption that the power price will be lower than forecasted, the plant would have to be run at times of very low power prices in order to fulfill the heat demand, since the storage is almost empty. In both situations, an inadequate anticipation of the power price leads to lower revenues on the power market because the plant cannot be operated with the necessary flexibility due to limited heat storage capacity.

Higher heat storage capacities can lead to a more flexible plant operation and effects of uncertainty on revenues can much better be exploited.

3 Simulation Study with Varying Heat Storage Capacities and Uncertain Power Prices

We present a simulation study with a small municipal energy provider who generates power and heat with a CHP plant with a back pressure steam turbine that can generate up to $215\,MWh_{el}$ of power with a corresponding heat output of $60\,MWh_{th}$. With this CHP plant, the operator has to guarantee the heat supply while power can be sold at the spot-market. The objective is to maximize the operational profit as described in section two. We analyze to what extent the capacity of heat storage influences the profitibility of CHP operation when power prices are uncertain. This indicates the importance of integrating short-term uncertainties into strategic planning.

In order to run the simulation in a reasonable amount of time, we developed a combined usage of two well-known algorithms. In [5] an efficient algorithm is presented that reshapes the three dimensional convex operation region of a CHP plant into a two dimensional efficient envelope pattern. This envelope pattern describes an adjusted cost-function for each feasible operation point of the plant dependent on the power price. A piecewise-linear cost-function is generated which is used as input for an efficient algorithm in [7] that determines the optimal storage policy for the capacitated economic lot sizing problem with piecewise-linear production costs. This combined use of the algorithm results in the optimal CHP planning with a heat storage device. To consider uncertainy in the model we generate power-price-anticipations that differ from the real power price by a deviation that follows a truncated normal distribution which is correlated and increases over time as described in [4].

In order to consider the process of rolling updates of the UC planning, forecasts are being updated every four hours during the planning horizon of two weeks on a hourly bases. By this we consider the possibility to react on new information about the future power price deployment, so that UC planning can be adjusted after any update, assuming a guaranteed heat demand and fixed prices.

The simulation was run without a heat storage and with five different heat storage capacities (fifty runs each). The additionally gained revenue due to the storage is defined as the added value. The impact of inaccurate forecasts leads to a decreased added value of up to 6 %. Figure 2 shows the expected additional revenues achieved by the integration of a heat storage. With the integrated storage the expected added value differs between 0.85 mio € and 1.07 mio € which is equal to an overall increase of revenues by 60 % up to 70 %. Therefore, the contribution of a heat storage to the revenue is significant. Figure 2 shows that the added value by an increased storage capacity diminishes, as e.g. the gap of the added value between a heat storage capacity of 250 and 300 MWh$_{th}$ is low. The investment costs for a higher storage capacity might exceed its additional value. It also shows the corresponding influences of uncertain power prices. For small to medium-sized capacities the impact of uncertainty on revenues stays constant as the dashed line that represents the standard deviation of the expected added value stays almost steady. Therefore, the decision maker is able to gain higher revenues without higher risk. In contrast, the heat storage with the highest capacity will lead to a significant deviation on the expected gained revenue, eventhough there is no significant higher expected added value. This is due to the fact that such a high storage level is rarely needed during the considered time horizon. Since there is no increase of risk on the added value between small and medium-sized heat storages, we can conclude that there are benefits of medium-sized storages because UC planning is able to react more flexible on inaccurate anticipated power prices. Further results confirm that the impact of uncertainty on UC, as described in Fig. 1, is less likely if storage capacity increases. But this effect holds true only to a certain level of capacity as it can be seen for 300 MW$_{th}$, where the impact of uncertainty strongly affects the revenues since the high capacity allows an anticipation of prices that are in the distant future and therefore likely to be inaccurate.

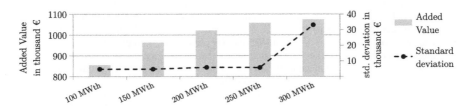

Fig. 2 Expected added value and impact of uncertainty for differing a heat storage capacities

4 Conclusion and Outlook

This contribution presents the impact of heat storage capacity on CHP unit commitment under power price uncertainties. Results show that power price uncertainties strongly affect revenues but also that the operational profit can be increased significantly by a power price oriented plant operation enabled by an increased heat storage capacity. For a certain range of heat storage capacity the influence of uncertainties remains constant due the higher flexibility described in section two and confirmed in section three. The effect of additionally gained revenues diminishes with a certain level of capacity while impact of uncertainty affects the revenues significantly. Information about the impact of short-term uncertainties have to be considered in the strategic planning of heat storage capacities, since too small or too large capacities either lead to a low added value or a high negative impact of uncertainty.

To confirm these insights different extensions to our model can be included, such as limiting the maximum charging capacity per period as well as a heat loss. Several power plants or additional power only units can be included in order to analyze the effect of uncertainties for a plant portfolio. Then, a concept to involve these short-term effects into a strategic optimization model for CHP units with heat storage devices can be developed.

References

1. Cerisola, S., Baillo, A., Fernndez-Lopez, J.M., Ramos, A., Gollmer, R.: Stochastic power generation unit commitment in electricity markets: a novel formulation and a comparison of solution methods. Oper. Res. **57**(1), 32–46 (2009)
2. Christidis, A., Koch, C., Pottel, L., Tsatsaronis, G.: The contribution of heat storage to profitable operation of combined heat and power plants in liberalized electricity markets. Energy **41**, 75–82 (2012)
3. Mitra, S., Sun, L., Grossman, I.E.: Optimal scheduling of industrial combined heat and power plants under time-sensitive electricity prices. Energy **54**, 194–211 (2013)
4. Ortega-Vazquez, M.A., Kirschen, D.S.: Economic impact assessment of load forecast errors considering the cost of interruptions. Power Eng. Soc. Gen. Meet. 2006 IEEE, 1–8 (2006)
5. Rong, A., Lahdelma, R.: Efficient algorithms for combined heat and power production planning under the deregulated electricity market. Eur. J. Oper. Res. **176**(2), 1219–1245 (2007)
6. Schacht, M., Schulz, K.: Kraft-Wärme-Kopplung in kommunalen Energieversorgungsunternehmen—Volatile Einspeisung erneuerbarer Energien als Herausforderung. Armborst, K. et al. (eds) Management Science—Festschrift zum 60. Geburtstag von Brigitte Werners, Dr. Kovac, Hamburg, pp. 337–363 (2013)
7. Shaw, D., Wagelmans, A.: An algorithm for single-item capacitated economic lot sizing withing piecewise linear production costs and general holding costs. Manage. Sci. **44**(6), 831–838 (1998)
8. Schulz, K., Schacht, M., Werners, B.: Influence of fluctuating electricity prices due to renewable energies on heat storage investments. To be published. In: Huisman, D. et al. (eds.) Proceedings of 2013 Operations Research. Springer, New York (2014)

Optimizing Information Security Investments with Limited Budget

Andreas Schilling and Brigitte Werners

Abstract The importance of information security is constantly increasing with technology becoming more pervasive every day. As a result, the necessity and demand for practical methods to evaluate and improve information security is particularly high. The aim of this paper is to apply mathematical optimization techniques tool improve information security. According to the identified problem structure, a combinatorial optimization model is established. The objective of the presented approach is to maximize system security by choosing the best combination of security controls limited by available budget. In addition, by performing a What-If analysis and systematic budget variations, the decision maker can get improved insights and thus determine an ideal budget proposition yielding the highest benefit among all possible control configurations. An exemplary case study demonstrates how this approach can be used as a tool within the risk management process of an organization.

1 Introduction

The importance of information security in information systems is constantly increasing with technology becoming more pervasive every day. It is in the very nature of risk that perfect security does not exist and, therefore, investments in information security only make sense up to a certain amount. As a consequence, it is crucial to ensure that available resources are spent as effectively as possible.

There exists a considerable market for security solutions which is why the problem is not finding security controls to protect a system, but identifying the *right ones* within a given budget. Possible controls have a wide variety of capabilities to protect against different threats and vulnerabilities and, in addition, they may be mutually exclusive or complement each other. To prioritize between controls it is necessary to

A. Schilling (✉) · B. Werners
Department of Management and Economics, Chair of Operations
Research and Accounting, Ruhr University Bochum, Bochum, Germany
e-mail: andreas.schilling@rub.de

B. Werners
e-mail: or@rub.de

© Springer International Publishing Switzerland 2016
M. Lübbecke et al. (eds.), *Operations Research Proceedings 2014*,
Operations Research Proceedings, DOI 10.1007/978-3-319-28697-6_69

measure their impact on security before they are deployed. A quantitative optimization model enables the decision maker to maximize the effect of deployed security controls prior to their deployment in a production environment.

In the following, a combinatorial nonlinear problem is described which minimizes expected losses caused by security incidents. We demonstrate in an exemplary case study how it can be used as a tool within different phases of the risk management process of an organization.

2 Security Optimization in the Context of Risk Management

Decisions on information security investments are usually taken as part of an organizational risk management process. There are different approaches and standards available which differ in detail but follow a similar overarching concept. The following description of the process is based on the widely used ISO 27005:2008 standard [6]. First, the context for the process is established in terms of scope and boundaries. The next step is a risk assessment to identify, estimate, and evaluate risk. The starting point for this assessment are threats, which may occur and cause damage to the organization. Threats may exploit different vulnerabilities and have a financial impact if successful. The goal of the risk identification phase is to identify what incidents may occur, what impact they could have, and what factors might be responsible for their success. In the risk estimation phase, consequences and the likelihood of incidents are determined by using quantitative or qualitative methods. The results are input to the evaluation phase to rank threats and prioritize mitigating actions. According to this prioritization, a risk treatment plan is defined and concrete security controls are selected. The whole process is supported by continuous monitoring and a learning feedback loop [6]. Figure 1 illustrates this process.

The problem with this procedure lies in the separation of risk evaluation and treatment. This separation results from the fact that the dependency structure of the incident and control side is too complex to be treated simultaneously by traditional

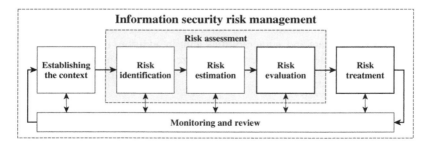

Fig. 1 Information security risk management process based on ISO 27005:2008

methods. This problem can be solved by using a quantitative optimization model which takes both sides into account and combines them into an integrated decision model. The goal is to find a selection of security controls that counteracts incidents as effectively as possible. Complex requirements like control incompatibilities and budget restrictions are integrated as constraints.

Using a quantitative optimization approach to solve this problem can improve the outcome of a risk management process significantly. The model can manage higher complexity and eliminates the necessity for decoupling the evaluation and treatment phase. Results can be produced faster and in a more consistent way. Changing circumstances in form of new vulnerabilities or threats can be considered in real-time without manually reevaluating the existing treatment plan.

3   Problem Description

The source of security risks are threats which have to be identified and their criticality has to be quantified to prioritize counter measures. A threat i, $i \in I$ results in an incident that causes loss l_i if it occurs. An incident may be induced by different threat agents including hackers, competitors, administrative staff, or malicious insiders. Loss typically results from damage to an asset order technical components like hardware, software, and data. To attack an asset, a threat agent can exploit one of many possible vulnerabilities j, $j \in J$. A vulnerability is a technical or organizational shortcoming of the system. This may be an insecure API endpoint [7], software backdoors [10], or even lack of, or insufficient, rules for employees [1]. To prevent an attacker from taking advantage of such vulnerabilities, an organization can deploy security controls k, $k \in K$ which affect the probability of a threat being successful [9].

In most cases, a threat can exploit a number of vulnerabilities and thus cause damage. For each threat, a set of potential vulnerabilities can be identified. Probability p_j^v denotes how likely it is that a vulnerability j is exploited by threat i. The same applies to vulnerabilities and security controls. p_{jk}^c indicates how likely it is that control k prevents vulnerability j from being exploited. This definition takes into account that most security controls do not only affect a single vulnerability, but multiple ones. Therefore, the interrelation of vulnerabilities and controls leads to a situation, where different combinations of controls have a completely different effect on security. If the selection of controls is not performed in an integrated way considering all vulnerabilities at the same time, possible positive interdependencies of controls remain unrealized. The following equation yields probability δ_i of an incident depending on the selection of controls:

$$\delta_i = p_i^T \left(1 - \prod_{j \in J} \left(1 - p_{ij}^V \prod_{k \in K} \left(1 - p_{jk}^c \cdot sc_k \right) \right) \right). \tag{1}$$

Variable $sc_k \in \{0, 1\}$ corresponds to the $|K|$ security controls available for selection and is defined as $sc_k = 1$ if control k is selected, and $sc_k = 0$ otherwise. Parameter p_i^T specifies the probability that a threat agent attacks the system once in a way corresponding to threat i. The probability δ_i of an incident is the joint probability that a threat occurs, at least one vulnerability is exploited, and all controls of this vulnerability fail.

The model is formulated with a nonlinear objective function and linear constraints. The optimal selection of controls minimizes the overall loss L of an organization. The objective function has the form

$$L = \sum_{i \in I} l_i \cdot \delta_i \cdot n_i, \tag{2}$$

where n_i is the number of occurrences of threat i. The structure of δ_i results in a nonlinear programming (NLP) problem. The solution space of possible control configurations is limited by a control compatibility constraint

$$sc_k + sc_l \leq 1 + \gamma_{kl} \quad \forall k \in K, l \in K, k < l, \tag{3}$$

where $\gamma_{kl} \in \{0, 1\}$ defines if control k is compatible to control l ($\gamma_{kl} = 1$) or not ($\gamma_{kl} = 0$). The budget constraint has the form

$$\sum_{k \in K} c_k \cdot sc_k \leq B, \tag{4}$$

where c_k is the cost of control k and B is the budget available for controls.

4 Computational Results

The model is implemented using the standard optimization software Xpress Optimization Suite [3]. The NLP problem is solved applying successive linear approximation provided by the Xpress-SLP solver module `mmxslp`. It allows modeling and solving of complex and large nonlinear problems [2]. To demonstrate the models application, an exemplary case study is developed. The data are based on the authors' practical experience in developing and maintaining real-world information systems. In the following, we analyze an exemplary cloud-based system to demonstrate how the model can be applied in a realistic setting.

During the risk identification phase 20 threats, 40 vulnerabilities, and 40 potential security controls were identified. All required model parameters were determined in the subsequent risk estimation phase by expert estimates. As shown in [8], expert judgement elicitation is a good approach to collect data in information security if no historical data are available.

Fig. 2 Optimal loss $L^*(B)$ depending on increasing budget B

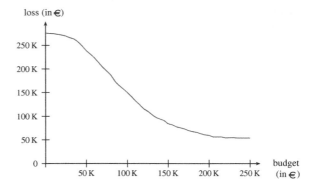

To solve the model, the decision maker has to provide a fixed budget limit as an input parameter. This budget limitation results from the practical requirement that information security budgets are preset based on a management decision. That means, security specialists involved in the risk treatment phase have to work with a fixed budget. It is, however, the case that security specialists advise the management on the size of the budget. Without quantitative estimation of losses and benefits this is an extremely difficult task. As a result, in practice, proposed budgets or even projected losses are normally reduced in some kind of a negotiation phase. To support this phase, quantifiable data are needed that can support decision making. Solving the proposed model with a systematic budget variation, an economically ideal budget proposition can be made. Figure 2 visualizes expected losses L of optimal solutions depending on budget B. The curve starts off with a slight negative gradient, drops steeply for mid-sized budgets, and levels off at some point. This shape is mainly caused by two factors: for small budgets there is a lack of flexibility in the solution, which means it is not possible to select controls that complement each other sufficiently. The flattening after the drop in the middle results from the fact that additional controls will only yield marginal improvements in loss reduction. This shape of the loss curve corresponds to a negative logistic function which is well suited to describe the effect of security investments [5]. This is an important extension of the widely accepted model proposed in [4] where marginal utility is constantly decreasing with higher investments.

The previous observations are particularly interesting because it is reasonable to assume that the ideal budget proposition succeeds the flat part in the beginning and precedes the flattening at the end. To verify this assumption, the actual benefit of each solution can be determined:

$$\text{Benefit}(B) = \overline{L} - L^*(B) - C(B). \tag{5}$$

$\overline{L}$ is the worst-case loss without any investment, $L^*(B)$ is the optimal loss according to budget B, and $C(B)$ are the costs of controls of the corresponding optimal solution. Figure 3 shows the benefit (5) in accordance with the loss curve depicted

Fig. 3 Benefit of optimal
solutions according to $L^*(B)$

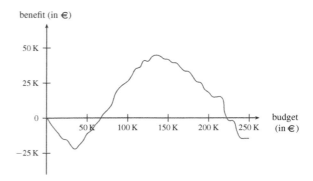

in Fig. 2. The figure reveals that the benefit takes its maximum of approximately
$44,800 €$ for $B = 135,000 €$. Initially the benefit is negative because the slowly
decreasing losses cannot compensate for the faster-rising costs of controls. When
the gradient of the loss curve decreases further, additional costs of controls are overcompensated by the resulting loss reduction. As a result, the benefit increases and
eventually becomes positive. When the loss curve declines more slowly, the benefit
curve decreases again.

Insights of such a What-If analysis can be used to determine and support a budget
proposition. Although an ideal budget can be found, it is not guaranteed that this
will be the budget the management team agrees on. In any case, using the presented
model, the decision maker can determine the optimal investment strategy according
to a given budget.

5   Conclusion

In this paper we presented a model to optimize information security expenditures as
part of the risk management process. The model considers how losses in an information system occur and how controls counteract threats by reducing the exploitability
of system vulnerabilities. We demonstrated that this approach can be used to determine and justify a budget proposition as well as to find an optimal selection of security
controls. This model is a first step towards sophisticated quantitative decision support in information security. The evaluation of first results shows the potential of
this approach and justifies more intensive research in this area. Future research may
include extending the scope of the model, taking uncertainty more explicitly into
account, and creating a multi-stage model to support mid- and long-term decision
making.

Acknowledgments This work was partially supported by the Horst Görtz Foundation.

References

1. Adams, A., Sasse, M.A.: Users are not the enemy. Commun. ACM **42**(12), 40–46 (1999)
2. FICO (2008) Xpress-SLP: Program Reference Manual
3. FICO (2009) FICO Xpress Optimization Suite: Getting Started with Xpress
4. Gordon, L.A., Loeb, M.P.: The economics of information security investment. ACM Trans. Inf. Syst. Secur. **5**(4), 438–457 (2002)
5. Hausken, K.: Returns to information security investment: the effect of alternative information security breach functions on optimal investment and sensitivity to vulnerability. Inf. Syst. Front. **8**(5), 338–349 (2006)
6. International Organization for Standardization (2008) ISO/IEC 27005 Information technology—Security techniques—Information security risk management
7. Mather, T., Kumaraswamy, S., Latif, S.: Cloud Security and Privacy: An Enterprise Perspective on Risks and Compliance. O'Reilly Media Inc (2009)
8. Ryan, J.J.C.H., Mazzuchi, T.A., Ryan, D.J., de la Cruz, J.L., Cooke, R.M.: Quantifying information security risks using expert judgment elicitation. Comput. Oper. Res. **39**(4), 774–784 (2012)
9. Schilling, A., Werners, B.: A quantitative threat modeling approach to maximize the return on security investment in cloud computing. In: Proceedings of the 1st International Conference on Cloud Security Management, ICCSM 13, Academic Conferences and Publishing International, Reading, pp. 68–78 (2013)
10. Schuster, F., Holz, T.: Towards reducing the attack surface of software backdoors. In: Proceedings of the ACM Conference on Computer and Communications Security, ACM, CCS'13, pp. 851–862 (2013)

Cut-First Branch-and-Price Second for the Capacitated Arc-Routing Problem

Claudia Schlebusch and Stefan Irnich

Abstract The basic multiple-vehicle arc-routing problem is called capacitated arc-routing problem (CARP) and was introduced by Golden and Wong (Networks 11:305–315, 1981 [9]). This paper presents a full-fledged branch-and-price (bap) algorithm for the CARP. In the first phase, the one-index formulation of the CARP is solved in order to produce strong cuts and an excellent lower bound. In the second phase, the master program is initialized with the strong cuts, CARP tours are iteratively generated by a pricing procedure, and branching is required to produce integer solutions. Furthermore, different pricing problem relaxations are analyzed and the construction of new labeling algorithms for their solution is presented. The cut-first branch-and-price second algorithm provides a powerful approach to solve knowingly hard instances of the CARP to proven optimality.

1 Capacitated Arc-Routing Problem

The capacitated arc-routing problem (CARP) is the fundamental multiple-vehicle arc-routing problem. It was introduced by Golden and Wong [9] and has applications in waste collection, postal delivery, winter services and more. The edited book by Corberán and Laporte [8] reports the increased attention since then.

For a formal definition of the CARP, we assume an undirected graph $G = (V, E)$ with node set V and edge set E. Non-negative integer demands $q_e \geq 0$ are located on edges $e \in E$. Those edges with positive demand form the subset $E_R \subset E$ of required edges that have to be serviced exactly once. A fleet K of $|K|$ homogeneous vehicles with capacity Q stationed at depot $d \in V$ is given. The problem is to find minimum cost vehicle tours which start and end at the depot d, service all required edges exactly

C. Schlebusch (✉) · S. Irnich (✉)
Chair of Logistics Management, Johannes Gutenberg University Mainz,
Jakob-Welder-Weg 9, 55128 Mainz, Germany
e-mail: claudia.bode@uni-mainz.de

S. Irnich
e-mail: irnich@uni-mainz.de

© Springer International Publishing Switzerland 2016   501
M. Lübbecke et al. (eds.), *Operations Research Proceedings 2014*,
Operations Research Proceedings, DOI 10.1007/978-3-319-28697-6_70

once, and respect the vehicle capacity Q. The tour costs consist of service costs c_e^{serv} for required edges e that are serviced and deadheading cost c_e whenever an edge e is traversed without servicing.

2   Cut-First Branch-and-Price Second

We developed a full-fledged cut-first branch-and-price second algorithm to solve the CARP. Its three key components are the cut generation procedure, the pricer, and the branching scheme. These components will be summarized in the following. For a detailed description the reader is referred to [5–7].

2.1   Cutting

In the first phase, the one-index formulation of the CARP is solved in order to produce strong cuts and an excellent lower bound. This formulation, first considered independently by Letchford [11] and Belenguer and Benavent [2], solely uses aggregated deadheading variables. However, the integer polyhedron of the one-index formulation is a relaxation of the CARP and can therefore contain infeasible integer solutions (see [3], p. 709). The cutting plane procedures are presented in [5]. At the end of phase one, binding cuts (odd cuts, capacity cuts, disjoint-path inequalities dp1, dp2, dp3) and odd/capacity cuts are identified and form the set $\mathscr{S}$. It suffices to know that the general form of all valid inequalities of the one-index formulation is

$$\sum_{e \in E} d_{es} y_e \geq r_s \quad s \in \mathscr{S}, \tag{1}$$

where s is the index referring to a particular inequality, d_{es} is the coefficient of edge e in the inequality, and $\mathscr{S}$ the set of all valid inequalities.

2.2   Master Problem

Applying Dantzig-Wolfe to the two-index formulation of [2] and aggregation over $k \in K$ identical subproblems leads to an aggregated integer master program. Let c_r indicate the cost of a route $r \in \Omega$ and let $\bar{x}_{er} \in \{0, 1\}$ and $\bar{y}_{er} \in \mathbb{Z}_+$ be the number of times that route r services or deadheads through edge e. There are binary decision variables λ_r for each route $r \in \Omega$. Furthermore, a variable $z_e \geq 0$ that represents a cycle $C_e = (e, e)$ along an edge e is added. This cycle corresponds to an extreme ray of the pricing polyhedron. The master problem (MP) reads then:

$$\min \sum_{r \in \Omega} c_r \lambda_r + \sum_{e \in E} (2c_e) z_e \tag{2}$$

$$\text{s.t.} \sum_{r \in \Omega} \bar{x}_{er} \lambda_r = 1 \quad \text{for all } e \in E_R \tag{3}$$

$$\sum_{r \in \Omega} d_{sr} \lambda_r + \sum_{e \in E} (2d_{es}) z_e \geq r_s \quad \text{for all } s \in \mathscr{S} \tag{4}$$

$$\mathbf{1}^\top \lambda = |K| \tag{5}$$

$$\lambda \geq \mathbf{0}, z \geq \mathbf{0} \tag{6}$$

The objective (2) minimizes over the costs of all tours. Equalities (3) ensure that every required edge is covered exactly once. The reformulated cuts of the first phase are given by (4). Herein, d_{sr} is the coefficient of the transformed cut $s \in \mathscr{S}$ for route r, which is $d_{sr} = \sum_{e \in E} d_{es} \bar{y}_{er}$. Equalities (5) are convexity constraints and require each vehicle to perform a CARP tour.

2.3 Branching

Let $\bar{\lambda}$ be a fractional solution to MP at a branch-and-bound node with associated values $\bar{x}$ and $\bar{y}$ ($x_e^k = \sum_{r \in \Omega} \bar{x}_{er} \lambda_r^k$, $y_e^k = \sum_{r \in \Omega} \bar{y}_{er} \lambda_r^k$). To obtain an integer solution, a branching scheme has to be devised. Our hierarchical branching scheme consists of three levels of decisions:

1. branching on node degrees
2. branching on edge flows
3. branching on followers and non-followers

The first two branching decisions are straight forward: Two branches are created whenever there exists a node $i \in V$ with non even node degree or an edge with fractional edge flow. The third branching decision is more intricate to implement. We define the follower information by

$$f_{ee'} = \sum_{r \in \Omega} f_{ee'r} \lambda_r \in \{0, 1\} \quad \text{for all } e, e' \in E_R$$

where $f_{ee'r} = |\{1 \leq q < p^r : \{e, e'\} = \{e_q^r, e_{q+1}^r\}\}|$ counts how often the two edges e and e' are serviced in succession by route $r \in \Omega$. If then for any two required edges e and e' this follower information is fractional, one can create two branches with constraints $f_{ee'} = 0$ and $f_{ee'} = 1$. The first branch implies that e and e' must not be serviced consecutively. This constraint can be implemented using the concept of task-2-loop free paths as presented in [10]. The second branch can be implemented by modifying the underlying graph on which pricing is carried out. This network modification does not destroy the structure of the pricing problem. A detailed description can be found in [5, Sect. 2.4.3].

2.4 Pricing Problem and Relaxations

The task of the pricing problem is to generate one or several variables with negative reduced cost or prove that no such variable exists. Let dual prices $\pi = (\pi_e)_{e \in E_R}$ to the partitioning constraints (3), $\beta = (\beta_s)_{s \in \mathscr{S}}$ to the cuts (4), and $\mu = (\mu^k)_{k \in K}$ to the convexity constraints (5) be given. Omitting the index k of the vehicle, the pricing problem is the two-index formulation of [2] with the following modified objective function:

$$z_{PP} = \min \tilde{c}^{serv,\top} x + \tilde{c}^\top y - \mu$$

where reduced costs for service and deadheading can be associated to the edges:

$$\tilde{c}_e^{serv} = c_e^{serv} - \pi_e \text{ for all } e \in E_R \quad \text{and} \quad \tilde{c}_e = c_e - \sum_{s \in \mathscr{S}} d_{es}\beta_s \text{ for all } e \in E.$$

Applying the suggested hierarchical branching scheme with branching on non-follower constraints means that any pricing problem relaxation must be able to handle two sets of tasks:

- tasks $\mathscr{T}^E$ for modeling the elementary routes
- tasks $\mathscr{T}^B$ for respecting non-follower constraints imposed by branching (2-loop-free tours)

In essence, a shortest-path problem where paths are elementary w.r.t. $\mathscr{T}^E$ and 2-loop-free w.r.t. $\mathscr{T}^B$ must be solved. We adopted the proposed labeling algorithms by Letchford and Oukil [12] to price out new routes that can handle two sets of tasks. This algorithm works on the original CARP graph G and exploits the sparsity of the network. A feasible path P ending at $i = i(P)$ can be extended along an edge either deadheaded or serviced. Any deadheading extension along an edge $e = \{i, j\} \in \delta(i)$ with associated reduced cost $\tilde{c}_e$ is feasible. On the other hand, a service extension along an edge $e = \{i, j\} \in \delta_R(i)$ with associated reduced cost $\tilde{c}_e^{serv}$ is feasible if $q(P) + q_e \leq Q$ holds. Moreover, in the ESPPRC case, the task sequences $\mathscr{T}^E(P)$ and $\mathscr{T}^E(i, j)$ are not allowed to share a common task, and $\mathscr{T}^B(P) \neq \mathscr{T}^B(i, j)$ needs to be fulfilled.

In [6, 7], we analyzed several pricing relaxations, and a detailed description of dominance rules and implementation details can be found there. Our basic branch-and-price approach [5] made use of just one relaxation producing 2-loop free tours [4]. This relaxation is particularly beneficial because it is compatible and at the same time indispensable for branching on followers. Stronger relaxations are k-loop elimination $k \geq 3$. In [7] we develop an efficient labeling algorithm for the combined $(k, 2)$-loop elimination in order to respect the two task sets resulting from branching. During experimental studies, we found out that reasonable parameters are $k \in \{2, 3, 4\}$. The ng-route relaxation by Baldacci et al. [1] has been successfully applied for solving several VRP variants using cut-and-column generation approaches. The principle of the ng-route relaxation is that the full sequence $\mathscr{T}^E(P)$ of served tasks associated with a path P is replaced by a subset $\mathscr{T}_{NG}^{E}(P)$ of the tasks $\mathscr{T}^{E}(P)$ in the sequence. It means that some of the tasks from the sequence $\mathscr{T}^{E}(P)$ are disregarded and also the ordering of the tasks is disregarded. The size of the set $\mathscr{T}_{NG}^{E}$ is a parameter where experimental studies show that the maximum size of neighborhoods in ng-route relaxations should be 5, 6 or 7. A description of the combined ng-route relaxation with 2-loop elimination is given in [7].

3   Computational Results

We tested the basic algorithm [5] on four standard benchmark sets (kshs, gdb, bccm and egl).[1] The advanced algorithms [6, 7], including more types of pricing relaxations and additional acceleration techniques, were tested on the egl benchmark set and additionally on the bmcv benchmark set.

All instances of kshs and gdb are often solved to optimality with the basic algorithm often in less than one second. For all bccm instances we can prove optimality either by computing an optimal solution or due to the tree lower bound. We can prove optimality for five out of 24 egl instances. Moreover, we can match our lower bound with a known upper bound from the literature for one more instance. Detailed computational results can be found in [5].

Comparing different pricing relaxations in [7], we obtain that for the egl instances, the k-loop relaxations ($k \in \{2, 3, 4\}$) are able to find more integer solutions, while for the bmcv the ng-route relaxation and the k-loop relaxations produce approximately the same number of optima. There is the tendency that the 2-loop relaxation in beneficial for problems of groups with higher vehicle capacity (54 best lower bounds out of 58), while the best ng-route relaxation performs worse on these instances (only 29 best lower bounds out of 58). On the other hand, for instances with lower capacity, the 2-loop-free relaxation results in 37 out of 58 best lower bounds, while ng-route relaxation gives 48 out of 58 best results.

4   Conclusion

We proposed a cut-first branch-and-price-second algorithm for the CARP. The strength of the new column-generation formulation results from strong lower bounds, symmetry elimination, efficient pricing, and an effective branching scheme. Computational experiments show that the proposed cut-first branch-and-price-second algorithm gives considerable results for all standard benchmark sets. Several earlier exact approaches proved optimality of known heuristic solutions by matching lower and upper bounds, but were not able to deliver optimal CARP routes. Our branching

---

[1] All instances can be downloaded from http://logistik.bwl.uni-mainz.de/benchmarks.php.

scheme, however, enables us to compute feasible integer solutions and in many cases even optimal ones.

Different relaxations known from the node-routing context were adapted to solve the CARP with a branch-and-price approach. The adaption is non-trivial when pricing is still performed on the original sparse graph: The result is a more intricate branching scheme, where two sets of tasks must be handled. We adapted and compared the ng-route relaxation and the relaxation with k-loop elimination. For instances with high capacity, k-loop elimination often outperforms ng-route relaxations. The opposite can generally be observed for instances with low capacity. Overall, several new best lower bounds were presented and some knowingly hard instances were solved to proven optimality for the first time.

References

1. Baldacci, R., Mingozzi, A., Roberti, R.: New route relaxation and pricing strategies for the vehicle routing problem. Oper. Res. **59**(5), 1269–1283 (2011)
2. Belenguer, J., Benavent, E.: The capacitated arc routing problem: valid inequalities and facets. Comput. Optim. Appl. **10**(2), 165–187 (1998)
3. Belenguer, J., Benavent, E.: A cutting plane algorithm for the capacitated arc routing problem. Comput. Oper. Res. **30**, 705–728 (2003)
4. Benavent, E., Campos, V., Corberán, A., Mota, E.: The capacitated arc routing problem: lower bounds. Networks **22**, 669–690 (1992)
5. Bode, C., Irnich, S.: Cut-first branch-and-price-second for the capacitated arc-routing problem. Oper. Res. **60**(5), 1167–1182 (2012)
6. Bode, C., Irnich, S.: In-depth analysis of pricing problem relaxations for the capacitated arc-routing problem. Transp. Sci. **49**(2), 369–383 (2015)
7. Bode, C., Irnich, S.: The shortest-path problem with resource constraints with k-loop elimination and its application to the capacitated arc-routing problem. Eur. J. Oper. Res. **238**(2), 415–426 (2014)
8. Corberán, Á., Laporte, G. (eds.): Arc routing: problems, methods and applications. MOS-SIAM Series on Optimization, SIAM, Philadelphia (2014)
9. Golden, B., Wong, R.: Capacitated arc routing problems. Networks **11**, 305–315 (1981)
10. Irnich, S., Villeneuve, D.: The shortest path problem with resource constraints and k-cycle elimination for $k \geq 3$. INFORMS J. Comput. **18**(3), 391–406 (2006)
11. Letchford, A.: Polyhedral results for some arc routing problems. Ph.D. dissertation, Department of Management Science, Lancaster University (1997)
12. Letchford, A., Oukil, A.: Exploiting sparsity in pricing routines for the capacitated arc routing problem. Comput. Oper. Res. **36**(7), 2320–2327 (2009)

Solving a Rich Position-Based Model for Dairy Products

Karl Schneeberger, Michael Schilde and Karl F. Doerner

Abstract By considering the lot-sizing problem and the detailed sequencing and scheduling problem simultaneously in a very realistic model formulation we aim to improve the overall performance of the entire production process for a dairy producing company. For this purpose we extended the Position-Based Model introduced by Lütke et al. (Int. J. Prod. Res. 43, 5071–5100, 2005 [5]). Based on a set of real-world production data, we used our model to determine exact solutions to very small problem settings. As even for small instances the time required for obtaining exact solutions is too long in general, we developed a fix&optimize inspired construction heuristic to obtain a feasible solution for several products. The results of our approach show that the solutions obtained are competitive compared to the optimal solution with less time consumption.

1 Introduction

Usually the lot-sizing problem and the detailed sequencing and scheduling problem are treated separately in the production planning process, especially for perishable goods. Instead, these two problems should be treated simultaneously to obtain better solutions (see, e.g., [1]). We developed a very realistic model formulation aiming to improve the overall performance of the entire production process. For this purpose we extended the Position-Based Model introduced by Lütke et al. [5]. The extensions include explicit product transfers via product pipes (i.e., pipes are used to transfer products between aggregates; no two transfers can be performed at the same time),

K. Schneeberger (✉) · M. Schilde (✉) · K.F. Doerner
Production and Logistics Management, Johannes Kepler University Linz,
Altenberger Str. 69, 4040 Linz, Austria
e-mail: karl.schneeberger@univie.ac.at

M. Schilde
e-mail: michael.schilde@univie.ac.at

K. Schneeberger · M. Schilde · K.F. Doerner
Department of Business Administration, University of Vienna,
Oskar-Morgenstern-Platz 1, 1090 Vienna, Austria

© Springer International Publishing Switzerland 2016
M. Lübbecke et al. (eds.), *Operations Research Proceedings 2014*,
Operations Research Proceedings, DOI 10.1007/978-3-319-28697-6_71

product-dependent durability during the production process (e.g., after fermentation the product has to be chilled within a certain time limit), cleaning and sterilization pipes which prevent simultaneous treatment of specific aggregates, maximum and minimum capacity of aggregates, sequence-dependent setup times, product loss caused by transfers, a product specific production speed for each aggregate, and cleaning intervals (i.e., the time between two consecutive cleaning procedures is limited). Gellert et al. [3] studied a related model considering only the filling lines, reasoning that they are the bottleneck of the entire production system. This however leads to infeasible solutions in practical use. Based on a set of real-world production data, we used our model to determine exact solutions to very small problem settings. As even for small instances the time required for obtaining exact solutions is too long in general (even finding a first feasible solution for one product on all available aggregates takes many hours), we developed a fix&optimize inspired construction heuristic to obtain a feasible solution for several products. This means, that the overall problem is first decomposed and iteratively solved while adding one product per iteration. Sahling et al. [6] and Helber and Sahling [4] used the fix&optimize heuristic to solve a dynamic multi-level capacitated lot sizing problem and extensions. Amorim et al. [2] give an overview of perishability issues in production and distribution.

2 Problem Description

The production process in our data set consists of eight aggregate levels (see Fig. 1; we use the term aggregate for both, machines and tanks). Each product has to be processed by a specific subset of aggregates; some aggregate levels can be omitted for some products (e.g., spoonable yoghurt skips the fermentation tanks whereas creamy yoghurt does not). Depending on the degree of processing the products are labeled differently. At the fillers and sorters the products are denoted as final products, at the heaters, fermentation tanks, coolers, and filling tanks as base mass, and at the mixers and mixing tanks as blended milk. Thus, generally speaking, blended milk is transformed into base mass, which is then transformed into final products by filling it into different types of bins (potentially together with some sort of fruit topping or cereals, which is neglected in our model).

Each machine is assigned a set of production speeds for all products that need to be processed by it, each tank has a minimum and a maximum capacity information. Each aggregate has a set of ingoing and a set of outgoing product pipes that can be used. Each such pipe is assigned a volume specifying the product loss due to cleaning after a transfer.

Additionally, each aggregate has a specific pipe for cleaning-in-place (CIP) and a pipe for sterilization-in-place (SIP) assigned to it. These pipes can be shared between different aggregates and cannot be used to service multiple aggregates simultaneously. The requirement of performing a CIP and SIP operation after processing a lot is sequence-dependent, as is the duration of the setup process required before

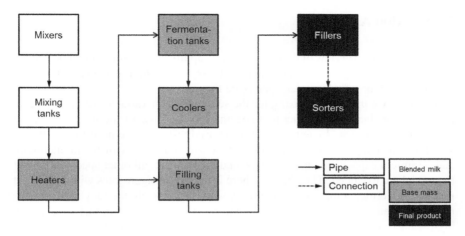

Fig. 1 Overview of the aggregate levels available for producing dairy products. Each product requires to be processed by a subset of aggregates; some aggregate levels can be omitted for some products. The transfer from an aggregate to the next level has to be performed using a product pipe (sorters need no pipe system for transportation; dummy pipes are used instead)

beginning the processing of a lot. Additionally, every aggregate must be cleaned after a specific period of time without cleaning (even when no lot is being processed during this time; cleaning interval). Coolers can voluntarily process at a slower speed to avoid cleaning due to the cleaning interval.

Each blended milk has a specific time determining how long the ingredients need to soak in a mixing tank (soaking time). Each base mass has a specific time determining how long the fermentation in a fermentation tank needs to be (fermentation time) and how soon after completing the fermentation the product must be chilled (tolerance time). Furthermore, blended milk and base mass may not spend more than a specific amount of time in any tank (maximum survival time).

If a lot processed by the mixer is smaller than a specific small batch size, the lots processing time has to be increased by a certain penalty time. This is due to the requirement to perform laboratory tests before moving the product into the mixing tank. The detailed mathematical formulation and description can be found in [7].

The main objective (1) is to minimize all starting times of all lots l_1 on all aggregates a_1 ($b_{a_1 l_1}$), of all lots l_1 on all product transfer pipes p_1 ($p_{p_1 l_1}$), of all lots l_1 on all CIP-pipes p_1 ($c_{p_1 l_1}$), and of all lots l_1 on all SIP-pipes p_1 ($s_{p_1 l_1}$). This ensures that the makespan of each aggregate is as short as possible and has also a positive influence on the overall makespan.

$$Min \sum_{a_1, l_1} b_{a_1 l_1} + \sum_{p_1 \in APIP, l_1} p_{p_1 l_1} + \sum_{p_1 \in ACIP, l_1} c_{p_1 l_1} + \sum_{p_1 \in ASIP, l_1} s_{p_1 l_1} \quad (1)$$

3 Solution Approach

Figure 2 gives an overview of the entire solution approach. To reduce the time needed to achieve an appropriate solution quality we separated the procedure of finding a solution into four interdependent steps (see Fig. 2).

In Step 1 we try to find a fixing for the setup indicator variables (if an aggregate is setup to produce one product in a specific lot). Because of the large number of aggregates (more than 40 aggregates for the used test case) we need to predefine a set of the most appropriate aggregates. We have to consider the different maximum and minimum capacities, pipe transfers, and the demand; all other specifications of the model are excluded. The objective here is to find the aggregates with the lowest processing, fermentation, or soaking time, while considering the lot position of the used aggregates. As the complexity of the remaining reduced model is still very high we also use a fix&optimize inspired procedure to solve it: First, we obtain a first feasible solution of the reduced model. Second, we use the determined setup indicators and fix them for all aggregates but the mixers. We re-run the model to obtain a new solution. We then fix the setup indicators for all aggregates but the mixing tanks and re-run the model. We repeat this procedure for each aggregate level to obtain a reasonably good solution quality.

In Step 2 we fix the setup indicators to the values found in Step 1, obtain a first feasible solution for the complete model and continue with Step 3.

In Step 3 we fix all binary variables except the CIP assignments. An overview of the solution process for a problem with two aggregates and two different products is presented in Fig. 3. Simultaneous usage of the same CIP and SIP pipe is not possible, so AGG1 has to wait before the cleaning process of AGG2 is finished. This is only the first feasible solution, so the solution quality can be improved. Aggregate AGG1 has finished processing earlier than AGG2, so it would reduce the overall makespan if the CIP-pipe first services aggregate AGG1 instead of AGG2. When fixing all binary variables except the CIP assignment, the CIP lots can be swapped.

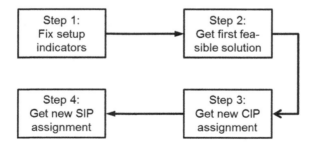

Fig. 2 Overview of the solution approach used to quickly obtain a first high-quality solution. First, we fix the setup indictors to predefine the needed aggregate lots. Second, we try to get a first feasible solution with the fixed setup indicators. Third, we fix all binary variables except the CIP assignments of the first feasible solution to get a better solution quality and last, we use this solution to fix again all binary variables except the SIP assignments to get the final solution

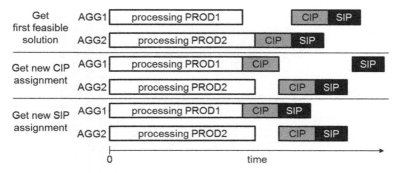

Fig. 3 Overview of an example with two aggregates and two products over the entire solution approach; including manufacturing process, CIP, and SIP. The aggregates (AGG1 and AGG2) share the same CIP and SIP pipe for cleaning and sterilization

In Step 4 we fix all binary variables, which includes the changed CIP assignment, except the SIP assignment. In this small example in Fig. 3 the SIP lots for aggregates AGG1 and AGG2 are swapped and the CIP assignment remains the same, because it was fixed. Thus we obtain a high quality solution with respect to the makespan on each aggregate.

4 First Solutions

Table 1 shows the results of 10 test cases. We obtained the objectives of each test case with CPLEX, CPLEX with 1 h time restriction, and with our presented solution approach. To get a feasible solution in CPLEX we used the first step of our solution approach and fixed the setup indicators (see Fig. 2); without this, it would not be possible to obtain a solution. The time measurement started after step one for each test case.

The solution quality of our fix&optimize solution approach seems to be very good (see Table 1). Our solution approach finds the best known solution for each single product test case. The objective obtained by the solution approach is 0.01 percent worse than the optimal solution obtained by CPLEX for products 3 + 5, and products 1 + 2.

5 Summary and Outlook

We presented a rich position-based model for dairy products by considering the lot-sizing problem and the detailed sequencing and scheduling problem simultaneously. Our model includes many specifications to handle real-world test cases. Because of

Table 1 Solutions of 10 test cases with one or two different products

Product	CPLEX [C]		CPLEX 1h [C1h]		Fix&optimize [F&O]		Gap [%]	
	Time	Objective	Time	Objective	Time	Objective	C-F&O	C1h-F&O
1	23.06	18,463.93	23.06	18,463.93	4.72	18,463.93	0.00	0.00
2	571.75	75,494.04	571.75	75,494.04	11.01	75,494.04	0.00	0.00
3	366,910.44	27,628.08	3,601.74	27,628.08	77.19	27,628.08	0.00	0.00
4	20.80	35,427.95	20.80	35,427.95	4.04	35,427.95	0.00	0.00
3 + 5	65,541.00	29,494.70	3,601.13	29,494.70	75.34	29,651.39	−0.01	−0.01
6	21.64	17,674.41	21.64	17,674.41	4.13	17,674.41	0.00	0.00
1 + 2	593,411.00	126,733.27	3,601.64	128,362.52	1,855.20	127,306.21	−0.01	0.01
8	20.69	29,164.85	20.69	29,164.85	3.80	29,164.85	0.00	0.00
9	22.87	20,275.47	22.87	20,275.47	3.97	20,275.47	0.00	0.00
10	21.07	37,918.23	21.07	37,918.23	3.72	37,918.23	0.00	0.00

The CPLEX columns indicate the total time needed to obtain the optimal solution and the objective value; CPLEX 1h shows the results with an abort criterion of 1h; fix&optimize reports the time needed to obtain the best available solution with our approach; Gap shows the difference in the objective function value between the CPLEX and fix&optimize, and CPLEX 1h and fix&optimize solution. For the fix&optimize heuristic, we used IBM ILOG CPLEX Optimization Studio (Version 12.5.1) and Microsoft Visual C# 2012 for Windows x86-64, and an Intel i5-2400 Quad-Core Processor (3.1 GHz) and 4 GB of memory. For exact solutions, we used the same CPLEX version for Linux x86-64, and 32 cores of an Intel Xeon E7-8837 (WestmereEX, 2.66 GHz) and 256 GB of memory.

the very detailed model a feasible solution cannot be found within reasonable time. We developed a fix&optimize based solution approach to reduce the time needed to obtain a good solution quality.

We plan to extend our solution approach to optimize larger test cases of five or more products by iterative product insertion. The method will start with optimizing the first product, fixing all variables adding the next product, and so on. So we first use the fix&optimize approach as a construction heuristic and also to improve the solution after each insertion step.

Acknowledgments Financial support from the Austrian Research Promotion Agency (FFG, Bridge) under Grant #838560 is gratefully acknowledged.

References

1. Amorim, P., Antunes, C.H., Almada-Lobo, B.: Multi-objective lot-sizing and scheduling dealing with perishability issues. Ind. Eng. Chem. Res. **50**, 3371–3381 (2011)
2. Amorim, P., Meyr, H., Almeder, C., Almada-Lobo, B.: Managing perishability in production-distribution planning: a discussion and review. Flex. Serv. Manuf. J. **25**, 389–413 (2013)
3. Gellert, T., Höhn, W., Möhring, R.H.: Sequencing and scheduling for filling lines in dairy production. Optim. Lett. **5**, 491–504 (2011)
4. Helber, S., Sahling, F.: A fix-and-optimize approach for the multi-level capacitated lot sizing problem. Int. J. Prod. Econ. **123**, 247–256 (2010)
5. Lütke-Entrup, M., Günther, H.-O., van Beek, P., Grunow, M., Seiler, T.: Mixed integer linear programming approaches to shelf life integrated planning and scheduling in yogurt production. Int. J. Prod. Res. **43**, 5071–5100 (2005)
6. Sahling, F., Buschkühl, L., Tempelmeier, H., Helber, S.: Solving a multi-level capacitated lot sizing problem with multi-period setup carry-over via a fix-and-optimize heuristic. Comput. Oper. Res. **36**, 2546–2553 (2009)
7. Schneeberger, K., Schilde, M., Doerner, K.: Solving a rich position-based model for dairy products with a fix&optimize based solution approach. Technical Report UNIVIE-PLIS-2015-01, University of Vienna, Department of Business Administration (2015)

A European Investment and Dispatch Model for Determining Cost Minimal Power Systems with High Shares of Renewable Energy

Angela Scholz, Fabian Sandau and Carsten Pape

Abstract In this paper a multistage combined investment and dispatch model for long-term unit commitment problems of large-scale hydrothermal power systems is presented. It is based on a combination of a continuous and mixed integer programming algorithm as well as Lagrangian relaxation. First, the required capacities of power generation and storage units are determined by a continuous linear program (LP). Second, in an optional stage the unit commitment problem for all investigated market areas, i.e. those of Europe, is solved by a mixed integer linear program (MILP). At last, a MILP solves the same problem with a higher level of detail for a focused subarea.

1 Introduction

In times of a proceeding electrification of the energy system (ES) in Europe and Germany, especially in the sectors mobility and heating, secure and reliable load coverage becomes increasingly important. At the same time, due to the foreseeable scarcity of fossil fuels as well as the climate policy objectives of the European Commission, the use of renewable energy sources (RES) increases. The volatile feed-in from RES results in a fluctuating residual load, wherefore the present power system has to be adapted. To identify a future ES at minimal overall costs, various generation technologies and constraints must be considered by the optimization.

This paper presents the most important aspects of the combined investment and dispatch model of power supply systems. The model intends to overcome difficulties in long-term optimization of large-scale unit commitment problems.

A. Scholz (✉) · F. Sandau · C. Pape
Fraunhofer-Institut für Windenergie Und Energiesystemtechnik, Königstor 59,
34119 Kassel, Germany
e-mail: angela.scholz@iwes.fraunhofer.de

© Springer International Publishing Switzerland 2016
M. Lübbecke et al. (eds.), *Operations Research Proceedings 2014*,
Operations Research Proceedings, DOI 10.1007/978-3-319-28697-6_72

2 Mathematical Programming

The optimization is based on a multistage procedure. For a cross-border investigation this includes a long-term LP in the first and a MILP in the optional second stage. In the third stage a subproblem is solved by an MILP but based on a more detailed model of a focused subarea as well as on results of previous stages. Moreover, it includes a rolling planning algorithm (Fig. 1).

Each of the programming is aimed at covering the residual load and heat demand for each hour per year at minimum costs. Therefore, several technologies for the conversion, storage and transport of energy are options for investment. Additional flexibility is provided by a given hydro-thermal power plant portfolio, for which only the operating costs are considered. For each investigated market area, the hourly residual load coverage is given, if it holds for all t in $\{1, \ldots, 8760\}$

$$D(t) = \sum_{i=1}^{I} b^i(t) \cdot P^i(t) - P^i_{con}(t) + \sum_{h=1}^{H} P^h_{out}(t) - P^h_{in}(t). \tag{1}$$

In (1) I indicates the number of thermal power plants and H that of hydro-storage units. The variables $P^i(t)$ and $P^h_{out}(t)$ represent the generated power of the thermal or hydraulic thermal units in hour t as well as $P^i_{con}(t)$ and $P^h_{in}(t)$ their additional power consumption, for instance by electric heaters, electrolyzers or hydraulic pumps. The binary variable $b^i(t) \in \{0, 1\}$ determines the operation mode of unit i and is set to 1 for all hours only in the investment model.

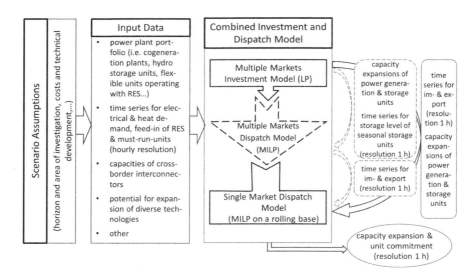

Fig. 1 Configuration of the multistage model

2.1 Multiple Markets Investment Model—MMIM

The programming in the first stage is a deterministic, continuous LP for determining the required capacities Γ of those technologies, whose expansion is within scope of the scenario. The objective function includes operating costs of all committed plants as well as annual investment costs for additional capacities. So, the function for total costs TC can be written as

$$TC(\Gamma, P) = \sum_{j=1}^{J} IC(\Gamma^j) + \sum_{t=1}^{8760} [\sum_{i=1}^{K} OC(P^i(t), P_{con}^i(t)) + \sum_{j=1}^{J} OC(P^j(t), P_{con}^j(t))] \qquad (2)$$

Within (2) the function for the variable operating costs of unit k is labeled with $OC(P^k(t), P_{con}^k(t))$. The numbers of existing or additional units respectively are denoted by K and J. As a simplification, only the aggregated capacity expansions Γ^j are modeled instead of individual units, while their potential $\widehat{\Gamma}^j$ can be limited, whereas Γ is limiting the maximal power generation or consumption per hour.

$$0 \le P^j(t) \le \Gamma^j \cdot \Delta t, \qquad \forall t = \in \{1, \ldots, 8760\}. \qquad (3)$$

The interaction between the different market areas is modeled by cross-border interconnectors.

$$0 \le Imp^j(t), \widehat{Imp}^j(t) \le NTC_{Imp}^j(t), \quad 0 \le Exp^j(t), \widehat{Exp}^j(t) \le NTC_{Exp}^j(t),$$

$$Imp^j(t) = \upsilon \cdot \widehat{Exp}^j(t), \quad \widehat{Imp}^j(t) = \upsilon \cdot Exp^j(t), \qquad \forall j \in J. \qquad (4)$$

Within this modeling, the transmission losses $\upsilon \in [0, 1]$ depend on the length of the power line. However, transport restrictions within market areas have been neglected and the results of this model stage are based on merged input data [3].

2.2 Multiple Markets Dispatch Model—MMDM

To make up for the inaccuracy in the first stage of the combined model, in the second stage the optimization is based on a detailed power plant portfolio, as a result of the MMIM. Besides considering single power units, this programming takes a more realistic and detailed model of the different generation technologies into account. As a result, more accurate time series for cross-border power exchange can be derived. For the high level of complexity in this step, standard mixed integer algorithms are not a pragmatic approach. Hence, the most complex constraints, those for load coverage (1), are relaxed by Lagrangian relaxation. In consequence, these constraints are added to the objective function by multiplication with the Lagrangian multipliers $\lambda_{a \in A} \in \mathbb{R}^T$, with A representing the number of regarded market areas.

Unlike the operating costs function in the investment model, OC includes costs for starting processes of thermal units. This term of OC is only different from zero in those time steps, if $b^i(t+1) - b^i(t) > 0$ holds for the binary operating mode variable of unit i. Hence, the objective function of MMDM is defined by

$$L(b^{i \in I}, P^{i \in I}, P_{con}^{i \in I}, P_{out}^{h \in H}, P_{in}^{h \in H}, \lambda_{a \in A}) = \tag{5}$$

$$= \sum_{t=1}^{8760} [\sum_{i=1}^{I} OC(b^i(t), P^i(t), P_{con}^i(t)) + \sum_{h=1}^{H} OC(P_{out}^h(t), P_{in}^h(t))] +$$

$$\sum_{a=1}^{A} \lambda_a(t)[D_a(t) - \sum_{i=1}^{I} b^{i,a}(t) \cdot P^{i,l}(t) - P_{con}^{i,a}(t) - \sum_{h=1}^{H} P_{out}^{h,a}(t) - P_{in}^{h,a}(t)]$$

Accordingly to the Lagrangian method, L has to be minimized with respect to $b := b^{i \in I}$, $P := P^{i \in I}$, $P_{con} := P_{con}^{i \in I}$, $P_{out} := P_{out}^{h \in H}$ and $P_{in} := P_{in}^{h \in H}$ and maximized with respect to $\lambda := \lambda_{a \in A}$ so that the dual problem is defined by

$$\max_{\lambda} \tilde{L}(\lambda) := \max_{\lambda} \min_{b, P, P_{con}, P_{out}, P_{in}} L(b, P, P_{con}, P_{out}, P_{in}, \lambda), \tag{6}$$

whereby the dual function, due to the $\tilde{L}(\lambda)$ is concave but nonsmooth, just like the primal function. Even so, the duality gap is supposed to be small, due to the large number of variables [1].

For solving the relaxed problem (6) iteratively, $\lambda_a(t)$ is initialized by the marginal costs for load coverage in market area a and time step t based on the merit order of fuel costs. Therefore, $L^*(b, P, P_{con}, P_{out}, P_{in}) := L(b, P, P_{con}, P_{out}, P_{in}, \lambda^*)$, λ^* as determined by the initialization, can be decomposed into several independent subproblems, each regarding just a few units. Those are solved by a branch and cut algorithm.

The maximization of the dual function $\tilde{L}$ is based on the subgradient method [2]. For each iteration k, the update of λ^k is determined by

$$\lambda_a^{k+1}(t) = max\{0, \lambda_a^k(t) + s_a(t) \frac{d_a(t)}{\|d_a(t)\|}\}. \tag{7}$$

Within this equation, the step size $s_a(t)$ is selected depending on the subgradient $d_a(t) = D_a(t) - b^a(t) \cdot P^a(t) + P_{con}^a(t) - P_{out}^a(t) + P_{in}^a(t)$, which is the current rate of load coverage in the market area a. This approach results in a faster convergence than updating the Lagrangian multipliers for each time step and market area in the same way. Besides this, oscillation of $(\lambda^k)_{k=1}^{\infty}$ decreases. If the stopping criteria is reached, a feasible solution of the primal problem is determined by an economic dispatch. Thus, the solutions for the binary variables b are fixed and the resulting (LP) is smooth, so that it can be solved by the barrier algorithm.

The presented method of the MMDM for the optimized power plant portfolio simultaneously takes into account the interaction of market areas and detailed technical unit restrictions.

2.3 Single Market Dispatch Model—SMDM

In the final stage, the optimized power plant portfolio, the result of the MMIM (cf. Sect. 2.1), and a fitted residual load time series are considered as input data. The fitted time series is based on the hourly feed-in of RES and the power demand of the focused market area as well as the hourly import and export, determined by the MMDM (cf. Sect. 2.2). That way, the interaction between the market areas is considered, despite the following deterministic MILP is an analysis for an isolated area. Besides load coverage (cf. (1)) and the technical restrictions (cf. Sect. 2.2), coverage of all types of balancing power is modeled in this programming. Such a cross-unit constraint can be described for each time step t and type of reserve control by:

$$R(t) \leq \sum_{i=1}^{\tilde{I}} b^i(t) \cdot (P_{max}^i - P^i(t)) + \sum_{h=1}^{\tilde{H}} min\{P_{max}^h - P^h(t), \frac{S^h(t)}{\eta_h} - P^h(t)\} \quad (8)$$

where $\tilde{I}$ is the set of those thermal power plants, which participate in this type of control power market. It includes units with a minimum power restriction as well as those without such a constraint (then $b^i(t) = 1$). Furthermore, $\tilde{H}$ defines the set of storage units. To simulate balancing power demand, a probability $\mathscr{P}^i$ is assumed, which depends on the technology. This leads to an additional term in the objective function

$$EC(b^{i \in \tilde{I}}, P^{i \in \tilde{I}}, P_{con}^{i \in \tilde{I}}, P_{out}^{h \in \tilde{H}}, P_{in}^{h \in \tilde{H}}) = \quad (9)$$

$$= \sum_{t=1}^{8760} [\sum_{i=1}^{\tilde{I}} [OC(b^i(t), P^i(t), P_{con}^i(t)) + \mathscr{P}^i \cdot b^i(t) \cdot (P_{max}^i - P^i(t))]$$

$$+ \sum_{h=1}^{\tilde{H}} [OC(P_{turb}^h(t), P_{in}^h(t)) + \mathscr{P}^h \cdot min\{P_{max}^h - P^h(t), \frac{S^h(t)}{\eta_h} - P^h(t)\}]]$$

In addition, the effect of RES forecast is simulated as well as the related costs. Therefore, the optimization is based on a rolling planning system with an optimization horizon equaling the forecast horizon of fluctuating RES. The forecast is updated continuously, i.e. every fourth hour, resulting in an updated residual load [4].

3 Case Study

The model described in Sect. 2 was developed within the framework of the research project "Roadmap Speicher".[1] To address the questions named in the introduction, the investigations focused on the development of power storage requirements. The program modeling has been done in MATLAB and the optimization by IBM's MIP-solver CPLEX. The optimization problems referring to each scenario or sensitivity, considered about 32.7 mill. variables in the MMIM, and 46.4 mill. in the MMDM, there of 4.3 mill. binaries. Additionally to the upper and lower bounds of each variable, consideration has also to be given to about 22.7 mill. constraints in the MMIM and 31.7 mill. in the MMDM.

One fundamental finding in this project has been a relative low storage demand in an energy supply system with a high share of renewable energies, based on the given assumptions. For more conclusions and the considered assumptions cf. [3].

4 Examplary Results

The first applications have shown that the presented model is suited analyzing problems as listed in Sect. 3. Computation times of 1–3 days for the MMIM and about 5 days for the SMDM are acceptable. The programming of the MMIM finds a near-optimal solution after 12 h and reaches the required accuracy for a primal feasible solution after 35 h (Fig. 2). For the study case this results in a load deficit of about 0.2 $\frac{MWh}{a}$ for the MMIM.

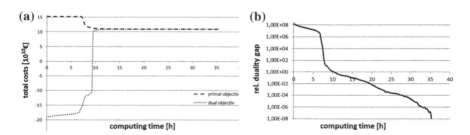

Fig. 2 Convergence behavior of the MMIM. **a** Primal and dual function. **b** Relative duality gap

[1]Funded by the Federal Ministry for Economic Affairs and Energy, funding code 0325327.

References

1. Dentcheva, D., Römisch, W.: Optimal power generation under uncertainty via stochastic programming. In: Lecture Notes in Economics and Mathematical Systems, vol. 458, pp. 22–56 (1998)
2. Fan, W., et al.: Evaluation of two Lagrangian dual optimization algorithms for large-scale unit commitment problems. J. Electr. Eng. Technol. (2012). doi:10.5370/JEET.2012.7.1.17
3. Pape, C., et al.: Roadmap Speicher, Speicherbedarf für erneuerbare Enegien - Speicheralternativen - Speicheranreiz - Überwingung von rechtlichen Hemmnissen - Endbericht (2014). http://publica.fraunhofer.de/documents/N-316127.html
4. von Oehsen, A.: Entwicklung und Anwendung einer Kraftwerks- und Speichereinsatzoptimierung für die Untersuchung von Energieversorgungsszenarien mit hohem Anteil erneuerbarer Energien in Deutschland (2012). http://d-nb.info/1038379601/34

Pre-selection Strategies for Dynamic Collaborative Transportation Planning Problems

Kristian Schopka and Herbert Kopfer

Abstract To improve the competitiveness, small and mid-sized carriers may ally in coalitions for request exchange. One main barrier is the "carrier-fear" of losing autonomy. A decentralized pre-selection that allows carriers to preserve own transportation requests for the private fleet may limit the information shared within the coalition and increase the autonomy. Several heuristic pre-selection strategies are presented. A computational study analyzes which of those are most qualified.

1 Introduction

Nowadays, freight carriers are confronted with customers demanding for quick execution of transportation requests and they have to plan the execution dynamically, i.e. new customer orders appear and have to be planned. Especially for small and mid-sized carriers (SMCs), it is difficult to create efficient transportation plans in such dynamic environments. By building horizontal coalitions SMCs may find a way to conquer their cost disadvantage against large forwarding companies. A facility to organize coalitions is the request exchange with competitors, which enables an improved request clustering and results in cost savings [6]. One main barrier of building coalitions is the "carrier-fear" for abandoning autonomy. Thereby, potential members are not willing to give all information about the request structure to their partners. However, to drive the building of coalitions, mechanisms have to be developed that increase the autonomy. The autonomy may be increased by using an independent decision making process (DMP), where each member decides if a request is released for the exchange or preserved for the private fleet. The contribution of this paper

K. Schopka (✉) · H. Kopfer
Chair of Logistics, University of Bremen, Wilhelm-Herbst-Str. 5, 28359 Bremen, Germany
e-mail: schopka@uni-bremen.de
URL: http://www.logistik.uni-bremen.de

H. Kopfer
e-mail: kopfer@uni-bremen.de

© Springer International Publishing Switzerland 2016
M. Lübbecke et al. (eds.), *Operations Research Proceedings 2014*,
Operations Research Proceedings, DOI 10.1007/978-3-319-28697-6_73

523

is to discuss adequate mechanisms (pre-selection strategies). The basic problem is introduced in Sect. 2. Section 3 presents the pre-selection strategies. A computational study is carried out in Sect. 4. Section 5 concludes the paper.

2 Selection of Collaborative Requests

To increase the autonomy in carrier coalitions for request exchange, each involved SMC has to decide independently, which transportation requests of his own request pool are preserved for the private fleet (self-fulfillment) and which are to be offered for the request exchange process (collaborative). Strategies to solve this assignment problem include preserving the requests with the highest marginal profit for the self-fulfillment [2] or performing a request clustering and releasing the clusters with the highest distance to the depot to the exchange process [11]. Other possibilities are approaches for the selective vehicle routing problem (SVRP) that include only the most profitable requests in the tour planning [1, 3, 7]. The remaining requests may be forwarded to subcontractors or exchanged with coalition partners. The basic SVRP-strategy is to perform a "cherry-picking", which tends to build solutions with a high efficiency for own vehicles. Against, in the considered problem a sequential two stage DMP exists, where the preserving of requests for the self-fulfillment has an immediate influence on the quality of the following request exchange process. Hence, a high capacity utilization may block the efficiency of the request exchange process. Another handicap of the SVRP is that the solution approaches require long computing time that is often not available in dynamic environments. Pre-selection strategies requiring less computing time have to be developed.

For the evaluation of the pre-selecting strategies, a multi vehicle dynamic collaborative transportation planning problem is considered, where some SMCs collaborate and exchange requests. Every request is exchangeable and assigned to one member. It generates earnings and has a time window. Furthermore, some requests appear during the planning horizon resulting in a dynamic planning environment. To handle this dynamic aspect, a periodic re-optimization within a rolling horizon planning is used [9]. At each planning update several problems related to the multi depot vehicle routing problem with time windows (MDVRPTW) have to be solved. The autonomy of each coalition member is increased by a postulated decentralized pre-selection of requests that are included in the request exchange. To solve the problem, a two step framework (TSF) depicted in Fig. 1 is used. The TSF can be classified in two DMPs, where each is followed by an optimization phase (OP1/2). In each optimization phase an MDVRPTW is solved by a modified adaptive large neighborhood search (ALNS). That the ALNS produce good results for diverse vehicle routing problems is shown by Pisinger and Ropke [10]. On the decentralized stage (DMP1) the members decide, which of their requests are preserved for the private fleet. On the centralized stage (DMP2) the request exchange is performed. It has to be clarified which request exchange mechanism is most qualified for solving the

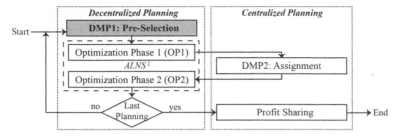

Fig. 1 Sequence of the two step framework

DMP2 of the TSF. Based on the periodic re-optimization, the sequence is repeated until the last planned update is performed. Finally, a profit sharing is conducted. The paper focuses on the pre-selection strategies for solving the DMP1.

3 Pre-selection Strategies

The pre-selection strategies are classified into request potential valuation strategies (RPVs) and tour potential valuation strategies (TPVs). Both types try to get a look forward and estimate the potential of preserving a request for the self-fulfillment.

Request Potential Valuation Strategies: The RPVs analyze characteristics of requests, where the idea is to identify those with the highest potential for preserving. For the evaluation, function (1) is introduced that calculates a potential-value (PV_j) for each request (j) in the carriers request pool (U). The requests with the highest PV_j are preserved for the self-fulfillment. Function (1) is divided in four parts, where each rates one specific request characteristic. $b_1(j)$ considers the ratio between the earnings (e_j) of j and the overall maximum of earnings per request. Separately, $b_2(j)$ rates the quantity of demand (d_j) proportionally. $b_3(j)$ analyzes whether a clustering of the requests is auspicious. Here, the sum of the distance $(dist(u, j))$ of the n nearest requests to j is built, where $Ubest$ represents the set of the nearest requests. For a normalization the minimal distance overall is identified and multiplied by the cardinal number of $Ubest$. $b_4(j)$ identifies the best insertion positions $(ins(i, j))$ for j, which seems sensible when a tour scheduling already exists. This situation exhibits in dynamic environments. The m best insertion positions are identified, where V represents the set of all possible insertion positions and $Vbest$ stores the m best positions. This term is also normalized. By varying the weights $c_1, \ldots, c_4$ different RPVs can be generated. $RPV1(c_1 = 1, c_2 = 0, c_3 = 0, c_4 = 0)$ and $RPV2(2, 1, 0, 0)$ relate to a greedy procedure, where the requests with the highest earnings are identified. $RPV3(0, 0, 0, 1)$ considers only the best insertion positions. In contrast, $RPV4(2, 1, 3, 1)$ and $RPV5(2, 1, 1, 3)$ represent a combination of the previous RPVs and consider all function parts. Each of those strategies preserves a percentage of the requests with the highest values of PV_j for the self-fulfillment. The remaining requests are reallocated to the request exchange process.

$$PV_j = c_1 \cdot b_1(j) + c_2 \cdot b_2(j) + c_3 \cdot b_3(j) + c_4 \cdot b_4(j); \quad \forall j \in U, \text{with:} \tag{1}$$

$$b_1(j) = \frac{e_j}{\max_{u \in U} e_u}; \, b_2(j) = \frac{\min_{u \in U} d_u}{d_j}; \, b_3(j) = \frac{|Ubest| * \min_{u \in U \setminus \{j\}} dist(u,j)}{\sum_{u' \in Ubest} dist(u',j)}; \, b_4(j) = \frac{|Vbest| * \min_{i \in V} ins(i,j)}{\sum_{i' \in Vbest} ins(i',j)}$$

Tour Potential Valuation Strategies: A drawback of the RPVs is that the incidental traveling costs per request cannot be calculated, if no current tour scheduling exists. This situation occurs especially in static problems or the first planning period of dynamic environments. Thereby, it is hard to give a forecast if preserving a request for the self-fulfillment will increase the overall profit. The idea of the TPVs is to build promising tours. The costs of those tours can be calculated easily, which results in a more realistic cost evaluation per request. For building promising tours, the TPVs use heuristic constructive strategies of the vehicle routing. *TPV*1 relates to the saving algorithm, where the request with the highest savings is identified and inserted into the current tour scheduling [8]. This procedure is repeated until all requests are inserted. For the considered problem, *TPV*1 is supplemented with the request earnings, so that the request j with the highest profit $(e_j - ins(best,j))$ is chosen first. *TPV*2 is a modification of the sweep algorithm, where tours are build based on the polar coordinates of the requests [4]. The procedures of both TPVs insert all potential requests in the tour schedule. Afterward, the DMP1 is executed previous to the OP1. To solve the DMP1, a defined percentage of the most profitable tours are preserved for the self-fulfillment. The requests of the residual tours are reallocated to the request exchange process. For the performance of *TPV*3, an initial tour scheduling is generated by *TPV*1. Afterward, the OP1 is performed, before the DMP1 selects the requests for the self-fulfillment. Because of this procedure, *TPV*3 relates to the SVRP and may also block the following request exchange.

4 Computational Experiments

For the computational studies instances are generated that base on Gehring and Homberger [5]. The instances are organized into three classes, where the number of coalition members, requests per member and planning periods vary.[1] For the evaluation, each instance is solved ten times on a Windows 7 PC with Intel Core i7-2600 processor (3.4 GHz, 16 GB) and the average of the overall profits is considered. To get a better comparability, the allowed number of preserved requests for the self-fulfillment is limited to 50 % of the requests for all strategies. The DMP2 of the

---

[1]http://www.logistik.uni-bremen.de/english/instances/.

TSF is solved by a combinatorial auction with a central clustering and the maximal number of iterations is limited to 1,000 for the OP1 and to 20,000 for the OP2.

The first study analyzes the ability of the RPVs for solving the DMP1. As experiments demonstrate, $RPV4$ generates auspicious solution, if no tour scheduling exists. In contrast, $RPV5$ gets good results, when an initial scheduling exists. To realize both benefits a combination of $RPV4$ and $RPV5$ is performed, where the strategy is switched at the first planning update. Table 1 gives an overview of the results. It can be derived, that $RPV4$ and $RPV5$ are preferable for random instances (R). The combination of both strategies is able to increase the results for some instances. If a random/cluster instance (RC) is considered, $RPV4$ generates best solutions.

As the first test case shows, to increase the results in dynamic environments, it seems sensible to switch the pre-selection strategy, if a tour scheduling already exists. The TPVs give opportunities to build an initial tour scheduling. In a second case study, it is analyzed, which TPV is qualified to generate an initial tour scheduling and to solve the DMP1. For each scenario, one TPV is used to build the initial tour scheduling. Afterward, the strategy is changed to $RPV1$ or $RPV5$. Table 2 presents the results. The test case indicates, that building an initial tour scheduling via $TPV1$ or $TPV2$ increases the overall results for many test scenarios. Furthermore, a switch to $RPV5$ is preferable, which confirms the results of the first test case. To get a better similarity of TPV3 to an approach for the SVRP, the number of iterations of the advanced OP1 is set to 20,000. Despite this increased computing time, $TPV3$ is seldom able to improve the results of the initial tour scheduling against $TPV1$ or $TPV2$. In this context, the assumption that strategies related to approaches of the SVRP block the request exchange process, may be confirmed. The computational

Table 1 Results of the RPV-evaluation (average profit of ten runs in $)

	RPV1	RPV2	RPV3	RPV4	RPV5	RPV4/RPV5
R1-2x400	23,588.91	23,802.43	21,833.02	24,288.59	24,024.13	**24,380.06**
R2-2x400	**25,477.64**	24,327.61	23,967.01	25,070.36	24,029.87	25,017.73
R3-2x400	34,450.53	34,337.16	34,706.58	35,279.20	**35,300.60**	34,775.81
RC-2x400	23,334.24	23,923.38	22,894.95	**25,247.74**	23,407.79	23,239.92
R1-4x250	10,427.84	8,167.35	7,676.21	10,348.10	9,702.55	**10,547.34**
R2-4x250	13,982.38	14,379.51	15,265.93	16,599.38	**16,912.63**	16,624.47
R3-4x250	20,355.05	21,059.65	20,428.07	**24,457.71**	22,569.13	24,059.09
RC-4x250	5,584.20	8,336.73	9,766.94	**9,860.16**	6,152.13	8,994.35
R1-6x200	65,987.04	66,722.61	68,574.71	76,967.42	68,278.78	**77,587.27**
R2-6x200	68,333.91	64,832.97	66,713.48	**79,013.88**	67,946.56	78,685.05
R3-6x200	57,325.79	56,669.78	57,729.00	65,208.92	61,839.19	**65,647.80**
RC-6x200	59,175.69	57,437.22	61,718.80	**64,716.57**	61,495.80	64,084.60
Average	34,001.94	33,666.37	34.272.89	**38,088.17**	35,138.26	37,803.62

528   K. Schopka and H. Kopfer

Table 2 Results of the TPV-evaluation (average profit of ten runs in $)

	TPV1/RPV1	TPV1/RPV5	TPV2/RPV1	TPV2/RPV5	TPV3/RPV1	TPV3/RPV5
R1-2x400	24,434.01	**24,905.10**	24,252.39	24,867.42	24,197.20	24,401.31
R2-2x400	25,624.52	**25,626.29**	25,372.58	25,403.97	24,828.47	24,715.57
R3-2x400	35,168.44	35,681.30	35,204.24	**35,711.52**	35,175.10	35,701.66
RC-2x400	**23,334.23**	22,721.61	21,855.05	22,605.62	21,914.74	22,249.02
R1-4x250	10,103.43	12,301.97	11,794.60	13,273.51	12,438.35	**14,083.01**
R2-4x250	13,506.78	16,926.41	17,531.03	**18,588.50**	17,078.39	18,235.03
R3-4x250	19,472.33	21,854.53	21,621.25	**22,929.51**	21,491.22	22,766.36
RC-4x250	9,982.99	**11,537.56**	10,697.20	10,006.35	9,802.38	10,173.42
R1-6x200	69,978.81	75,025.08	72,744.08	**78,247.19**	72,209.90	76,838.80
R2-6x200	69,904.03	**73,326.25**	67,035.82	69,627.33	66,857.91	69,649.16
R3-6x200	61,516.83	**65,748.06**	62,539.61	65.089.83	62,676.53	63,872.33
RC-6x200	62,593.82	**67,518.84**	60,511.64	65,571.43	61,083.16	65,239.06
Average	35,468.35	**37,764.42**	35,929.96	37,660.18	35,812.78	37,327.06

studies determine, that the RPVs as well as the TPVs have the ability to solve the DMP1. For the considered test case the best results are achieved by using *RPV4*, although advances that switch the pre-selection strategy generate good results.

5 Conclusion

In this paper, the pre-selection strategies of the RPV and the TPV are presented, that give an advice for preserving requests for the self-fulfillment. The RPVs and TPVs may be possibilities to increase the autonomy for all coalition members and result in a long-term, stable coalition process. Computational studies identify that both types are qualified to solve the DMP1. In contrast, strategies which are related to approaches for the SVRP are less effective since they block the following request exchange process of our TSF. Future research should focus on powerful request exchange mechanisms for solving the DMP2. Because of the identified potential, it has to be analyzed if a modified RPV within a continuous re-optimization is able to generate an auspicious tour scheduling for problems with a higher grade of dynamic.

Acknowledgments The research was supported by the German Research Foundation (DFG) as part of the project "Kooperierende Rundreiseplanung bei rollierender Planung".

References

1. Archetti, C., Speranza, M.G., Vigo, D.: Vehicle routing problems with profits. Technical Report, Department of Economics and Management, University of Brescia, Italy (2012)
2. Berger, S., Bierwirth, C.: Solutions to the request reassignment problem in collaborative carrier networks. Transp. Res. Part E **46**, 627–638 (2010)
3. Butt, S.E., Cavalier, T.M.: A heuristic for the multiple tour maximum collection problem. Comput. Oper. Res. **21**, 101–111 (1994)
4. Clarke, G., Wright, J.W.: Scheduling of vehicles from a central depot to a number of delivery points. Oper. Res. **12**, 368–581 (1964)
5. Gehring, H., Homberger, J.: A parallel hybrid evolutionary metaheuristic for the vehicle routing problem with time windows. Proc. EUROGEN **99**, 57–64 (1999)
6. Krajewska, M.A., Kopfer, H.: Collaborating freight forwarding enterprise. OR Spectr. **28**, 301–317 (2006)
7. Laporte, G., Martello, S.: The selective travelling salesman problem. Discrete Appl. Math. **26**, 193–207 (1990)
8. Gillett, B.E., Miller, L.R.: A heuristic for the vehicle-dispatch problem. Oper. Res. **22**, 340–349 (1974)
9. Pillac, V., Gendreau, M.: Guret C., Medaglia A.L.: A review of dynamic vehicle routing problems. Eur. J. Oper. Res. **225**, 1–11 (2013)
10. Pisinger, D., Ropke, S.: A general heuristic for vehicle routing problems. Comput. Oper. Res. **34**, 2403–2435 (2007)
11. Schwind, M., Gujo, O., Vykoukal, J.: A combinatorial intra-enterprise exchange for logistics services. Inf. Syst. E-Business Manage. **7**, 447–471 (2009)

Optimal Operation of a CHP Plant for the Energy Balancing Market

Katrin Schulz, Bastian Hechenrieder and Brigitte Werners

Abstract For energy companies with a combined heat and power (CHP) plant and a heat storage device, the provision of balancing power provides additional revenue potential. Balancing power is needed to ensure a stable frequency if current generation differs from demand. In Germany, the responsible transmission system operator procures the needed type of reserve energy through an auction. To participate in the balancing market for minute reserve, energy companies have to submit a bid for a certain time frame of the following day that comprises a price and the amount of electricity at which power generation can be increased or decreased considering the expected load. Therefore, capacity allocation for the balancing market has to be considered simultaneously with the uncertain next day's unit commitment of the CHP plant and the heat storage device. To support energy companies planning their bids, an optimization model is developed to determine the optimal bidding amount based on a forecasted market clearing price.

1 Introduction

As a climate protection measure, the European Union approved the Climate Change and Energy Package in 2008 containing i.e. the reduction of greenhouse gas emissions aiming at a sustainable energy system. Conventional power plants shall be replaced by renewable energies—in Germany to an extent of 80 % until 2050. Since power generation of renewables is feature-dependent, a share of conventional, preferably efficient power plants is still needed to ensure system stability. CHP plants generate heat and electricity simultaneously and excel in a high fuel utilization rate as well

K. Schulz (✉) · B. Hechenrieder · B. Werners
Faculty of Management and Economics, Chair of Operations Research
and Accounting, Ruhr University Bochum, Bochum, Germany
e-mail: katrin.schulz@rub.de

B. Hechenrieder
e-mail: bastian.hechenrieder@rub.de

B. Werners
e-mail: or@rub.de

© Springer International Publishing Switzerland 2016
M. Lübbecke et al. (eds.), *Operations Research Proceedings 2014*,
Operations Research Proceedings, DOI 10.1007/978-3-319-28697-6_74

as an increased efficiency by 10–40 % compared to separate generation. Thus, these plants are promoted by the German government in order to increase the share of CHP in the generation of electricity to 25 % by 2020. Given the current and stagnating share of merely 16 %, the implemented government support is apparently not sufficient. In the following, it is analyzed whether participating in the energy balancing market can facilitate further revenue potential for gas-fired CHP plants which allow a flexible adaption of operation.

2 Participation of a CHP Plant in the Energy Balancing Market

Since electricity can only be stored to a limited extent, a stable balance between demand and supply has to be ensured what is typically done by the transmission system operator (TSO). In any case of deviation from the setpoint frequency of 50 Hz in the European integrated network, balancing energy is needed to compensate for imbalances. As balancing deviations can, for instance, occur due to fluctuating energy generation from renewables, deviations from demand predictions or a forced power plant outage, the sufficient provision of so called electricity reserve power is crucial to ensure system stability. Since Germany is part of the European Network of Transmission System Operators for Electricity (ENTSO-E), it is distinguished between primary (PCP) and secondary control power (SCP) as well as minute reserve power (MRP). The latter two are separated into positive and negative reserve power. If demand exceeds supply, positive reserve power is needed and if supply surpasses demand, negative reserve power is used [2]. The three qualities of reserve power differ in their delivery times which are 30 s for PCP, 5 min for SCP and 15 min for MRP. Thus, a power plant operator has to ensure that the reserved capacity is provided in time. In the following, a CHP plant with a start-up time of less than 15 min is considered enabling the provision of reserve capacity as MRP even if the plant is shut down. To prevent a must-run condition, the CHP plant takes part only in the energy balancing market for minute reserve.

On the German energy balancing market, the four German TSOs are interconnected and coordinate their operations allowing power plants to offer MRP in all four control areas with decreased prices compared to the previous market design [2]. On a common web-based tendering platform, generation companies compete for the published demand. This single-stage demand auction for minute reserve is carried out daily and day-ahead as multi-unit auction, i.e. it is differentiated between positive and negative minute reserve and different time slots. As a multi-part auction, the two-part tariff consists of a capacity price for the stand-by provision and an energy price for the actual call. The last selected bid for the capacity price establishes the market clearing price and the dispatched generation companies receive their bidding price following a pay-as-bid pricing mechanism. For the actual call of reserve power, energy prices are sorted in a merit order [5].

In principle, a bid for positive as well as for negative minute reserve power comprises a time frame of 4 h. Throughout this period, the generation company has to ensure that the offered amount of reserve power is available. Thus, the unit commitment of the CHP plant, i.e. the economic dispatch, has to be considered simultaneously with planning the submission of a tender.

3   Optimization Model for the Operation of a CHP Plant

To support generation companies in planning their bids for the energy balancing market, different modeling methods have been developed. Optimization models for a single generation company focus on one specific market participant while making simplifying assumptions about the other market participants and influencing factors. In this respect, the generation company is assumed to be a price-taker. In contrast to a price-maker who influences the market clearing price with his bidding strategy, the market clearing price can be regarded as an exogenous variable for a price-taker [4]. Thus, the generation company treats the market clearing price as a random variable which has to be forecasted like in [1] or [8]. Based on the forecast, a bidding strategy is introduced in [1] considering the optimal self-scheduled generation. In the following, the optimal bidding of reserve capacity is determined assuming the bidding price to be equal to the forecasted clearing price.

Since the bidding process is divided into time slices of 4 h, bidding for a certain block b comprises a bidding price for positive BPP_b or negative reserve power BPN_b accompanied by an offered capacity for positive op_b or negative reserve capacity on_b, respectively. Whereas the stand-by provision of reserve capacity is remunerated with the company's bidden price BPN_b or BPP_b, respectively, the energy price is only paid in case of delivery and reflects those costs that arise due to the adapted operation of the CHP plant. The generation company has to decide 1 day ahead which capacity to offer as positive op_b or negative reserve capacity on_b, respectively, in each block b focussing on negative reserve power in the following. Since priority is given to the regular power and heat demand, the available capacity is restricted. However, the demand of the following day as well as the power price on the spot market are subject to uncertainty. To exploit the full revenue potential of the energy balancing market, the maximum reserve capacity that can be offered is determined with regard to the unit commitment of the CHP plant considering different scenarios for the supply situation, trading conditions on the spot market and the heat storage device. It has to be ensured that the offered negative reserve capacity on_b for a certain block b can be provided in all scenarios and time periods t within the corresponding block T_b. For each period t and scenario s the available capacity for negative reserve power is determined as difference r_{st}^- between the actual and the minimum power generation at this operation point. The offered negative reserve capacity on_b is restricted by the

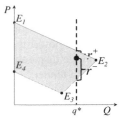

Fig. 1 Shows all possible combinations of heat Q and power P that can be generated by a CHP extraction condensing turbine. The generation costs vary with the chosen operation point. Hence, every extreme point E_i (with I as a set of extreme points) is complemented with its corresponding costs C_i, power generation P_i and heat generation Q_i [3]. q^* gives the minimal heat generation to satisfy the demand. Point • gives the current operation of the CHP plant, r^+ and r^- mark the available capacity for adjustments in the power generation for positive respectively negative reserve power

minimum of the available negative reserve capacity of all time periods t within the corresponding block T_b (1).

$$on_b \leq r_{st}^- \qquad\qquad \forall b \in B \text{ and } t \in T_b, \forall s \in S \qquad (1)$$

Based on the optimization model developed in [6], a flexible CHP plant with an extraction condensing turbine and a heat storage device is considered. Taking into account the characteristic diagram of the CHP plant, the available reserve capacity is determined considering one operation point (see Fig. 1). If this indicates a certain heat generation q^* to the right of E_3, r^- is determined using the slope of the line between E_3 and E_2 given as $m_{32} := (P_2 - P_3)/(Q_2 - Q_3)$. Since the available capacity for reserve power depends on the necessary heat generation, it is calculated whether an adjustment of the operation point to the right of E_3 is possible if the whole storage content is used (given as the storage level of last period $\ell_{s(t-1)}$ minus a proportional heat loss V). The binary variable $n_{st} = 1$ indicates a heat generation q^* located to the right of E_3 as Eqs. (3) and (4) compare the heat demand HD_{st} in scenario s and period t with the storage level and the heat generation Q_3 of E_3. The variable v_{st} specifies the necessary heat generation in period t and scenario s if the scheduled operation point is located right to E_3. The available capacity for negative reserve power r_{st}^- is determined in (6) using the power generation p_{st} of the current operation point and the minimum power generation P_3 at the extreme point E_3 if the CHP plant is operated ($y_{st} = 1$) subtracting the heat generation to the right of E_3 ($v_{st} - Q_3 \cdot n_{st}$) multiplied with the slope m_{32}.

$$z_s = \sum_{t \in T} \left(\sum_{i \in I} C_i \cdot x_{ist} + SUC \cdot u_{st} + SP_{st} \cdot (bsm_{st} - ssm_{st}) \right)$$
$$- \sum_{b \in B} \left(BPP_b \cdot op_b - BPN_b \cdot on_b \right) \qquad\qquad \forall s \in S \qquad (2)$$

$$HD_{st} - (1 - V) \cdot \ell_{s(t-1)} - Q_3 - n_{st} \cdot M \leq 0 \qquad \forall s \in S, t \in T \tag{3}$$

$$HD_{st} - (1 - V) \cdot \ell_{s(t-1)} - Q_3 + (1 - n_{st}) \cdot M \geq 0 \qquad \forall s \in S, t \in T \tag{4}$$

$$HD_{st} - (1 - V) \cdot \ell_{s(t-1)} - (1 - n_{st}) \cdot M \leq v_{st} \qquad \forall s \in S, t \in T \tag{5}$$

$$r_{st}^- = p_{st} - P_3 \cdot y_{st} - m_{32} \cdot (v_{st} - Q_3 \cdot n_{st}) \qquad \forall s \in S, t \in T \tag{6}$$

In Eq. (2), the net acquisition costs for each scenario z_s are defined as generation costs for heat and power plus the difference between purchase costs and revenues from power trading minus revenues from the energy balancing market. For each period t and scenario s it has to be decided how to operate the CHP plant x_{ist} and whether a start-up ($u_{st} = 1$) is necessary. The power demand can be met by own generation or purchase bsm_{st} and excess power ssm_{st} can be sold at a charge of SP_{st} on the spot market. The trading results, the hourly generation costs $\sum_{i \in I} C_i \cdot x_{ist}$ and the start-up costs SUC for each start-up are considered. Moreover, the aspired revenues for reserve capacity are deducted from the bidding prices (BPP_b and BPN_b) multiplied by the offered reserve capacity (op_b and on_b). With regard to the risk attitude of the decision maker, the objective function is formed as expected value of the net acquisition costs for each scenario in the first case and contains a minimax regret approach in the second case. While the optimization model seeks to determine the maximum reserve capacity, first priority is given to the fulfilment of the regular heat and power demand. However, due to unit flexibility and especially use of the heat storage device, generation can be shifted between different periods according to the revenue potential on the reserve market still taking into account trade options on the spot market.

4 Illustrative Results

The results for a CHP plant with heat storage according to [7] are presented for negative reserve power. Exemplarily, three different scenarios for the power and heat demand and spot prices are assumed and the scenario-independent optimal bidding amount of negative reserve power is determined using the expected-value (EV) and the minimax regret (MR) approach. Whereas the optimal bidding amounts for the blocks 2 to 6 (hours 5 to 24) come close together in both approaches, negative reserve power is only offered for block 1 (hours 1 to 4) in the MR approach as shown in Fig. 2. This means that the bidding strategy is more aligned to scenario 1 in the EV approach. A 2-h shut-down of the CHP plant in this scenario is more beneficial than operating the CHP plant for the participation in the energy balancing market. The scenario-individual costs for scenario 1 are about 6 % lower in the EV approach than in the MR approach whereas the latter leads to lower costs in scenario 2 and 3 compared to the

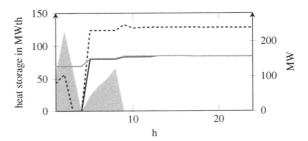

Fig. 2 Negative reserve power in the EV approach (—) and the MR approach (—) considering the power generation in scenario 1 (---) and heat storage level (■) for the EV approach

EV approach. Since the three scenarios are assumed to be equiprobable, the possible savings in scenario 1 without bidding for the first block on the reserve market exceed the savings in scenario 2 and 3 with bidding for this block in the EV approach. Hence, bidding negative reserve power prevents an optimal operation of the CHP plant in a certain scenario according to the scenario-specific heat and power demand and spot market prices. Depending on the risk attitude of the decision maker and the trading conditions, a single scenario-optimal operation mode may dominate. Overall, the participation in the energy balancing market leads to additional revenues regarding average costs for the three scenarios independent of the risk attitude.

5 Conclusion

For the participation in the energy balancing market available capacities for positive and negative reserve power are maximized simultaneously considering unit commitment. The herein examined market for minute reserve offers additional revenue potential for a CHP plant. However, such an adapted operation for the balancing market considering different scenarios can prevent the optimal operation of the CHP plant in a certain scenario. According to the risk attitude of the decision maker, an expected value approach and a minimax-regret approach are compared. The two approaches lead to different results for the optimization of the available reserve capacity and the operation of the CHP plant. Assuming a successful auction, the operation of the CHP plant is optimized again for the case of delivery. Resulting costs are reflected by the energy price. In order to examine the effect of a specific bidding strategy for different scenarios, herein the bidding price equals the market clearing price. Since the revenue potential depends on the realizable bidding price, further research will focus on the optimal price bidding strategy taking into account the corresponding optimal bidding amount.

References

1. Conejo, A.J., Nogales, F.J., Arroyo, J.M.: Price-taker bidding strategy under price uncertainty. IEEE Trans. Power Syst. **17**(4), 1081–1088 (2002)
2. Haucap, J., Heimeshoff, U., Jovanovic, D.: Competition in Germany's minute reserve power market: an econometric analysis. Energy J. **35**(2), 137–156 (2014)
3. Lahdelma, R., Hakonen, H.: An efficient linear programming algorithm for combined heat and power production. Eur. J. Oper. Res. **148**(1), 141–151 (2003)
4. Li, G., Shi, J., Qu, X.: Modeling methods for GenCo bidding strategy optimization in the liberalized electricity spot market—a state-of-the-art review. Energy **36**(8), 4686–4700 (2011)
5. Müller, G., Rammerstorfer, M.: A theoretical analysis of procurement auctions for tertiary control in Germany. Energy Policy **36**(7), 2620–2627 (2008)
6. Schulz, K., Schacht, M., Werners, B.: Influence of fluctuating electricity prices due to renewable energies on heat storage investments. In: Huisman, D., et al. (eds.) Operations Research Proceedings 2013. Springer, New York (2014)
7. Schacht, M., Schulz, K.: Kraft-Wärme-Kopplung in kommunalen Energieversorgungsunternehmen - Volatile Einspeisung erneuerbarer Energien als Herausforderung. In: Armborst, K., et al. (eds.) Management Science - Festschrift zum 60. Geburtstag von Brigitte Werners, Dr. Kovac, 337–363. Hamburg (2013)
8. Valenzuela, J., Mazumdar, M.: Commitment of electric power generators under stochastic market prices. Oper. Res. **51**(6), 880–893 (2003)

Gas Network Extension Planning for Multiple Demand Scenarios

Jonas Schweiger

Abstract Today's gas markets demand more flexibility from the network operators which in turn have to invest into their network infrastructure. As these investments are very cost-intensive and long-living, network extensions should not only focus on one bottleneck scenario, but should increase the flexibility to fulfill different demand scenarios. We formulate a model for the network extension problem for multiple demand scenarios and propose a scenario decomposition. We solve MINLP single-scenario sub-problems and obtain valid bounds even without solving them to optimality. Heuristics prove capable of improving the initial solutions substantially. Results of computational experiments are presented.

1 Introduction

Recent changes in the regulation of the German gas market are creating new challenges for gas network operators. Especially the unbundling of gas transport and trading reduces the influence of network operators on transportation requests. Greater flexibility in the network is therefore demanded. Traditional, deterministic planning approaches focus on one bottleneck scenario. Stochastic or robust approaches, in contrast, can consider a set of scenarios and therefore lead to more flexible network extensions.

Gas transmission networks are complex structures that consist of passive pipes and active, controllable elements such as valves and compressors. For planning purposes, the relationship of flow through the element and the resulting pressure difference is appropriately modeled by nonlinear functions and the description of the active elements involves discrete decisions (e.g., whether a valve is open or closed) (see [4, 5] for the details of our model and algorithmic approach to solve deterministic models). The resulting model is thus an *Mixed-Integer Nonlinear Program (MINLP)*.

J. Schweiger (✉)
Konrad-Zuse-Zentrum für Informationstechnik Berlin,
Takustraße 7, 14195 Berlin, Germany
e-mail: schweiger@zib.de

© Springer International Publishing Switzerland 2016   539
M. Lübbecke et al. (eds.), *Operations Research Proceedings 2014*,
Operations Research Proceedings, DOI 10.1007/978-3-319-28697-6_75

In this presentation, we focus on additional pipes as extension candidates. A new pipe allows flow but also couples the pressures at the end nodes, possibly rendering previously feasible transport requests (also known as *nominations*) infeasible. An additional valve retains all possibilities of the original network. Opening the valve, corresponds to building the extension pipe and is therefore penalized the cost for the extension. Closing the valve forbids flow over the pipe which effectively removes the pipe from the system. Details on the approach for topology optimization for a single-scenario can be found in [1].

To approach the optimization over a finite set of scenarios (i.e., transport requests), we propose a scenario decomposition. Section 2 describes the model. The decomposition method is presented in Sect. 3 together with some details about primal and dual bounds and results on the ability to reuse solutions from previous optimization runs over the same scenario. Section 4 presents the results of computational experiments. Section 5 provides an outlook on planned future work on the topic.

2 Planning for Multiple Demand Scenarios

Assume a gas network, a set of scenarios $\omega \in \Omega$, i.e., nominations, and a set of extensions $\mathscr{E}$ (each extension consisting of a pipe and a valve) is given. We denote the set of characteristic vectors of feasible extension sets for scenario ω with

$$\mathscr{F}^{\omega} = \left\{ \chi_E \in \{0, 1\}^{\mathscr{E}} \mid \text{Extensions } E \subseteq \mathscr{E} \text{ make } \omega \text{ feasible} \right\}$$

In our situation, a closed form description of $\mathscr{F}^{\omega}$ is not at hand. However, we assume monotonicity in the sense that adding extensions to an element of the set is still feasible:

$$x_1 \in \mathscr{F}^{\omega}, \ x_2 \in \{0, 1\}^{\mathscr{E}}, \ x_2 \geq x_1 \implies x_2 \in \mathscr{F}^{\omega}.$$

Especially in the context of gas network planning this property cannot be taken for granted but adding valves to all extensions ensures monotonicity in our application.

For a specific scenario ω the extension planning problem can now be stated as

$$\min c^T x^{\omega} \qquad\qquad \text{(SSTP)}$$

$$\text{s.t. } x^{\omega} \in \mathscr{F}^{\omega}$$

This formulation hides the difficulties in describing and optimizing over the set $\mathscr{F}^{\omega}$. Our approach uses problem (SSTP) as sub-problem and assumes a black-box solver to be available (e.g., from [1]).

In the multi-scenario extension planning problem we seek for a set of extensions of minimal cost such that the resulting network allows a feasible operation in *all* scenarios. We stress that in the different scenarios not all extensions that have been built have to be actually used; in fact, using them might not even be feasible. The multi-scenario problem can then be stated as:

$$\min c^T y \qquad\qquad\qquad\text{(MSTP\_TS\_Node)}$$

$$\text{s.t. } x^\omega \in \mathscr{F}^\omega \qquad\qquad \text{for all }\ \omega \in \Omega \qquad\qquad (1)$$

$$x^\omega \leq y \qquad\qquad\qquad \text{for all }\ \omega \in \Omega \qquad\qquad (2)$$

$$y \in \{0, 1\}^{\mathscr{E}} \qquad\qquad\qquad\qquad\qquad\qquad (3)$$

This is a two-stage stochastic program. y are the first stage variables which indicate which extensions are built. Finding a feasible operational mode for the scenarios given the extensions selected by y is the second stage problem.

3 Scenario Decomposition

The algorithmic idea is scenario decomposition. First, we solve the scenario sub-problems (SSTP) independently and in parallel. If one scenario sub-problem is infeasible, the multi-scenario problem is infeasible.

Branching on the y variables is used to coordinate the scenarios. To this end, we identify extensions that are selected in some but not all scenarios. Two sub-problems, i.e., nodes in the Branch&Bround tree, are created: one with the condition $y_e = 0$ and one with the condition $y_e = 1$. In the two nodes, sub-problems have to be modified accordingly. For $y_e = 0$, variable x_e^ω is fixed to zero. For $y_e = 1$, extension e is built and using it does not incur additional cost.

Each node of the Branch&Bround tree is identified by the sets E_0 and E_1 of extensions that are fixed to 0 and 1, respectively. The modified single-scenario problem for scenario ω then reads:

$$\min \sum_{e \notin E_1} c_e x_e^\omega + \sum_{e \in E_1} c_e \qquad\qquad (SingleScen_\omega)$$

$$\text{s.t. } x^\omega \in \mathscr{F}^\omega$$

$$x_e^\omega = 0 \qquad\qquad \text{for all }\ e \in E_0$$

The following lemma states that adding more elements to E_0 and E_1 might only deteriorate the objective function value.

Lemma 1 *Let $E_0^1 \subseteq E_0^2$ and $E_1^1 \subseteq E_1^2$ and c_i^* the optimal value of ($SingleScen_\omega$) with respect to (E_0^i, E_1^i). Then $c_1^* \leq c_2^*$.*

Dual bounds for the single-scenario problems can be instantly translated into dual bounds for the multi-scenario problem.

Lemma 2 *Let the objective function coefficients be non-negative, i.e., $c \geq 0$. Then any dual bound for problem ($SingleScen_\omega$) for any scenario is also a dual bound for problem* (MSTP_TS_Node).

We propose three ways to get or enhance feasible solutions: First, by construction the union of all extensions used in the different scenarios constitutes a primal solution for the multi-scenario problem. Therefore, we construct a solution to (MSTP_TS_Node) in every node by setting $y = \max_{\omega \in \Omega} x_e^\omega$ where x_e^ω is taken as the best solution for scenario ω.

Second, we observed that checking if a certain subset of extensions is feasible is typically very fast. This observation is used by a 1-opt procedure that takes the best current solution to (MSTP_TS_Node), removes one extension, and checks all scenarios for feasibility.

Third, in stochastic programming optimal single-scenario solutions often lack flexibility and do not occur in optimal solutions to the stochastic program (e.g., [7]). To benefit from all solutions the solver provides, we access its solution pool and store all sub-optimal solutions for the scenarios. This has two benefits. The solver might be able to use them as start solutions in the next node. On the other hand, we construct the "best known" solution so far by solving an auxiliary MIP.

3.1 Reusing Solutions

The Branch&Bround procedure solves slight modifications of the same problem over and over again. In some important cases, not all scenarios need to be solved again since we already know the optimal solution. As an example, take the extreme case where a scenario is found to be feasible without extensions. Clearly, the procedure should never touch this scenario again.

In order to decide whether a solution from a previous node can be reused, we need to take the fixations under which the solution was computed and the current fixations into account. In addition to the current fixations E_0 and E_1, we define the sets E_0^S and E_1^S as the extensions that were fixed to the respective values when solution S was computed. We assume $E_i^S \subseteq E_i$, i.e., currently more extensions are fixed than when solution S was computed. Abusing notation, we identify the solution with the set of extensions it builds.

We start with the simple observation, that if all the extensions in a solution are already built (i.e., y_e is fixed to 1), then the solution is optimal for the restricted problem:

Lemma 3 *If $S \in \mathscr{F}^\omega$ and $S \subseteq E_1$, then S an optimal solution for ($SingleScen_\omega$) for fixings E_0 and E_1.*

If a solution is optimal for (E_0^S, E_1^S) and all extensions in E_1 are part of the solution, then the solution is still optimal.

Lemma 4 *Let $S \in \mathcal{F}^\omega$ be an optimal solution to $(SingleScen_\omega)$ given the fixations E_0^S and E_1^S. If $E_1 \subseteq S$ and $S \cap E_0 = \emptyset$, then S is an optimal solution to $(SingleScen_\omega)$ for fixings E_0 and E_1.*

This is the situation, for example, after branching in the root node. In the 1-branch, a scenario which uses this extension does not need to be recomputed. In the 0-branch, solutions that did not use the extension remain optimal.

The situation becomes tricky if a solution does not use extensions that are already built but still uses unfixed extensions. The following lemma generalizes Lemma 4.

Lemma 5 *Let $S \in \mathcal{F}^\omega$ be an optimal solution given the fixations E_0^S and E_1^S. If $E_1 \setminus E_1^S \subseteq S$ and $S \cap E_0 = \emptyset$, then S is an optimal solution to $(SingleScen_\omega)$ for fixings E_0 and E_1.*

4 Computational Results

We tested our approach on realistic instances from the gaslib-582 testset of the publicly available GASLIB [2, 5]. The GASLIB contains network data and flow scenarios that are distorted versions of the real data from the German gas network operator Open Grid Europe GmbH. The approach is implemented in the framework Lamatto++ [3]. Methods to solve the single-scenario problems and to generate suitable extension candidates were developed in the FORNE project. We used a time limit of 600s for the sub-problems which is reduced to 300s in the 1-opt heuristic. The total timelimit for set to 10h. The experiments were performed on Linux computers with two 64 bit Intel Xeon X5672 CPUs at 3.20 GHz having 4 cores each such that up to 8 threads were used to solve the single-scenario problems in parallel.

Instances are composed from a pool of 126 infeasible instances that in single-scenario optimization find feasible solutions in the first 10 min. Table 1 summarizes the results. All but 3 instance are solved to proven optimality. The 3 instances that run into timeout each solve 3 nodes and then arrive to a point, where all extensions are fixed, but the single-scenario subproblem can neither find a feasible solution nor prove infeasibility.

Table 1 Summary of computational results

Scenarios	Instances	Status		Nodes		Time (s)	
		Optimal	Timelimit	Avg	Max	Avg	Max
4	186	184	2	1.1	3	475	36017
8	90	89	1	1.3	3	660	36001
16	42	42	0	2.0	15	676	4598
32	18	18	0	3.9	14	1890	6080

5   Outlook

We presented a method for capacity planning with multiple demand scenarios. The computational experiments show the potential of our approach. Even though developed in the context of gas network planning, the few assumptions on the problem structure suggest the generalization to other capacity planning problems.

In the future, we also want to consider active elements (compressors, which can increase the pressure, and control valves, which can reduce it) as extension candidates. They possess 3 states: active, bypass, and closed. In case the element is not used in active mode, the abilities needed can be covered by a much cheaper valve. Then the binary "build"-"not build" decision is replaced by the three possibilities "build as active element", "build as valve", and "do not build".

Last, we want to mention that Singh et al. [6] present an approach for capacity planning under uncertainty based on Dantzig-Wolfe decomposition. A comparison of our approach to theirs is future work.

Acknowledgments The author is grateful to Open Grid Europe GmbH (OGE, Essen/Germany) for supporting this work. Furthermore, the author wants to thank all collaborators in the FORNE project and all developers of Lamatto++.

References

1. Fügenschuh, A., Hiller, B., Humpola, J., Koch, T., Lehman, T., Schwarz, R., Schweiger, J., Szabó, J.: Gas network topology optimization for upcoming market requirements. In: IEEE Proceedings of the 8th International Conference on the European Energy Market (EEM), pp. 346–351 (2011)
2. Gaslib: A library of gas network instances (2013). http://gaslib.zib.de
3. Geißler, B., Martin, A., Morsi, A.: Lamatto++ (2014). http://www.mso.math.fau.de/edom/projects/lamatto.html
4. Koch, T., Hiller, B., Pfetsch, M.E., Schewe, L. (eds.): From Simulation to Optimization: Evaluating Gas Network Capacities. MOS-SIAM Series on Optimization. SIAM—Society for Industrial and Applied Mathematics (2015)
5. Pfetsch, M.E., Fügenschuh, A., Geißler, B., Geißler, N., Gollmer, R., Hiller, B., Humpola, J., Koch, T., Lehmann, T., Martin, A., Morsi, A., Rövekamp, J., Schewe, L., Schmidt, M., Schultz, R., Schwarz, R., Schweiger, J., Stangl, C., Steinbach, M.C., Vigerske, S., Willert, B.M.: Validation of nominations in gas network optimization: models, methods, and solutions. Optim. Methods Softw. (2014)
6. Singh, K.J., Philpott, A.B., Wood, R.K.: Dantzig-wolfe decomposition for solving multistage stochastic capacity-planning problems. Oper. Res. **57**(5), 1271–1286 (2009)
7. Wallace, S.W.: Stochastic programming and the option of doing it differently. Ann. Oper. Res. **177**(1), 38 (2010)

Impacts of Electricity Consumers' Unit Commitment on Low Voltage Networks

Johannes Schäuble, Patrick Jochem and Wolf Fichtner

Abstract Todays electricity consumer tend to become small businesses as they invest in their own decentralized electricity generation and stationary electricity storage as well as in information technology (IT) to connect and organize these new devices. Furthermore, the installed IT allows them at least technically to establish local markets. The variety of consumers and their characteristics implies numerous ways of how they optimize their individual unit commitment. This paper aims to analyze the impact of the individual consumers decisions on a future electricity demand and feed-in on low voltage network level. Therefore, in a first step the different unit commitment problems of the different small businesses have been modeled using linear programming (LP). In a second step these consumers are modeled as learning agents of a multi-agent system (MAS). The MAS comprises a local electricity market in which participants negotiate supply relationships. Finally, using scenarios with different input parameters the resulting impact is studied in detail. Amongst others, the simulations' results show major changes in electricity demand and feed-in for scenarios with high market penetration of storages.

1 Introduction

The design of a likewise sustainable, climate-friendly, safe, and efficient energy supply presents both current and future society with great challenges. In order to meet this requirement, the energy sector, driven by political, economical, and social decisions, is changing continuously. This evolution thereby affects all areas of energy supply, namely provision, transport, distribution and demand. Induced by expan-

J. Schäuble (✉) · P. Jochem · W. Fichtner
Chair of Energy Economics, Karlsruhe Institute of Technology,
Kaiserstr. 12, 76131 Karlsruhe, Germany
e-mail: johannes.schaeuble@kit.edu; schaeuble@kit.edu

P. Jochem
e-mail: patrick.jochem@kit.edu

W. Fichtner
e-mail: wolf.fichtner@kit.edu

© Springer International Publishing Switzerland 2016
M. Lübbecke et al. (eds.), *Operations Research Proceedings 2014*,
Operations Research Proceedings, DOI 10.1007/978-3-319-28697-6_76

sion of decentralized electricity generation by renewable energy sources (RES), use of storage, new load characteristics such as electric vehicles [1], and market liberalization [2], as well as increased involvement of society on climate protection and market participation [3], the growing number and heterogeneity of actors and elements particularly increase the complexity of the electricity sector.

Apart from the implied problems, these developments offer great potential within a new design of a future power supply: More and more consumers will generate electricity (e.g. by using photovoltaic (PV) systems), and will apply storages [1, 4], therefore becoming in most hours less dependent on centralized conventional power generation. More frequently those electricity producing consumers will be situated in electricity networks which changed from a top-down to a bottom-up cell structure [3]. Moreover they might be organized in local markets [5] with simple access for individual actors using new IT appliances. These local systems offer an incentive to locally balance power supply and consumption, and hence reduce the degree of grid capacity utilization.

To estimate the potential of such a new design of a future power supply system, its elements and their impact on the system must be analyzed in detail. This paper therefore aims to examine individual households and their cost optimized scheduling of power consumption, generation, and storage, as well as the implications for the local system. Therefore, in a first step the different unit commitment problems of the above described households are described and modeled using LP. In a second step they are modeled as learning agents of a MAS although retaining their individual LP. Following a local electricity market in which participants negotiate supply relationships is integrated into the MAS. Finally, using different scenarios with several input parameters the resulting impacts are studied in detail.

2 Scheduling of Consumers' Generation, Demand, and Storage

With an increasing complexity of the households and their options of configuration, the demand for ways to optimize the scheduling of the system elements of the household rises. A connection to the electricity network (EN) and an electricity demand (ED) is thereby assumed for each household and each point in time t. The ED in this paper includes no load shifting potential (e.g. via delayed charging of electric vehicles). This paper focuses on three household configurations. Two of the configurations include an electric generation (EG), here a generation using PV is assumed exemplary. One of these configurations includes an electrical storage (ES). Figure 1 displays nodes and flows from source to sink ($F_{Source,Sink,t}$) in the last-mentioned household configurations. Whereat all nodes can function as sink. Moreover, nodes EN, EG, and ES equally can act as a source. Depending on their configuration the types of households are referred to as ND (Nodes EN and ED), NDG (Nodes EN, ED and EG), and NDGS (Nodes EN, ED, EG and ES).

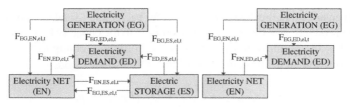

Fig. 1 Nodes and flows of the household configurations NDGS (*left*) and NDG (*right*)

Because households with the configuration NDG have to decide about the amount and the ratio of the feed-in ($F_{EG,EN,t}$) and own consumption ($F_{EG,ED,t}$) for the points in time t ($0 < t \leq T$) with a positive activity level of EG ($X_{EG,t} > 0$) and an positive absolute value ED ($|D_t| > 0$) these households are modeled via LP. For the configuration NDGS the flows into the storage ($F_{EG,ES,t}$) from a positive own generation ($X_{EG} > 0$) have to be introduced into the above mentioned ratio. Moreover, decisions have to be taken on the level and the ratio of the feed-in ($F_{ES,EN,t}$) and coverage of ED ($F_{ES,ED,t}$) respectively by the ES in points of times t with a positive state of charge of the ES ($X_{EG,t} > 0$). For the points in time when maximum charge of the ES level is not reached ($X_{ES,t} \leq \bar{L}_{ES,t}$) additionally the decision on the amount of electricity to store from the EN ($F_{EN,ES,t}$) has to be taken. Table 1 lists relevant parameters and indices for the mathematical formulation of the model.

Objective function of the optimization problem of the household configuration NDG is the minimization of the system costs C_{NDG}

$$min\ C_{NDG} = \sum_{t \in T} \left(F_{EN,ED,t} \cdot p_t - F_{EG,EN,t} \cdot r_{fi,t} - F_{EG,ED,i} \cdot r_{oc,t} \right) \qquad (1)$$

Subject to (selection of constraints)

$$D_t = F_{EG,ED,t} + F_{EN,ED,t}, \forall t \in T \qquad (2)$$

$$\bar{X}_{EG} \cdot \alpha_t \cdot j_t = F_{EG,ED,t} + F_{EG,EN,t}, \forall t \in T \qquad (3)$$

Table 1 Description of selected model variables and parameters

$X_{ES,t}$	State of charge of the ES [kWh]
p_t	Electricity price in t [EUR/kWh]
$r_{fi,t}$	Price for feed-in electricity in t [EUR/kWh]
$r_{oc,t}$	Price for the own consumption in t [EUR/kWh]
D_t	ED of the household in t [kWh]
$\bar{X}_{EG}$	Maximal power of EG [kWp]
α_t	Capacity factor of the EG in t, $\alpha_t \in [0, 1]$
$\bar{L}_{ES,t}$	Storage capacity of the ES [kWh]
$\bar{X}_{ES}$	Maximal power of ES [kWp]
j_t	Length of the time step t [h]

where (2) ensures that the demand D_t is met for each t and (3) takes into account the electricity production of the PV system according to the capacity factor and installed capacity. For the optimization problem of the household configuration NDGS the objective function is the minimization of the system costs C_{NDGS}

$$
\begin{aligned}
min\ C_{NDGS} = C_{NDG} + \sum_{t \in T} \Big(& F_{EN,ES,t} \cdot p_t \\
& - \frac{F_{ES,ED,t}}{F_{ES,ED,t} + F_{ES,EN,t}} \cdot F_{EG,ES,t} \cdot r_{fi,t} \\
& - \frac{F_{ES,EN,t}}{F_{ES,ED,t} + F_{ES,EN,t}} \cdot F_{EG,ES,t} \cdot r_{oc,t} \Big)
\end{aligned}
\tag{4}
$$

Subject to (in addition to constraints (2) and (3))

$$
\bar{L}_{ES,t} \geq X_{ES,t-1} + F_{EG,ES,t} + F_{EN,ES,t}, \forall t \in T \tag{5}
$$

$$
\bar{X}_{ES,t} \cdot j_t \geq F_{ES,ED,t} + F_{ES,EN,t}, \forall t \in T \tag{6}
$$

$$
F_{ES,ED,t} + F_{ES,EN,t} + X_{ES,t} = F_{EG,ES,t} + F_{EN,ES,t} + X_{ES,t-1}, \forall t \in T \tag{7}
$$

where (5) takes into account the maximal amount of energy that can be stored in the ES and (6) the maximal power of the ES. Equation (7) ensures balanced flows to and from the ES. Apart from the costs all variables are subject to the non-negativity constraint.

3 A Local Electricity System Modeled as Multi-agent System

To investigate the effects of the households actions on the local electricity system, the respective households are placed in a MAS as agents. The MAS is formed by the structuring low-voltage network, the agents (households, distribution system operator (DSO), and power supply company), and the environment (geographical and political system and market design). The modeled MAS is then implemented to run simulations with different input parameters.

The low voltage network is modeled in order to map the load flow between consumers and higher-level networks (medium voltage). In the model, the network can thus restrict the load flow and network bottlenecks and capacity requirements can be identified. The low voltage network is composed of the individual line sections (modeled as agents), the connection points of the generating and consuming agents, and their respective connections.

The environment describes the context in which the agents act. With the necessary IT installed agents can (at least technically) take action in a local market. Such a market is introduced into the model and modeled as regional spotmarket in which agents can negotiate their offers for sale or buy. The parameters, design and processes of this market are defined within this context. The environment also determines the geographical context and thereby has an influences on generation and demand of electricity (e.g. insolation and load characteristics).

The agents of the system represent different actors with different objectives (see Sect. 2). The characteristic of the agent population is determined within their context before ($t = 0$) each simulation run. Because no investment decisions are modelled the population of the agents does not change during simulation runtime. However, at each simulation step agents will readjust their decisions based on the newly gained information (e.g. update projections for electricity prices using precedent local spot market prices). Several agents have particular tasks such as the DSO which has to perform load flow calculations and the power supply company which operates as well as dispatches power plants and therefore provides the necessary operating reserve.

The implementation of the described MAS has been performed in Repast Simphony[1] which allows to include the optimization calculations of the respective agents. Anytime a recalibration of the unit commitment problems is necessary during a simulation run a connection to either GAMS[2] or Matlab®[3] is established to recalculate the agents individual LP.

4 Input Parameters and Results

In this analysis the simulation runs are limited to one typical day (for reasons of simplicity) and do start with several agent population scenarios (different load characteristics and degree of technology diffusion) as initial points which are based on the energy system of the French island of Guadeloupe. Figure 2 shows the average costs of the power supply company for a typical weekday in October in different scenarios whereat A(08) serves as baseline as it denotes the situation at this day in 2008 without any local market transactions and ES diffusion. A(20,1) denotes a scenario with a low (5 %) and A(20,2) with a high (45 %) diffusion of ES in the households (storage capacity and maximal power depend on household size). B(20,1) and B(20,2) show respective scenarios though with a higher number of household agents

---

[1] Open source, Java-based simulation system, see http://repast.sourceforge.net/ [20.01.2015].

[2] Modeling system for mathematical optimization, see http://www.gams.com/ [20.01.2015].

[3] Numerical computation system, see http://www.mathworks.com/products/matlab/ [20.01.2015].

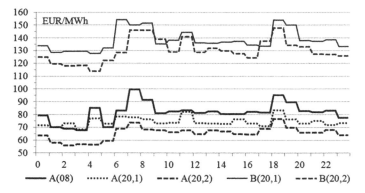

Fig. 2 Average costs of the system's power supply company during a typical weekday day for different scenarios

(which reflects the situation for a corresponding day in 2020) and thus a relatively higher electricity demand and peak load. The baseline A(08) shows price peaks in the morning and evening which are typical for the island. Those peaks are flattened and lowered for A(20,1) as well as A(20,2). For B(20,1) and B(20,2) price fluctuations are higher which is mainly due to the estimated power plant park in 2020 which comprises expensive peak power plants to handle higher demand. The diagramed prices in Fig. 2 demonstrate that, depending on the chosen environment and agent population, local electricity networks with high market penetration of ES and a local electricity market can lead to lower average costs for the power supply companies due to the balanced power consumption and generation. In the case of the island of Guadeloupe this additionally could help to avoid capacity bottlenecks in both grid and production which will most probably occur more frequently as power plant and grid expansion are not proportional to the increase in demand.

5 Conclusion

The applied bottom-up modeling process from individual customer to a local power network allows to integrate a multitude of details, but also reduces complexity where necessary. This made it possible to focus on the modeling of consumer with own PV electricity generation and storage and their actions on a local market. The composed multi agent system optimizes an overall cost minimizing objective for the local grid cell although agents do individually optimize their systems. Furthermore, simulations' results show major changes in electricity demand and feed-in for scenarios with high market penetration of storages. Nevertheless, simulation runs with present or pessimistic technology diffusion scenarios (such as A(08) and A(20,1) in Fig. 2) still show a high dependence of consumers and consequently grid cells on a connection to a greater power network to balance consumption and production.

References

1. Flath, C.M., Ilg, J.P., et al.: Improving electric vehicle charging coordination through area pricing. Trans. Sci. (2013)
2. Markard, J., Truffer, B.: Innovation processes in large technical systems: market liberalization as a driver for radical change? Res. Policy **35**(5), 609–625 (2006)
3. Clastres, C.: Smart grids: another step towards competition, energy security and climate change objectives. Energy Policy **39**(9), 5399–5408 (2011)
4. International Energy Agency (IEA): World Energy Outlook 2012. OECD, Paris (2012)
5. Hvelplund, F.: Renewable energy and the need for local energy markets. Energy **31**(13), 2293–2302 (2006)

Congestion Games with Multi-Dimensional Demands

Andreas Schütz

Abstract Weighted congestion games are an important and extensively studied class of strategic games, in which the players compete for subsets of shared resources in order to minimize their private costs. In my Master's thesis (Congestion games with multi-dimensional demands. Master's thesis, Institut für Mathematik, Technische Universität Berlin, 2013, [17]), we introduced congestion games with multi-dimensional demands as a generalization of weighted congestion games. For a constant $k \in \mathbb{N}$, in a congestion game with k-dimensional demands, each player is associated with a k-dimensional demand vector, and resource costs are k-dimensional functions $c : \mathbb{R}_{\geq 0}^k \to \mathbb{R}$ of the aggregated demand vectors of the players using the resource. Such a cost structure is natural when the cost of a resource depends not only on one, but on several properties of the players' demands, e.g., total weight, total volume, and total number of items. We obtained a complete characterization of the existence of pure Nash equilibria in terms of the resource cost functions for all $k \in \mathbb{N}$. Specifically, we identified all sets of k-dimensional cost functions that guarantee the existence of a pure Nash equilibrium for every congestion game with k-dimensional demands. In this note we review the main results contained in the thesis.

1 Introduction

Game theory provides mathematical tools to model and analyze real-world situations, in which multiple participants interact and mutually affect one another by their decisions. Road networks and communication systems are two examples of a wide range of applications that can be viewed as *strategic games* with finitely many players. We assume that these players act selfishly and make their decisions to maximize their private wealth. Thus, a fundamental objective in game theory is to study the existence of equilibrium states in such decentralized systems. The most famous notion

Work was done while the author was at Technische Universität Berlin, Berlin, Germany

A. Schütz (✉)
d-fine GmbH, Opernplatz 2, 60313 Frankfurt, Germany
e-mail: andreas.schuetz@d-fine.de

© Springer International Publishing Switzerland 2016
M. Lübbecke et al. (eds.), *Operations Research Proceedings 2014*,
Operations Research Proceedings, DOI 10.1007/978-3-319-28697-6_77

553

of an equilibrium state is the *pure Nash equilibrium* as introduced by Nash [13]. In a pure Nash equilibrium, no player has an incentive to switch its chosen pure strategy because unilateral deviations would not benefit any of the players.

Rosenthal [15] addressed the existence problem of pure Nash equilibria in *congestion games*. In this well-known and extensively studied class of strategic games the players compete for subsets of shared resources. The cost of each resource depends on the number of players using it simultaneously. Rosenthal [15] showed that congestion games always admit a pure Nash equilibrium. A natural extension of congestion games are *weighted congestion games*, in which an unsplittable demand $d_i > 0$ is assigned to each player i and the cost function $c_r : \mathbb{R}_{\geq 0} \to \mathbb{R}$ of a resource r depends on the cumulated demands of all players sharing r rather than the number of players. In contrast to congestion games, weighted congestion games do not always possess a pure Nash equilibrium. This was demonstrated by Fotakis et al. [6] and Libman and Orda [11], who constructed weighted network congestion games with two players and non-decreasing costs, which do not admit a pure Nash equilibrium. Goemans et al. [7] used continuous quadratic resource cost functions with the same effect.

Consequently, many recent studies have focused on subclasses of weighted congestion games with *restricted* strategy spaces. In *singleton weighted congestion games* the strategies of all players correspond to single resources only. Fabrikant et al. [5] showed in the final section of their paper the existence of pure Nash equilibria in singleton congestion games where the resource costs are non-decreasing functions depending on the set of players using them. As a direct implication, every singleton weighted congestion game with non-decreasing resource costs admits a pure Nash equilibrium. Moreover, Andelman et al. [3] and Harks et al. [9] showed the existence of a *strong equilibrium* for every such game. Introduced by Aumann [4], a strong equilibrium is stronger than a pure Nash equilibrium and protects against deviations of coalitions, which may include more than one player. Rozenfeld and Tennenholtz [16] gave the complementary result for singleton weighted congestion games with non-increasing cost functions.

Furthermore, even broader strategy spaces exist, which give rise to a pure Nash equilibrium. Ackermann et al. [1] showed that weighted congestion games with non-decreasing costs always admit a pure Nash equilibrium if the strategy space of each player is identical to the collection of bases of a matroid over the resources. These games are referred to as *matroid weighted congestion games*. The matroid property has been shown to be the maximal condition on the strategy spaces ensuring the existence of a pure Nash equilibrium.

Other studies examine resource cost functions that ensure the existence of a pure Nash equilibrium in weighted congestion games with *arbitrary* strategy spaces. Fotakis et al. [6] showed that every weighted congestion game with affine cost functions always admits a pure Nash equilibrium. Moreover, Panagopoulou and Spirakis [14] proved the existence of pure Nash equilibria in every instance with exponential resource costs of type $c_r(x) = e^x$. Later, Harks et al. [8] additionally confirmed the existence of pure Nash equilibria in weighted congestion games with exponential cost functions of type $a_c e^{\phi x} + b_c$, where $a_c, b_c \in \mathbb{R}$ may depend on c while $\phi \in \mathbb{R}$ is independent of c. In a recent study, Harks and Klimm [8] introduced the notion of a *consistent* set $\mathscr{C}$ as a set of cost functions ensuring a pure Nash equilibrium in every weighted congestion game with resource cost functions in $\mathscr{C}$. The authors gave a complete characterization of consistent sets of cost functions. They proved that a set $\mathscr{C}$ of continuous cost functions is consistent if and only if $\mathscr{C}$ only contains affine cost functions $c(x) = a_c x + b_c$ or $\mathscr{C}$ only consists of exponential cost functions of type $c(x) = a_c e^{\phi x} + b_c$, where $a_c, b_c \in \mathbb{R}$ may depend on c while $\phi \in \mathbb{R}$ is independent of c.

For illustration, we give a typical example of a weighted congestion game. The game is represented by the directed network shown in Fig. 1a. The resources correspond to the arcs of the network and the cost of each arc is a function depending on the cumulated demand x of all players using it. The corresponding cost function is written next to each arc. Furthermore, there are two players 1 and 2 with demands $d_1 = 1$ and $d_2 = 2$, respectively. Each player has the same strategy set corresponding to all directed paths from node t to node w. This weighted congestion game can be interpreted as a network congestion game where the demand of each player corresponds to the amount of data it wants to send through the network. The resource costs then can be interpreted as delay functions. Hence, the delay of an arc depends on the total amount of data that is sent through it and the delay of a t-w-path corresponds to the cumulated delays associated with all arcs on the path. Thus, the aim of each player is to choose a t-w-path with minimum delay.

However, what if the delay of an arc does not only depend on the total amount of data, but additionally on the number of files that are sent through it? This might be modeled by the directed network shown in Fig. 1b. In this network the delay of each arc is a function of two variables x_1 and x_2, where x_1 and x_2 might represent the aggregated amount of data and the total number of files that are sent through the corresponding arc, respectively. Additionally, the demand of each player now is a 2-dimensional vector consisting of its amount of data and its number of files. In this example, the players have demands $\mathbf{d}_1 = (2, 3)^\top$ and $\mathbf{d}_2 = (4, 1)^\top$, respectively. We call a game of this kind a *congestion game with* 2-*dimensional demands*. Moreover, if the demand of each player i is a strictly positive k-dimensional vector $\mathbf{d}_i \in \mathbb{R}^k_{\geq 0}$ for a fixed number $k \in \mathbb{N}_{\geq 1}$, the game is called a *congestion game with k-dimensional demands*. The class of *congestion games with multi-dimensional demands* then contains all congestion games with k-dimensional demands for all $k \in \mathbb{N}_{\geq 1}$. These games

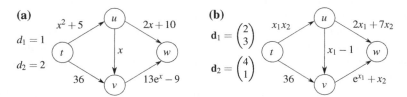

Fig. 1 Directed network congestion game with **a** 1-dimensional demands and cost functions $c : \mathbb{R}_{\geq 0} \to \mathbb{R}$ of one variable x, and **b** 2-dimensional demands and cost functions $c : \mathbb{R}^2_{\geq 0} \to \mathbb{R}$ of two variables x_1 and x_2

generalize the class of weighted congestion games and enable us to model real-world applications where the cost of each resource r is a multivariable function $c_r : \mathbb{R}^k_{\geq 0} \to \mathbb{R}$.

However, to date there is no research on congestion games with multi-dimensional demands. While previous research focused on weighted congestion games with 1-dimensional demands, the equilibrium existence problem in congestion games with multi-dimensional demands has remained unsolved. The purpose of the thesis is to investigate the existence of pure Nash equilibria in both congestion games with multi-dimensional demands and *arbitrary* strategy spaces as well as in games of different subclasses with *restricted* strategy spaces. Specifically, we characterize resource cost functions that guarantee the existence of a pure Nash equilibrium in every congestion game with multi-dimensional demands using two different approaches.

Firstly, we study the existence of pure Nash equilibria in games with arbitrary strategy spaces. Let $k \in \mathbb{N}_{\geq 1}$, we follow the notion of [8] and introduce k-*consistency* as a property of a set $\mathscr{C}$ of cost functions $c : \mathbb{R}^k_{\geq 0} \to \mathbb{R}$ such that every congestion game with k-dimensional demands and resource costs in $\mathscr{C}$ possesses a pure Nash equilibrium. Specifically, we determine all types of cost functions, which form a k-consistent set $\mathscr{C}$.

Secondly, we investigate k-consistency for different subclasses of congestion games with k-dimensional demands and restricted strategy spaces. More specifically, we examine whether the results of [2, 16] for weighted congestion games with 1-dimensional demands can be extended to the class of congestion games with multi-dimensional demands. Thus, we determine necessary and sufficient conditions on k-consistent sets of cost functions for singleton and matroid congestion games with k-dimensional demands.

2 Characterizations of k-Consistency

We formally capture the class of congestion games with k-dimensional demands for any fixed $k \in \mathbb{N}_{\geq 1}$. We find that for any instance the demand dimension k may be assumed to be upper bounded by the number of its players n.

In order to address the equilibrium existence problem in games with arbitrary strategy spaces, we initially consider congestion games with 2-dimensional demands. Our first main result is a complete characterization of 2-consistent sets of cost functions. For a set $\mathscr{C}$ of continuous cost functions $c : \mathbb{R}^2_{\geq 0} \to \mathbb{R}$, we show that $\mathscr{C}$ is 2-consistent if and only if there are $\phi_1, \phi_2 \geq 0$ or $\phi_1, \phi_2 \leq 0$ such that $\mathscr{C}$ only contains functions of type $c(x_1, x_2) = a_c(\phi_1 x_1 + \phi_2 x_2) + b_c$ or $\mathscr{C}$ only contains functions of type $c(x_1, x_2) = a_c e^{\phi_1 x_1 + \phi_2 x_2} + b_c$, where $a_c, b_c \in \mathbb{R}$ may depend on c.

We then extend this characterization to congestion games with k-dimensional demands for any $k \in \mathbb{N}_{\geq 3}$. Our second main result states that a set $\mathscr{C}$ of continuous cost functions $c : \mathbb{R}^k_{\geq 0} \to \mathbb{R}$ is k-consistent if and only if there is a vector $\Phi \in \mathbb{R}^k_{\geq 0}$ or $\Phi \in \mathbb{R}^k_{\leq 0}$ such that $\mathscr{C}$ only contains functions of type $c(\mathbf{x}) = a_c \Phi^\top \mathbf{x} + b_c$

or $\mathscr{C}$ only contains functions of type $c(\mathbf{x}) = a_c e^{\Phi^\top \mathbf{x}} + b_c$, where $a_c, b_c \in \mathbb{R}$ may depend on c. A set $\mathscr{C}$ with this property is called *uniformly degenerate*. Provided that $\mathscr{C}$ is uniformly degenerate, our results imply that every congestion game with k-dimensional demands and cost functions in $\mathscr{C}$ is isomorphic to a congestion game with 1-dimensional demands and the same sets of players, resources, strategies and private costs. Our contributions to k-consistency for $k \in \mathbb{N}_{\geq 1}$ generalize the complete characterization of 1-consistent sets for weighted congestion games derived by Harks and Klimm [8].

Key to the proof of these results are three Lemmas that provide necessary conditions on the k-consistency of cost functions. The first of these lemmas, the *Extended Monotonicity Lemma for k-dimensional demands*, which is a generalization of the Extended Monotonicity Lemma introduced by Harks and Klimm [8] for weighted congestion games with 1-dimensional demands, states that every function $c \in \mathscr{C}$ and every integral linear combination $\lambda_1 c_1 + \lambda_2 c_2$ of any two functions $c_1, c_2 \in \mathscr{C}$ must be monotonic. Second, the *Hyperplane Restriction Lemma* implies that, given $i \in \{1, \ldots, k\}$, a set of restrictions of functions $c \in \mathscr{C}$ to hyperplanes of type $H_{\hat{x}_i}^i = \{\mathbf{x} \in \mathbb{R}_{\geq 0}^k : x_i = \hat{x}_i\}$ for some $\hat{x}_i \geq 0$ is $(k-1)$-consistent. Finally, we prove the *Line Restriction Lemma*. Given a vector $\mathbf{v} \in \mathbb{R}_{>0}^k$, the lemma states that a set containing restrictions of functions $c \in \mathscr{C}$ to the line $L_\mathbf{v} = \{\mathbf{z} \in \mathbb{R}_{\geq 0}^k : \mathbf{z} = x\mathbf{v}$ for some $x \geq 0\}$ is 1-consistent. The latter two necessary conditions establish a connection between k-consistency, $(k-1)$-consistency and 1-consistency for every $k \in \mathbb{N}_{\geq 2}$.

Subsequently, we discuss the existence of pure Nash equilibria in congestion games with k-dimensional demands and restricted strategy spaces. We start by analyzing games with matroid strategy spaces and derive necessary conditions on k-consistency. Let $\mathscr{C}$ be a set of continuous cost functions $c : \mathbb{R}_{\geq 0}^k \to \mathbb{R}$, which is k-consistent for matroid congestion games with k-dimensional demands. We then show that $\mathscr{C}$ only contains monotonic functions. Additionally, we derive that every set $\mathscr{C}$ containing either non-decreasing cost functions or non-increasing cost functions $c : \mathbb{R}_{\geq 0}^k \to \mathbb{R}$ is k-consistent for matroid congestion games with k-dimensional demands. Matroid strategy spaces were firstly introduced by Ackermann et al. [1]. They analyzed the existence of pure Nash equilibria in weighted matroid congestion games with 1-dimensional demands and their work forms the basis of our considerations.

Finally, we study the equilibrium existence problem in singleton congestion games with k-dimensional demands. As singleton congestion games are special matroid congestion games, necessary and sufficient conditions on k-consistency for matroid congestion game carry over to k-consistency for singleton congestion games with k-dimensional demands. Additionally, we show that a set $\mathscr{C}$ containing either non-decreasing or non-increasing cost functions does not only imply the existence of a pure Nash equilibrium, but it also guarantees the existence of a strong equilibrium in every singleton congestion game with k-dimensional demands and resource costs in $\mathscr{C}$. This result is based on the work of Harks et al. [9].

Acknowledgments I would like to express my gratitude to Prof. Dr. Rolf H. Möhring who offered continuing support and constant encouragement during the course of my studies. Special thanks also go to my advisor Dr. Max Klimm whose constructive comments and insightful advice made an enormous contribution to my work.

References

1. Ackermann, H., Röglin, H., Vöcking, B.: Pure Nash equilibria in player-specific and weighted congestion games. Theor. Comput. Sci. **410**(17), 1552–1563 (2009)
2. Ackermann, H., Skopalik, A.: On the complexity of pure Nash equilibria in player-specific network congestion games. In: Deng, X., Graham, F. (eds.) Proceedings 3rd International Workshop on Internet and Network Economics, LNCS, vol. 4858, pp. 419–430 (2007)
3. Andelman, N., Feldman, M., Mansour, Y.: Strong price of anarchy. Games Econ. Behav. **65**(2), 289–317 (2009)
4. Aumann, R.: Acceptable points in general cooperative n-person games. In: Luce, R.D., Tucker, A.W. (eds.) Contributions to the Theory of Games IV, pp. 287–324. Princeton University Press, Princeton (1959)
5. Fabrikant, A., Papadimitriou, C.H., Talwar, K.: The complexity of pure Nash equilibria. In: Proceedings 36th Annual ACM Symposium Theory Computing, pp. 604–612 (2004)
6. Fotakis, D., Kontogiannis, S., Spirakis, P.G.: Selfish unsplittable flows. Theor. Comput. Sci. **348**(2–3), 226–239 (2005)
7. Goemans, M.X., Mirrokni, V.S., Vetta, A.: Sink equilibria and convergence. In: Proceedings 46th Annual IEEE Symposium Foundations Computing Science, pp. 142–154 (2005)
8. Harks, T., Klimm, M.: On the existence of pure Nash equilibria in weighted congestion games. Math. Oper. Res. **37**(3), 419–436 (2012)
9. Harks, T., Klimm, M., Möhring, R.H.: Strong equilibria in games with the lexicographical improvement property. Int. J. Game Theory **42**(2), 461–482 (2012)
10. Held, S., Korte, B., Rautenbach, D., Vygen, J.: Combinatorial optimization in VLSI design. In: Chvtal, V. (ed.) Combinatorial Optimization: Methods and Applications, pp. 33–96. IOS Press, Amsterdam (2011)
11. Libman, L., Orda, A.: Atomic resource sharing in noncooperative networks. Telecommun. Syst. **17**(4), 385–409 (2001)
12. Müller, D., Radke, K., Vygen, J.: Faster minmax resource sharing in theory and practice. Math. Programm. Comput. **3**(1), 1–35 (2011)
13. Nash, J.F.: Equilibrium points in n-person games. Proc. Natl. Acad. Sci. USA **36**, 48–49 (1950)
14. Panagopoulou, P.N., Spirakis, P.G.: Algorithms for pure Nash equilibria in weighted congestion games. ACM J. Exp. Algorithmics **11**, 1–19 (2006)
15. Rosenthal, R.W.: A class of games possessing pure-strategy Nash equilibria. Int. J. Game Theory **2**(1), 65–67 (1973)
16. Rozenfeld, O., Tennenholtz, M.: Strong and correlated strong equilibria in monotone congestion games. In: Mavronicolas, M., Kontogiannis, S. (eds.) Proceedings 2nd International Workshop on Internet and Network Economic, LNCS, vol. 4286, pp. 74–86 (2006)
17. Schütz, A.: Congestion games with multi-dimensional demands. Master's thesis, Institut für Mathematik, Technische Universität Berlin (2013)

Unit Commitment by Column Generation

Takayuki Shiina, Takahiro Yurugi, Susumu Morito and Jun Imaizumi

Abstract The unit commitment problem is to determine the schedule of power generating units and the generating level of each unit. The decisions involve which units to commit at each time period and at what level to generate power to meet the electricity demand. We consider the heuristic column generation algorithm to solve this problem. Previous methods used the approach in which each column corresponds to the start–stop schedule and output level. Since power output is a continuous quantity, it takes time to generate the required columns efficiently. In our proposed approach, the problem to be solved is not a simple set partitioning problem, because the columns generated contain only a schedule specified by 0–1 value. It is shown that the proposed heuristic approach is effective to solve the problem.

1 Introduction

In this paper, we consider the unit commitment problem to determine the schedule of power generating units and the generating level of each unit. The decisions involve which units to commit at each time period and at what level to generate power to meet the electricity demand [6]. In early research, deterministic models [2, 5] in which deterministic power demand is given, was considered. Later, models considering the variation in power demand were developed [9]. Shiina et al. [8] modified these models to reflect the operation of a real system and described a stochastic programming

T. Shiina (✉)
Chiba Institute of Technology, 2-17-1 Tsudanuma, Narashino, Chiba 275-0016, Japan
e-mail: shiina.takayuki@it-chiba.ac.jp

T. Yurugi · S. Morito
Waseda University, 3-4-1 Okubo, Shinjuku-ku, Tokyo 169-8555, Japan
e-mail: yurugi@fuji.waseda.jp

S. Morito
e-mail: morito@waseda.jp

J. Imaizumi
Toyo University, 5-28-20 Hakusan, Bunkyo-ku, Tokyo 112-8606, Japan
e-mail: jun@toyo.jp

© Springer International Publishing Switzerland 2016 559
M. Lübbecke et al. (eds.), *Operations Research Proceedings 2014*,
Operations Research Proceedings, DOI 10.1007/978-3-319-28697-6_78

model which takes into account the uncertainty of the power demand. The method for solving this problem decomposes the original problem into each of the power generation by the Lagrangian relaxation method and generates a schedule efficiently to meet the power demand at the same time. This approach can calculate a feasible solution in a short time, but has the disadvantage that a good solution cannot be obtained, since the same schedule is always created for each generator having the same characteristics.

In this paper, we propose a heuristic method to the unit commitment problem using the column generation method, and the performance of the proposed method is evaluated by numerical experiments. The column generation method [1] has not been applied frequently to solve the unit commitment problem. The only research that uses a column generation technique is a paper by Shiina-Birge [7]. They used the column generation approach in which each column corresponds to the start–stop schedule and output level. Since power output is a continuous quantity, it takes time to generate the required column efficiently.

2 Unit Commitment Problem

The mathematical formulation of the stochastic unit commitment problem (UC) is described as follows. We assume that there are I generating units. The status of unit i at period t is represented by the 0–1 variable u_{it}. Unit i is on at time period t if $u_{it} = 1$, and off if $u_{it} = 0$. The power generating level of the unit i at period t is represented by x_{it} (≥ 0). The fuel cost function $f_i(x_{it})$ is given by a convex quadratic function of x_{it}. The start-up cost function $g_i(u_{i,t-1}, u_{it})$ satisfies the condition $g_i(0, 1) > 0$, $g_i(0, 0) = 0$, $g_i(1, 0) = 0$, $g_i(1, 1) = 0$. Fuel cost $f_i(x_{it})$ must be 0 in the case of not performing the activation. Since $f_i(0) \neq 0$ for some i, the exact fuel cost is described as $f_i(x_{it})u_{it}$.

$$\text{(UC): min} \sum_{i=1}^{I} \sum_{t=1}^{T} f_i(x_{it})u_{it} + \sum_{i=1}^{I} \sum_{t=1}^{T} g_i(u_{i,t-1}, u_{it}) \tag{1}$$

$$\text{subject to} \sum_{i=1}^{I} x_{it} \geq d_t, t = 1, \ldots, T \tag{2}$$

$$u_{it} - u_{i,t-1} \leq u_{i\tau}, \tau = t+1, \ldots, \min\{t + L_i - 1, T\} \tag{3}$$
$$i = 1, \ldots, I, t = 2, \ldots, T$$

$$u_{i,t-1} - u_{it} \leq 1 - u_{i\tau}, \tau = t+1, \ldots, \min\{t + l_i - 1, T\} \tag{4}$$
$$i = 1, \ldots, I, t = 2, \ldots, T$$

$$q_i u_{it} \leq x_{it} \leq Q_i u_{it}, i = 1, \ldots, I, t = 1 \ldots, T \tag{5}$$

$$u_{it} \in \{0, 1\}, i = 1, \ldots, I, \quad t = 1, \ldots, T \tag{6}$$

The objective function of minimization (1) is the sum of the fuel cost and the start-up cost. The sum of the levels of generation must be greater than the demand (2). When unit i is switched on, it must continue to run for at least a certain period L_i. These minimum up-time constraints are described in (3). Similarly, when unit i is switched off, it must continue to be off for at least period l_i. These constraints are called minimum down-time constraints (4). Let $[q_i, Q_i]$ be an operating range of the generating unit i as shown in (5).

3 Application of Column Generation

In order to apply the column generation technique for (UC), the problem is reformulated based on the decomposition principle of Danzig-Wolfe [3]. Let the number of the schedules generated for unit i be K_i. The new binary variable v_i^k indicates whether the kth schedule of unit i is selected. In the reformulated problem, the state of generator i in period t in the kth schedule is given as a constant u_{it}^k. The reformulation of (UC) is given as (RUC).

$$(\text{RUC}): \min \quad \sum_{i=1}^{I}\sum_{k=1}^{K_i}\left\{\sum_{t=1}^{T}f_i(x_{it})u_{it}^k\right\}v_i^k + \sum_{i=1}^{I}\sum_{k=1}^{K_i}\left\{\sum_{t=1}^{T}g_i(u_{i,t-1}^k, u_{i,t}^k)\right\}v_i^k \quad (7)$$

$$\text{subject to} \quad \sum_{k=1}^{K_i}v_i^k = 1, \quad i = 1,\ldots,I \quad (8)$$

$$\sum_{i=1}^{I}x_{it} \geq d_t, \quad t = 1,\ldots,T \quad (9)$$

$$\sum_{k=1}^{K_i}q_iu_{it}^kv_i^k \leq x_{it} \leq \sum_{k=1}^{K_i}Q_iu_{it}^kv_i^k, \quad i = 1,\ldots,I, \quad t = 1,\ldots,T \quad (10)$$

$$v_i^k \in \{0, 1\}, i = 1,\ldots,I, k = 1,\ldots,K_i \quad (11)$$

The constraint (8) is added to (RUC) to express the choice of schedule. In addition, the output level pair of constraints (5) is transformed into (10). The constants u_{it}^k and $g_i(u_{i,t-1}^k, u_{i,t}^k)$ involved in (RUC) are generated sequentially in the column generation procedure described later. The minimum uptime (3) and downtime (4) constraints are taken into account when the schedule is generated.

Since the first term of the objective function (7) becomes the product of a convex quadratic function and a binary variable, we consider the transformation of the fuel cost term. The convex quadratic function $f_i(x_{it})$ is decomposed into the constant $f_i(0)$ and the term $\bar{f}_i(x_{it}) = f_i(x_{it}) - f_i(0)$. Thus, the master problem (IPM) transformed from (RUC) is shown as follows.

(IPM): min $\displaystyle\sum_{i=1}^{I}\sum_{t=1}^{T}\bar{f}_i(x_{it}) + \sum_{i=1}^{I}\sum_{k=1}^{K_i}\sum_{t=1}^{T}\left\{g_i(u_{i,t-1}^k, u_{i,t}^k) + f_i(0)u_{it}^k\right\}v_i^k$

subject to $(8)-(11)$

Here, defining $\bar{f}_i(x_{it}) = (1/2)a_i x_{it}^2 + b_i x_{it}$, the dual problem for the continuous relaxation of (IPM) becomes the following (DPM).

$$(\text{DPM}): \max \sum_{i=1}^{I}\mu_i + \sum_{t=1}^{T}d_t\pi_t - \sum_{i=1}^{I}\sum_{t=1}^{T}\frac{(b_i - \pi_t - \underline{\lambda}_{it} + \overline{\lambda}_{it} - \nu_{it})^2}{2a_i} \tag{12}$$

$$\text{subject to} \sum_{t=1}^{T}(-q_i u_{it}^k \underline{\lambda}_{it} + Q_i u_{it}^k \overline{\lambda}_{it}) + \mu_i \leq \sum_{t=1}^{T}\left\{g_i(u_{t,t-1}^k, u_{it}^k) + f_i(0)u_{it}^k\right\},$$
$$\tag{13}$$

$$i = 1,\ldots,I, \quad k = 1,\ldots,K_i$$

$$\pi_t, \underline{\lambda}_{it}, \overline{\lambda}_{it}, \nu_{it} \geq 0, \quad i = 1,\ldots,I, t = 1,\ldots,T \tag{14}$$

Let the optimal solution for the dual problem (DPM) be $(\pi_t^*, \underline{\lambda}_{it}^*, \overline{\lambda}_{it}^*, \mu_i^*)$. Then, we consider the column generation problem which generates the schedule of generator i.

There is a possibility that the optimal solution of (DPM) when all of the schedules are not be enumerated is not dual feasible for the continuous relaxation problem of (RUC) with all columns enumerated. The constraint (13) should be satisfied for all schedules. The new column is generated when the optimal objective function value satisfies $\zeta_i < 0$.

$$\zeta_i = \min \quad \sum_{t=1}^{T}\left\{(g_i(u_{i,t-1}, u_{it}) + f_i(0)u_{it} + q_i\underline{\lambda}_{it}^* u_{it} - Q_i\overline{\lambda}_{it}^* u_{it}\right\} - \mu_i^* \tag{15}$$

subject to $(3), (4), (6)$

This column generation problem results in the shortest path problem, and thus it is possible to seek the schedule of the generator u_{it}^k by dynamic programming. Details of the dynamic programming are shown in [7, 8].

4 Numerical Experiments

To evaluate the performance of this solution method, problem instances are based on Shiina-Birge [7]. The numbers of generation units and periods are varied in the experiments to seek objective value, computation time, and duality gap. The following three strategies are compared.

- (a) adding columns by solving the continuous relaxation problem of (IPM) and then solving (IPM)
- (b) after removing of columns with a large reduced cost, solving (IPM)
- (c) solving (UC) using the Shiina–Birge method [7]

Experiments were carried out using AMPL [4]-Gurobi 5.5.0 on a Xeon E5507 2.00 GHz (2 processors, memory: 12.0 GB). Accuracy of the solution is evaluated by the duality gap. In Tables 1 and 2, LB and UB denote the lower and upper bound values, respectively cols is the number of columns generated (after deleting columns for strategy (b)). Abbreviations ite, c-time, u-time, and t-time represent the number of iterations in the column generation procedure, time spent for the column generation, time of the upper bound calculation, and total computation time (in seconds), respectively. The maximum computation time is limited to within 7200 s. If the total computation time exceeds this upper limit, then the time is shown as *. Performance evaluation is performed by varying the number of units with a fixed T (Table 1) and by varying the number of periods with a fixed I (Table 2).

Generally speaking, the column generation method is able to calculate a lower bound at high speed. The time required for the calculation of the upper bound is longer than the time required for the column generation procedure. Solving (IPM) took much longer than solving the column generation when the size of the problem is large. The duality gap between our proposed methods (a) and (b) is approximately 3 % or less, and the solution method is accurate and precise. Comparing strategies (a) and (b), it can be seen that strategy (b) can seek a solution effectively in terms of calculation time without raising the obtained objective function value.

Table 1 Computational results ($T = 24$)

I	Strategy	LB	UB	Gap	Cols	Ite	c-time	u-time	t-time
10	a	530366	542773	2.33	236	19	3	19	22
	b	530366	542773	2.33	188	19	3	2	5
	c	529356	569763	7.63	1071	168	27	3626	3673
20	a	1054294	1081470	1.95	476	21	4	7196	7200*
	b	1054294	1081470	1.95	386	21	5	132	137
	c	1053712	1108573	5.21	2161	168	72	7128	7200*

Table 2 Computational results ($I = 10$)

T	Strategy	LB	UB	Gap	Cols	Ite	c-time	u-time	t-time
24	a	530366	542773	2.33	236	19	3	19	22
	b	530366	542773	2.33	188	19	3	2	5
	c	529356	569763	7.63	1071	168	27	3626	3673
48	a	1081966	1081966	2.62	563	50	12	5948	5948
	b	1081966	1081966	2.62	455	51	16	39	55
	c	1071395	1116367	4.20	3090	436	147	7053	7200*

From Table 1, the number of iterations for strategies (a) and (b) in the column generation procedure does not vary as the number of generators increases. In addition, the accompanying increase in the time required for the column generation procedure is very small. Furthermore, compared with strategy (c), strategies (a) and (b) produced good solutions in terms of the obtained objective function value. In particular, in the case of $T = 24, I = 20$, strategy (b) provided a solution in 137 s.

As shown in Table 2, the number of iterations in the column generation procedure increases proportionally with the number of periods, and the time of the column generation also grows. This is because the calculation step of dynamic programming is proportional to the value of T. Generation of good schedules takes a long time when the number of periods is large, especially for strategy (c). Strategy (c) uses column generation in which each column corresponds to the start–stop schedule and the output level. Since power output is a continuous quantity, it takes considerable time to generate the required column. In addition, compared with the calculation results of strategy (a) and (c), strategy (b) provides the solution efficiently in terms of the total computation time. Especially for the case with $T = 72$, the calculation required long time using either strategy (a) or (c), but strategy (b) solved the problem in 55 s.

5   Concluding Remarks

In this study, a heuristic algorithm based on the column generation method for the unit commitment problem is proposed. We have developed a procedure to obtain a solution with a certain level of accuracy within a short computation time. Improvement of the method to seek feasible solutions and to raise the lower bound by adding a family of valid inequalities is left as future work. As for application to real power systems, the coordination of the operation of hydroelectric generation plants must be considered.

References

1. Barnhart, C., Johnson, E.L., Nemhauser, G.L., Savelsbergh, M.W.P., Vance, P.H.: Branch-and-price: column generation for solving huge integer programs. Oper. Res. **46**, 316–329 (1998)
2. Bard, J.F.: Short-term scheduling of thermal-electric generators using Lagrangian relaxations. Oper. Res. **36**, 756–766 (1988)
3. Dantzig, G.B., Wolfe, P.: Decomposition principle for linear programs. Oper. Res. **8**, 101–111 (1960)
4. Fourer, R., Gay, D.M., Kernighan, B.W.: AMPL: a modeling language for mathematical programming, Scientific Press (1993)
5. Muckstadt, J.A., Koenig, S.A.: An application of Lagrangian relaxation to scheduling in power-generation systems. Oper. Res. **25**, 387–403 (1977)
6. Sheble, G.B., Fahd, G.N.: Unit commitment literature synopsis. IEEE Trans. Power Syst. **11**, 128–135 (1994)

7. Shiina, T., Birge, J.R.: Stochastic unit commitment problem. Int. Trans. Oper. Res. **11**, 19–32 (2004)
8. Shiina, T., Watanabe, I.: Lagrangian relaxation method for price-based unit commitment problem. Eng. Optim. **36**, 705–719 (2004)
9. Takriti, S., Birge, J.R., Long, E.: A stochastic model for the unit commitment problem. IEEE Trans. Power Syst. **11**, 1497–1508 (1996)

Parallel Algorithm Portfolio with Market Trading-Based Time Allocation

Dimitris Souravlias, Konstantinos E. Parsopoulos and Enrique Alba

Abstract We propose a parallel portfolio of metaheuristic algorithms that adopts a market trading-based time allocation mechanism. This mechanism dynamically allocates the total available execution time of the portfolio by favoring better-performing algorithms. The proposed approach is assessed on a significant Operations Research problem, namely the single-item lot sizing problem with returns and remanufacturing. Experimental evidence suggests that our approach is highly competitive with standard metaheuristics and specialized state-of-the-art algorithms.

1 Introduction

Algorithm portfolios (APs) emerged the past two decades as a promising framework that combines different algorithms or copies of the same algorithm to efficiently tackle hard optimization problems [3, 4]. Recently, they have gained increasing attention as a general framework for incorporating different population-based algorithms to solve continuous optimization problems [7, 13]. Significant effort has been paid on the selection of the constituent algorithms of the APs [11], which may run in an independent [10] or in a cooperative way [7]. The selection is usually based on a preprocessing phase, where the constituent algorithms are selected according to their performance from a wide range of available optimization algorithms.

In parallel implementations, the minimization of the total execution time is crucial due to the limited (or expensive) resources allocated to the users in high-performance

D. Souravlias (✉) · K.E. Parsopoulos (✉)
Department of Computer Science and Engineering,
University of Ioannina, Ioannina, Greece
e-mail: dsouravl@cs.uoi.gr

K.E. Parsopoulos
e-mail: kostasp@cs.uoi.gr

E. Alba
Department of Languages and Computer Science,
University of Malaga, Malaga, Spain
e-mail: eat@lcc.uma.es

© Springer International Publishing Switzerland 2016
M. Lübbecke et al. (eds.), *Operations Research Proceedings 2014*,
Operations Research Proceedings, DOI 10.1007/978-3-319-28697-6_79

computer infrastructures. The AP's framework offers inherent parallelization capability that stems from the ability of concurrently using its constituent algorithms. In practice, the algorithms that are used to solve a problem in parallel are typically assigned equal execution time or function evaluation budgets that remain constant throughout the optimization process [7]. Also, it is frequently observed that different algorithms perform better in different phases of the optimization procedure or problem instances [7, 10].

Motivated by this observation, we propose an AP where the algorithms are rewarded additional execution time on a performance basis. Specifically, the portfolio adopts a trading-based mechanism that dynamically orchestrates the allocation of the total available execution time among the AP's constituent algorithms. Better-performing algorithms are assigned higher fractions of execution time compared to worse-performing ones, without modifying the AP's total execution time. The core idea behind the attained mechanism is inspired by stock trading models and involves a number of algorithms-investors that invest on elite solutions that act as stocks, using execution time as currency.

The performance of the proposed AP is evaluated on a well studied Operations Research (OR) problem, namely the single-item dynamic lot sizing problem with returns and remanufacturing [9, 12]. Its performance is compared to other meta-heuristics [6] as well as state-of-the-art heuristics for the specific problem [9]. The rest of the paper is structured as follows: Sect. 2 briefly describes the problem, while Sect. 3 presents the proposed AP model. The experimental setting and results are exposed in Sect. 4 and the paper concludes in Sect. 5.

2 Problem Formulation

The considered problem constitutes an extension of the well-known Wagner-Whitin dynamic lot sizing problem [14]. It employs the dynamic lot sizing model with separate manufacturing and remanufacturing setup costs as it was introduced in [12] and further studied in [9]. The problem assumes a manufacturer that sells a single type of product over a finite planning horizon of T time periods. In each time period $t = 1, 2, \ldots, T$, the consumers state their demand denoted by D_t, along with a number of used products that are returned to the manufacturer. The fraction R_t of returned products in period t that can be recovered and sold as new is stored at a recoverables inventory with a holding cost h^R per unit time. To satisfy the demand, a number of z_t^R and z_t^M products are remanufactured and manufactured, respectively, in period t and then brought to a serviceables inventory with a holding cost h^M per unit time. Naturally, the manufacturing and remanufacturing process incur setup costs denoted by K^R and K^M, respectively.

The target is to minimize the incurring setup and holding costs by determining the exact number of manufactured and remanufactured items per period under a number of constraints. The corresponding cost function is defined as follows [9]:

$$C = \sum_{t=1}^{T} \left(K^R \gamma_t^R + K^M \gamma_t^M + h^R y_t^R + h^M y_t^M \right), \tag{1}$$

where γ_t^R and γ_t^M are binary variables denoting the initiation of a remanufacturing or manufacturing lot, respectively. The inventory levels of items that can be remanufactured or manufactured in period t are denoted by y_t^R and y_t^M, respectively. The operational constraints of the model are defined as follows:

$$y_t^R = y_{t-1}^R + R_t - z_t^R, \quad y_t^M = y_{t-1}^M + z_t^R + z_t^M - D_t, \quad t = 1, 2, \ldots, T, \tag{2}$$

$$z_t^R \leq Q \, \gamma_t^R, \quad z_t^M \leq Q \, \gamma_t^M, \quad t = 1, 2, \ldots, T, \tag{3}$$

$$y_0^R = y_0^M = 0, \quad \gamma_t^R, \gamma_t^M \in \{0, 1\}, \quad y_t^R, y_t^M, z_t^R, z_t^M \geq 0, \quad t = 1, 2, \ldots, T. \tag{4}$$

Equation (2) guarantees the inventory balance, while Eq. (3) assures that fixed costs are paid whenever a new lot is initiated. In [9] the value of Q is suggested to be equal to the total demand of the planning horizon. Finally, Eq. (4) asserts that inventories are initially empty and determines the domain of each variable. The decision variables of the optimization problem are z_t^M and z_t^R for each period t. Thus, for a planning horizon of T periods the corresponding problem is of dimension $n = 2T$. More details about the considered problem can be found in [6, 9, 12].

3 Proposed Algorithm Portfolio

We propose an AP that consists of metaheuristic algorithms that operate in parallel. We denote with N the number of algorithms. The AP employs a typical *master-slave* parallelization model, where each algorithm runs on a single slave node. Each algorithm invests a percentage of its assigned running time to buy solutions from the other algorithms of the AP. The remaining time is used for its own execution. We assign equal initial execution time budgets, T_{tot}, and investment time budgets, $T_{\text{inv}} = \alpha T_{\text{tot}}$, for all algorithms. The parameter $\alpha \in (0, 1)$ tunes each algorithm's investment policy. Clearly, high values of α indicate a risky algorithm-investor, while lower values characterize a more conservative one.

The master node retains in memory a solution archive that is asynchronously accessed by the slaves via a message passing communication mechanism. The archive holds the elite solution found by each algorithm. For their pricing, the solutions are sorted in descending order with respect to their objective values. If p_i is the position of the ith solution after sorting, then its cost is defined as $CS_i = (p_i \times SBC)/N$, where $SBC = \beta T_{\text{inv}}$ is a fixed base cost. The parameter $\beta \in (0, 1)$ tunes each algorithm's elitism. High values of β limit the number of the best elite solutions each algorithm can buy throughout the optimization process.

Whenever an algorithm cannot improve its elite solution for an amount of time, it requests to buy a solution from another algorithm. The master node acts as a trading broker that applies a solution selection policy to help the buyer-algorithm make the most profitable investment. In particular, the master node proposes to the buyer-algorithm elite solutions that are better than its own and cost less or equal to its current investment budget. Among the possible solutions, the algorithm opts to buy the solution that maximizes the *Return On Investment* (ROI) index, defined as $ROI_j = (C - C_j)/CS_j$, $j \in \{1, 2, \ldots, N\}$, where C is the objective value of the algorithm's own elite solution, C_j is the objective value of the candidate buying solution and CS_j is its corresponding cost. If the buyer-algorithm decides to buy the jth elite solution, it pays its price of CS_j running time to the seller-algorithm (the one that found this solution). The seller algorithm adds this time to its total execution time budget. Thus, better-performing algorithms sell solutions more often, gaining longer execution times. Yet, the total execution time of the AP remains constant.

In the present work, the proposed AP consists of 4 algorithms, namely Particle Swarm Optimization (PSO) [1], Differential Evolution (DE) [8], Tabu Search (TS) [2], and Iterated Local Search (ILS) [5].

4 Experimental Results

The proposed approach was evaluated on the established test suite used in [9]. It consists of a full factorial study of various problem instances with common planning horizon $T = 12$. Table 1 summarizes the configuration of the problem parameters

Table 1 Parameters of the considered problem and the employed algorithms

Problem parameter	Value(s)		Algorithm parameter	Value(s)
Dimension	$n = 24$	AP	Number of slave algorithms	$N = 4$
Setup costs	$K^M, K^R \in$ $\{200, 500, 2000\}$		Per algorithm execution time	$T_{tot} = 75000\,\text{ms}$
Holding costs	$h^M = 1, h^R \in$ $\{0.2, 0.5, 0.8\}$		Constants α, β	$\alpha = 0.1, \beta = 0.05$
Demand for period t	$D_t \sim N(\mu_D, \sigma_D^2)$	PSO	Model	lbest (ring topology)
	$\mu_D = 100$		Swarm size	60
	$\sigma_D^2 = 10\%$ of μ_D (small variance)		Constriction coefficient	$\chi = 0.729$
	$\sigma_D^2 = 20\%$ of μ_D (large variance)		Cognitive/social constants	$c_1 = c_2 = 2.05$
Returns for period t	$R_t \sim N(\mu_R, \sigma_R^2)$	DE	Population size	60
	$\mu_R \in \{30, 50, 70\}$		Operator	DE/rand/1
	$\sigma_R^2 = 10\%$ of μ_R (small variance)		Differential/crossover constants	$F = 0.7, CR = 0.3$
	$\sigma_R^2 = 20\%$ of μ_R (large variance)	TS	Size of tabu list	24

Table 2 Percentage % error of the compared algorithms for different problem parameters

Param	Alg.	Avg	StD	Max	Param	Alg.	Avg	StD	Max	Param	Alg.	Avg	StD	Max
All	SM_4^+	2.2	2.9	24.3	$h^R = 0.2$	SM_4^+	1.7	2.5	21.1	$\mu_R = 30$	SM_4^+	1.2	1.8	12.1
	PSO	4.3	4.5	49.8		PSO	4.5	5.2	49.8		PSO	3.5	3.1	45.5
	DE	3.3	5.1	31.9		DE	3.0	5.3	30.9		DE	3.3	5.0	28.2
	TS	51.6	33.4	255.5		TS	45.0	26.4	255.5		TS	37.2	23.9	255.5
	ILS	80.3	54.3	450.8		ILS	94.8	67.2	450.8		ILS	70.6	46.5	336.8
	AP	1.9	2.8	35.6		AP	1.5	2.5	35.6		AP	1.6	2.5	25.7
$\sigma_D^2 = 10\%$	SM_4^+	2.1	2.8	18.9	$h^R = 0.5$	SM_4^+	2.3	3.0	24.3	$\mu_R = 50$	SM_4^+	2.3	2.7	16.2
	PSO	4.4	4.6	49.8		PSO	4.3	4.5	45.5		PSO	4.1	4.0	34.0
	DE	3.4	4.8	31.7		DE	3.3	5.0	31.9		DE	3.5	5.2	31.9
	TS	50.9	33.2	200.2		TS	50.8	32.1	202.1		TS	50.8	27.3	153.6
	ILS	79.7	54.2	450.8		ILS	77.6	48.2	261.1		ILS	83.7	53.0	364.2
	AP	1.8	2.6	26.8		AP	1.9	2.8	27.4		AP	2.0	2.9	27.4
$\sigma_D^2 = 20\%$	SM_4^+	2.4	3.0	24.3	$h^R = 0.8$	SM_4^+	2.8	3.0	20.6	$\mu_R = 70$	SM_4^+	3.3	3.5	24.3
	PSO	4.1	4.5	48.3		PSO	4.0	3.9	42.9		PSO	5.1	5.9	49.8
	DE	3.3	5.2	31.9		DE	3.7	4.5	31.4		DE	3.3	4.6	31.7
	TS	52.4	33.5	255.5		TS	59.1	39.0	235.2		TS	66.9	39.8	235.2
	ILS	80.9	54.4	421.5		ILS	68.4	40.8	211.4		ILS	86.5	61.1	450.8
	AP	2.0	2.9	35.6		AP	2.2	3.0	21.6		AP	2.0	2.9	35.6
$K^M = 200$	SM_4^+	2.3	2.6	13.5	$K^R = 200$	SM_4^+	1.9	2.1	11.8	$\sigma_R^2 = 10\%$	SM_4^+	2.2	2.9	21.1
	PSO	4.0	3.1	45.5		PSO	5.7	5.5	49.8		PSO	4.3	4.6	46.7
	DE	3.2	3.9	24.0		DE	3.8	4.0	24.0		DE	3.4	5.0	31.4
	TS	39.1	27.3	255.5		TS	75.2	38.0	203.3		TS	52.1	34.3	233.8
	ILS	62.6	64.0	450.8		ILS	63.0	45.4	260.4		ILS	80.4	54.4	450.8
	AP	2.4	3.0	21.6		AP	3.0	3.3	21.6		AP	1.8	2.7	35.6

(continued)

Table 2 (continued)

	Alg.	Avg	StD	Max
$K^M = 500$	SM_4^+	2.1	2.5	**12.8**
	PSO	4.5	4.1	27.5
	DE	2.5	2.6	15.2
	TS	67.9	33.1	197.1
	ILS	62.0	40.6	278.1
	AP	**1.8**	**2.4**	17.6
$K^M = 2000$	SM_4^+	2.3	3.4	**24.3**
	PSO	4.4	5.9	49.8
	DE	4.3	7.1	31.9
	TS	47.9	32.7	235.2
	ILS	116.3	34.2	260.4
	AP	**1.4**	**2.8**	35.6

	Alg.	Avg	StD	Max
$K^R = 500$	SM_4^+	3.4	3.2	19.1
	PSO	3.8	4.1	37.4
	DE	1.8	2.0	11.2
	TS	50.8	23.8	235.2
	ILS	62.4	37.8	244.8
	AP	**1.3**	**1.7**	**11.6**
$K^R = 2000$	SM_4^+	1.4	2.9	**24.3**
	PSO	3.3	3.5	45.5
	DE	4.4	7.1	31.9
	TS	29.0	16.4	255.5
	ILS	115.5	59.2	450.8
	AP	**1.3**	**2.8**	35.6

	Alg.	Avg	StD	Max
$\sigma_R^2 = 20\%$	SM_4^+	2.3	2.9	**24.3**
	PSO	4.2	4.5	49.8
	DE	3.3	4.9	31.9
	TS	51.2	32.5	255.5
	ILS	80.1	54.2	399.1
	AP	**2.0**	**2.9**	25.6

as well as the employed algorithm parameters for the AP. Further details on the problem setting can be found in [9]. The proposed AP was compared against the best-performing variant (SM_4^+) of the state-of-the-art Silver-Meal heuristic [9], as well as against the sequential versions of its constituent algorithms. The goal of the experiments was to achieve the lowest possible percentage error [9] from the global optimum within a predefined budget of total execution time T_{tot}. The global optimum per problem was computed by CPLEX and was provided in the test suite.

Table 2 shows the average (Avg), standard deviation (StD), and maximum (Max) value of the percentage error for the different values of the problem parameters. A first inspection of the results reveals superiority of the proposed AP, which achieves the best overall mean percentage error (1.9 %). The second lowest value was achieved by SM_4^+ (2.2 %), followed by the sequential versions of DE (3.3 %) and PSO (4.3 %). Specifically, AP prevails in 14 out of 17 considered parameter cases, while in the rest 3 cases SM_4^+ is the dominant algorithm. The results of SM_4^+ and PSO were directly adopted from [9] and [6], respectively.

The results indicate that population-based algorithms (DE and PSO) outperform (by far) the trajectory-based ones (TS and ILS). Moreover, when all algorithms are integrated into the AP, the overall performance with respect to solution quality is further enhanced. This can be attributed to the dynamics of the trading among the algorithms. In particular, we observed that the population-based algorithms were mainly the seller ones, using their exploration capability to discover high-quality solutions. On the other hand, trajectory-based algorithms employed their exploitation power to further fine-tune the vast number of acquired solutions. From this point of view, the employed algorithms of the AP exhibited complementarity, which is a desired property in APs [7, 13]. Also, we observed that between the two population-based algorithms, PSO acquired a higher number of solutions than DE during the optimization, whereas the solutions of the latter were of better quality.

5 Conclusions

We proposed an Algorithm Portfolio (AP) of metaheuristic algorithms that operate in parallel and exchange solutions via a sophisticated trading-based time allocation mechanism. This mechanism favors better-performing algorithms with more execution time than the others. Also, it combines the exploration/exploitation dynamics of each individual constituent algorithm in an efficient way. We assessed our approach on a well studied OR problem. The experimental results were promising, indicating that the AP is highly competitive against its constituent algorithms, individually, as well as against a state-of-the-art algorithm of the considered problem.

References

1. Eberhart, R.C., Kennedy, J.: A new optimizer using particle swarm theory. In Proceedings Sixth Symposium on Micro Machine and Human Science, pp. 39–43, Piscataway, NJ (1995)
2. Glover, F.: Future paths for integer programming and links to artificial intelligence. Comput. Oper. Res. **13**(5), 533–549 (1986)
3. Gomes, C.P., Selman, B.: Algorithm portfolio design: theory vs. practice. In: Proceedings Thirteenth conference on Uncertainty in artificial intelligence, pp. 190–197 (1997)
4. Huberman, B.A., Lukose, R.M., Hogg, T.: An economics approach to hard computational problems. Science **27**, 51–53 (1997)
5. Lourenco, H.R.: Job-shop scheduling: computational study of local search and large-step optimization methods. Eur. J. Oper. Res. **83**(2), 347–364 (1995)
6. Moustaki, E., Parsopoulos, K.E., Konstantaras, I., Skouri, K., Ganas, I.: A first study of particle swarm optimization on the dynamic lot sizing problem with product returns. In: Proceedings BALCOR 2013, pp. 348–356 (2013)
7. Peng, F., Tang, K., Chen, G., Yao, X.: Population-based algorithm portfolios for numerical optimization. IEEE Trans. Evol. Comput. **14**(5), 782–800 (2010)
8. Price, K.: Differential evolution: a fast and simple numerical optimizer. In: Proceedings NAFIPS'96, pp. 524–525 (1996)
9. Schulz, T.: A new silver-meal based heuristic for the single-item dynamic lot sizing problem with returns and remanufacturing. Int. J. Prod. Res. **49**(9), 2519–2533 (2011)
10. Shukla, N., Dashora, Y., Tiwari, M., Chan, F., Wong, T.: Introducing algorithm portfolios to a class of vehicle routing and scheduling problem. In: Proceedings OSCM 2007, pp. 1015–1026 (2007)
11. Tang, K., Peng, F., Chen, G., Yao, X.: Population-based algorithm portfolios with automated constituent algorithms selection. Inf. Sci. **279**, 94–104 (2014)
12. Teunter, R.H., Bayindir, Z.P., Van den Heuvel, W.: Dynamic lot sizing with product returns and remanufacturing. Int. J. Prod. Res. **44**(20), 4377–4400 (2006)
13. Vrugt, J.A., Robinson, B.A., Hyman, J.M.: Self-adaptive multimethod search for global optimization in real-parameter spaces. IEEE Trans. Evol. Comput. **13**(2), 243–259 (2009)
14. Wagner, H.M., Whitin, T.M.: Dynamic version of the economic lot size model. Manag. Sci. **5**(1), 88–96 (1958)

Global Solution of Bilevel Programming Problems

Sonja Steffensen

Abstract We discuss the global solution of Bilevel Programming Problems using their reformulations as Mathematical Programs with Complementarity Constraints and/or Mixed Integer Nonlinear Programs. We show that under suitable assumptions the Bilevel Program can be reformulated and globally solved via MINLP refomulation. We also briefly discuss some simplifications and suitable additional constraints.

1 Introduction

In this paper, we are interested in the global solution of the following Bilevel Program

$$
(BP) \qquad \min_{x,y} \quad F(x, y)
$$
$$
\text{s.t.} \quad c(x, y) \leq 0 \qquad\qquad (1)
$$
$$
y \in \Sigma(x) := \operatorname{argmin}_y \{ f(x, y) \mid g(x, y) \leq 0 \},
$$

where $(x, y) \in \mathbb{R}^n \times \mathbb{R}^p$, x denotes the upper-level variable (i.e. the leader) and y is the lower-level variable (i.e. the follower). All inequalities are meant componentwise and throughout, we assume that $F : \mathbb{R}^n \times \mathbb{R}^p \to \mathbb{R}$, $f : \mathbb{R}^n \times \mathbb{R}^p \to \mathbb{R}$, $c : \mathbb{R}^n \times \mathbb{R}^p \to \mathbb{R}^m$ and $g : \mathbb{R}^n \times \mathbb{R}^p \to \mathbb{R}^q$ are twice continuously differentiable functions. As can be deduced from (1), we use the optimistic approach (i.e. if $\Sigma(x)$ is not single-valued, we assume that $y \in \Sigma(x)$ is chosen such that $F(x, y)$ is smallest).

As indicated by the names of the variables, the Bilevel Program is a hierarchical optimization problem that is mostly used to model a so-called Stackelberg game. However, other applications are parameter identification problems, where the variable x then denotes the vector of parameters that parametrizes the optimization problem:

$$
\min_y f(x, y) \qquad \text{s.t.} \quad g(x, y) \leq 0 . \qquad\qquad (2)
$$

S. Steffensen (✉)
Department of Mathematics, RWTH Aachen University, Templergraben 55,
D-52062 Aachen, Germany
e-mail: steffensen@igpm.rwth-aachen.de

© Springer International Publishing Switzerland 2016
M. Lübbecke et al. (eds.), *Operations Research Proceedings 2014*,
Operations Research Proceedings, DOI 10.1007/978-3-319-28697-6_80

575

Bracken and McGill [4, 5] first considered such kind of problems in the 1970th having applications in the military field in mind. Recent applications comprise problems from economic sciences, such as problems arising from designing and planning transportation networks (e.g. network design, toll-setting problem or the problem of designing parking facilities for parkn ride trips [12]) or revenue management (e.g. planning the pricing and seat allocation policies in the airline industry [7]), but also engineering problems, where they appear as optimization problems that involve chemical or physical equilibria.

For more information on bilevel programs, we refer the interested reader e.g. to the two monographs [1, 8] and the survey paper [6].

Even in the case where, all functions involved are convex, i.e. both the upper- and the lower-level problem are convex, the global solution of the resulting Bilevel Problem is not trivial as the solution mapping $\Sigma(x)$ might be nonconvex (it might even be not compact). In this paper we will first describe the reformulations of (1) as Mathematical Program with Complementarity Constraints (MPCC), as well as its reformulation as Mixed Integer Nonlinear Program (MINLP) [2, 13, 20]. We will then relate the feasible sets and give two results concerning the relationship between the global solutions in the convex case of the original problem and its reformulations. Furthermore, we briefly discuss some simplifications for some special type of (1) and additional constraints for the nonconvex case.

2 Reformulations

The reformulations we propose here are such that the two-level problem (1) is replaced by a single-level problem. The most often used reformulation of (1) is to replace the lower-level problem (2) by its stationarity conditions [9]

$$\nabla_y \mathcal{L}(x, y, \mu) = 0 \tag{3}$$

$$g(x, y) \leq 0, \quad \mu \geq 0, \quad \mu^T g(x, y) = 0, \tag{4}$$

where $\mathcal{L}(x, y, \mu) = f(x, y) + g(x, y)^T \mu$ denotes the Lagrangian function associated with (2). The conditions (4) are also referred to as complementarity conditions. Hence the resulting problem becomes a so-called Mathematical Program with Complementarity Constraints

$$
(MPCC) \qquad
\begin{aligned}
\min_{x,y,\mu} \quad & F(x, y) \\
\text{s.t.} \quad & c(x, y) \leq 0 \\
& \nabla_y \mathcal{L}(x, y, \mu) = 0 \\
& g(x, y) \leq 0, \quad \mu \geq 0, \quad \mu^T g(x, y) = 0.
\end{aligned}
\tag{5}
$$

These problems are inherently nonconvex, due to the complementarity constraints and it is known, that they do not admit the MFCQ at any feasible point. There exist a variety of theoretical results (see e.g. [17, 21]), as well as local solution methods for MPCC in the literature, ranging from relaxation and smoothing methods to SQP methods and interior point methods (see e.g. [10, 11, 14, 16, 18, 19]). However, since we are interested in the global solution of (1), we introduce another reformulation which directly takes the decision structure of the complementarity constraints into account. Note that (4) can be reformulated by

$$g(x, y) \leq 0, \quad \mu \geq 0 \quad \text{and} \quad (\mu_i = 0 \quad \text{or} \quad g_i(x, y) = 0, \quad \forall \, i = 1, .., q). \quad (6)$$

Assuming that $g(x, y)$ and μ are bounded (6) can equivalently be replaced by

$$g(x, y) \leq 0, \quad \mu \geq 0 \quad \text{and} \quad \mu \leq Ms, \quad g(x, y) \geq M(1 - s), \quad s \in \{0, 1\}^q, \quad (7)$$

where $\mathbf{1} := (1, 1, \ldots, 1) \in \mathbb{R}^q$. This reformulation then yields the following Mixed Integer Nonlinear Programming Problem as a reformulation of (1)

$$
\begin{aligned}
(MINLP) \qquad \min_{x, y, \mu, s} \quad & F(x, y) \\
\text{s.t.} \quad & c(x, y) \leq 0 \\
& \nabla_y \mathcal{L}(x, y, \mu) = 0 \\
& g(x, y) \leq 0, \quad \mu \geq 0 \\
& \mu \leq Ms, \qquad g(x, y) \geq -M(1 - s) \\
& s \in \{0, 1\}^q,
\end{aligned}
$$

$$(8)$$

where M is a positive scalar that is supposed to be larger than the corresponding upper bounds on μ and $|g(x, y)|$.

Remark 1 If the lower-level problem is simply a convex, box-constrained problem, we can omit the multiplier variable. Since either $\alpha_i \leq y_i$ or $y_i \leq \beta_i$ might be satisfied with equality but not both of them either $\mu_{\alpha,i} = 0$ or $\mu_{\beta,i} = 0$ and the conditions (3) and (4) can be reduced to

$$
\begin{aligned}
& \alpha \leq y, \quad y \leq \beta, \quad y - \alpha \leq Ms_1, \quad \beta - y \leq Ms_2 \\
& \nabla_y f(x, y) \geq -M(1 - s_2), \quad \nabla_y f(x, y) \leq M(1 - s_1) \quad s_1, s_2 \in \{0, 1\}^p.
\end{aligned}
$$

3 Relations of Global Solutions

In the following, we compare the set of global solutions of the original problem (1) to the set of global solution of its reformulations. In [9] the global solutions of (5) are compared to the ones of (1). In this paper, we therefore restrict ourselves to the comparison of the global solutions of (8) to the ones of (1).

As starting point, we compare the set of feasible upper- and lower-level variables. Thus, define the following sets:

$$\mathscr{Z}_{BP} := \{(x, y) \in \mathbb{R}^n \times \mathbb{R}^p \mid c(x, y) \le 0, \; y \in \Sigma(x)\}$$

$$\mathscr{Z}_{MPCC} := \{(x, y) \in \mathbb{R}^n \times \mathbb{R}^p \mid c(x, y) \le 0,$$
$$\exists \mu \in \mathbb{R}_+^q : \nabla_y \mathscr{L}(x, y, \mu) = 0, \; g(x, y) \le 0, \; \mu^T g(x, y) = 0\}$$

$$\mathscr{Z}_{MINLP} := \{(x, y) \in \mathbb{R}^n \times \mathbb{R}^p \mid c(x, y) \le 0,$$
$$\exists \mu \in \mathbb{R}_+^q, \; s \in \{0, 1\}^q : \nabla_y \mathscr{L}(x, y, \mu) = 0$$
$$g(x, y) \le 0, \; \mu \le Ms, \; g(x, y) \ge -M(1 - s) \}$$

Moreover, define the Slater CQ for the lower-level problem.

Definition 1 The **Slater CQ** is said to hold in $x \in \mathbb{R}^n$ for the lower-level problem (2), if there exists $z(x)$ such that $g_i(x, z(x)) < 0$ for all $i = 1, .., q$.

Lemma 1 *Let f and g be convex. Then it holds:*

$$(\bar{x}, \bar{y}) \in \mathscr{Z}_{MINLP} \quad \Leftrightarrow \quad (\bar{x}, \bar{y}) \in \mathscr{Z}_{MPCC} \quad \Rightarrow \quad (\bar{x}, \bar{y}) \in \mathscr{Z}_{BP}.$$

Furthermore, if the Slater CQ holds in $(\bar{x}, \bar{y})$ then it also holds

$$(\bar{x}, \bar{y}) \in \mathscr{Z}_{BP} \quad \Rightarrow \quad (\bar{x}, \bar{y}) \in \mathscr{Z}_{MPCC}.$$

Proof First, assume that $(\bar{x}, \bar{y}) \in \mathscr{Z}_{MINLP}$, then there exists $\bar{\mu}$ and s such that (7) is satisfied so that either $\bar{\mu}_i = 0$ (if $s_i = 0$) or $g_i(\bar{x}, \bar{y}) = 0$ (if $s_i = 1$) holds. Hence (4) is satisfied and therefore $(\bar{x}, \bar{y}) \in \mathscr{Z}_{MPCC}$.

On the other hand, if $(\bar{x}, \bar{y}) \in \mathscr{Z}_{MPCC}$, then there exists a multiplier $\bar{\mu}$ such that (3) and (4) hold. Now choose $s \in \{0, 1\}^q$ such that $s_i = 0$ if $\bar{\mu}_i = 0$ and $g_i(\bar{x}, \bar{y}) < 0$ and $s_i = 1$ if $g_i(\bar{x}, \bar{y}) = 0$ and $\bar{\mu}_i > 0$. In case that both $\bar{\mu}_i = 0$ and $g_i(\bar{x}, \bar{y}) = 0$ one might choose $s_i \in \{0, 1\}$. Thus $(\bar{x}, \bar{y}, \bar{\mu}, s)$ satisfies (7), i.e. $(\bar{x}, \bar{y}) \in \mathscr{Z}_{MINLP}$.

Next the last direction of the first part is clear since, due to the convexity assumption, the stationarity conditions (3) and (4) are sufficient for $\bar{y}$ to be a global minimizer.

Finally, if $(\bar{x}, \bar{y}) \in \mathscr{Z}_{BP}$, then $\bar{y} \in \Sigma(\bar{x})$. Hence by convexity of f and g and the Slater CQ, we know there exists a multiplier $\bar{\mu}$ such that $(\bar{x}, \bar{y}, \bar{\mu})$ satisfies (3) and (4), i.e. $(\bar{x}, \bar{y}) \in \mathscr{Z}_{MPCC}$.   □

Remark 2 Note that for the first part of the Lemma the assumptions of convexity and Slater CQ are crucial. However, they might obviously be replaced by the assumption that f and g are linear or by a stronger regularity condition as e.g. MFCQ or LICQ, respectively.

Theorem 1 *Assume that f and g are convex and let $(\bar{x}, \bar{y})$ be a global solution of (BP) such that the Slater CQ holds in $\bar{x}$ for the lower-level problem (2). Then for* *any* $\bar{\mu}, \bar{s}$, *so that* $(\bar{x}, \bar{y}, \bar{\mu}, \bar{s})$ *is feasible for the associated (MINLP),* $(\bar{x}, \bar{y}, \bar{\mu}, \bar{s})$ *is a global optimal solution of (MINLP).*

Proof Since $(\bar{x}, \bar{y}) \in \mathscr{Z}_{BP}$, it follows by Lemma 1 that $(\bar{x}, \bar{y}) \in \mathscr{Z}_{MINLP}$, i.e. there exist $\bar{\mu}, \bar{s}$, such that $(\bar{x}, \bar{y}, \bar{\mu}, \bar{s})$ is feasible for the associated (MINLP). Assume that there exists $(\hat{x}, \hat{y}) \in \mathscr{Z}_{MINLP}$ with $F(\hat{x}, \hat{y}) < F(\bar{x}, \bar{y})$ then by Lemma 1 $(\hat{x}, \hat{y}) \in \mathscr{Z}_{BP}$ which contradicts the global optimality of $(\bar{x}, \bar{y})$ for (BP), which concludes the proof. $\square$

For the opposite direction we have to slightly strengthen the assumption in the way that the Slater CQ is supposed to hold for (2) for any feasible $\bar{x}$.

Theorem 2 *Assume that f and g are convex and let* $(\bar{x}, \bar{y}, \bar{\mu}, \bar{s})$ *be a global solution of (MINLP). Furthermore assume that the Slater CQ is satisfied for (2) for any feasible x (i.e. there exist* $y \in \mathbb{R}^p$ *so that* $c(x, y) \leq 0$ *and* $g(x, y) \leq 0$*). Then* $(\bar{x}, \bar{y})$ *is a global optimal solution of (BP).*

Proof Since $(\bar{x}, \bar{y}) \in \mathscr{Z}_{MINLP}$, it follows by Lemma 1 and the assumptions that $(\bar{x}, \bar{y}) \in \mathscr{Z}_{BP}$, i.e. $(\bar{x}, \bar{y})$ is feasible for (BP). Assume that there exists a feasible $(\hat{x}, \hat{y})$ for (BP) so that $F(\hat{x}, \hat{y}) < F(\bar{x}, \bar{y})$ then by Lemma it also holds that Lemma 1 $(\hat{x}, \hat{y}) \in \mathscr{Z}_{MINLP}$ and hence there exist $\hat{\mu}, \hat{s}$, so that $(\hat{x}, \hat{y}, \hat{\mu}, \hat{s})$ is feasible for the associated (MINLP) which contradicts the global optimality of $(\bar{x}, \bar{y}, \bar{\mu}, \bar{s})$ for (BP). This concludes the proof. $\square$

Remark 3 Note that these results do not transfer to local solutions as the discussion in [9] already shows.

Remark 4 Taking into account that in general the global solution of MINLP can be guaranteed to be found under the convexity assumption of the objective function and all constraint functions [2, 3, 13], by the results presented, it becomes clear, that a general purpose MINLP solver (e.g. DICOPT) can guarantee to solve the MINLP (and hence the BP), if f is quadratic and convex in y for all feasible x (e.g. $f(x, y) = y^T A y + x^T B y + b(x)$), F and c are convex, g is linear and the Slater CQ is satisfied for (2) for any feasible x.

Remark 5 In the case of a convex upper-level program, however, a nonconvex lower-level program the discussed approach might not be able to find the correct global solution. In particular, if the lower-level problem is nonconvex, the reformulation of the BP as an MPCC (and hence as a MINLP) is not exact anymore (in the sense of Lemma 1), as the stationarity conditions are not sufficient anymore. Thus even if the lower-level problem satisfies a regularity assumption (as e.g. LICQ), the MPCC reformulation is in fact a relaxation. However, if further constraints on the curvature of the function f are included in the reformulation, the computation of a global solution $(\bar{x}, \bar{y})$ where y is a saddlepoint or even worse a local maximum of (2) can be prevented. Another option for such cases are specialized algorithms as presented e.g. in [15].

References

1. Bard, J.F.: Practical Bilevel Optimization: Algorithms and Applications. Kluwer Academic Publishers, Dordrecht (1998)
2. Belotti, P., Kirches, Ch., Leyffer, S., Linderoth, J., Luedtke, J., Mahajan, A.: Mixed-Integer nonlinear optimization. Acta Numerica 22, 1–131 (2013)
3. Bonami, P., Biegler, L.T., Conn, A.R., Cornujols, G., Grossmann, I.E., Laird, C.D., Lee, J., Lodi, A., Margot, F., Sawaya, N., Wchter, A.: An algorithmic framework for convex mixed integer nonlinear programs. Discret. Optim. 5(2), 186–204 (2008)
4. Bracken, J., McGill, J.: Mathematical programs with optimization problems in the constraints. Oper. Res. 21, 3744 (1973)
5. Bracken, J., McGill, J.: Defense applications of mathematical programs with optimization problems in the constraints. Oper. Res. 22, 1086–1096 (1974)
6. Colson, B., Marcotte, P., Savard, G.: An overview of bilevel optimization. Ann. Oper. Res. 153, 235–256 (2007)
7. Ct, J.-P., Marcotte, P., Savard, G.: A bilevel modeling approach to pricing and fare optimization in the airline industry. J. Revenue Pricing Manage. 2, 23–36 (2003). Dempe, S.: A necessary and a sufficient optimality condition for bilevel programming (1992a)
8. Dempe, S.: Foundations of Bilevel Programming. Kluwer Academic Publishers, Dordrecht (2002)
9. Dempe, S., Dutta, J.: Is Bilevel programming a special case of mathematical programs with complementarity constraints?. Math. Program. Series A, 131, 37–48 (2012)
10. de Miguel, A.V., Friedlander, M.P., Nogales, F.J., Scholtes, S.: A two-sided relaxation scheme for mathematical programs with equilibrium constraints. SIAM J. Optim. 16, 587–609 (2005)
11. Fletcher, R., Leyffer, S., Ralph, D., Scholtes, S.: Local convergence of SQP-methods for mathematical programs with equilibrium constraints. SIAM J. Optim. 17(1), 259–286 (2006)
12. Garcia, R., Marin, A.: Parking Capacity and pricing in parkn ride trips: a continuous equilibrium network design problem. Ann. Oper. Res. 116, 153–178 (2002)
13. Grossmann, I.E.: Review of nonlinear mixed-integer and disjunctive programming techniques. Optim. Eng. 3, 227–252 (2002)
14. Hu, X.M., Ralph, D.: Convergence of a penalty method for mathematical programming with complementarity constraints. J. Optim. Theory Appl. 123(2), 365–390 (2004)
15. Mitsos, A., Lemonidis, P., Barton, P.I.: Globla solution of bilevel programs with a nonconvex inner program. J. Optim. Theory Appl. 123(2), 365–390 (2004)
16. Raghunathan, A.U., Biegler, L.T.: Interior point methods for Mathematical Programs with Complementarity Constraints (MPCCs). SIAM J. Optim. 15(3), 720–750 (2005)
17. Scheel, H., Scholtes, S.: Mathematical programs with complementarity constraints: stationarity. Optim. Sensit. Math. Oper. Res. 25, 1–22 (2000)
18. Scholtes, S.: Convergence properties of a regularization scheme for mathematical programs with complementarity constraints. SIAM J. Optim. 11(4), 918–936 (2001)
19. Steffensen, S., Ulbrich, M.: A new relaxation scheme for mathematical programs with equilibrium constraints. SIAM J. Optim. (2010)
20. Tawarmalani, M., Sahinidis, N.V.: Global optimization of mixed-integer nonlinear programs: a theoretical and computational study. Math. Program. 99(3), 563–591 (2004)
21. Ye, J.J.: Necessary and sufficient optimality conditions for mathematical programs with equilibrium constraints. J. Math. Anal. Appl. 307, 350–369 (2005)

Robustness to Time Discretization Errors in Water Network Optimization

Nicole Taheri, Fabian R. Wirth, Bradley J. Eck, Martin Mevissen
and Robert N. Shorten

Abstract Water network optimization problems require modeling the progression of flow and pressure over time. The time discretization step for the resulting differential algebraic equation must be chosen carefully; a large time step can result in a solution that bears little relevance to the physical system, and small time steps impact a problem's tractability. We show that a large time step can result in meaningless results and we construct an upper bound on the error in the tank pressures when using a forward Euler scheme. We provide an optimization formulation that is robust to this discretization error; robustness to model uncertainty is novel in water network optimization.

1 Introduction

Addressing uncertainty is an important aspect of many optimization applications. Most robust optimization work focuses on data uncertainty, e.g., inaccurate parameters or neglected terms [1, 2], including for water network applications [3, 4]. Uncertainty in the problem modeling receives little attention, even though the discretization can add uncertainty to a perfect model.

Water network optimization requires modeling flow and pressure in discrete time. The time step chosen is a critical part of this model: a large time step may result in a

N. Taheri (✉) · B.J. Eck · M. Mevissen · R.N. Shorten
IBM Research Ireland, Dublin, Ireland
e-mail: nictaher@ie.ibm.com; nicole.taheri@ie.ibm.com

F.R. Wirth
University of Passau, Passau, Germany
e-mail: fabian.wirth@uni-passau.de

B.J. Eck
e-mail: bradley.eck@ie.ibm.com

M. Mevissen
e-mail: martmevi@ie.ibm.com

R.N. Shorten
e-mail: robshort@ie.ibm.com

© Springer International Publishing Switzerland 2016
M. Lübbecke et al. (eds.), *Operations Research Proceedings 2014*,
Operations Research Proceedings, DOI 10.1007/978-3-319-28697-6_81

simulation that is feasible for the model, but unsuitable for the real system, while a small time step can result in a large intractable problem. We show an example where errors in the simulated tank levels lead to a solution that inaccurately describes the physical system, and we provide a method to compute the upper bound on the error introduced by a forward Euler scheme. We construct an optimization formulation that is robust to the discretization error. The robust counterpart of the optimization formulation can drastically reduce the energy cost given the worst-case error.

2 Problem Description

A water network consists of a set of nodes N and links L. There are 3 types of nodes: demand nodes, reservoirs and tanks. A link j connects 2 nodes $i_1 \neq i_2 \in N$ and a subset of links are pumps. Table 1 defines the terminology.

Define $A \in \mathbb{R}^{|N| \times |L|}$ as the network's directed node-link incidence matrix and partition $A^\top = \begin{bmatrix} A_D^\top & A_T^\top & A_R^\top \end{bmatrix}$, where rows correspond to demand nodes, tanks and reservoirs. Similarly, partition $p = (p_D, p_T, p_R)$, where p_R is assumed to be constant. For a pump $j \in P$, let $\widehat{S}_j \in \{0, 1\}^{t_{\text{end}}}$ be a pump schedule. For ease in explanation, we define a diagonal matrix $S(t) \in \{0, 1\}^{|L| \times |L|}, \forall t \in [0, t_{\text{end}}]$, where $S_{jj}(t) := \widehat{S}_j(t)$ if $j \in P$, and $S_{jj}(t) := 1$ if $j \in L \backslash P$.

At a given time t, the dynamics of a network can be described by:

$$(I - S(t))q(t) \qquad\qquad = 0 \tag{1a}$$

$$S(t)(H(q(t)) + A_D^\top p_D(t)) \quad = -S(t)\left(A_T^\top p_T(t) + A_R^\top p_R(t) + A^\top e\right) \tag{1b}$$

$$A_D q(t) \qquad\qquad\qquad = d(t), \tag{1c}$$

where the unknowns $q(t)$ and $p_D(t)$ are on the left-hand side. Equation (1a) ensures there is flow in a pump only if it is on. Equations (1b) and (1c) describe energy and

Table 1 List of constants and sets

Sets		Constants	
Name	Description	Name	Description
N	Nodes	$t_{\text{end}} \in \mathbb{R}$	Time horizon
T	Tanks, $T \subset N$	$h \in \mathbb{R}$	Size of time Step ($h = 1/t_{\text{end}}$)
D	Demand nodes, $D \subset N$	$e \in \mathbb{R}^{|N|}$	Node elevation (m)
L	Links	$\bar{q}, \underline{q} \in \mathbb{R}^{|L|}$	Bounds on flow (m³/day)
P	Pumps, $P \subset L$	$\bar{p}, \underline{p} \in \mathbb{R}^{|N|}$	Bounds on pressure (m)
Variables		$p_0 \in \mathbb{R}^{|R|}$	Initial tank pressure (m)
$q(t) \in \mathbb{R}^{|L|}$	Flow at time t (m³/day)	$\lambda \in \mathbb{R}^{|R|}$	Tank areas (m²)
$p(t) \in \mathbb{R}^{|N|}$	Pressure at time t (m)	$d \in \mathbb{R}^{|D| \times T}$	Demand (m³/day)
$\widehat{S}(t) \in \mathbb{R}^{|P|}$	Pump schedule at time t ($\{0,1\}$)	$A \in \mathbb{Z}^{|N| \times |L|}$	Directed incidence matrix

mass balance, respectively. Given $d(t)$, $S(t)$ and $p_\tau(t)$, under mild conditions there is a unique solution to (1) [5, 6]. However, uniqueness does not necessarily imply that solutions are physically meaningful.

Define the change in tank pressure with the function:

$$\dot{p}_\tau(t) = f(p_\tau(t), S(t), d(t)) := \Lambda^{-1} A_\tau q(p_\tau(t), S(t), d(t)), \tag{2}$$

or equivalently, by the integral equation,

$$p_\tau(t + h) = p_\tau(t) + \int_t^{t+h} f(p_\tau(s), S(s), d(s))ds. \tag{3}$$

To stay consistent with the field standard [7], we consider a discretization of (3) with forward Euler, which gives the discretization:

$$\widehat{p}_\tau(t + h) = \widehat{p}_\tau(t) + h \cdot f(\widehat{p}_\tau(t), S(t), d(t)). \tag{4}$$

The choice of h adjusts the trade-off of approximation to the differential and computational burden.

Define the set $\Omega(S, d)$ of feasible pressures and flows for a given network:

$$\Omega(S, d) = \left\{ (p, q) \in \mathbb{R}^{|N| \times |L|} \;\middle|\; \begin{array}{l} \text{(1a)–(1c) are satisfied} \\ p \in [\underline{p}, \overline{p}], \; q \in [\underline{q}, \overline{q}] \end{array} \right\}. \tag{FEAS}$$

And define the following sets of feasible points given a subset of the values: $\Omega_\tau(S, d) = \{ \widehat{p}_\tau \,|\, \exists (\widehat{p}, q) \in \Omega(S, d) \}$, $\Omega_q(S, d) = \{ \widehat{q} \,|\, \exists (p, \widehat{q}) \in \Omega(S, d) \}$. Let $\phi(p, q, S, d) \in \mathbb{R}$ be a chosen convex objective function, and define the general water network optimization formulation:

$$
\begin{array}{ll}
\underset{p, q, S}{\text{minimize}} & \phi(p, q, S, d) \\
\text{subject to} & p_\tau(t + h) = p_\tau(t) + h \cdot f(p_\tau(t), S(t), d(t)) \\
& (p(t), q(t)) \in \Omega(S(t), d(t)) \\
& p_\tau(0) = p_0,
\end{array}
\tag{OPT}
$$

where the constraints hold for all $t \in \{0, h, \ldots, t_{\text{end}}\}$.

3 Time Discretization

To differentiate solutions of the ODE (2) and discretization (4), denote by $p_\tau(\cdot; t_0, p_0)$ solutions of (2) with initial tank pressure p_0, and $\widehat{p}_\tau(\cdot; t_0, p_0)$ by solutions of (4) with the same initial pressures. The local error describes the error due to a single time step:

$$E^L(h) := \sup \left\{ \left\| p_T(h:0, p_0) - \widehat{p}_T(h; 0, p_0) \right\|_\infty | p_0 \in \Omega_T(S, d) \right\} . \tag{5}$$

The global error gives the cumulative error at time $kh \in [0, t_{\text{end}}]$, $k > 1$:

$$E(h, t_0, t_{\text{end}}) := \sup_{p_0 \in \Omega_T(S,d)} \max_{k \in \mathbb{N}} \left\{ \left\| p_T(kh; 0, p_0) - \widehat{p}_T(kh; 0, p_0) \right\|_\infty \right\} . \tag{6}$$

If this global error is too large, a modeled pump schedule may be unsuitable in the physical system. For our test network, we found an example where a pump schedule is feasible for a larger (30-min) time step, but results in tank overflow for smaller, more accurate time steps (Fig. 1).

We now find bounds on the local and global discretization errors. Our bound on the local error $E^L(h)$ depends on the maximum value of $f(\cdot)$ and the Lipschitz constant of $f(\cdot)$ with respect to the tank pressures. An upper bound on the Lipschitz constant, $\mathcal{L}$, of $f(\cdot, S, d)$, with respect to p_T, bounds the relative change in $f(\cdot)$. We use the ∞-norm here to get the maximum error in pressure for all tanks. The Lipschitz constant $\mathcal{L}$ is defined as

$$\mathcal{L}(S, d) := \sup \left\{ \frac{\| f(p_{T1}, S, d) - f(p_{T2}, S, d) \|_\infty}{\| p_{T1} - p_{T2} \|_\infty} \Big| p_{T1}, p_{T2} \in \Omega_T(S, d) \right\} .$$

A bound on $f(\cdot)$ is found by norm maximization over the feasible set:

$$c^\star := \sup_{(p,q) \in \Omega(S,d)} \left\{ \| f(p_T(t), S(t), d(t)) \|_\infty \right\} = \left\{ \begin{array}{l} \text{maximize } \| \Lambda^{-1} A q \|_\infty \\ \hphantom{\text{maximize }}_{p,q,S,d} \\ \text{subject to } (p, q) \in \Omega(S, d) \end{array} \right\} ,$$

where $c^\star$ can be used to bound the difference in tank pressures in time h,

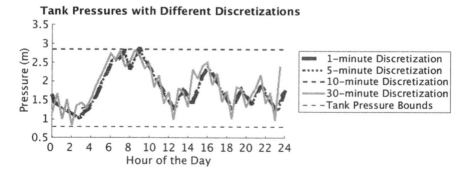

Fig. 1 Tank pressures using different time discretizations with the same schedule

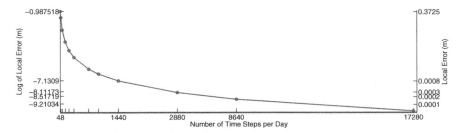

Fig. 2 Comparison of errors using different time steps h

$$\|p_T(t+h) - p_T(t)\|_\infty \leq \left\| \int_t^{t+h} f(p_T(s), S(s), d(s))ds \right\|_\infty \leq h \cdot c^\star .$$

Lemma 1 *Let S, d be fixed on $[0, t_{\text{end}}]$ and $\mathcal{L} = \mathcal{L}(S, d)$. Then the local discretization error $E^L(h)$ at any tank is bounded above by $(h^2 \cdot \mathcal{L} \cdot c^\star)$.*

Proof

$$E^L(h) = \|(p_T(h) - p_0) - (\widehat{p}_T(h) - p_0)\|_\infty \leq \int_0^h \|f(p_T(s)) - f(p_T(0))\|_\infty ds \leq h^2 \mathcal{L} c^\star .$$

☐

Figure 2 shows the bound on $E^L(h)$ for different values h on a test network.

Lemma 1 gives an upper bound on the local error, which is only relevant at $t = 0$; the differential (4) and algebraic equations (1) lead to propagating errors over time. The global error analysis follows established principles [8] that are used to find the bound in Theorem 1. We omit the proof for brevity.

Theorem 1 *Let S, d be fixed on $[0, t_{\text{end}}]$, $\mathcal{L} = \mathcal{L}(S, d)$, and $0 = h_0 < h_1 < \ldots < t_{\text{end}} = t_{\text{end}}$. The global discretization error $E(h)$ is bounded above by*

$$E(h, t_0, h_k) \leq hc^\star \left(e^{\mathcal{L}h_k} - 1\right) .$$

Comment: This is a theoretical upper bound that is much larger than the practical bound. A more realistic bound can be found by simulating the water network with different hydraulic time steps; the error can be measured by the difference in tank pressures from those of the finest possible discretization.

4 Robustness to Discretization Error

A discretization error may lead to errors in the objective, which can result in a suboptimal solution implemented in the physical system. We present a robust counterpart to (OPT) that takes into account the effect of the worst-case modeling error on the objective:

$$
\begin{aligned}
&\underset{\widehat{p},\widehat{q},S}{\text{minimize}} \; \underset{p,q}{\text{maximize}} \quad \phi(p, q, S, d) \\
&\text{subject to } (p, q) \in \Omega(S, d),\ (\widehat{p}, \widehat{q}) \in \Omega(S, d) \\
&\qquad\qquad \widehat{p}_T(t + h) = \widehat{p}_T(t) + h \cdot f(\widehat{p}_T(t), S(t), d(t)) \\
&\qquad\qquad \widehat{p}(0) = p(0) = p_0 \\
&\qquad\qquad p_T(t) \in [\widehat{p}_T(t) - E(h, t_0, t),\ \widehat{p}_T(t) + E(h, t_0, t)].
\end{aligned}
\tag{ROB}
$$

The robust counterpart of a convex optimization problem is intractable in general, but is tractable when the optimality of the objective can be easily checked [1, 9]. As shown below, this will be true for a number of practical cases in water network optimization. We consider two simple but common cases that result in a tractable robust counterpart (ROB):

Case 1: $\phi(p, q, S, d) = \psi(p)$, $\quad \psi$ is convex and increasing in p $\qquad$ (C1)

Case 2: $\phi(p, q, S, d) = \psi(q)$, $\quad \psi$ is convex and increasing in q. $\qquad$ (C2)

Many water network optimization problems will have such objectives, such as pump scheduling [10] or valve settings problems [11].

Lemma 2 *If either* (C1) *or* (C2) *is true, then solving* (ROB) *is equivalent to solving* (OPT) *with a different objective; the robust counterpart evaluates* $\psi(\cdot)$ *at the pressure or flow added to the respective error bound.*

Proof Assume (C2) is true. Define the upper bound on the flow error at time t with the optimization problem:

$$
\begin{aligned}
E_Q(t) := &\underset{p,q,\widehat{p},\widehat{q},S}{\text{maximize}} \; \|\widehat{q}(t) - q(t)\|_\infty \\
&\text{subject to } (p, q), (\widehat{p}, \widehat{q}) \in \Omega(S, d) \\
&\qquad\qquad \|\widehat{p}_T(t) - p_T(t)\| \le E(h, t_0, t).
\end{aligned}
\tag{7}
$$

Because the value of $E_Q(t)$ increases with the value of $E(h, t_0, t)$, and $\psi(\cdot)$ is convex and increasing, (ROB) is equivalent to

$$
\begin{aligned}
&\underset{p,q,S}{\text{minimize}} \quad \psi(q + E_Q) \\
&\text{subject to } p_T(t + h) = p_T(t) + h \cdot f(p_T(t), S(t), d(t)) \\
&\qquad\qquad (p(t), q(t)) \in \Omega(S(t), d(t)), \qquad \forall t \in \{0, h, \dots, t_{\text{end}}\} \\
&\qquad\qquad p(0) = p_0.
\end{aligned}
\tag{8}
$$

Similarly, if (C1) is true and the maximum flow error at time t is $E_D(t)$ (with a definition analogous to (7)), replacing the objective in (8) with $\psi((p_D + E_D, p_T + E(h), p_R))$ gives similar results. $\qquad\square$

Fig. 3 Comparison of worst-case cost of (ROB) versus (OPT) pump schedules

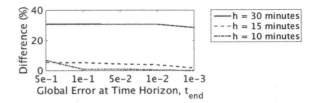

We tested (ROB) for a pump scheduling optimization problem, where the cost-minimization objective is convex and increasing in q. In our test network (of 1 reservoir, 2 pumps, 2 tanks and 1 demand node) the pumps flow from a reservoir directly into tanks, where the pump flow increases as tank pressure decreases. Thus, the worst-case flow error accounted for in (ROB) occurs when $p_T(t) = \left(\widehat{p}_T(t) - E(h, t_0, t)\right)$, for all $t \in [0, t_{\text{end}}]$.

We compared the energy cost in the scenario of the worst-case error for pump schedules optimal for (ROB) and (OPT). We tested 3 different time steps with 5 different possible global error values that increase linearly with time; results are shown in Fig. 3. The average difference in worst-case energy cost comparing optimal schedules of (ROB) and (OPT) was 9.25 %, which shows the potential benefit of taking into account discretization-based modeling error in the optimization formulation.

5   Conclusion

The time step chosen in water network optimization problems is critically important; a large time step can result in a simulation that is inaccurate or physically unsuitable. We provide an upper bound on the time discretization error caused by using forward Euler to update tank pressures and a formulation to find the optimal solution that is robust to the worst-case error.

References

1. Bertsimas, D., Brown, D., Caramanis, C.: Theory and applications of robust optimization. SIAM Rev. **53**(3), 464–501 (2011)
2. Ben-Tal, A., Nemirovski, A.: Robust optimization methodology and applications. Math. Program. **92**(3), 453–480 (2002)
3. Marques, J., Cunha, M.C., Sousa, J., Savić, D.: Robust optimization methodologies for water supply systems design. Drink. Water Eng. Sci. **5**, 31–37 (2012)
4. Goryashko, A., Nemirovski, A.: Robust energy cost optimization of water distribution system with uncertain demand (2011)
5. Collins, M., Cooper, L., Helgason, R., Kennington, J., LeBlanc, L.: Solving the pipe network analysis problem using optimization techniques. Manage. Sci. **24**(7), 747–760 (1978)
6. Berghout, B.L., Kuczera, G.: Network linear programming as pipe network hydraulic analysis tool. J. Hydraul. Eng. **123**(6), 549–559 (1997)

7. United States Environmental Protection Agency. Epanet: Software that models the hydraulic and water quality behavior of water distribution piping systems. http://www.epa.gov/nrmrl/wswrd/dw/epanet.html. Accessed 25 Oct 2013
8. Griffiths, D.F., Higham. D.J.: Numerical Methods for Ordinary Differential Equations. Springer (2010)
9. Ben-Tal, A., El Ghaoui, L., Nemirovski, A.: Robust Optimization. Princeton University Press, Princeton (2009)
10. Burgschweiger, J., Gnádig, B., Steinbach, M.C.: Nonlinear programming techniques for operative planning in large drinking water networks. The Open Appl. Math. J. **3**, 14–28 (2009)
11. Giacomello, C., Kapelan, Z., Nicolini, M.: Fast hybrid optimization method for effective pump scheduling. J. Water Resour. Plan. Manage. **139**(2), 175–183 (2013)

Forecasting Intermittent Demand with Generalized State-Space Model

Kei Takahashi, Marina Fujita, Kishiko Maruyama, Toshiko Aizono
and Koji Ara

Abstract We propose a method for forecasting intermittent demand with generalized state-space model using time series data. Specifically, we employ mixture of zero and Poisson distributions. To show the superiority of our method to the Croston, Log Croston and DECOMP models, we conducted a comparison analysis using actual data for a grocery store. The results of this analysis show the superiority of our method to the other models in highly intermittent demand cases.

1 Introduction

Accurately forecasting intermittent demand is important for manufacturers, transport businesses, and retailers [3] because of the diversification of consumer preferences and the consequent small production lots of the highly diversified products. There are many models for forecasting intermittent demand. Croston's model [2] is one of the most popular and has many variant models, including log-Croston and modified Croston. However, Croston's model has an inconsistency in its assumptions as pointed out by Shenstone and Hyndman [9]. Further, Croston's model generally needs round-up approximation on the inter-arrival time to estimate the parameters from discrete time-series data.

We employ non-Gaussian nonlinear state-space models to forecast intermittent demand. Specifically, we employ a mixture of zero and Poisson distributions because the occurrence of an intermittent phenomenon generally implies low average demand. As in DECOMP [5, 6], time series are broken down into trend, seasonal, auto-regression, and external terms in our model. Therefore, we cannot obtain parameters via ordinal maximum likelihood estimators because the number of parameters exceeds the number of data items owing to non-stationary assumptions on the para-

K. Takahashi (✉)
The Institute of Statistical Mathematics, 10-3 Midori-cho, Tachikawa-shi,
Tokyo 190-8562, Japan
e-mail: k-taka@ism.ac.jp

M. Fujita · K. Maruyama · T. Aizono · K. Ara
Central Research Laboratory, Hitachi Ltd., Tokyo, Japan

© Springer International Publishing Switzerland 2016
M. Lübbecke et al. (eds.), *Operations Research Proceedings 2014*,
Operations Research Proceedings, DOI 10.1007/978-3-319-28697-6_82

meters. Therefore, we adopt the Bayesian framework, which is similar to DECOMP. We employ a particle filter [7] for our filtering method instead of the Kalman filter in DECOMP because of the non-Gaussianness of the system, and the observation noises and nonlinearity in these models. To show the superiority of our method to other typical intermittent demand forecasting methods, we conduct a comparison analysis using actual data for a grocery store.

2 Model

2.1 Mixture Distribution and Components

Let the observation of a time series for discrete product demand be y_n ($n = 1, 2, \ldots, N$). We assume that demand for a product at arbitrary time step n follows a mixture distribution, considering the non-negativity of product demand. We do not need to conduct any approximating operations as in Croston's model. This mixture distribution is composed of a discrete probability distribution with a value of 0 with weight w_n and a Poisson distribution that has parameter λ_n with weight $1 - w_n$:

$$y_n \sim w_n \cdot 0 + (1 - w_n)y'_n, \tag{1}$$

$$y'_n \sim \frac{e^{\lambda_n} \lambda_n^y}{y!}. \tag{2}$$

From the expectation property of the Poisson distribution, the expected value of the mixture distribution becomes $(1 - w_n)\lambda_n$.

Now assume that parameter λ_n has trend component t_n, seasonal component s_n, steady component d_n, and external component e_n:

$$\lambda_n = \exp(t_n + s_n + d_n + e_n). \tag{3}$$

Specifically, the fluctuations in each component are as follows:

$$\Delta^k w_n = v_{0,n}, \tag{4}$$

$$\Delta^l t_n = v_{1,n}, \tag{5}$$

$$\Delta_q^m s_n = v_{2,n}, \tag{6}$$

$$d_n = \sum_{i=1}^{I} a_i d_{n-i} + v_{3,n}, \tag{7}$$

$$e_n = \sum_{j=1}^{J} \left(\gamma_j c_{k,n} + v_{e,j,n} \right). \tag{8}$$

Here, Δ_q^m indicates the difference between cycle q and degree m in the trend term. k and l are the degrees of differences in the weight and the seasonal component, respectively. I is the auto-regression order and J is the number of external variables. a_i is the jth auto-regression coefficient. $v_{0,n}$, $v_{1,n}$, $v_{2,n}$, $v_{3,n}$ and $v_{e,j,n}$ are the noise terms for the components, and they follow Gaussian distributions:

$$v_{i,n} \sim N(0, \tau_i^2) \quad \forall \, i = 0, 1, \ldots, 3, \tag{9}$$

$$v_{e,j,n} \sim N(0, \tau_{e,j}^2), \tag{10}$$

where $N(0, \sigma^2)$ is a Gaussian distribution with mean 0 and variance σ^2. In the seasonal component, we can employ multiple components simultaneously. However, the introduction of plural components often leads to mistakes in practice. Therefore, we employ singular components in the seasonal component. The external component corresponds to variables and parameters such as price and promotion variables. In addition, we utilize the external component to consider the holiday effect via dummy variables.

2.2 State-Space Expression

It is meaningless to estimate the time-varying parameter λ_n via ordinary maximum likelihood estimators. In the simplest setting, the number of unknown variables λ_n equals the number of data items y_n. Furthermore, we cannot estimate the parameters in our settings via ordinary maximum likelihood estimators because the number of unknown variables t_n, s_n, d_n, and e_n exceeds the number of data items y_n.

We introduce the state-space expression to resolve the above formulation. Let the model be expressed as a state-space model. When $k = 2, l = 2, m = 1, q = 7$, $I = 2$, and $J = 1$, we can write the state vector as

$$x_n = [w_n, w_{n-1}, t_n, t_{n-1}, s_n, \ldots, s_{n-5}, d_n, d_{n-1}, e_n]^T. \tag{11}$$

Therefore, the system and observation models are described as

[System Model] $x_n = F_n(x_{n-1}) + G_n v_n,$ (12)

$$F_n = \begin{bmatrix} 2 & -1 & 0 & 0 & 0 & 0 & \ldots & 0 & 0 & 0 & 0 \\ 1 & 0 & 0 & 0 & 0 & 0 & \ldots & 0 & 0 & 0 & 0 \\ 0 & 0 & 2 & -1 & 0 & 0 & \ldots & 0 & 0 & 0 & 0 \\ 0 & 0 & 1 & 0 & 0 & 0 & \ldots & 0 & 0 & 0 & 0 \\ 0 & 0 & 0 & 0 & -1 & -1 & \ldots & -1 & 0 & 0 & 0 \\ 0 & 0 & 0 & 0 & 1 & 0 & \ldots & 0 & 0 & 0 & 0 \\ \vdots & \vdots & \vdots & \vdots & & \ddots & \ddots & \vdots & \vdots & \vdots & \vdots \\ 0 & 0 & 0 & 0 & 0 & & 1 & 0 & 0 & 0 & 0 \\ 0 & 0 & 0 & 0 & 0 & 0 & \ldots & 0 & a_1 & a_2 & 0 \\ 0 & 0 & 0 & 0 & 0 & 0 & \ldots & 0 & 1 & 0 & 0 \\ 0 & 0 & 0 & 0 & 0 & 0 & \ldots & 0 & 0 & 0 & 1 \end{bmatrix}, \, G_n = \begin{bmatrix} 1 & 0 & 0 & 0 & 0 \\ 0 & 0 & 0 & 0 & 0 \\ 0 & 1 & 0 & 0 & 0 \\ 0 & 0 & 0 & 0 & 0 \\ 0 & 0 & 1 & 0 & 0 \\ 0 & 0 & 0 & 0 & 0 \\ \vdots & \vdots & \vdots & \vdots & \vdots \\ 0 & 0 & 0 & 0 & 0 \\ 0 & 0 & 0 & 1 & 0 \\ 0 & 0 & 0 & 0 & 0 \\ 0 & 0 & 0 & 0 & 1 \end{bmatrix}, \tag{13}$$

[Observation Model] $y_n \sim$ Zero–inflated Poisson$(\cdot | x_n)$, (14) where $\boldsymbol{v}_n$ is an independent and identically distributed noise term vector corresponding to $v_{0,n}, v_{1,n}, v_{2,n}, v_{3,n}$, and $v_{e,j,n}$. The observation model is not linear, and therefore, we cannot employ a Kalman filter and have to go with a particle filter.

2.3 Parameter Estimation

In the above setting, the elements in $\boldsymbol{F}_n$ (with the exception of a_i) are given; however, we need to estimate the other (hyper) parameters, τ_i, $\tau_{e,j}$, and a_i. Let $R(y_n|\boldsymbol{x}_n)$ be the likelihood at arbitrary time n; then, the likelihood with all data $(y_1, \ldots, y_N)$ is given by

$$L = \prod_{n=1}^{N} R(y_n|\boldsymbol{x}_n). \tag{15}$$

We estimate the parameters by maximizing Eq. (15).

We employ a grid search algorithm to maximize Eq. (15), because of the existence of Monte Carlo errors in calculating the likelihood via particle filters, which varies in each trial. Therefore, we cannot employ gradient methods such as the Newton method. Within the particle filter, we use residual resampling [8] and sequential importance sampling [4] to update the particles.

3 Comparison Analysis

3.1 Analyzing Data, and Models for Comparison

To show the superiority of our method, we conduct a comparison analysis of our method and typical intermittent demand forecasting and other relevant methods, including Croston, log-Croston [10], and DECOMP. The estimation methods used in the Croston and log-Croston methods are those shown in Syntetos and Boylan [11]. The smoothing parameter in the Croston model is set as $\alpha = 0.5$. The data analyzed here comprise fifty days of daily retail data for four SKU-level products in a Japanese grocery store. Further details of the data are shown in Table 1. #1 and #4 have relative higher intermittent demand than #2 and #3. Owing to differences in the estimation schemes, we compare forecast accuracy among these models by root mean squares (RMS).

To shorten the calculation time, the number of particles in each time step is fixed as $10,000$ in this paper. The setting of the degrees and cycles in our model and DECOMP are $k = 2, l = 2, m = 1, q = 7, I = 2$, and $J = 1$ (the external variable is the daily

Table 1 Data details

Product		Data for estimation				Data for forecasting			
#	Category	Term	Tot. amt. of purchase	Daily mean	Var.	Term	Tot. amt. of purchase	Daily mean	Var.
1	Yogurt	1 Jan.–19 Feb.	36	0.72	4.16	20 Feb.–24 Feb.	0	0.00	0.00
2	Ice cream	1 Oct.–19 Nov.	110	2.20	3.68	20 Nov.–24 Nov.	8	1.60	1.04
3	Milk	1 Apr.–20 May	120	2.40	2.04	21 May–25 May	10	2.00	1.60
4	Pudding	11 Jul.–29 Aug.	19	0.38	0.48	30 Aug.–3 Sep.	1	0.20	0.16

Table 2 Main results of the comparison analysis

Product #	1-ahead forecast	Actual	Prediction				RMS			
			Croston	log-Croston	DECOMP	Our model	Croston	log-Croston	DECOMP	Our model
1	1	0	0.92	0.66	1.77	0.66	0.92	0.66	1.77	0.66
	2	0	0.92	0.66	2.01	0.41	0.92	0.66	2.01	0.41
	3	0	0.92	0.66	1.86	0.71	0.92	0.66	1.86	0.71
	4	0	0.92	0.66	3.75	0.39	0.92	0.66	3.75	0.39
	5	0	0.92	0.66	1.61	0.07	0.92	0.66	1.61	0.07
	Product total						4.62	3.29	11.00	2.24
2	1	2	1.96	1.76	2.17	1.49	0.04	0.24	0.17	0.51
	2	0	1.96	1.76	1.80	1.66	1.96	1.76	1.80	1.66
	3	1	1.96	1.76	0.97	1.40	0.96	0.76	0.03	0.40
	4	3	1.96	1.76	2.33	2.08	1.04	1.24	0.67	0.92
	5	2	1.96	1.76	2.25	2.15	0.04	0.24	0.25	0.15
	Product total						4.04	4.24	2.93	3.64
3	1	2	2.01	2.11	1.21	1.52	0.01	0.11	0.79	0.48
	2	2	2.01	2.11	1.68	2.59	0.01	0.11	0.32	0.59
	3	0	2.01	2.11	1.92	2.52	2.01	2.11	1.92	2.52
	4	2	2.01	2.11	2.28	3.46	0.01	0.11	0.28	1.46
	5	4	2.01	2.11	1.60	3.47	1.99	1.89	2.40	0.53
	Product total						4.04	4.33	5.71	5.57
4	1	0	0.49	1.33	0.06	0.26	0.49	1.33	0.06	0.26
	2	0	0.49	1.33	0.35	0.24	0.49	1.33	0.35	0.24
	3	0	0.49	1.33	0.03	0.24	0.49	1.33	0.03	0.24
	4	1	0.49	1.33	0.08	0.31	0.51	0.33	0.92	0.69
	5	0	0.49	1.33	−0.11	0.33	0.49	1.33	0.11	0.33
	Product total						2.46	5.66	1.47	1.77
Overall							15.15	17.51	21.10	13.22

price for an objective product, which is not used in DECOMP). In the grid search, each hyperparameter of the error term has five nodes ($v_i = 0.003125, 0.00625, \ldots, 0.05$) and each auto-regression coefficient has 10 nodes ($a_i = -1.0, -0.8, \ldots, 1.0$).

3.2 Results

Table 2 shows the results for each data set. The overall RMS for our model is less than those of the other three models. For #1, the RMS for our model is the lowest. Thus, our method is superior to the other three models. For #4, our method is superior to Croston and log-Croston, but inferior to DECOMP. However, DECOMP predicts negative demand that never happens (five-ahead forecast). Therefore, our method can be concluded to be superior to the other models in highly intermittent demand situations.

In contrast to the highly intermittent demand situation, we cannot show substantial superiority of our model to the other three models for #2 and #3. It is conceivable that the degree of non-Gaussianness in the data influences these differences. If the data have high Gaussianness, Croston and DECOMP are suitable. On the other hand, log-Croston and our model are suitable if the data have low demand (namely low Gaussianness).

4 Conclusions

This paper proposed a method to forecast intermittent demand with non-Gaussian nonlinear state-space models using a particle filter. To show the superiority of our method to other typical intermittent demand forecasting methods, we conducted a comparison analysis using actual data for a grocery store. The results of this comparison analysis show the superiority of our method to the Croston, log-Croston, and DECOMP models in highly intermittent demand cases. In the furture, we intend to shorten the calculation time, and the MCMC filter [1] is a promising method by which to overcome the problem.

Acknowledgments We acknowledge the support of JSPS Grant Number 23730415.

References

1. Andrieu, C., Doucet, A., Holenstein, R.: Particle Markov chain Monte Carlo methods. J. Roy. Stat. Soc. B. **72**(3), 269–342 (2010)
2. Croston, J.D.: Forecasting and stock control for intermittent demands. Oper. Res. Q. **23**(3), 289–303 (1972)

3. Doucet, A.: On sequential simulation-based methods for Bayesian filtering. Technical Report CUED/F-INFENG/TR. 310, Cambridge University Department of Engineering (1998)
4. Fildes, R., Nikolopoulos, K., Crone, S.F., Syntetos, A. A.: Forecasting and operational research: a review. J Oper. Res. Soc. **59**, 1150–1172 (2008)
5. Kitagawa, G., Gersch, W.: A smoothness priors-state space approach to the modeling of time series with trend and seasonality. J. Am. Stat. Assoc. **79**(386), 378–389 (1984)
6. Kitagawa, G.: Decomposition of a nonstationary time series—an introduction to the program DECOMP—. Proc. Inst. Stat. Math. **34**(2), 255–271 (1986). in Japanese
7. Kitagawa, G.: Monte Carlo filter and smoother for non-Gaussian nonlinear state space models. J. Comput. Graph. Stat. **5**(1), 1–25 (1996)
8. Liu, J.S., Chen, R.: Sequential Monte Carlo methods for dynamic systems. J. Am. Stat. Assoc. **93**(443), 1032–1044 (1998)
9. Shenstone, L., Hyndman, R.J.: Stochastic models underlying Croston's method for intermittent demand forecasting. J. Forecast. **24**, 389–402 (2005)
10. Syntetos, A.A., Boylan, J.E.: On the bias of intermittent demand estimates. Int. J. Prod. Econ. **71**, 457–466 (2001)
11. Syntetos, A.A., Boylan, J.E.: Intermittent demand: estimation and statistical properties. In: Altay, N., Litteral, L.A. (eds.) Service Parts Management, pp. 1–30. Springer, London (2010)

Optimal Renewal and Electrification Strategy for Commercial Car Fleets in Germany

Ricardo Tejada and Reinhard Madlener

Abstract In this paper we model the uncertainty inherent in oil, electricity, and battery prices, in order to find the optimal renewal strategy for transport fleets in Germany from a car fleet operator's perspective. We present a comprehensive statistical model of total operating costs for the usage of light duty vehicles in the transport industry. The model takes into consideration current and future power train technologies, such as internal combustion and electric engines. The framework allows for the calculation of sensitivities of the relevant explanatory variables (fuel price, interest rate, inflation rate, economic lifetime, subsidy/tax policies, and economic development). We also calculate and evaluate relevant diffusion scenarios for commercially used e-vehicles.

1 Introduction

In Germany, internal combustion engine vehicles (ICEV) directly cause over 14 % of the yearly CO_2 emissions [1]. The increasing usage of fossil fuels, such as oil and gas, is one of the major driving forces of climate change [2, 3]. Increasing renewable energy supply, combined with the proliferation of electric vehicles (EV), might help to alleviate some of these problems. However, network externalities of the incumbent technology and the low degree of internalization of its external costs present a great obstacle. The required investments in electric vehicle supply equipment (EVSE)

R. Tejada
RWTH Aachen University, Templergraben 55, 52056 Aachen, Germany
e-mail: ricardo.tejada@rwth-aachen.de

R. Madlener (✉)
School of Business and Economics / E.ON Energy Research Center,
Institute for Future Energy Consumer Needs and Behavior (FCN),
RWTH Aachen University, Mathieustrasse 10, 52074 Aachen, Germany
e-mail: RMadlener@eonerc.rwth-aachen.de

© Springer International Publishing Switzerland 2016   597
M. Lübbecke et al. (eds.), *Operations Research Proceedings 2014*,
Operations Research Proceedings, DOI 10.1007/978-3-319-28697-6_83

infrastructure for the usage of battery electric vehicles (BEV) can only be financed for a high number of BEV, and the mass adoption of BEV can only be triggered by large investments in the charging infrastructure. Still, some niche markets exist already today where BEV are full substitutes for ICEV. Especially interesting are commercial applications, where scheduled routes and planned working hours allow the unreserved usage of BEV.

Our study aims at modeling the optimal diffusion of electric light duty vehicles (LDV), depending on various endogenous factors (e.g. a company's driving profile and fleet size) and exogenous factors (e.g. energy and gasoline prices and government subsidies). In our study, we take a closer look into the main cost factors and the way they influence the viability of BEV usage in commercial applications. Further, we perform an extensive sensitivity analysis of the explanatory variables and forecast the yearly sales and fleet structure of a company using various scenarios.

2 Methodology

The main aim of this study is to determine the optimal strategy for the introduction of BEV in the commercial sector in Germany from a fleet operator's perspective. The economic comparison of ICEV and EV is challenging, since there are direct and indirect factors that influence the benefits and costs of each technology. Similar to [4], we concentrate on the measurable and quantifiable factors. This means that we ignore the effects of non-monetizable gains since these are company-specific. For the sake of simplicity, we assume that the benefits (B) from owning a vehicle are constant independently of the engine technology used. Furthermore, we assume that the supply side of the BEV market is able to provide fitting solutions for each requirement set, i.e. the matching level of requirements and performance (M) is also assumed constant. In case that these assumptions hold, the value comparison of both technologies is reduced to a comparison of the total cost of ownership (TCO) of the competing technologies.

The resulting TCO model is an integrated analytical model to determine the economic viability of the usage of EV in the commercial sector. The model concentrates solely on the monetizable factors that influence the vehicle's TCO. Neither the external costs nor the network externalities of ICEVs have been considered. This design decision was taken to enhance the robustness of the results, since there is currently no generally accepted guideline in place for the quantification of the above-mentioned externalities, especially considering the lack of policies for their internalization. The TCO model consists of five independent statistical models (oil price model, battery cost model, electricity price model, inflation rate model, and interest rate model), a deterministic model (TCO calculation module), two integrated databases (COST and CIA), and a managing handler algorithm (the Fleet Renewal Algorithm). The

Fig. 1 TCO model
architecture

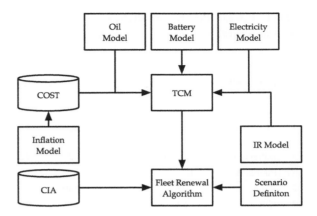

overall model structure is shown in Fig. 1; for a description of the individual model
components and their interrelatedness see [5].

3 The Data

A major challenge in performing this analysis arises from the limited information
available regarding technical characteristics and pricing of BEV. Although some electric LDV are already commercially available today, the current models do not match
all the possible requirements of the industry. In order to overcome this obstacle we
have defined artificial vehicles by using the available information of similar ICEV
and extrapolated known characteristics of already commercially available electric
LDV. Even though the presented model allows for the analysis of vehicles independently of their technical characteristics to obtain reasonable results, we have defined
three vehicle sizes in two versions. The first version of the vehicles has an electric
engine (EE), the latter an ICE. The complete set of characteristics relevant to our
model is listed in Table 1. The data used has been collected from various studies (e.g.
[6–8]), EEX, Datastream, and vehicle manufacturers (Renault, Iveco).

In order to find the most relevant factors for the diffusion of BEV we performed
an extensive analysis of the collected data. The analysis consists of three parts: (1) a
normalized sensitivity analysis of the TCO of ICEV and BEV and their difference;
(2) an analysis of the evolution of the TCO over time while parameterizing the results
for different variations of the describing variables; and (3) an analysis of the changes
in the TCO when varying the most relevant factors simultaneously (for a complete
list of tested variables and parameters see Table 2).

Table 1 BEV and ICEV base model characteristics

N^o	Characteristic	Unit	BEV			ICEV		
			Large	Medium	Small	Large[a]	Medium[b]	Small
1	Price[c]	[1000 €]	50	40	20	42	35	16
2	Fuel type[d]	[–]	E	E	E	D	D	D
3	Diesel consumption	[l/100 km]	–	–	–	13	9	6
4	Electricity consumption	[kWh/100 km]	30	22	15.5	–	–	–
5	Battery size	[kWh]	42	32	22	0	0	0
6	Fuel tank	[l]	–	–	–	60	70	60
7	Range	[km]	140	140	140	500	500	500
8	ICE performance	[kW(hp)]	–	–	–	93(126)	78(106)	66(90)
9	EE performance	[kW]	84	60	44	–	–	–
10	Trunk (min./max.)	[m^3]	15/17	7/10	3/3.5	15/17	7/10	3/3.5
11	Weight	[t]	3.7	2.4	1.48	3.5	2.2	1.35

[a] IVECO Daily 35C 11V; [b] IVECO Daily 29L 11V; [c] excl. battery; [d] E = electric, D = diesel

Table 2 Parameter definition

N^o	Variable	Name	Lower limit	Upper limit	Step size	Unit
1	L	Economic life	4	10	2	[a]
2	G_{sub}	Subsidies	0	7000	1000	[€]
3	C_{oil}	Oil price	50	+150	5	[%]
4	C_{ele}	Electricity price	50	+150	5	[%]
5	C_{bat}	Battery price	50	+150	10	[%]
6	c_{fuel}	Fuel consumption	80	+180	20	[%]
7	Bat	Battery size	4	10	1	[kWh]
8	km_Y	Annual mileage	10	40	5	[1000 km]
9	i_t	Discount rate	2	8	1	[%]

4 Results

4.1 Normalized Sensitivity and Scenario Analysis

The sensitivity analysis is based on a small-sized LDV, the Renault Kangoo, currently the only mass-produced commercially available vehicle with diesel and electric engines. Unless stated otherwise, the assumed values of the variables match the ones of the scenario described further below. First, we systematically vary one variable value *cet. par.* and test for the impact of this change on the TCO normalized

(a)   **(b)**

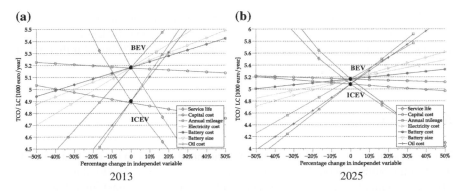

                            2013   2025

Fig. 2 Normalized sensitivity of the TCO of a small LDV in electric and diesel version. **a** 2013. **b** 2025

to the economic life of the vehicle. Figure 2a, b show the results for the years 2013 and 2025. As expected, the annual mileage and the duration of the service life of the vehicles are the most influential factors. Furthermore, the TCO of the BEV is more sensitive to the electricity price changes than to battery prices. For the analogous analysis for differences in the TCO, see [5].

Based on the results of the sensitivity analysis, we selected three variables to define plausible scenarios (oil, electricity, and battery prices) for predicting their joint impact on overall cost. *Scen. 1* represents the *base case*, in which current trends are projected into the future. The environmental awareness of the customers is reflected in lower average CO_2 emissions for new vehicles. Oil prices rise in accordance with the statistical average to about US$110 per barrel in 2020. No new policies for the reduction of CO_2 emissions are introduced, but already passed bills are enforced. No revolutionary innovation takes place in battery technology. The base case scenario is founded on the results of statistical analysis of the three key variables (omitting any cross correlations of the variables). It is the development expected when neither major policy change nor technical breakthrough take place. *Scen. 2* represents the *acceleration case*, in which oil prices rise to US$180 in 2020. The environmental problems are defining governmental policy-making; as a result, new and stricter regulations of CO_2 emissions are put in place. Governmental research funding has triggered a technological breakthrough, enabling a substantial battery cost reduction. The large-scale usage of BEV has allowed for the increased integration of volatile renewable energy sources, thereby reducing the electricity price. *Scen. 3* represents the *deceleration case*, in which environmental concerns are assumed to diminish in the transport industry. Sizable new oil fields are being discovered. The price of oil is only US$60 per barrel in 2020. CO_2 emission reduction efforts come to a standstill due to low prices of CO_2 certificates. The low incentives to purchase BEV reduce the forecasted market size for batteries. We assume that the business requirements remain constant during the time of observation, and that the decisions on the vehicle acquisitions are solely based on vehicle replacement requirements. Furthermore, note that we assume that these decisions are taken annually and that the vehicles' age is uniformly distributed over the economic lifetime of the vehicles. The economic lifetime is assumed to be six years and the annual milage 30,000 km. Further details can be found in [5].

5 Conclusions

The results show that the oil prices are the dominating factor in the calculation of the TCO, contributing almost 40 % of the total costs of an ICEV. The total mileage, as a product of annual mileage and the service life, is the second-most important factor. Higher total milages allow the BEV to reduce the monetary gap incurred by its higher TCA through lower operating costs. However, the results of our analysis on future oil and electricity prices show that the electricity prices will increase at a higher rate than the oil prices, thereby reducing the attractiveness of BEV. While BEV are able to match the requirements of certain commercial applications in range, security, and reliability, the slow market penetration limits the leveraging of economies of scale. Despite of this, the predicted gap between the TCO of ICEV and BEV is relatively small (5–10 %, depending on the framework conditions). The battery accounts for almost 40 % of the BEV purchase costs (ca. 20 % of the total lifecycle vehicle cost). Battery price development remains the biggest unknown for the TCO for electric LDV. Further R&D investment could trigger the technological breakthrough necessary to render BEV economical. Investment in EVSE is not considered to be crucial for the diffusion of electric LDV, but would enable the wider usage of BEV, thus increasing the battery market size and lowering manufacturing costs due to economies of scale.

We can conclude that the economic introduction of small electric LDV cannot be achieved in the near future without incentives in the form of subsidies or new pollution regulation policies. The current tax advantages of BEV are totally insufficient for accelerating the diffusion of electric LDV. Furthermore, the results show that low capital costs positively influence the diffusion of BEV. Government loans at low interest rates could have the politically desired effect.

There remains scope for further research. For instance, the model presented can be expanded to take into consideration the actual driving profiles of the companies, in order to present company-specific results. Furthermore, by assessing the overall structure of the LDV market in Germany, the current model could be used to forecast the adoption rate of the BEV technology on the country level.

References

1. German Federal Government: German Federal Government's National Electromobility Development Plan. Berlin (2009)
2. UNEP: The Emissions Gap Report 2012. Nairobi, Kenia (2012a)
3. UNEP: GEO-5: Global Enviromental Outlook. Nairobi, Kenia (2012b)
4. Kleindorfer, P.R., Neboian, A., Roset, A., Spinler, S.: Fleet renewal with electric vehicles at La Poste. Interfaces **42**(5), 465–477 (2012)
5. Tejada, R., Madlener, R.: Optimal renewal and electrification strategy for commercial car fleets in Germany. FCN Working Paper No. 7/2014, Institute for Future Energy Consumer Needs and Behavior (FCN), RWTH Aachen University, June 2014
6. Delucchi, M.A., Lipman, T.E.: An analysis of the retail and lifecycle cost of batterypowered electric vehicles. Transp. Res. Part D: Transp. and Environ. **6**(6), 371–404 (2001). doi:10.1016/S1361-9209(00)00031-6
7. Eaves, S., Eaves, J.: A cost comparison of fuel-cell and battery electric vehicles. J. Power Sources **130**(130) 208–212 (2004). doi:10.1016/j.jpowsour.2003.12.016
8. Offer, G.J., Howey, D., Contestabile, M., Clague, R., Brandon, N.P.: Comparative analysis of battery electric, hydrogen fuel cell and hybrid vehicles in a future sustainable road transport system. Energy Policy **38**(1) 24–29 (2010). doi:10.1016/j.enpol.2009.08.040

A Time-Indexed Generalized Vehicle Routing Model for Military Aircraft Mission Planning

Jorne Van den Bergh, Nils-Hassan Quttineh, Torbjörn Larsson and Jeroen Beliën

Abstract We introduce a time-indexed mixed integer linear programming model for a military aircraft mission planning problem, where a fleet of cooperating aircraft should attack a number of ground targets so that the expected effect is maximized. The model is a rich vehicle routing problem and the direct application of a general solver is only practical for scenarios of very moderate sizes. Therefore, a Dantzig–Wolfe decomposition and column generation approach is considered. A column here represents a specific sequence of tasks for one aircraft, and to generate columns, a longest path problem with side constraints is solved. We compare the column generation approach with the time-indexed model with respect to upper bounding quality and conclude that the Dantzig–Wolfe decomposition yields a much stronger formulation of the problem.

1 Introduction

We study a military aircraft mission planning problem (MAMPP) which was introduced by Quttineh et al. [5]. In general, a military mission might involve various tasks, such as surveillance, rescue assistance, or an attack. We only consider the situation where a set of ground targets needs to be attacked with a fleet of aircraft. The studied problem can be classified as a generalized vehicle routing problem with synchronization, in space and time between aircraft, and precedence relations.

Synchronization in vehicle routing problems (VRPs) might be exhibited with regard to spatial, temporal, and load aspects. A recent survey of VRPs with synchronization constraints is given by Drexl [2] and shows that this topic is emerging and challenging. Following the definitions in that paper, the synchronization in MAMPP can be classified as operation synchronization, in which one has to decide about time and location of some interaction between vehicles. A general framework for VRPs

J. Van den Bergh (✉) · J. Beliën
KU Leuven, Campus Brussels, Warmoesberg 26, 1000 Brussels, Belgium
e-mail: jorne.vandenbergh@kuleuven.be

N.-H. Quttineh · T. Larsson
Linköping University, 58183 Linköping, Sweden

© Springer International Publishing Switzerland 2016
M. Lübbecke et al. (eds.), *Operations Research Proceedings 2014*,
Operations Research Proceedings, DOI 10.1007/978-3-319-28697-6_84

605

with time windows and temporal dependencies, including exact synchronization, is given by Dohn et al. [1]. In the context of generalized vehicle routing, the time windows extension is considered by Moccia et al. [3]. Their work concerns an application to the design of home-to-work transportation plans. We believe to contribute to the literature by taking into account multiple non-standard characteristics of the generalized VRP, such as operation synchronization and precedence relations.

2 Problem Setting

The geographical area of interest for an aircraft mission is known as the target scene, and it includes the targets that need to be attacked and other objects such as enemy defense positions and protected objects, like hospitals and schools. The diameter of a target scene is typically of the order of 100 km, while the target distances are of the order of a few kilometers. Typically, a mission involves 6–8 targets and 4–6 aircraft. The timespan of a mission is of the order of a quarter of an hour.

The objective of a mission is to gain maximal expected effect against the targets within short timespan. The mission time is defined by the time the first aircraft passes an entry line of the target scene and the time the last aircraft passes an exit line. Each aircraft has an armament capacity, limiting the number of attacks it can perform. It can also be equipped with an illumination laser pod to guide weapons. Each target needs to be attacked exactly once, and requires one aircraft that illuminates the target with a laser beam and one aircraft that launches the weapon. Since an attack requires continuous illumination from the launch of the weapon until its impact, both aircraft need to team up. This rendez-vous shall take place both in time and space.

Fig. 1 The feasible attack space defined by inner and outer radii, and divided into six sectors, each with three attack and two illumination alternatives. A pair of compatible attack and illumination positions is marked, where the arrows indicate the flight directions

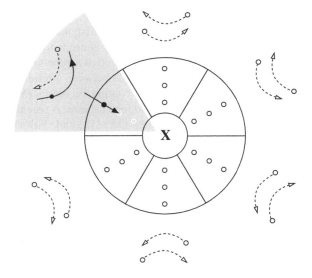

Figure 1 shows how a target is modeled. The feasible attack space is represented by inner and outer radii and divided into six sectors, which each holds at most three discretized attack positions and two compatible illumination positions. For any attack position, the expected effect on the target can be calculated. The expected effect depends mainly on the kind of weapon being used, which is decided in advance, and its kinetic energy. Depending on the wind direction and the proximity between targets, dust and debris might reduce the visibility of the illumination. Hence, precedence constraints are given, specifying which targets are not allowed to be attacked before other targets.

A detailed report on the problem setting is found in [5]. Below we present a network-based time-indexed mixed integer linear programming model for the MAMPP. It is derived from a continuous-time model in the same reference.

3 Mathematical Model

We divide the nomenclature into indices and sets, parameters, and decision variables, given in Tables 1, 2 and 3, respectively.

The goal is to maximize the expected effect against all the targets, while also minimizing the mission timespan. We choose to optimize a weighted combination of these conflicting objectives, which yields a solution that is Pareto optimal.

The time-indexed mathematical model for the MAMPP is given below.

$$\max \quad \sum_{r \in \mathbf{R}} \sum_{(i,j) \in \mathbf{A}} c_{ij}^r x_{ij}^r - \mu t_{end} \qquad (1)$$

Table 1 Indices and sets

$\mathbf{R}$	Fleet of aircraft, r
$\mathbf{M}$	Set of targets, m, to be attacked
$\mathbf{N}$	Set of nodes in the network, excluding the origin ($orig$) and destination ($dest$) nodes
$\mathbf{G}, \mathbf{G}_m$	Set of all sectors for all targets and for target m, respectively
$\mathbf{N}_m^A, \mathbf{N}_m^I$	Set of feasible attack (A) and illumination (I) nodes, respectively, for target m
$\mathbf{A}, \mathbf{A}_g, \mathbf{I}_g$	Set of arcs in the network (including from $orig$ and to $dest$) and sets of arcs (i, j) such that node j is an attack (A) node or illumination (I) node in sector g, respectively
$\mathbf{P}$	Set of ordered pairs (m, n) of targets such that target m cannot be attacked before target n
$\mathbf{S}$	Set of time periods within a discretized planning horizon, each of step length Δt

Table 2 Parameters

c_{ij}^r	For arcs (i, j) with $i \in N_m^A$, that is, for arcs leaving attack nodes, the value of c_{ij}^r is the expected effect of the attack, and otherwise the value is zero
S_{ij}^r	The time needed for aircraft r to traverse arc (i, j), expressed in number of time periods; equals actual time to traverse the arc divided by Δt, rounded upwards
T_s	The ending time of period s, which equals $s \cdot \Delta t$, $s = 0, 1, \ldots, |S|$
Γ^r	Armament capacity of aircraft r
q_m	Weapon capacity needed towards target m
μ	Positive parameter, used to weigh mission timespan against expected effect on targets

Table 3 Decision variables

x_{ij}^r	Routing variable, equals 1 if aircraft r traverses arc (i, j), and 0 otherwise
y_{is}^r	Equals 1 if node i is visited by aircraft r in time period s, and 0 otherwise
t_{end}	Time the last aircraft passes the exit line

subject to

$$\sum_{(orig,j) \in A} x_{orig,j}^r = 1, \qquad r \in R \qquad (2)$$

$$\sum_{(i,dest) \in A} x_{i,dest}^r = 1, \qquad r \in R \qquad (3)$$

$$\sum_{(i,k) \in A} x_{ik}^r = \sum_{(k,j) \in A} x_{kj}^r, \qquad k \in N, r \in R \qquad (4)$$

$$\sum_{r \in R} \sum_{g \in G_m} \sum_{(i,j) \in A_g} x_{ij}^r = 1, \qquad m \in M \qquad (5)$$

$$\sum_{r \in R} \sum_{g \in G_m} \sum_{(i,j) \in I_g} x_{ij}^r = 1, \qquad m \in M \qquad (6)$$

$$\sum_{r \in R} \sum_{(i,j) \in A_g} x_{ij}^r = \sum_{r \in R} \sum_{(i,j) \in I_g} x_{ij}^r, \qquad g \in G \qquad (7)$$

$$\sum_{m \in M} \sum_{g \in G_m} \sum_{(i,j) \in A_g} q_m x_{ij}^r \leq \Gamma^r, \qquad r \in R \qquad (8)$$

$$y_{orig,0}^r = 1, \qquad r \in R \qquad (9)$$

$$\sum_{t=s+S_{ij}^r}^{|S|} y_{jt}^r \geq x_{ij}^r + y_{is}^r - 1, \qquad (i, j) \in A, \ s \in \{0\} \cup S, \ r \in R \qquad (10)$$

$$\sum_{s \in S} y_{ks}^r = \sum_{(k,j) \in A} x_{kj}^r, \qquad k \in N, r \in R \qquad (11)$$

$$\sum_{r \in R} \sum_{i \in N_m^A} y_{is}^r = \sum_{r \in R} \sum_{i \in N_m^I} y_{is}^r, \qquad m \in M, \ s \in S \qquad (12)$$

$$\sum_{r \in \mathbf{R}} \sum_{t=s}^{|S|} \sum_{i \in \mathbf{N}_m^A} y_{it}^r \geq \sum_{r \in \mathbf{R}} \sum_{i \in \mathbf{N}_n^A} y_{is}^r, \qquad\qquad (m,n) \in \mathbf{P}, \; s \in \mathbf{S} \quad (13)$$

$$\sum_{s \in \{0\} \cup \mathbf{S}} T_s y_{dest,s}^r \leq t_{end}, \qquad\qquad\qquad\qquad\qquad r \in \mathbf{R} \quad (14)$$

$$x_{ij}^r \in \{0,1\}, \qquad\qquad\qquad\qquad (i,j) \in \mathbf{A}, \; r \in \mathbf{R} \quad (15)$$

$$y_{is}^r \in \{0,1\} \qquad\qquad i \in \mathbf{N} \cup \{orig, dest\}, \; s \in \{0\} \cup \mathbf{S}, \; r \in \mathbf{R} \quad (16)$$

Constraints (2)–(3) force each aircraft to leave and enter the target scene via the origin and destination nodes, respectively, and constraint (4) is flow conservation. The requirement that each target shall be attacked and illuminated exactly once is implied by constraints (5) and (6), respectively, while constraint (7) synchronizes these tasks to the same sector. Constraint (8) is armament limitation. Constraint (9) states that each aircraft is leaving the origin at time zero. Constraint (10) ensures that if aircraft r is visiting node j directly after node i, then the time of visiting node j cannot be earlier than the time of visiting node i plus the time needed to traverse arc (i, j). Constraint (11) enforces that if node i is not visited by an aircraft, no outgoing arc (i, j) from that node can be traversed by the aircraft. Constraint (12) states that the attack and the illumination of a target need to be synchronized in time. Constraint (13) imposes the precedence restrictions on the attack times of pairs of targets. Constraint (14) defines the total mission time, since all aircraft end up at the destination node. Finally, (15)–(16) are definitional constraints.

4 Dantzig–Wolfe Decomposition

As indicated by Quttineh et al. [5], solving the continuous time version of MAMPP to optimality takes a general solver several hours already for problem instances of moderate sizes. To meet the expectations of this application in a real-life decision support setting, with limited planning time available, more efficient algorithms are needed. Especially the upper bounding quality of the continuous-time MAMPP is very poor; we therefore propose a Dantzig–Wolfe decomposition and column generation approach for the time-indexed model given above, as a means for constructing a stronger formulation of MAMPP and thereby find stronger upper bounds.

Our decomposition approach follows the well known route generation paradigm for the solution of vehicle routing problems. Constraints associated with the targets and the mission timespan, that is, (5)–(7) and (12)–(14), are coupling and thus included in the master problem. The remaining constraints hold for individual aircraft and are included in $|\mathbf{R}|$ column generation subproblems. Each column represents a route for an aircraft, and is defined by a target sequence, timing, the task to be performed (attack or illumination) at each target, and from which position.

The restricted master problem optimally combines the available routes, while the pricing problem, which amounts to a side constrained longest path problem, finds profitable new routes for individual aircraft. Upper bounds on the optimal value of the master problem are calculated in a standard way. To improve the practical performance of the column generation scheme, a stabilization, see e.g. [4], is used.

5 Results and Conclusion

We have made a preliminary assessment of the time-indexed model of MAMPP and the column generation approach by using a few small problem instances that are identical to, or slight modifications of, instances used in [5].

Table 4 shows problem characteristics and results obtained with the continuous-time model of MAMPP in [5] and the above time-indexed model. Even for rather large time steps, the optimal solutions found by the continuous-time and time-indexed models are very similar, with respect to attack sequences and to attack and illumination nodes. Although not reported in the table, we also observe that the solution times of the continuous-time and time-indexed models are similar for large time steps while the latter is much more demanding when the steps are small. Further, the upper bounds given by the linear programming relaxations of the continuous time and time-indexed version of MAMPP are very similar, independent of the sizes of the time steps, and very weak. Table 5 shows a comparison between the time-indexed model and the column generation approach. Clearly, the column generation approach provides vastly superior upper bounds.

Our main conclusion is that the Dantzig–Wolfe decomposition gives rise to a very strong formulation of the MAMPP. The solution times of our first implementation of the column generation approach are not competitive compared to direct methods. There are however many opportunities for tailoring and streamlining the computations. A great advantage of the column generation approach to MAMPP in a real-life planning situation would be its creation of many possible routes for the aircraft. This is of practical interest since a real-life MAMPP can never be expected to include

Table 4 Problem characteristics and comparison of the continuous-time and time-indexed models

	Problem			Cont.		$\Delta t = 60$		$\Delta t = 45$		$\Delta t = 30$	
No.	$|M|$	Prec.	Γ	Eff.	t_{end}	Eff.	t_{end}	Eff.	t_{end}	Eff.	t_{end}
1	3	–	3	0.974	333	0.808	420	0.974	405	0.974	390
2	3	–	2	0.974	338	0.808	420	0.974	405	0.974	390
3	3	{1|23}	3	0.863	352	0.808	420	0.863	405	0.808	390
4	4	{1|2|3|4}	3	0.917	628	1.000	840	0.917	720	0.917	720
5	4	{1|2|3|4}	2	0.917	638	1.000	840	0.917	720	0.917	720

Here, $\mu = 0.005$ and all instances include two aircraft. The notation {1|23} means that target 1 is attacked before targets 2 and 3. The maximal possible total effect on targets is 1.000

Table 5 Comparison of the time-indexed model and column generation

No.	Time-indexed			CG: $\Delta t = 45$			CG: $\Delta t = 30$		
	LP	IP45	IP30	LP	IP	Iter.	LP	IP	Iter.
1	23.173	1.933	2.683	1.933	1.933	16	2.683	2.683	22
2	23.173	1.887	2.674	1.887	1.887	11	2.674	2.674	15
3	22.813	0.346	0.080	0.346	0.346	22	1.271	–	22
4	30.117	−7.677	−7.730	−6.532	–	37	−4.744	–	37
5	30.115	−7.730	−7.730	−7.083	–	29	−6.002	–	60

The LP optimal values of the time-indexed model vary very little with the step size; we give the value for $\Delta t = 60$. The columns IP45 and IP30 are the optimal values of the time-indexed model with different time steps. Further, IP are the objective values obtained when solving the integer version of the final master problem (if a feasible solution exists), and Iter. is the number of column generation iterations needed to reach optimality

all possible aspects of the mission to be planned, and because of the multi-objective nature of the problem. The access to multiple aircraft routes can then be exploited in an interactive decision support system.

References

1. Dohn, A., Rasmussen, M.S., Larsen, J.: The vehicle routing problem with time windows and temporal dependencies. Networks **58**, 273–289 (2011)
2. Drexl, M.: Synchronization in vehicle routing—A survey of VRPs with multiple synchronization constraints. Transp. Sci. **46**, 297–316 (2012)
3. Moccia, L., Cordeau, J.-F., Laporte, G.: An incremental tabu search heuristic for the generalized vehicle routing problem with time windows. J. Oper. Res. Soc. **63**, 232–244 (2012)
4. du Merle, O., Villeneuve, D., Desrosiers, J., Hansen, P.: Stabilized column generation. Discrete Math. **194**, 229–237 (1999)
5. Quttineh, N.-H., Larsson, T., Lundberg, K., Holmberg, K.: Military aircraft mission planning: a generalized vehicle routing model with synchronization and precedence. EURO J. Transp. Logistics **2**, 109–127 (2013)

Adapting Exact and Heuristic Procedures in Solving an NP-Hard Sequencing Problem

Andreas Wiehl

Abstract The paper on hand focuses on the development of computerized solutions for a particular class within the area of shunting yard optimization. Our aim is to determine a humping sequence to shunt freight cars from inbound trains to outbound trains. The objective is to minimize the total weighted tardiness. We present a simple mixed integer problem formulation, two heuristic approaches and an implementation in CPLEX for the study. In addition, we compare the CPU and objective value of the proposed algorithms with the results of CPLEX optimizer in a computational study.

1 Introduction

Shunting yards are used to separate freight cars on to several tracks and the associated new directions. They represent important nodes in freight rail networks.

Figure 1 shows a schematic layout of a hump yard. At the receiving tracks, there are inbound trains that contain freight cars going from one origin to one destination. In order to transport each freight car to its destination, the inbound trains are decoupled and disassembled into individual freight cars. Then an inbound train is humped as a whole over a hump track, whose inclination accelerates the freight cars by gravity. Via a system of tracks and switching points the freight cars are humped to given classification tracks, where they can be reassembled such that homogeneous outbound trains with freight cars in a shunting-dependent order can be generated. Finally, a train is pulled as a whole from the classification tracks to the departure tracks.

The paper on hand focuses on the development of computerized solutions for a particular class within the area of railway optimization. Our aim is to determine a humping sequence to shunt freight cars from inbound trains to outbound trains. The objective is to minimize the total weighted tardiness. We consider the sum of all priority values assigned to the outbound trains multiplied by the time units that have

A. Wiehl (✉)
Sustainable Operations and Logistics, University of Augsburg,
Universitaetsstr. 16, 86159 Augsburg, Germany
e-mail: andreas.wiehl@wiwi.uni-augsburg.de

© Springer International Publishing Switzerland 2016   613
M. Lübbecke et al. (eds.), *Operations Research Proceedings 2014*,
Operations Research Proceedings, DOI 10.1007/978-3-319-28697-6_85

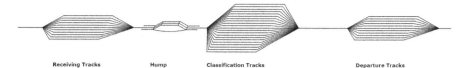

Receiving Tracks   Hump   Classification Tracks   Departure Tracks

Fig. 1 A schematic layout of a typical hump yard [3]

exceeded the given due date. It should be noted that this article is a brief review of the paper 'Minimizing delays in a shunting yard' by Jaehn et al. submitted in 2015 [7]. In the context of shunting yard operations, the objective of total weighted tardiness has, despite of its high practical relevance, only partially been investigated by Kraft [8–10]. In the last two decades, literature on operational problems at shunting yards has significantly grown. Just recently, a broad survey on shunting yard operations has been presented by Boysen et al. [3]. A situation in which the freight cars can be humped again, is analyzed by Bohlin et al. [1] and [2]. We call this multi stage shunting. The paper on hand focuses on a single stage shunting problem, for a detailed literature survey on other shunting problems, the reader may refer to the work of Gatto et al. [4], Hansmann and Zimmermann [5] and the seminal paper of Jacob et al. [6].

2 Modeling the Problem

In this section we provide a detailed description and the mathematical program of our single stage shunting problem considering weighted tardiness ($SSSWT$). Our problem restricts on forming single-destination trains with arbitrary freight car order. The freight cars of each outbound train $o \in \{1 \ldots O\}$ are served by one or many inbound trains t ($t = 1 \ldots T$). Trains can only leave the yard after an inbound train has been fully processed and all freight cars dedicated to this outbound train have been shunted. Hence, the possible departure times are represented by steps $s = 1 \ldots T$. The time of each step is determined by the time for shunting the according inbound train. Hence, outbound trains may leave the yard simultaneously leading to the same departure time. The processing time of one inbound train consists of the constant setup time su and processing times pt of each freight car assigned to train t. Each freight car has a fixed dedicated outbound train.

We have to find a humping sequence of all inbound trains $t = 1 \ldots T$ so that we can determine the departure time of the outbound trains. The sequence of freight cars within an in- or outbound train is not relevant. At the starting point of the shunting process all inbound trains are already located on the receiving tracks of the shunting yard, i.e. we do not consider release dates. The parameter g_t represents the number of freight cars of each inbound train t ($t = 1 \ldots T$). The outbound trains have to be formed on the classification tracks. We do not allow further rearranging of freight cars using the switches between classification tracks and departure tracks. There is a sufficient number of tracks, i.e. at least T receiving and O classification tracks

Table 1 Notation

T	Number of inbound trains (indices t, s and q)
O	Number of outbound trains (index o)
su	Setup time for each train in time units
pt	Processing time for one freight car in time units
$a_{t,o}$	Number of freight cars from inbound train t to outbound train o
g_t	Number of freight cars of inbound train t ($g_t := \sum_{o=1}^{O} a_{t,o}$)
f_o	Number of freight cars of outbound train o ($f_o := \sum_{t=1}^{T} a_{t,o}$)
w_o	Sum of all priority values assigned to outbound Train o
d_o	Due date of outbound train o in time units
$x_{s,t}$	Binary variable: 1, if inbound train t is shunted in step s; 0, otherwise
$y_{s,o}$	Binary variable: 1, if outbound train o has not left the yard until step s; 0, otherwise
$z_{t,o}$	Binary variable: 1, if inbound train t is processed before the departure of o; 0, otherwise

and no length restrictions. Hence, every in- and outbound train can be located on a separate track. The hump represents the only bottleneck of this optimization problem. Only one inbound train can be pushed over the hump at once and all freight cars of an inbound train are shunted before the next train is shunted. Additionally each freight car receives a non-negative weight, depending on priority aspects of the cargo. The weight w_o of an outbound train $o \in \{1 \ldots O\}$ then corresponds to the sum of the freight car weights. An inbound train can serve many outbound trains and an outbound train may receive freight cars from several inbound trains. The departure time of an outbound train is the point in time in which an outbound train is completely build up on a classification track and is ready to be moved to a departure track. For the departure time of every outbound train $o \in \{1 \ldots O\}$, a due date d_o is given. If the train departs later than its due date, the tardiness (measured in time units) is multiplied with the train's weight w_o so that we receive a value for the impact of this train's tardiness. Summing up these values for all outbound trains defines our objective function.

Using the notation summarized in Table 1, (*SSSWT*) consists of constraints (2) to (6) and objective function (1):

Minimize

$$\sum_{o=1}^{O} \left(w_o \cdot \max \left\{ 0, \sum_{t=1}^{T} \left(z_{t,o} \cdot (g_t \cdot pt + su) \right) - d_o \right\} \right) \tag{1}$$

subject to

$$\sum_{t=1}^{T} x_{s,t} = 1 \qquad \forall s = 1 \ldots T \tag{2}$$

$$\sum_{s=1}^{T} x_{s,t} = 1 \quad \forall t = 1 \dots T \tag{3}$$

$$\sum_{t=1}^{T}\sum_{q=1}^{s-1} \frac{x_{q,t} \cdot a_{t,o}}{f_o} \geq 1 - y_{s,o} \quad \forall o = 1 \dots O; \ \forall s = 1 \dots T \tag{4}$$

$$y_{s,o} + x_{s,t} - 1 \leq z_{t,o} \quad \forall t = 1 \dots T; \ \forall o = 1 \dots O; \ \forall s = 1 \dots T \tag{5}$$

$$x_{s,t}, y_{s,o}, z_{t,o} \in \{0, 1\} \quad \forall t = 1 \dots T; \ \forall o = 1 \dots O; \ \forall s = 1 \dots T \tag{6}$$

Objective function (1) minimizes the weighted tardiness of all outbound trains within an instance of (SSSWT). Therefore, we sum up the time units ($g_t \cdot pt + su$) of all inbound trains t that are processed before the departure of outbound train o and subtract it by the due date d_o. Since we only consider tardiness (and not lateness), this value must not be negative. Obviously, this equation can easily be linearized.

Equalities (2) and (3) ensure that an inbound train can only be processed once and only one train can be assigned to each step s. The assignment is defined by the binary variable $x_{s,t}$, which receives the value one, whenever train t is sorted in step s, otherwise 0. Constraints (4) force binary variables $y_{s,o}$ to receive the value one, whenever outbound train o is not fully sorted before step s and is still waiting on one of the classification tracks (0, if a train already left the yard by step s). This holds true whenever the number of sorted freight cars until step s, divided by the total number of freight cars assigned to train o, is smaller than one. Thus, no outbound train can depart before the first step is completed and thus, the departure time of every outbound train is greater zero. Constraints (5) ensure that binary variables $z_{t,o}$ receive value 1, whenever $y_{s,o}$ and $x_{s,t}$ both receive value 1 in a single step $s = 1 \dots T$. Hence, $z_{t,o} = 1$ indicates a train t that is processed before the departure of train o and thus affecting the departure time of o (0, otherwise). Finally, constraints (6) force $x_{s,t}$, $y_{s,o}$, and $z_{t,o}$ to be binary.

3 Algorithm

We generate an effective sorting for the decision variable $x_{s,t}$. After testing several different sorting procedures, the following SORT approach has the most promising results: The priority value consists of the sum of all penalty costs assigned to the outbound trains o that are served by a train t, if all outbound trains o would leave the yard at the same time step h. Additionally, we sum up the values of each time step h between 0 to the processing time for an instance. We sort the priority values non-increasingly. This sorting presents a far better solution than the simple First Come

First Serve (FCFS) approach. In the next step we try to improve the initial solution as obtained with SORT using a hill climbing algorithm combined with tabu search. Our neighborhood solution consists of all possible interchanges of two arbitrary trains within the sequence of inbound trains Υ. As a result, we obtain a set of feasible sorted solutions Φ with $|\Phi| = T \cdot (\frac{T-1}{2})$. After computing all sequences of Φ, we select the one with the lowest objective function score Υ', which is not part of the tabu list Ψ and add it to the list. Afterwards we start the next iteration with the new solution Υ' in step one. Once we reached 100 iterations with no improvement ($noimp = 100$) to the current best solutions score UB, the procedure ends and delivers UB as the result. Obviously, a solution with objective function score UB can easily be determined in the course of the algorithm. The tabu search algorithm offers a good solution in very short time. Usually the heuristic stops after 200 iterations on average. Further, we managed to improve the Tabu Search (TS) algorithm by letting it run with various initial solutions.

4   Solution

In this section we present a numerical study that investigates the efficiency of the presented algorithms. The parameters for the number of in- and outbound trains $\{10, 20, 30, 40\}$, the departure times $\{early, middle, late\}$ and the maximum number of different trains that an incoming train can deliver $\{5, 10, 15\}$, are linked in a fully factorial design. With 30 instances per combination, we obtain 1080 different instances for the study. The maximum number for the tracks (40) arises from the fact that European shunting yards usually never have more than 40 directional tracks. We have implemented ($SSSWT$) in ILOG using model (1)–(6) and used CPLEX $v12.3$ for solving the instances with the standard MIP-Gap. Unfortunately, the CPU rises sharply for only ten trains, with a maximum computation time of 1878 s for only one instance. Hence, we used CPLEX with a time limit of 30 s in our study. Table 2 shows the objective value and CPU in seconds for all procedures on the 1080 instances. The total objective function score of CPLEX is on average 49.57 % above the result of the TS approach, which is close to the least efficient 'first come first serve' method with 80.07 % (FCFS).

SORT delivers a significantly better score in a negligible time with an average gap of 28.84 % above the optimum. Interestingly, the tabu search (TS) delivers a very promising result in a relatively short amount of time. Further, the gap of SORT in regard to TS rises with the number of in- and outbound trains from 14.90 to 28.84 %.

Table 2 Comparison of the algorithms on 4 · 270 instances

		Number of in- and outbound trains				Sum	Mean	Gap[c]
		10	20	30	40			
FCFS	Obj.	14929910	55419180	128829480	221467892	420646462	389487.46	80.07
	CPU[a]	0.17	0.19	0.23	0.45	1.04	0.00	
CPLEX[b]	Obj.	10226544	55917690	105405938	177845428	349395600	323514.44	49.57
	CPU[a]	8035.76	8166.43	8078.12	8202.73	32483.04	30.08	
SORT	Obj.	11331556	40247527	92639939	156741360	300960382	278667.02	28.84
	CPU[a]	0.18	0.16	0.23	0.33	0.90	0.00	
TS	Obj.	9861899	32190418	71496633	120052123	233601073	216297.29	0.00
	CPU[a]	21.52	288.57	1545.61	5399.09	7254.79	6.72	

[a] CPU in seconds
[b] CPLEX with a time limit of 30 s
[c] Gap to the TS approach in %

References

1. Bohlin, M., Flier, H., Maue, J., Mihalák, M.: Track allocation in freight-train classification with mixed tracks. In: 11th Workshop on Algorithmic Approaches for Transportation Modelling, Optimization, and Systems, vol. 20, pp. 38–51. Dagstuhl, Germany (2011)
2. Bohlin, M., Dahms, F., Flier, H., Gestrelius, S.: Optimal freight train classification using column generation. In: 12th Workshop on Algorithmic Approaches for Transportation Modelling, Optimization, and Systems, vol. 25, pp. 10–22. Dagstuhl, Germany (2012)
3. Boysen, N., Fliedner, M., Jaehn, F., Pesch, E.: Shunting yard operations: theoretical aspects and applications. Eur. J. Oper. Res. **220**, 1–14 (2012)
4. Gatto, M., Maue, J., Mihalk, M., Widmayer, P.: Shunting for dummies: an introductory algorithmic survey, pp. 310–337. In: Robust and Online Large-Scale Optimization. Springer, Berlin (2009)
5. Hansmann, R.S., Zimmermann, U.T.: Optimal Sorting of rolling stock at hump yards. In: Mathematics—Key Technology for the Future, pp. 189–203 (2008)
6. Jacob, R., Marton, P., Maue, J., Nunkesser, M.: Multistage methods for freight train classification. Networks **57**, 87–105 (2011)
7. Jaehn, F., Rieder, J., Wiehl, A.: Minimizing Delays in a Shunting Yard. Submitted (2015)
8. Kraft, E.: A hump sequencing algorithm for real time management of train connection reliability. J. Transp. Res. Forum **39**, 95–115 (2000)
9. Kraft, E.: Priority-based classification for improving connection reliability in railroad yards: part i. Integration with car scheduling. J. Transp. Res. Forum **56**, 93–105 (2002)
10. Kraft, E.: Priority-based classification for improving connection reliability in railroad yards: part ii. Dynamic block to track assignment. J. Transp. Res. Forum **56**, 107–119 (2002)

Strategic Deterrence of Terrorist Attacks

Marcus Wiens, Sascha Meng and Frank Schultmann

Abstract Protection against terrorist threats has become an integral part of organisational and national security strategies. But research on adversarial risks is still dominated by approaches which focus too much on historical frequencies and which do not sufficiently account for the terrorists motives and the strategic component of the interaction. In this paper we model the classical risk analysis approach using a specific variant of adaptive play and compare it with a direct implementation approach. We find that the latter allows for a more purposeful use of security measures as defenders avoid to get caught in a "hare-tortoise-trap". We specify the conditions under which the direct implementation outperforms adaptive play in the sense that it lowers the cost of defence at a given rate of deterrence. We analyse the robustness of our results and discuss the implications and requirements for practical application.

1 Introduction

Terrorist attacks frequently happen. Since 9/11, organisations and countries ("defenders") have heavily adapted their security measures to it. Analysing terrorist risks ("adversarial risks") has been a largely unknown terrain for security authorities. Inevitably, well-known standard risk analysis approaches have initially been used as blueprints [5, 6]. The idea is to use historical data, interpreting risks as random outcomes based on historical distributions. However, different risk types call for different "logics of protection". While risks arising from natural events or accidents can be modelled using historical frequencies, adversarial risks have a strong strategic component which need to be considered. Protection measures certainly can be anticipated by terrorists ("offenders"). Thus, pure history-orientated approaches play into

M. Wiens (✉) · S. Meng · F. Schultmann
KIT, Hertzstraße 16, 76187 Karlsruhe, Germany
e-mail: marcus.wiens@kit.edu

S. Meng
e-mail: sascha.meng@kit.edu

F. Schultmann
e-mail: frank.schultmann@kit.edu

© Springer International Publishing Switzerland 2016
M. Lübbecke et al. (eds.), *Operations Research Proceedings 2014*,
Operations Research Proceedings, DOI 10.1007/978-3-319-28697-6_86

offenders hands. As a result, the systems vulnerability and the offenders intention are both partly endogenous.

The remainder of the paper is organised as follows. In Sect. 2 we present the model and the basic assumptions. In Sect. 3 we analyse the dynamic behaviour of the model, derive the stationary solution and compare the results with a direct implementation mechanism. In Sect. 4 we discuss the implications of our result and suggest some needs for future research.

2 The Basic Model

We set up a simple defender-offender game. The defender wants to protect his assets and the offender seeks to attack them. Keeping it simple, we restrict the model to three targets $j \in \{1, 2, 3\}$. Target $j = 1$ represents a human target where an attack predominantly implies personal injuries and fatalities; $j = 2$ is associated with high economic and immaterial loss; and $j = 3$ implies minor material damage which can be overcome within a short period of inconvenience. The game is a sequential move game, with the defender as the first-mover and the offender as the second-mover.

The defender has incomplete information about the offender's motivation. According to [7], we consider two offenders $i \in \{1, 2\}$. Offender T1 ($i = 1$) wants to change the political order or societal system and, thus, regards persons or institutions which fulfil systemic key functions or symbolizes the systems strength as enemies. He values the three targets according to his pay-offs (u_{ij}) as follows: $u_{12} > u_{11} > u_{13} > 0$. Offender T2 ($i = 2$) views society fundamentally different and is characterized by deep hatred with regard to peoples way of life and value systems. The latter wants to destroy the prevailing system, maximising damage and causing widespread fear. He values the three targets according to his pay-offs as follows: $u_{21} > u_{22} > u_{23} > 0$.

The defender faces T1 with probability θ and T2 with probability $(1 - \theta)$. His preferences (v_j) over the targets (if the attack succeeds) are ordered as follows: $0 > v_3 > v_2 > v_1$. If an attack is successful, the defender realises losses which are highest for fatalities, lower for high material damages and lowest for marginal damages. Table 1 summarises the respective pay-offs and gives an example.

The defender's objective is to protect all targets by allocating defence-units $d_j \in D$, with D being the maximum amount of defence-units available to him. The defence-units represent the costs of protection. That are the expenditures to install, maintain

Table 1 Target-dependent pay-offs of defender and terrorist in the case of a successful attack

Target	Defender	Anti-system T1	Fanatic T2
(1) High number of fatalities	$v_1 = -6$	$u_{11} = +3$	$u_{21} = +7$
(2) High material loss	$v_2 = -3$	$u_{12} = +5$	$u_{22} = +3$
(3) Low material loss	$v_3 = -1$	$u_{13} = +1$	$u_{31} = +1$

and operate security measures. The defence-units act as barriers (obstacles) for the offender, reducing the attractiveness of particular attack strategies. The offender as a second-mover observes the installed defence-units and accordingly decides. His net pay-off (u_{ij}^{net}) is a combination of his pay-off minus the defence-units: $u_{ij}^{net} = u_{ij} - d_j$. The offender only attacks if $u_{ij}^{net} > 0$. By using defence-units the defender can prevent an attack, but he cannot attenuate the impact of an attack. He seeks to minimize his expected loss according to Eq. (1). The outcomes are evaluated by VNM-utility functions.

$$\min_{d} \sum_{j=1}^{3} v_j \text{ subject to } 0 < \sum_{j=1}^{3} d_j \leq D \tag{1}$$

3 Traditional Risk Analysis as Adaptive Play

We aim at identifying weaknesses of risk analysis approaches for adversarial risks which are exclusively history-orientated. A way to model such an approach is to apply adaptive or fictitious play [3, 4]. These are essentially learning procedures where the players update their beliefs along with empirical frequency distributions of past strategies. This is the continuation of our previous work presented in [7].

In a first step we set up the dynamic equations for the offender and the defender, getting a system of recurrence equations. It is impossible to derive a particular solution in closed form. Thus, we focus on the stationary (steady-state) solution. This solution can be interpreted as long-term equilibrium where all learning rests. We show that a static allocation mechanism can outperform the adaptive play from the defenders point of view.

The offender's attack-strategy is represented by a_{ij}. For the sake of simplicity we allow just one attack per period of either T1 or T2. We define a_{ij} as a binary strategy variable; it takes the value $a_{ij} = 1$ if an attack occurs, and $a_{ij} = 0$ if no attack occurs. The offender will attack if and only if two conditions are fulfilled: $u_{ij}^{net} > 0$ and $u_{ij}^{net} > max$. Equation (2) states the dynamic attack-strategy for the subsequent period $t + 1$, given a distribution of defence-units $\mathbf{d}[t] = \{d_1[t], d_2[t], d_3[t]\}$ in t.

$$a_{ij}[t + 1] = \begin{cases} 1, & \text{for } u_{ij} - d_j = \max\{u_{i1} - d_1[t], u_{i2} - d_2[t], u_{i3} - d_3[t], \varepsilon\} \\ 0, & \text{else} \end{cases} \tag{2}$$

The auxiliary parameter $\varepsilon > 0$ represents the smallest possible value for $u_{ij} - d_j$, assuring a strictly positive utility for any attack. Let $\bar{a}_{ij}$ be the average rate of attack until period t, executed by the offender against a particular target. The defender then allocates defence-units with respect to the experienced damage according to equation (3).

$$d_j[t + 1] \cdot D = \frac{\bar{a}_{ij}[t] \cdot S_j}{\sum_{j=1}^{3} \bar{a}_{ij}[t] \cdot S_j} \tag{3}$$

The defenders reaction causes the adaptive play property of the model in the first place. Although the offender reacts to the defenders preceding decision, he is not aggregating historical data in a statistical sense. In contrast to "pure" fictitious play the defenders strategy is not based on the offenders strategy distribution alone, but also on the distribution of the expected damage (average attack rate and target-related damage). The dynamic interaction between offender and defender can be described by Eqs. (2) and (3). Figure 1 illustrates a simulation of this dynamic interaction for the parameter constellation of our example (see Table 1) and $\theta = 0$.

Average and expected damage follow an iterating damped oscillation with one curve permanently overtaking the other. In Fig. 1, regions of excess damage are shaded. These are the critical periods where the defender is caught off guard due to the sequential and adaptive procedure, corresponding to the well-known 'hare-tortoise-trap'. Too adaptive defence strategies run the risk of just 'reacting', forcing the defender into the inferior role.

The process involves adaptive learning with steadily declining expectation errors. In the long run the system approaches a stationary equilibrium where the proportion of defence-units for j converges to Eq. (4) and the attack rate to Eq. (5).

$$d_j^* \cdot D = \frac{\bar{a}_{ij} \cdot S_j}{\sum\limits_{k=1}^{3} \bar{a}_{ik} \cdot S_k} \tag{4}$$

$$\bar{a}_{ij}^* = \frac{\prod\limits_{\substack{k \neq j}}^{2} (d_k \cdot D - u_{ik})}{\left(\prod\limits_{\substack{k \neq j}}^{2} (d_k \cdot D^2 - u_{ik}) + d_j \cdot D^2 \sum\limits_{\substack{k \neq j}}^{2} d_k + u_{ij} \sum\limits_{\substack{k \neq j}}^{2} u_{ik} \right) - \left[\sum\limits_{j=1}^{3} d_j \cdot \sum\limits_{j=1}^{3} u_{ij} - \sum\limits_{j=1}^{3} u_{ij} \right] \cdot D} \tag{5}$$

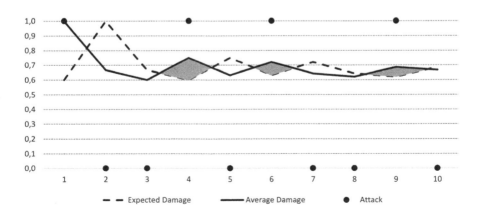

Fig. 1 Sequence of expected and realized damage (Adaptive Play)

Evaluating Eqs. (4) and (5), we get $d^* = \{d_1^* = 0.67, d_2^* = 0.27, d_3^* = 0.06\}$ and $a^* = \{a_1^* = 0.42, a_2^* = 0.35, a_3^* = 0.23\}$ as optimal long run strategies for the defender and the offender, respectively. In the long run the defender will assign two third of all defence-units to the most vulnerable target, but he cannot fully prevent attacks by this. In 42 % of all cases the offender chooses $j = 1$. With 35 % the offender goes for $j = 2$, mainly due to its increased vulnerability. The defender just assigns 27 % of his defence-units to $j = 2$. The same argument applies for $j = 3$ which is least attractive for the offender, but also unprotected and therefore an easy task.

In the example the defender assigns all 10 defence-units and the average damage amounts to 3.8 units (long run). But can the damage further be reduced by concentrating on the most vulnerable target? The defender could allocate the necessary defence-units to $j = 1$, just to inhibit all attacks once and for all. He should deter T2 because $j = 1$ is most vulnerable to him. The defender can achieve this by allocating $d_1 \cdot D = u_{21}$ defence-units to $j = 1$. In our example, he should spend 7 units to $j = 1$ to fully deter both types. The remaining defence-units should be assigned to $j = 2$. By doing this he can deter T2 again but not T1 in this case, because T1 has a higher utility in attacking $j = 2$. The expected damage $3\theta + (1 - \theta)$ is strictly less than 3.8 (average damage of the adaptive play scenario). It becomes apparent that a direct implementation mechanism which starts at the offenders motivation structure can lead to a more purposeful allocation of defence resources. In a last step we derive a condition under which the direct implementation mechanism outperforms the adaptive play approach. In order to avoid the complex allocation problem for lower-range targets, we set the maximum amount of resources equal to the T2-utility for target 1 ($D = u_{21}$). This leads to a unique optimal implementation of defence-units, given by the vector $d^* = \{d_1^* = 1, d_2^* = 0, d_3^* = 0\}$ and the optimal (unique) attack rates $a^* = \{a_1^* = 0, a_2^* = 1, a_3^* = 0\}$ for T1 and T2. This leads to a certain damage of S_2 units. We now need to look for critical conditions which have to be fulfilled so that the implementation mechanism outperforms the adaptive play approach. This is formally expressed through inequality (6).

$$a_1^* S_1 + a_2^* S_2 + a_3^* S_3 > S_2 \text{ with } D = u_{21} \tag{6}$$

This inequality holds if both u_{21} and S_1 are sufficiently high, but also when u_{23} and S_3 are sufficiently low. Thus, if there is an extremely critical target in the sense that the defender is enormously hurt by an attack, but also if the offender puts a very high priority on it, then the direct implementation should be preferred. Under these conditions the adaptive play approach is too risky, because even a minimal frequency of attack against $j = 1$ leads to a sharp increase in average damage.

4 Discussion

The adaptive play model allows for a precise analysis of history-orientated risk analysis strategies vis–vis human adversaries which has been criticized by [1, 2] and others. It highlights the main weakness of this approach: The defender permanently reacts and lags behind which leads to recurrent episodes of excess damage. This can quickly lead to a severe loss of resources, but also to significant damages of reputation. We showed that this aspect is highly relevant if very critical targets are in the crosshair of fanatic terrorists. By contrast, the direct implementation mechanism takes some kind of "illusion of control" from the defending institution. It compels that there are inevitably natural limits to risk reduction due to limited resources. The most vulnerable parts are the systems limits, urging the defender to establish clear priorities. However, the direct implementation mechanism requires sufficient knowledge about possible motives, targets and restrictions of offenders. They should thoroughly be analysed to get a clearer picture of the offenders' incentive structures. In this sense, historical data are an important input. It is essential to investigate possible intentions and motives of offenders by carving out clear motivation structures of potential offender-types. We conclude that it is more important to concentrate on profiling activities with regard to *current* threats rather than to reproduce and analyse bygone menace.

References

1. Anthony Jr. Cox, L.: Game theory and risk analysis. Risk Anal. **29**(8), 1062–1068 (2009b)
2. Anthony Jr. Cox, L.: Improving risk-based decision making for terrorism applications. Risk Anal. **29**(3), 336–341 (2009a)
3. Brown, G.W.: Iterative solution of games by fictious play. In: Koopmans, T.C. (ed.) Activity analysis of production and allocation, pp. 374–376. Wiley, New York (1951)
4. Crawford, Vincent P.: Learning behavior and mixed-strategy Nash equilibria. J. Econ. Behav. Organ. **6**(1), 69–78 (1985)
5. Deisler, Paul F.: A perspective: risk analysis as a tool for reducing the risks of terrorism. Risk Anal. **22**(3), 405–413 (2002)
6. Dillon, Robin L., Liebe, Robert M., Bestafka, Thomas: Risk-based decision making for terrorism applications. Risk Anal. **29**(3), 321–335 (2009)
7. Meng, S., Wiens, M., Schultmann, F.: A game theoretic approach to assess adversarial risks. In: Brebbia, C.A. (ed.) Risk Anal. IX. WIT Transactions on Information and Communication Technologies, vol. 47, pp. 141–152. WIT Press, Southampton (2014)

Optimal Airline Networks, Flight Volumes, and the Number of Crafts for New Low-Cost Carrier in Japan

Ryosuke Yabe and Yudai Honma

Abstract Recently, numerous low-cost carriers (LCCs) have been established and have become popular as a new style of airline service. In Japan, Peach Aviation began business as an LCC in 2012. However, while it is true that some airline companies are suffering from a slump in business, Peach Aviation has succeeded because it set up a hub airport at Kansai International Airport and runs many cash-cow routes. To establish a new LCC, consideration of airline networks is most important for success. Therefore, in this study, we propose a mathematical model to optimize the airline network, flight volume, and number of aircrafts for maximizing a new LCC's profit supposing a hub-spoke style network, and solve the model as a mathematical programming problem. First, we investigate the case of a single-hub network, and subsequently consider a two-hub network. It was determined that, when both Narita and Kansai International Airports are chosen as hub airports, a new LCC's profit is maximized.

1 Introduction

Recently, many airlines that offer low fares by simplifying services have been established and are referred to as low-cost carriers (LCCs); this has resulted in a greater number of transportation choices for travelers. When a new LCC is founded, it is essential for the LCC to consider optimum routes. From this viewpoint, in this study, we consider an LCC that has newly entered the domestic airline network in Japan. We evaluate the optimal airline network with simultaneous consideration of

R. Yabe (✉)
Faculty of Science and Engineering, Waseda University, Okubo 3-4-1,
Shinjuku, Tokyo 169-8555, Japan
e-mail: ryabe@toki.waseda.jp

Y. Honma (✉)
Institute of Industrial Science, The University of Tokyo, Komaba 4-6-1,
Meguro, Tokyo 153-8505, Japan
e-mail: yudai@iis.u-tokyo.ac.jp

© Springer International Publishing Switzerland 2016
M. Lübbecke et al. (eds.), *Operations Research Proceedings 2014*,
Operations Research Proceedings, DOI 10.1007/978-3-319-28697-6_87

flight volume and number of aircrafts that will maximize the new LCC's profit. In particular, we consider a hub-spoke network used by many LCCs.

An earlier study by Sohn and Park [1, 2] is considered as a fundamental basis for the present study. Sohn discussed the means to formulate and obtain an exact solution for the problem of which link between a spoke airport and a hub airport should be connected when the location of the hub airport is given to minimize transportation cost. Sohn's model indicated that every spoke airport must be connected to a hub airport. In our model, however, it is possible, under optimum conditions, that some spoke airports are not connected to any hub airport. Furthermore, Sohn considered a comparatively simple objective function involving strictly the minimization of transportation cost; the model did not include the parameters of airline volume or number of aircrafts. Many differences exist therefore between Sohn's model and the model proposed in this study. In what follows, we propose an optimal airline network to maximize a new LCC's profit, and apply the model to Japan's domestic airlines.

2 Proposed Model

In this section, we assume a new LCC that gains access to Japan's domestic airline network, and simultaneously determine the optimal network, flight volume, and number of aircrafts that maximize the new LCC's profit.

First, we suppose a total of I hub airports indexed by $i(i = \{1, 2, \cdots, I\})$ and a total of $J - I$ spoke airports indexed by $j(j = \{I + 1, I + 2, \cdots, J\})$. In the proposed network model, the hub airport is given and each spoke airport can be arbitrarily connected to other hub airports resulting in a hub-spoke network (which suggests that we assume a multiple allocation model). Within the proposed network, hub-hub connections are possible but spoke-spoke connections are not allowed. As such, in this network model there are

$$I \times (J - I) + \binom{I}{2} \tag{1}$$

possible flight links.

In this study, we propose flight volumes R_{ij} from a given hub airport to each hub-spoke airport and number of aircrafts A as decision variables. Additionally, we suppose a hub-spoke network, so that $R_{ij} = 1$ indicates a single service trip both outward and homeward between airports i and j. Furthermore, in this study, we suppose that the same aircraft can be used in any flight.

In this study, we consider both income and cost to maximize the profit of the new LCC. First, fare income and incidental business income are considered for the total income. Incidental business income is earned by charging food/drink or seat assignments. This income, given by the parameter β (%), is attached to the fare income. When the fare for link i is Y_i (Yen) and the number of embarkations between airports i and j is m_{ij}, the total income REV is formulated as given by Eq. (2).

$$REV = (1 + \beta) \sum_{i=1}^{I} \sum_{j=i+1}^{J} \left\{ Y_{ij} \left(m_{ij} + m_{ji} \right) \right\} \tag{2}$$

By calculating the summation in (2), every link in (1) is considered.

Next, the costs of airport use (C_i^H and C_j^S, Yen/year), sales management (K, Yen/year), aircraft (M, Yen/year), fuel and employment (f, Yen/flight·minute), and traveling, maintenance, and shipping (L, Yen/flight) are considered. Here, the superscripts H and S respectively indicate hub and spoke. The parameter L also includes the cost to employees for airport use.

When t_{ij} is the traveling time between airports i and j, and Q is 365 (days), the total expenditures $COST$ is formulated as given in Eq. (3).

$$COST = \sum_{i=1}^{I} C_i^H + \sum_{j=I+1}^{J} b_j C_j^S + \sum_{i=1}^{I} \sum_{j=i+1}^{J} e_{ij} K + AM + Q \sum_{i=1}^{I} \sum_{j=i+1}^{J} R_{ij} \left\{ f \left(t_{ij} + t_{ji} \right) + 2L \right\} \tag{3}$$

Furthermore, e_{ij}, in the above equation, indicates whether there is more than one flight between airports i and j, where $e_{ij} = 0$ for no flight and $e_{ij} = 1$ for more than one flight. Additionally, b_j indicates whether a spoke airport is placed at airport j, where $b_j = 0$ when airport j is not used and $b_j = 1$ if airport j is placed as a spoke airport.

Under the above assumptions, the proposed formulation for maximizing the new LCC profit is given by Eqs. (4)–(16).

$$\max. \; REV - COST \tag{4}$$

$$\sum_{i=1}^{I} \sum_{j=i+1}^{J} \left\{ R_{ij} \left(t_{ij} + t_{ji} + 2\theta \right) \right\} \leq TA \tag{5}$$

$$m_{ij} \leq n_{ij} \; \forall i, j \tag{6}$$

$$m_{ij} \leq QGR_{ij} \; \forall i, j \tag{7}$$

$$m_{ji} \leq n_{ji} \; \forall i, j \tag{8}$$

$$m_{ji} \leq QGR_{ji} \; \forall i, j \tag{9}$$

$$R_{ij} \leq 99e_{ij} \; \forall i, j \tag{10}$$

$$\sum_{i=1}^{I} e_{ij} \leq Ib_j \; \forall i, j \tag{11}$$

$$0 \leq m_{ij} \forall i, j \tag{12}$$

$$R_{ij} \in \{0, 1, 2, \cdots \} \tag{13}$$

$$A \in \{0, 1, 2, \cdots \} \tag{14}$$

$$e_{ij} \in \{0, 1\} \tag{15}$$

$$b_j \in \{0, 1\} \tag{16}$$

In the above equations, T indicates the maximum operation time (minutes.), θ indicates the shuttle time (min.), n_{ij} indicates the predicted number of people using traveling in an airplane between airports i airport and j airport, and G indicates the number of seats of one in an aircraft. Incidentally, the coefficients of determination of this mathematical programming are R_{ij}, A, e_{ij}, and b_j.

3 Method of Parameter Assignments

To calculate revenues and costs, we must assign a value to each parameter. In this section, we describe the parameter setting method employed.

To assign a value to n_{ij}, we use data from the "Domestic Passenger Record per each route and month" [3] and "Inter-Regional Travel Survey" [4] produced by the Ministry of Land, Infrastructure, Transport and Tourism.

The Domestic Passenger Record contains data regarding the number of people for each route among existing airlines and the total number of all airlines for each route in 2012. The Inter-Regional Travel Survey contains the number of people traveling by airplane between arbitrary prefectures in 2010. We use these data based on the rule above to set n_{ij}.

If the new LCC obtains a route wherein an existing LCC is already in service, we ascertain the number of people using that route and divide that number by the number of competing LCCs. In this case, we do not consider an increase in demand.

When only a legacy carrier is in service, we ascertain the number of legacy carrier customers on that route and consider an increase in demand owing to the new LCC service. We subsequently employ a logit model to calculate the rate of LCC use. We incorporated travel fee and travel time as utility functions. Finally, we multiply the number of legacy carrier customers including the assumed increase in demand by the obtained rate of LCC use.

Because Inter-Regional Travel Survey data indicate air traffic between arbitrary prefectures, we first allocate airports to each prefecture. We already calculated the rate of LCC use for each route by the logit model, and we calculate an average for each airport. Then, we multiply the number of travelers derived from the Inter-Regional Travel Survey by the average from the logit model for each airport and arrive at a value for n_{ij}.

If an existing LCC is in service between airport i and j, we use this value for Y_{ij}. We ascertain this value from each LCC's homepage and use an average over one month. If an existing LCC is not in service, we employ the relationship given by Eq. (17) below, where $distance$ is the air travel distance between airports.

$$Y_{ij}(calc.) = 872.96 \times distance^{-0.731}. \qquad (17)$$

We omit details about rental fees C_i^H and C_j^S, the cost of sales management K, cost of aircraft M, and traveling, maintenance, and shipping cost L, but these data are mainly derived from Skymark Airlines Inc. [5], one of Japan's largest LCCs.

4　Example Calculations

In this section, we provide a specific calculation using our profit maximization model for a new LCC given in Sect. 2, and the method of parameter assignment given in Sect. 3 in the case of a domestic airline network in Japan.

There are more than 100 airports in Japan, but in this study we restrict our investigation to the Narita Airport and 15 other airports, and attempt to obtain an exact solution by employing a branch and bound approach using Mathematica 9. It is not possible to consider all airports owing to limitations in calculational resources. Therefore, we chose airports that is ranked in the top 16 based upon the number of incoming and outgoing passengers. Furthermore, we chose Narita, Shinchitose, Fukuoka, Kansai, Naha, and Chubu from these 16 airports as possible hub airports.

First, we list the results of single-hub calculations, where $I = 1$, in Table 1.

From Table 1, the new LCC's profit is maximized when Narita Airport is the chosen hub airport.

Subsequently, we list the results of two-hub calculations, where $I = 2$, in Table 2 when Narita-Kansai, Narita-Chubu, Narita-Shinchitose, Narita-Fukuoka, or Shinchitose-Kansai are chosen as the pair of hub airports.

Table 1　Results of single-hub calculations (Million Yen)

Hub	Profit	Revenue	Cost	Number of aircraft	Hub	Profit	Revenue	Cost	Number of aircraft
Narita	449	7,783	7,334	4	Fukuoka	−148	1,771	1,919	1
Kansai	71	2,045	1,975	1	Shinchitose	−149	1,690	1,839	1
Chubu	−58	1,778	1,836	1	Naha	−203	0	2,032	0

Table 2　Results of two-hub calculations (Million Yen)

Hub pair	Profit	Revenue	Cost	Aircraft	Hub pair	Profit	Revenue	Cost	Aircraft
Narita-Kansai	740	9,931	9,191	5	Narita-Fukuoka	433	9,727	9,293	5
Narita-Chubu	636	7,922	7,285	4	Shinchitose-Kansai	377	7,616	7,240	4
Narita-Shinchitose	465	9,637	9,172	5	Chubu-Kansai	283	5,815	5,532	3

Table 3 Flight volumes from Narita and Kansai International Airports (Flights)

	Narita	Kansai	Chubu	Shinchitose	Fukuoka	Naha	Kagoshima	Sendai
Narita	• •	0	0	0	0	0	1	0
Kansai	0	• •	0	0	0	0	1	1
	Hiroshima	Kumamoto	Miyazaki	Kobe	Matsuyama	Nagasaki	Komatsu	Oita
Narita	2	2	1	0	1	0	1	1
Kansai	0	0	1	0	0	0	0	0

As shown in Table 2, profit is maximized when Narita and Kansai International Airports are chosen as the hub airports. Flight volumes from Narita and Kansai International Airports are listed in Table 3 below.

Similar to the case of the single-hub calculations, numerous links connect to small airports in rural areas, and, the rental fees affect the results. In fact, the primary reason that the top two most profitable hub airport pairs are Narita-Kansai and Narita-Chubu is that these three airports are the top three airports based on the number of incoming and outgoing passengers with the lowest rental fees.

5 Conclusion

In this study, we consider a new LCC that has gained access to Japan's domestic airline network, and establish an airline network that maximizes the new LCC's profit. We consider a hub-spoke network that numerous existing LCCs presently use, and we investigate a multi-hub network using simultaneous decision of flight volume and number of aircrafts.

Furthermore, we found that rental fees profoundly affect connection choices. As such, it would be most beneficial to establish more accurate rental fees to increase the accuracy of our calculations because the top three airports based on the number of incoming and outgoing passengers with the lowest rental fees were chosen for hub airports as a result.

A source of inaccuracy in our calculations, which could be rectified in the future, is the use of 2012 data. Use of 2013 data would allow us the calculation which is closer to today's situation if the data were disclosed to the public.

References

1. Sohn, J., Park, S.: A linear program for the two-hub solution problem. Eur. J. Oper. Res. **100**, 617–622 (1997)
2. Sohn, J., Park, S.: The single allocation problem in the interacting three-hub network. Networks **35**, 17–25 (2000)

3. Ministry of Land, Infrastructure, Transport and Tourism, Domestic Passenger Record per each route and month (2014). http://www.mlit.go.jp/k-toukei/index.html. Accessed 30 Apr 2014
4. Ministry of Land, Infrastructure, Transport and Tourism, The Inter-Refional Travel Survey (2014). http://www.mlit.go.jp/sogoseisaku/soukou/sogoseisaku_soukou_fr_000016.html. Accessed 30 Apr 2014
5. Skymark Airlines Inc., *Skymark official HP* (2014). http://www.skymark.co.jp. Accessed 30 Apr 2014

Variable Speed in Vertical Flight Planning

Zhi Yuan, Armin Fügenschuh, Anton Kaier and Swen Schlobach

Abstract Vertical flight planning concerns assigning cruise speed and altitude to segments that compose a trajectory, such that the fuel consumption is minimized and the time constraints are satisfied. The fuel consumption over each segment is usually given as a black-box function depending on aircraft speed, weight, and altitude. Without time consideration, it is known that it is fuel-optimal to fly at a constant speed. If an aircraft is under time pressure to speed up, the industrial standard of cost index cannot handle it explicitly, while research literature suggest using a constant speed. In this work, we formulate the vertical flight planning with variable cruise speed into a mixed integer linear programming (MILP) model, and experimentally investigate the fuel saving potential over a constant speed.

1 Introduction and Motivation

Planning a fuel-efficient flight trajectory connecting a departure and an arrival airport is a hard optimization problem. The solution space of a flight trajectory is four-dimensional: a 2D horizontal space on the earth surface, a vertical dimension consisting of discrete altitude levels, and a time dimension controlled by aircraft speed. In practice, the flight planning problem is solved in two separate phases: a horizontal phase that finds an optimal 2D trajectory consisting of a series of segments; followed by a vertical phase that assigns optimal flight altitude and speed to

Z. Yuan (✉) · A. Fügenschuh (✉)
Professorship of Applied Mathematics, Department of Mechanical Engineering,
Helmut Schmidt Universität, Hamburg, Germany
e-mail: yuanz@hsu-hh.de

A. Fügenschuh
e-mail: fuegenschuh@hsu-hh.de

A. Kaier · S. Schlobach
Lufthansa Systems AG, Kelsterbach, Germany
e-mail: anton.kaier@lhsystems.com

S. Schlobach
e-mail: swen.schlobach@lhsystems.com

© Springer International Publishing Switzerland 2016
M. Lübbecke et al. (eds.), *Operations Research Proceedings 2014*,
Operations Research Proceedings, DOI 10.1007/978-3-319-28697-6_88

635

each segment. The altitude and speed can be changed only at the beginning of each segment. This work focuses on the vertical phase. A vertical flight profile consists of five stages: take-off, climb, cruise, descend, and landing. Here we focus on the cruise stage, since it consumes most of the fuel and time during a flight, while the other stages are relatively short and have relatively fixed procedures due to safety considerations. Lovegren and Hansman [5] considered assigning optimal speed and altitude for the cruise stage, and comparing the optimal vertical profile to the real operating vertical profiles in USA. A potential fuel saving of up to 3.6 % was reported by the vertical profile optimization. However, no time constraint is taken into account in their computation as in real life. Note that in such case, it is known that the fuel-optimal speed assignment is to use a constant optimal cruise speed throughout the flight.

A practical challenge in airline operations is to handle delays due to disruptions such as undesirable weather conditions and unexpected maintenance requirements. Such delays are typically recovered by increasing the cruise speed, such that the next connection for passengers as well as for the aircraft can be caught. Speeding up an aircraft may also be useful, for example, to enter a time-dependent restricted airspace before it is closed, or when an aircraft is reassigned to a flight which used to be served by a faster aircraft. The industrial standard *cost index* was introduced by aircraft manufacturers to input a value (e.g., between 0–999) that reflects the importance between time-related cost and fuel-related cost, such that optimal flight speed is controlled. However, this approach cannot handle explicitly hard time constraints such as the about-to-close airspace. Aktürk et al. [1] considered increasing cruise speed in the context of aircraft rescheduling, and handled time constraint explicitly for scheduling purpose. However, their mathematical model only considered assigning a constant speed for the whole flight. It leaves an open research question: given a flight to be accelerated from its optimal speed, is it more fuel-efficient to allow variable speed on each segment? We formulate this problem as a mixed integer nonlinear programming (MINLP) model, and present linearization techniques in Sect. 2, examine its computational scalability in Sect. 3 and empirically investigate the question above using data for various aircrafts.

2 Mathematical Model

The unit distance fuel consumption of an aircraft depends on its speed, altitude, and weight. Each aircraft's unit distance fuel consumption data is measured at discrete levels of each of the three factors. Given speed and weight, the optimal altitude can be precomputed by enumerating all possible altitudes. Thus the unit distance fuel consumption $F_{v,w}$ defined for a speed level $v \in V$ between optimal and maximal speed, and a weight level $w \in W$ can be illustrated in Fig. 1. Other input parameters include a set of n segments $S := \{1, \ldots, n\}$ with length L_i for all $i \in S$; the minimum and maximum trip duration $\underline{T}$ and $\overline{T}$; and the dry aircraft weight W^{dry}, i.e. the weight of a loaded aircraft without trip fuel (reserve fuel for safety is included in the dry

Fig. 1 The unit distance fuel
consumption (kg per nautical
mile) by aircraft speed (mach
number, from optimal speed
to maximal speed) and
weight (kg) for Airbus 320

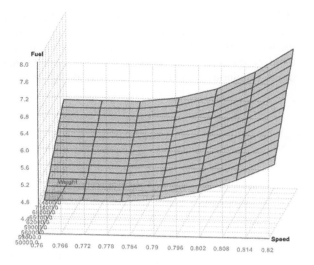

weight). The variables include the time vector t_i for $i \in S \cup \{0\}$, where t_{i-1} and t_i denote the start and end time of segment i; the travel time Δt_i spent on a segment $i \in S$; the weight vector w_i for $i \in S \cup \{0\}$ and w_i^{mid} for $i \in S$ where w_{i-1}, w_i^{mid}, and w_i denote the start, middle, and end weight at a segment i; the speed v_i on a segment $i \in S$; and the fuel f_i consumed on a segment $i \in S$. A general mathematical model for the vertical flight planning problem can be stated as follows:

$$\min \quad w_0 - w_n \tag{1}$$

$$\text{s.t.} \quad t_0 = 0, \quad \underline{T} \le t_n \le \overline{T} \tag{2}$$

$$\Delta t_i = t_i - t_{i-1} \quad \forall i \in S \tag{3}$$

$$L_i = v_i \cdot \Delta t_i \quad \forall i \in S \tag{4}$$

$$w_n = W^{dry} \tag{5}$$

$$w_{i-1} = w_i + f_i \quad \forall i \in S \tag{6}$$

$$w_{i-1} + w_i = 2 \cdot w_i^{mid} \quad \forall i \in S \tag{7}$$

$$f_i = L_i \cdot \tilde{F}(v_i, w_i^{mid}) \quad \forall i \in S. \tag{8}$$

Equation (1) minimizes the total fuel consumption; (2) enforces the flight duration within a given interval; (3) ensures the time consistency; the basic equation of motion on each segment is given in (4); the weight vector is initialized in (5) by assuming all trip fuel is burnt during the flight; weight consistency is ensured in (6), and the middle weight of each segment calculated in (7) will be used in the calculation of fuel consumption in (8), where $\tilde{F}(v, w)$ is a piecewise linear function interpolating F for all the continuous values of v and w within the given grid of $V \times W$. $\tilde{F}$ can be formulated as a MILP submodel using Danzig's convex combination method [3].

Here we present one of its variants, and drop the index i hereafter for simplification. The grids of $V \times W$ are first partitioned by a set of triangles K. The grid indices of the three vertices of each triangle $k \in K$ is stored in N_k. Each triangle is assigned a binary variable y_k, y_k equals 1 if (v, w) is inside triangle k. We further introduce three continuous variables for each triangle $\lambda_{k,n} \in \mathbb{R}^+$ for $k \in K$, $n \in N_k$ such that

$$\sum_{k \in K} y_k = 1 \tag{9a}$$

$$\sum_{n \in N_k} \lambda_{k,n} = y_k \quad \forall k \in K \tag{9b}$$

$$\sum_{k \in K, n \in N_k} \lambda_{k,n} \cdot V_{k,n} = v \tag{9c}$$

$$\sum_{k \in K, n \in N_k} \lambda_{k,n} \cdot W_{k,n} = w \tag{9d}$$

$$\sum_{k \in K, n \in N_k} \lambda_{k,n} \cdot F(V_n, W_n) = \tilde{F}(v, w) \tag{9e}$$

where (9a) ensures only one triangle is selected, (9b) sums λ of each triangle to 1 only if the triangle is selected, together with (9c) and (9d), the value of non-zero lambda is determined, such that the fuel estimation at a (v, w) is given by (9e) as a convex combination of λ and the grid value.

Another difficulty in the model is to handle the quadratic constraint in Eq. (4). It can be linearized by quadratic cone approximations. First we can rewrite the equality (4) into an equivalent inequality $L \leq v \cdot \Delta t$, since neither increasing v nor Δt leads to fuel saving. Applying the variable transformation $\alpha = \frac{1}{2}(v - \Delta t)$, $\tau = \frac{1}{2}(v + \Delta t)$, $\beta = \sqrt{L}$ yields $\sqrt{\alpha^2 + \beta^2} \leq \tau$, which defines a second-order cone, and thus can be approximated by linear inequality system as introduced by Ben-Tal and Nemirovski [2] and refined by Glineur [4]. We introduce continuous variables $\alpha_j, \beta_j \in \mathbb{R}$ for $j = 0, 1, \ldots, J$, and initialize by setting $\alpha_0 = \frac{1}{2}(v - \Delta t)$ and $\beta_0 = \sqrt{L}$. The *approximation level* parameter J controls the approximation accuracy. Then the following constraints can be added:

$$\alpha_{j+1} = \cos\left(\frac{\pi}{2^j}\right) \cdot \alpha_j + \sin\left(\frac{\pi}{2^j}\right) \cdot \beta_j, j = 0, 1, \ldots, J - 1, \tag{10a}$$

$$\beta_{j+1} \geq -\sin\left(\frac{\pi}{2^j}\right) \cdot \alpha_j + \cos\left(\frac{\pi}{2^j}\right) \cdot \beta_j, j = 0, 1, \ldots, J - 1, \tag{10b}$$

$$\beta_{j+1} \geq \sin\left(\frac{\pi}{2^j}\right) \cdot \alpha_j - \cos\left(\frac{\pi}{2^j}\right) \cdot \beta_j, j = 0, 1, \ldots, J - 1, \tag{10c}$$

$$\frac{1}{2}(v + \Delta t) = \cos\left(\frac{\pi}{2^J}\right) \cdot \alpha_J + \sin\left(\frac{\pi}{2^J}\right) \cdot \beta_J. \tag{10d}$$

3 Experiments and Results

Four different aircrafts are used for our study: Airbus 320, 380 and Boeing 737 and 777. The characteristics of these aircrafts are listed in Table 1. Our preliminary experiments for fuel estimation accuracy test confirmed that when dividing a longest possible 7500 nautical miles (NM) trip into equidistance segments of 100 NM, the total fuel estimation error is under 1 kg in 200 tons consumption (i.e. a relative error of under $5 \cdot 10^{-6}$). With the same 1 kg error threshold, we experimentally determine $J = 10$ for A320 and B737, $J = 11$ for B777 and $J = 12$ for A380.

We set up instances for each of the four aircraft types by considering different levels of speed-up and different travel distances. Two levels of speed-up are used: 2.5 and 5 %, since the maximum possible speed-up is around 7.2 to 8.5 % as shown in Table 1, higher speed-up settings also do not leave much room for speed variation. For each aircraft, different typical travel distances are tested, ranging from 800 NM for B737, which is around the distance from Frankfurt to Madrid, to 7500 NM for A380 and B777, which is around the distance from Frankfurt to west coast of Australia. Each trip is divided into equidistance segments of 100 NM each.

These instances were first tried to be solved without conic reformulation, and SCIP 3.1 was used as a MINLP solver. Each run was performed on a computing node with 12-core Intel Xeon X5675 at 3.07 GHz and 48 GB RAM. Only single thread was used for SCIP. These realistic instances cannot be solved by SCIP within 24 h. We reduced the number of segments and coarsened the weight grid, and found the largest instance solved is with 10 segments and 4 weight levels ($|W| = 4$).

With the conic reformulation, the MINLP model becomes MILP model, so commercial MILP solver such as CPLEX can be applied. We applied CPLEX 12.6, and each run was performed on the same computing node, with 12 threads per run. The computational results including the computation time and the gap (if cut off at 24 h) was shown in Table 2. All the real-world instances are solved to provable optimality or near-optimality (less than 0.05 %). Instances with no more than 25 segments can typically be solved within one minute. Increasing the number of segments seems to increase the computational difficulty noticeably.

Table 1 Four aircraft types, Airbus 320, 380, Boeing 737, 777, and their characteristics, such as optimal and maximal speed (in Mach number), dry weight and maximal weight (in kg), and maximal distance (in NM)

Type	Opt. speed	Max. speed	Dry weight	Max. weight	Max. distance	$|V|$	$|W|$	J
A320	0.76	0.82	56614	76990	3500	7	15	10
A380	0.83	0.89	349750	569000	7500	7	24	12
B737	0.70	0.76	43190	54000	1800	7	12	10
B777	0.82	0.89	183240	294835	7500	8	16	11

The number of speed grids $|V|$ (between optimal and maximal speed) and weight grids $|W|$, and the empirically determined conic approximation level J are also listed

Table 2 Instances and their computational results on four aircraft types, with two speed-up factors 2.5 or 5 %, and various numbers of segments |S|

Aircraft	|S|	2.5 % Speed up			5 % Speed up		
		Comp. time	Gap	Fuel saving (%)	Comp. time	Gap	Fuel saving (%)
A320	15	16	0	0.009	21	0	0.007
	20	28	0	0.009	32	0	0.009
	25	57	0	0.009	43	0	0.009
	30	191	0	0.009	23901	0	0.009
	35	735	0	0.005	34631	0	0.010
A380	30	525	0	0.008	276	0	0.053
	40	1691	0	0.005	360	0	0.093
	50	4952	0	0.012	14043	0	0.117
	60	86400	0.02 %	0.013	86400	0.02 %	0.121
	70	86400	0.02 %	0.013	86400	0.03 %	0.180
	75	86400	0.03 %	0.017	86400	0.03 %	0.177
B737	8	2	0	0.013	4	0	0.020
	12	9	0	0.015	11	0	0.014
	15	15	0	0.011	22	0	0.016
	18	17	0	0.013	31	0	0.023
B777	25	275	0	0.011	69	0	0.007
	35	4759	0	0.001	425	0	0.005
	45	86400	0.02 %	0.004	86400	0.03 %	0.015
	55	13552	0	0.001	86400	0.02 %	0.013
	65	86400	0.04 %	0.005	86400	0.05 %	0.020
	75	86400	0.05 %	0.023	86400	0.03 %	0.020

Each segment is 100 NM, so the total distance is $100 \times |S|$. Computation time (seconds), gap from optimality, and potential fuel saving are listed

We compared also the optimal value of using variable speed as computed above with an optimal constant speed. Since the fuel consumption is a monotone function of speed, the optimal constant speed can be computed as 2.5 or 5 % over the optimal speed, respectively. As shown in Table 2, the potential fuel savings of using variable speed compared to a constant speed are rather small for the relatively mature aircrafts A320, B737, and B777. The rather new A380 shows the highest potential fuel savings of up to 0.18 %. Although this number seems small, it means a lot in the highly competitive market of the airline industry. It also shows that there is room for Airbus to improve the performance of their new flagship airplane.

Our current experiments do not consider the influence of the weather, in particular, the wind. As also suggested in [5], a strong head wind may favor higher speed, while flying slower may be advantageous in a strong tail wind. The current MILP model can be easily extended to include wind influence. Practically, the current MILP approach may require undesirable long computation time, but its optimal solution may be used to assess the quality of further heuristic approaches.

Acknowledgments This work is supported by BMBF Verbundprojekt E-Motion.

References

1. Aktürk, M.S., Atamtürk, A., Gürel, S.: Aircraft rescheduling with cruise speed control. Operations Research (2014). To appear
2. Ben-Tal, A., Nemirovski, A.: On polyhedral approximations of the second-order cone. Math. Oper. Res. **26**(2), 193–205 (2001)
3. Dantzig, G.B.: On the significance of solving linear programming problems with some integer variables. Econometrica **28**(1), 30–44 (1960)
4. Glineur, F.: Computational experiments with a linear approximation of second-order cone optimization. Image Technical Report 0001, Faculté Polytechnique de Mons, Belgium (2000)
5. Lovegren, J.A., Hansman, R.J.: Estimation of potential aircraft fuel burn reduction in cruise via speed and altitude optimization strategies. Technical report, ICAT-2011-03, MIT International Center for Air Transportation (2011)

A Fast Greedy Algorithm for the Relocation Problem

Rabih Zakaria, Laurent Moalic, Mohammad Dib
and Alexandre Caminada

Abstract In this paper, we present three relocation policies for the relocation problem in one carsharing system. We implement these policies in a greedy algorithm to evaluate their performance. Compared with CPLEX, greedy algorithm proved that it is able to solve the most difficult system configurations in at most one second while providing good quality solutions. On the other side, greedy algorithm results show that relocation policies that do not rely on historical data, will not be very efficient in reducing rejected user demands, on the contrary they can contribute in increasing their number while increasing the total number of relocation operations.

1 Introduction

It's safe to say, vehicle-sharing systems are among the top modern-world up and coming transportation innovations. In our study, we focused on carsharing systems and more precisely, one-way carsharing system. Carsharing operators offer many cars for public use by system users, generally for short-term rentals. Cars are located in different stations all over urban areas; each station has a fixed number of parking spaces. Users can pick up a car from a station at time t and drop it off in any other station. Issues may arise however during the day, when many one-way trips are done, the system tends to become imbalanced [1]. Stations that are popular points of departure become empty, preventing users who want to make a ride at those stations, from using the system. Consequently, other stations are full. Thus, users who want

R. Zakaria (✉) · L. Moalic · A. Caminada
OPERA, UTBM, Belfort, France
e-mail: rabih.zakaria@utbm.fr; rabie_zakaria@hotmail.com

L. Moalic
e-mail: laurent.moalic@utbm.fr

A. Caminada
e-mail: alexandre.caminada@utbm.fr

M. Dib
GDF SUEZ - CEEME, Paris, France
e-mail: mohammad.dib@gmail.com

© Springer International Publishing Switzerland 2016
M. Lübbecke et al. (eds.), *Operations Research Proceedings 2014*,
Operations Research Proceedings, DOI 10.1007/978-3-319-28697-6_89

to return the car are compelled to look for another station or to wait until a place is free. Ultimately, users tend to lose interest in using this system that often appears to be unavailable and unreliable when they need it. To solve the imbalance problem, carsharing operators recruit employees to relocate cars from the saturated stations to the empty stations or to the stations that need more cars to satisfy user demands. We call these operations: Car Relocation and we refer to the employees that perform these operations by "Jockeys".

There are many related works in the literature dealing with this problem, both for carsharing and bikesharing. In [2], they proposed to use the client himself to contribute in the relocation operation; although this approach was successful in reducing 42 % of the overall number of relocation operations, this only works when 100 % of clients participate, which is obviously not always guaranteed. In another paper, [4] presented a decision support system for carsharing companies to determine a set near-optimal operating parameters for the vehicle relocation problem. Suggested parameters lead to a reduction in staff cost of 50 %, a reduction in zero-vehicle-time ranging between 4.6 and 13.0 %, a maintenance of the already low full-port-time and a reduction in number of relocations ranging between 37.1 and 41.1 % of staff and operating parameters for the car relocation problem. Furthermore, The importance of relocation operations in increasing the carsharing operator profit has been demonstrated in [3].

In next sections, we provide a brief system description. Then we present our relocation policies. After that, we present our results and analyze them. Finally, we sum up by a conclusion and perspectives.

2 System Description

We represent a carsharing system by a time-space network where a day is subdivided into 96 time steps, each time-step covers 15 min. For each time step, we have the number of arriving, departing and available cars at each station. We obtain the system data by using a platform developed by our team [5], which enables us to generate different carsharing system configurations by varying input parameters such as number of stations, number of parking spaces in each station and the average number of trips per car. This platform takes advantage of real mobility data collected by professional in regional planning, in addition to socio-economic information and GIS shape files for the region of the study. Using google maps, the platform calculates the time and distance needed to move from one station to another.

3 Greedy Algorithm

In a previous work, we modeled the relocation problem as an Integer Linear Programming model [6]. We solved the model using CPLEX. After running the model through different configurations, we noticed that the execution time tends to increase dramatically when we increase the number of jockeys involved in the relocation operations. For some configurations, CPLEX takes more than two days to deliver a solution and for other configurations, we could not get any results using this solver. The long execution time to solve the relocation problem using CPLEX, pushed us to think about a different approach that solves the relocation problem in a faster time. For this sake, we developed a greedy algorithm that tries to reduce the number of rejected demands using the minimum number of relocation operations. Our greedy algorithm proves that is able to deliver, in at most one second, good quality solutions when compared with CPLEX results.

4 Relocation Policies

In this section, we propose three relocation policies and then we implement these approaches using a greedy algorithm in a policy pattern to measure the effect of each policy on the total number of rejected demands. Each relocation operation consists of two steps: in the first step, the jockey chooses the station where he will take a car. Then in the second step, he chooses the station where he will drop that car in order to regain the balance of the system. The choice of relocation policy plays a major role in reducing the number of rejected user demands and therefore in increasing the client satisfaction.

4.1 Policy 1

In this approach, as the first step of the relocation operation, the jockey tries to find the nearest stations to his current location. If he finds many stations having the same distance, he chooses the station that has the maximum number of cars. Then, in the second step, the jockey looks for the nearest stations to his location. If he finds many stations having the same distance, he chooses the station that has the least number of cars. As we see in Fig. 1 each station is represented by a disc that contains the station name and the number of available cars at the specified time. In this example, during the first step, the jockey finds two stations S_2 and S_3 that are on the same distance from his current station, but he goes to station S_3 since it has a bigger number of cars. Then, during the second step, the jockey goes to station S_1, since it is the nearest station at first and because it has the minimum number of cars knowing that we have the matrix of distance between each pair of stations. In this policy, the priority is given to the distance between the stations with the aim of minimizing the total time of relocation operations.

Fig. 1 Simple relocation
operation using policy 1

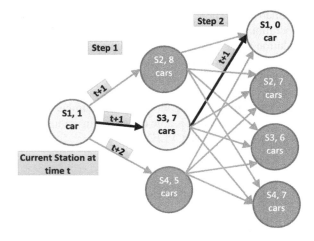

4.2 Policy 2

This approach is similar to the previous one, but here, we reverse the order of choosing
the stations at each step. As the first step of the relocation operation, the jockey tries
finding the list of stations having the maximum number of cars, and then chooses the
nearest station amongst this list. Then, in the second step, the jockey looks for the
list of stations having the minimum number of cars, and then he chooses the nearest
station amongst this list. In this policy, the priority is given to rebalance the number
of cars at each station instead of the distance as in the policy 1 described earlier, with
the aim of regain balance for the stations.

4.3 Policy 3

In this approach, the jockey tries to solve the maximum number of rejected demands
in each relocation operation. As the first step, the jockey looks for the list of stations
that will have the soonest expected rejected demands because stations are full and
the list of stations that can provide cars for other stations. Then, in a second step,
the jockey looks for the list of stations that will have the soonest expected rejected
demands because stations are empty and the list of stations that may need cars in
the future. When the jockey gets these lists, he plans the relocation operation in a
way that reduces the maximum number of rejected demand while avoiding that these
operations cause future rejected demands in the affected stations. In this policy, we
consider that the jockey has a perfect knowledge of what will happen in the future,
so he can avoid removing or adding cars to some stations when it can lead to cars
shortage or cars saturation respectively.

5   Results and Experimentation

In Fig. 2, we see a comparison of the three policies described earlier. Results are
shown for a carsharing system having 20 stations of 10 parking places for each, using
150 cars with 9 trips per day as an average. As we can see, the performance of policy 1
is slightly better than policy 2 at first, then when we increase the number of jockeys
over 19, Policy 2 performs better in reducing the number of rejected demands. This
difference is expected, since policy 2 prioritizes the relocation operations that tends to
rebalance the systems vehicle inventory. In addition, when using both of policy 1 and
policy 2, we observe that when we increase the number of jockeys over 20, the number
of remaining rejected demands increases. This is due to bad relocation decisions that
cause new rejected demands to appear in the future. However, policy 3 appears to
be the best. We can explain this by the fact that the jockey has perfect knowledge
of the future. Which enables him to take better relocation decisions that reduce the
maximum number of rejected demands without causing new rejected demands in
the future. Using this policy, the jockey relocates cars only when he is sure that
the relocation operation will reduce the number of rejected demands, otherwise he
waits for the right moment for the relocation operations. This is clear in Fig. 2, we
see that the number of relocation operations using policy 1 and policy 2 is kind off
constant, but policy 2 uses less relocation operations than policy 1. However, the
number of relocation operations in Policy 3, tends to decrease along with the number
of remaining rejected demands.

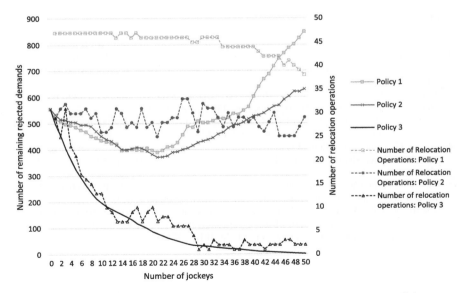

Fig. 2 Comparison of the performance of the relocation operations using the three policies

6 Conclusion and Future Works

One-way carsharing systems are attractive to users who want to do one-way trips, since it is flexible and available on the go. However, the imbalance in cars inventory that occurs during the day due to the one-way trips in some stations, makes the system unavailable for users when they need it. Car relocation operations seems to be essential to mitigate this problem and thus to increase client satisfaction. In this paper, we compare three different car relocation policies. We found that the performance of policy 1 and policy 2 are near at first, and then policy 2 performs better after the number of jockeys surpasses a limit of 20. However, the performance of Policy 3 where the jockey has a perfect knowledge of the future is much better than the two other policies. We can conclude that applying policies that are based on intuitive decisions, such as distance and number of cars at stations, without taking into consideration the effect of these relocation operations on the whole system, will not be very efficient in reducing the number of rejected demands. In addition, these policies lead to bigger number of relocation operations, which increases the total operation cost. A relocation policy that takes into consideration historical data seems to be promising in reducing the number of rejected demands. In future works, we aim to develop a stochastic heuristic approach to solve the relocation problem based on historical data, in a simulation environment that takes real life factors into consideration.

References

1. Barth, M., Shaheen, S.A.: Shared-use vehicle systems: framework for classifying carsharing, station cars, and combined approaches. Transp. Res. Rec.: J. Transp. Res. Board 1791(1), 105–112 (2002)
2. Barth, M., Todd, M., Xue, L.: User-based vehicle relocation techniques for multiple-station shared-use vehicle systems. Transp. Res. Rec. 1887, 137–144 (2004)
3. Jorge, D., Correia, G.H., Barnhart, C.: Comparing optimal relocation operations with simulated relocation policies in one-way carsharing systems (2013)
4. Kek, A.G., Cheu, R.L., Meng, Q., Fung, C.H.: A decision support system for vehicle relocation operations in carsharing systems. Transp. Res. Part E: Logist. Transp. Rev. 45(1), 149–158 (2009)
5. Moalic, L., Caminada, A., Lamrous, S.: A fast local search approach for multiobjective problems. In: Learning and Intelligent Optimization, pp. 294–298. Springer (2013)
6. Zakaria, R., Dib, M., Moalic, L., Caminada, A.: Car relocation for carsharing service: comparison of cplex and greedy search. In: 2014 IEEE Symposium on Computational Intelligence in Vehicles and Transportation Systems (CIVTS), pp. 51–58. IEEE (2014)

Multicriteria Group Choice via Majority Preference Relation Based on Cone Individual Preference Relations

Alexey Zakharov

Abstract A multicriteria group choice problem that is considered in the article includes: a set of feasible decisions; a vector criterion reflecting general goals of a group of Decision Makers (DMs); asymmetric binary relations of DMs, which reflect individual preferences. Individual preferences are given by "quanta" of information, which indicate a compromise between two components of vector criterion. A majority preference relation is also considered. It is proved that such majority relation is a cone one, and the cone, generally speaking, is not convex. The property of convex is equivalent to transitivity of the corresponding relation. The goal of the research is to construct a convex part of a majority preference relation cone, which gives a transitive part of this relation. The case of group of three DMs and three components of criteria is considered. It is shown how to specify a convex part of a majority preference relation cone, and construct a set of nondominated vectors.

1 Introduction

A group of people makes decisions in various spheres of human life: economics, politics, engineering, science, etc. Generally problems are considered from different points of view, and each person, the Decision Maker (DM), has their own preferences. Here the following problem arises: how we should aggregate individual preferences to satisfy as many members of the group as possible. Also, preferences could be expressed in different ways: ranking, utility function, preference relation.

Bergson and Samuelson proposed a social welfare function which was considered as a rule of aggregation of individual ordering. The problem of constructing of a "reasonable" rule of aggregation was widely discussed by Arrow [3, 4]. He formulated axioms (5 conditions) for social welfare function, which strict rationality of individual and group behavior. Also, he proved that if there are at least three alternatives, then an aggregation ordering rule (social welfare function) satisfying

A. Zakharov (✉)
Saint-Petersburg State University Universitetskii pr., 35, Petrodvorets, St.,
Petersburg 198504, Russia
e-mail: a.zakharov@spbu.ru; zakh.alexey@gmail.com

© Springer International Publishing Switzerland 2016
M. Lübbecke et al. (eds.), *Operations Research Proceedings 2014*,
Operations Research Proceedings, DOI 10.1007/978-3-319-28697-6_90

649

these axioms does not exist. Having analyzed the construction of group choice function upon individual choice functions, Aizerman and Aleskerov proposed axioms of "rational" behavior for the choice functions [1, 2]. Nowadays there exist many methods to solve a multicriteria group choice problem [5–7, 12].

The simplest rule of individual preferences aggregation is the majority rule: a group chooses a decision if it is the "best" according to the preferences of half of the group. Arrow proved that a majority rule could be not of a weak order (not connected or not transitive) [3]. In this paper individual preferences are expressed in notation of cone preference relations with particular properties, and such individual preference relations are aggregated by the majority relation. It is proved that it is a cone relation. Some properties of such majority preference relation are considered, and the problem of nontransitivity is discussed. It is shown how to use the information about individual preference relations ("quanta" of information) in a group choice when the group consists of three members.

2 Model of Multicriteria Individual Choice Problem

Let the group of DMs consists of n members. Denote them by $DM_1, \ldots, DM_n$. The group of DMs should make the choice among some set of feasible alternatives, solutions $X \subseteq \mathbb{R}^k$. The goals of the group are reflected by components of the vector criterion $\mathbf{f}$, which is defined on X, $Y = \mathbf{f}(X)$. For example, a vector criterion has such components: profit, loss, ecological impact, etc. The tastes of each DM_l, $l \in \{1, \ldots, n\}$, are reflected using the individual preference relation $\succ_l$. Expression $\mathbf{y}^{(1)} \succ_l \mathbf{y}^{(2)}$ for any vectors $\mathbf{y}^{(1)}, \mathbf{y}^{(2)} \in \mathbb{R}^m$ means that for DM_l variant $\mathbf{y}^{(1)}$ is more preferable than variant $\mathbf{y}^{(2)}$. Thus, a multicriteria group choice model consists of the following objects:

- a set of feasible vectors Y;
- n preference relations $\succ_1, \ldots, \succ_n$, defined on Y, of $DM_1, \ldots, DM_n$.

In [9, 10] axioms of "reasonable" individual choice, which restrict the behavior of any DM, are introduced. Further, we assume that each preference relation $\succ_l$ satisfies these axioms. The Pareto set reduction axiomatic approach applied to this class of multicriteria individual choice problems restricted by this axioms [8–11] is developed by Noghin.

Definition 1 [9–11] It is said that we have a "quantum" of information about a DM_l's preference relation with groups of criteria A and B and with two sets of positive parameters $w_i^{(l)}$ for all $i \in A$ and $w_j^{(l)}$ for all $j \in B$ if for vector $\mathbf{y} \in \mathbb{R}^m$ such that

$$y_i = w_i^{(l)}, \ y_j = -w_j^{(l)}, \ y_s = 0 \ \forall i \in A, \forall j \in B, \forall s \in I \setminus (A \cup B) \,,$$

where $I = 1, \ldots, m$, A, $B \subset I$, $A \neq \varnothing$, $B \neq \varnothing$, $A \cap B = \varnothing$, the following relation is valid: $\mathbf{y} \succ_l \mathbf{0}$. In such case the group of criteria A is called more important, and the group B is called less important with the given positive parameters $w_i^{(l)}$ (profit), $w_j^{(l)}$ (loss).

Thus, the existence of "quantum" of information means that a vector $\mathbf{y} \in N^m$, $N^m = \mathbb{R}^m \setminus (\mathbb{R}_+^m \cup (-\mathbb{R}_+^m) \cup \{0_m\})$, is given such that $\mathbf{y} \succ_l \mathbf{0}$. Here, set N^m is the set of the vectors, which have at least one positive component and at least one negative component. The collection of "quantum" of information is defined by the sequence of vectors $\mathbf{y}^{(s)} \in N^m$, $s = 1, \ldots, p_l$. In papers [8–11] it is shown how to use such collections of different types in individual choice process that allows to reduce bounds of this choice. It lets a DM to choose the "best" alternative from more narrow set than the initial set (before using the information). Now it is of interest to consider how to use individual information in terms of "quantum" in group decision making process.

3　Majority Preference Relation and Its Properties

In this section the group relation, which aggregates individual relations, is introduced, and properties of such relation are investigated.

Definition 2 A binary relation $\mathfrak{R}$ defined on $\mathbb{R}^m$ is called a cone relation if there exists a cone K such that the following equivalence holds for any $\mathbf{y}^{(1)}$, $\mathbf{y}^{(2)} \in \mathbb{R}^m$: $\mathbf{y}^{(1)} \mathfrak{R} \mathbf{y}^{(2)} \Leftrightarrow \mathbf{y}^{(1)} - \mathbf{y}^{(2)} \in K$. Note, that the inclusion $\mathbf{y}^{(1)} - \mathbf{y}^{(2)} \in K$ is the same to $\mathbf{y}^{(1)} \in \mathbf{y}^{(2)} + K$.

According to [8, 9], the preference relation $\succ_l$ of DM_l for any $l \in \{1, \ldots, n\}$, which satisfies the axioms of "reasonable" choice, is a cone relation with a pointed convex cone K_l (without the origin $\mathbf{0}$) that contains the nonnegative orthant $\mathbb{R}_+^m$.

Definition 3 Let us call the group preference relation $\succ_{maj}$ defined on $\mathbb{R}^m$ the *majority preference relation* if for any vectors $\mathbf{y}^{(1)}$, $\mathbf{y}^{(2)} \in \mathbb{R}^m$ relation $\mathbf{y}^{(1)} \succ_{maj} \mathbf{y}^{(2)}$ is equivalent to the existence of subset $\{l_1, \ldots, l_p\}$ such that relations $\mathbf{y}^{(1)} \succ_{l_j} \mathbf{y}^{(2)}$ are valid for all $j \in \{1, \ldots, p\}$, where $p = [(n + 1)/2]$.

Here by $[a]$ we denote an integer part of number a. According to Definition 3, the majority preference relation $\succ_{maj}$ takes into account intentions of at least a half of the group. Consider the cone

$$K = \bigcup_{l=1}^{C_n^p} \bigcap_{j=1}^{p} K_{lj}, \qquad (1)$$

where $\{K_{l1}, \ldots, K_{lp}\}$ is the subset of the set of the cones $\{K_1, \ldots, K_n\}$. Note, that for any $l, k \in \{1, \ldots, C_n^p\}$, $l \neq k$, subsets $\{K_{l1}, \ldots, K_{lp}\}$ and $\{K_{k1}, \ldots, K_{kp}\}$ do not coincide, C_n^p is a p-combination of the set $\{1, \ldots, n\}$. It is easy to check that K is a cone as a union and intersection of cones.

Lemma 1 *The majority preference relation $\succ_{maj}$ is a cone relation with cone K.*

Proof To prove the lemma it is sufficient to establish the following equivalence: $\mathbf{y}^{(1)} \succ_{maj} \mathbf{y}^{(2)} \Leftrightarrow \mathbf{y}^{(1)} - \mathbf{y}^{(2)} \in K$.

Necessity. Due to Definition 3, if the relation $\mathbf{y}^{(1)} \succ_{maj} \mathbf{y}^{(2)}$ is valid, then there exists subset $\{l_1, \ldots, l_p\}$ such that the relations $\mathbf{y}^{(1)} \succ_{l_j} \mathbf{y}^{(2)}$ hold for all $j \in \{1, \ldots, p\}$, where $p = [(n + 1)/2]$. Since the preference relation $\succ_s$ of DM$_s$ for any $s \in \{1, \ldots, n\}$ is a cone relation with the appropriate cone K_s, there exists a subset $\{K_{l1}, \ldots, K_{lp}\}$ such that $\mathbf{y}^{(1)} - \mathbf{y}^{(2)} \in K_{lj}$ for all $j \in \{1, \ldots, p\}$. This implies the inclusion $\mathbf{y}^{(1)} - \mathbf{y}^{(2)} \in \bigcap_{j=1}^{p} K_{lj}$, and we obtain $\mathbf{y}^{(1)} - \mathbf{y}^{(2)} \in K$.

Sufficiency. Let the inclusion $\mathbf{y}^{(1)} - \mathbf{y}^{(2)} \in K$ holds. Using the inverse transformations we obtain that the relation $\mathbf{y}^{(1)} \succ_{maj} \mathbf{y}^{(2)}$ is valid. $\qquad\square$

Lemma 2 *The majority preference relation $\succ_{maj}$ is an irreflexive relation, invariant under a linear positive transformation, and its cone K contains the nonnegative orthant $\mathbb{R}_+^m$ and does not contain the origin $\mathbf{0}$.*

One can obtain that an arbitrary cone is convex, if and only if the corresponding cone relation is transitive [8]. In general, the cone K is not convex. Therefore, the majority relation $\succ_{maj}$ is not transitive. Let $\hat{K}$ be a convex cone, such that $\hat{K} \subseteq K$, $\mathbb{R}_+^m \subseteq \hat{K}$, call this cone $\hat{K}$ the *transitive part* of the cone K. And if a convex cone $\tilde{K}$, such that $\tilde{K} \subseteq K$, $\tilde{K} \subset \hat{K}$, $\tilde{K} \neq \hat{K}$, does not exist, then call this cone $\hat{K}$ the *maximum transitive part* of the cone K. In general, it is not unique for particular cone K. Similarly, let us call the corresponding cone relations $\succ_{tr}$ and $\succ_{maxtr}$ the *transitive part* and the *maximum transitive part* of the majority preference relation $\succ_{maj}$.

Let us denote by $Ndom_{\succ_{tr}}(Y)$ the set of nondominated vectors of the set Y according to the relation $\succ_{tr}$, i.e. $Ndom_{\succ_{tr}}(Y) = \{\mathbf{y}^* \in Y \mid \nexists \mathbf{y} \in Y : \mathbf{y} \succ_{tr} \mathbf{y}^*\}$. It forms a group choice as a set of unimprovable vectors.

The inclusion $\mathbb{R}_+^m \subseteq \hat{K}$ implies the inclusion $Ndom_{\succ_{tr}}(Y) \subseteq P(Y)$, where $P(Y)$ is the set of Pareto-optimal vectors (the Pareto set), $P(Y) = \{\mathbf{y}^* \in Y \mid \nexists \mathbf{y} \in Y : \mathbf{y} \geq \mathbf{y}^*\}$. Note, that if each DM does not have any "quantum" of information, then $K = \hat{K} = \mathbb{R}_+^m$, and, therefore, the equality $Ndom_{\succ_{tr}}(Y) = P(Y)$ holds, i.e. in such case the group choice is the Pareto set $P(Y)$. Thus, the inclusion $Ndom_{\succ_{tr}}(Y) \subseteq P(Y)$ reduces the bounds of group choice according to the majority rule and the given "quanta" of information.

4 Using "Quanta" of Information in Case of 3 DMs

Let the group of DMs consists of three members: DM$_1$, DM$_2$, DM$_3$. Each DM$_l$ has its individual preference relation $\succ_l$, which satisfies axioms of "reasonable" choice, and each relation $\succ_l$ is associated with cone K_l for any $l \in \{1, 2, 3\}$. Note, that cone K_l is pointed, convex, $\mathbb{R}_+^m \subseteq K_l$, $\mathbf{0} \notin K_l$ for any $l \in \{1, 2, 3\}$. According to Lemma 1 and formula (1), the majority preference relation $\succ_{maj}$ has cone $K = (K_1 \cap K_2) \cup (K_1 \cap K_3) \cup (K_2 \cap K_3)$.

Let $m = 3$. Consider information about individual preferences (I): each DM has two "quanta" of information as follows. The vectors $\mathbf{y}^{(l)}$, $\mathbf{u}^{(l)}$ by DM_l are given such that $\mathbf{y}^{(l)} \succ_l \mathbf{0}$, $\mathbf{u}^{(l)} \succ_l \mathbf{0}$ for any $l \in \{1, 2, 3\}$, where $\mathbf{y}^{(1)} = (w_1^{(1)}, -w_2^{(1)}, 0)^T$, $\mathbf{u}^{(1)} = (0, v_2^{(1)}, -v_3^{(1)})^T$, $\mathbf{y}^{(2)} = (0, w_2^{(2)}, -w_3^{(2)})^T$, $\mathbf{u}^{(2)} = (-v_1^{(2)}, 0, v_3^{(2)})^T$, $\mathbf{y}^{(3)} = (-w_1^{(3)}, 0, w_3^{(3)})^T$, $\mathbf{u}^{(3)} = (v_1^{(3)}, -v_2^{(3)}, 0)^T$.

For example, for DM_1 it means the following: (1) the first criterion is more important, the second criterion is less important with positive parameters $w_1^{(1)}$ (profit), $w_2^{(1)}$ (loss); (2) the second criterion is more important, the third criterion is less important with positive parameters $v_2^{(1)}$ (profit), $v_3^{(1)}$ (loss). Similar interpretations for DM_2 and DM_3 can be done.

Theorem 1 *Let information (I) is given. Then the inclusion*

$$\hat{P}_{12}(Y) \cap \hat{P}_{13}(Y) \cap \hat{P}_{23}(Y) \subseteq P(Y) \tag{2}$$

holds, where $\hat{P}_{lk}(Y) = \mathbf{f}(P_{\mathbf{g}^{lk}}(X))$, $\mathbf{g}^{lk} = \begin{pmatrix} \mathbf{A}_l \\ \mathbf{A}_k \end{pmatrix} \mathbf{f}$ *for any* $l, k \in \{1, 2, 3\}$, $l \neq k$. *The intersections* $\hat{P}_{12}(Y) \cap \hat{P}_{13}(Y) \cap \hat{P}_{23}(Y)$ *form a group choice according to information (I). Here matrices* $\mathbf{A}_1$, $\mathbf{A}_2$, *and* $\mathbf{A}_3$ *are the following*

$$\mathbf{A}_1 = \begin{pmatrix} 1 & 0 & 0 \\ w_2^{(1)} & w_1^{(1)} & 0 \\ w_2^{(1)} v_3^{(1)} & w_1^{(1)} v_3^{(1)} & w_1^{(1)} v_2^{(1)} \end{pmatrix}, \quad \mathbf{A}_2 = \begin{pmatrix} w_2^{(2)} v_3^{(2)} & w_3^{(2)} v_1^{(2)} & w_2^{(2)} v_1^{(2)} \\ 0 & 1 & 0 \\ 0 & w_3^{(2)} & w_2^{(2)} \end{pmatrix},$$

$$\mathbf{A}_3 = \begin{pmatrix} w_3^{(3)} & 0 & w_1^{(3)} \\ w_3^{(3)} v_2^{(3)} & w_3^{(3)} v_1^{(3)} & w_1^{(3)} v_2^{(3)} \\ 0 & 0 & 1 \end{pmatrix}.$$

Proof Consider cone $M_l = \{\mathbf{e}^1, \mathbf{e}^2, \mathbf{e}^3, \mathbf{y}^{(l)}, \mathbf{u}^{(l)}\} \setminus \mathbf{0}$ for any $l \in \{1, 2, 3\}$, and let $M = (M_1 \cap M_2) \cup (M_1 \cap M_3) \cup (M_2 \cap M_3)$. Obviously, $M_l \subseteq K_l$ for any $l \in \{1, 2, 3\}$, $M \subseteq K$.

It can be proved that the cone M_l coincides with the set of all nonzero solutions of the system $\mathbf{A}_l \mathbf{y} \geq \mathbf{0}$ for any $l \in \{1, 2, 3\}$. From here the inclusion $\mathbf{y} \in M_l \cap M_k$ holds, if and only if vector $\mathbf{y}$ satisfies the system $\begin{pmatrix} \mathbf{A}_l \\ \mathbf{A}_k \end{pmatrix} \mathbf{y} \geq \mathbf{0}$ for any $l, k \in \{1, 2, 3\}, l \neq k$.

Intersections $M_1 \cap M_2$, $M_1 \cap M_3$, and $M_2 \cap M_3$ are convex cones. Cone M is not convex, and we should specify the convex part of it. Consider this problem from another point of view. The set of nondominated vectors $Ndom_{M_l M_k}(Y) = \{\mathbf{y}^* \in Y \mid \nexists \mathbf{y} \in Y : \mathbf{y} - \mathbf{y}^* \in M_l \cap M_k\}$ according to cone $M_l \cap M_k$ for any $l, k \in \{1, 2, 3\}, l \neq k$. If the inclusion $\mathbf{y} - \mathbf{y}^* \in M_l \cap M_k$ is valid for some vectors $\mathbf{y}, \mathbf{y}^* \in Y$ and some indices $l, k \in \{1, 2, 3\}, l \neq k$, then $\begin{pmatrix} \mathbf{A}_l \\ \mathbf{A}_k \end{pmatrix} (\mathbf{y} - \mathbf{y}^*) \geq \mathbf{0}$. It implies $\hat{P}_{lk}(Y) = Ndom_{M_l M_k}(Y)$.

Then, due to $M = (M_1 \cap M_2) \cup (M_1 \cap M_3) \cup (M_2 \cap M_3)$, the equality $Ndom_M(Y) = Ndom_{M_1 M_2}(Y) \cap Ndom_{M_1 M_3}(Y) \cap Ndom_{M_2 M_3}(Y)$ is true, where $Ndom_M(Y)$ is the set of nondominated vectors according to cone M. And we obtain $Ndom_M(Y) = \hat{P}_{12}(Y) \cap \hat{P}_{13}(Y) \cap \hat{P}_{23}(Y)$. The inclusion $\mathbb{R}^3_+ \subseteq M_l \cap M_k$ for any $l, k \in \{1, 2, 3\}$, $l \neq k$ implies the inclusion $Ndom_{M_l M_k}(Y) \subseteq P(Y)$. As a result we obtain that inclusion (2) is valid. We can conclude that the set $\hat{P}_{12}(Y) \cap \hat{P}_{13}(Y) \cap \hat{P}_{23}(Y)$ is an "optimal" group choice according to the individual preferences (I) and the majority preference relation $\succ_{maj}$, as a principle of aggregation. $\qquad\square$

Theorem 1 shows how to aggregate information about individual preferences (I) using the majority rule. Note, that intersections in inclusion (2) should be nonempty.

5 Conclusion

The majority preference relation constructed upon cone preference relations is considered. Due to nontransitivity of this relation, one should specify a transitive part. The case of group of three DMs and criteria with three components is studied. It is shown how to aggregate "quanta" of information of each DM, and construct an "optimal" group choice upon this information. Moreover, this result reduces bounds of "optimal" group choice in comparison with the situation when DMs do not have a "quantum" of information.

Acknowledgments This work is supported by the Russian Foundation for Basic Research, project no. 14-07-00899.

References

1. Aizerman, M.A., Aleskerov, F.T.: Theory of Choice. Elsevier, North-Holland (1995)
2. Aleskerov, F.T.: Arrovian Aggregation Models. Kluwer Academic Publisher, Dordrecht (1999)
3. Arrow, K.J.: Social Choice and Individual Values. Wiley, and New Haven: Yale University Press, New York (1963)
4. Arrow, K.J., Raynaud, H.: Social Choice and Multicriterion Decision-Making. MIT Press, Cambridge (1986)
5. Ehrgott, M., Figueira, J.R., Greco S.: Trends in Multiple Criteria Decision Analysis. Springer (2010)
6. Hwang, Ch.-L, Lin, M.-J.: Group Decision Making under Multiple Criteria. Methods and Applicatoins. Springer (1987)
7. Kilgour, D.M., Eden, C.: Handbook of Group Decision and Negotiation. Springer, Dordrecht (2010)
8. Noghin, V.D.: Relative importance of criteria: a quantitative approach. J. Multi-Criteria Decis. Anal. **6**(6), 355–363 (1997)
9. Noghin, V.D.: Prinatie reshenii v mnogokriterial'noi srede: collichestvennyi podkhod (Decision Making in Multicriteria Sphere: Quantitative Approach). Fizmatlit, Moscow (2005) (in Russian)

10. Noghin V.D.: Axiomatic approach to reduce the pareto set: computational aspects. In: Moscow International Conference on Operational Research (ORM2013), pp. 58–60 (2013)
11. Noghin, V.D., Baskov, O.V.: Pareto set reduction based on an arbitrary finite collection of numerical information on the preference relation. Doklady Math. **83**(3), 418–420 (2011)
12. Petrovskii, A.B.: Teoriya prinyatiya reshenii (Theory of Decision Making). Akademia, Moscow (2009) (in Russian)

Author Index

© Springer International Publishing Switzerland 2016
M. Lübbecke et al. (eds.), *Operations Research Proceedings 2014*,
Operations Research Proceedings, DOI 10.1007/978-3-319-28697-6

657

9 783319 286952